AF598237

Gmelin Handbook of Inorganic Chemistry

8th Edition

Organometallic Compounds in the Gmelin Handbook

The following listing indicates in which volumes these compounds are discussed or are referred to:

Ag	Silber B5 (1975)
Au	Organogold Compounds (1980)
Bi	Bismut-Organische Verbindungen (1977)
Co	Kobalt-Organische Verbindungen 1 (1973), 2 (1973), Kobalt Erg.-Bd. A (1961), B1 (1963), B2 (1964)
Cr	Chrom-Organische Verbindungen (1971)
Cu	Organocopper Compounds 1 (1985), 2 (1983)
Fe	Eisen-Organische Verbindungen A1 (1974), A2 (1977), A3 (1978), A4 (1980), A5 (1981), A6 (1977), A7 (1980), Organoiron Compounds A8 (1986), B1 (partly in English; 1976), B2 (1978), Eisen-Organische Verbindungen B3 (partly in English; 1979), B4 (1978), B5 (1978), Organoiron Compounds B6 (1981), B7 (1981), B8 (1985), B9 (1985), B10 (1986), present volume, B11 (1983), B12 (1984), Eisen-Organische Verbindungen C1 (1979), C2 (1979), Organoiron Compounds C3 (1980), C4 (1981), C5 (1981), C7 (1986), and Eisen B (1929-1932)
Hf	Organohafnium Compounds (1973)
Nb	Niob B4 (1973)
Ni	Nickel-Organische Verbindungen 1 (1975), 2 (1974), Register (1975), Nickel B3 (1966), C1 (1968), C2 (1969)
Np, Pu	Transurane C (partly in English; 1972)
Pt	Platin C (1939) and D (1957)
Ru	Ruthenium Erg.-Bd. (1970)
Sb	Organoantimony Compounds 1 (1981), 2 (1981), 3 (1982)
Sc, Y, La to Lu	D6 (1983)
Sn	Zinn-Organische Verbindungen 1 (1975), 2 (1975), 3 (1976), 4 (1976), 5 (1978), 6 (1979), Organotin Compounds 7 (1980), 8 (1981), 9 (1982), 10 (1983), 11 (1984), 12 (1985)
Ta	Tantal B2 (1971)
Ti	Titan-Organische Verbindungen 1 (1977), 2 (1980) 3 (1984), 4 and Register (1984)
U	Uranium Suppl. Vol. E2 (1980)
V	Vanadium-Organische Verbindungen (1971), Vanadium B (1967)
Zr	Organozirconium Compounds (1973)

Gmelin Handbook of Inorganic Chemistry

8th Edition

Gmelin Handbuch der Anorganischen Chemie

Achte, völlig neu bearbeitete Auflage

Prepared and issued by

Gmelin-Institut für Anorganische Chemie
der Max-Planck-Gesellschaft
zur Förderung der Wissenschaften
Director: Ekkehard Fluck

Founded by — Leopold Gmelin

8th Edition — 8th Edition begun under the auspices of the Deutsche Chemische Gesellschaft by R. J. Meyer

Continued by — E.H.E. Pietsch and A. Kotowski, and by Margot Becke-Goehring

Springer-Verlag Berlin · Heidelberg · New York · Tokyo 1986

Gmelin Handbook of Inorganic Chemistry

8th Edition

Fe Organoiron Compounds

Part B 10

Mononuclear Compounds 10

With 33 illustrations

AUTHOR Adolf Slawisch

EDITOR Marlis Mirbach

FORMULA INDEX Edgar Rudolph

CHIEF EDITOR Adolf Slawisch

Springer-Verlag Berlin · Heidelberg · New York · Tokyo 1986

LITERATURE CLOSING DATE: 1983
IN SOME CASES MORE RECENT DATA HAVE BEEN CONSIDERED

Library of Congress Catalog Card Number: Agr 25-1383

ISBN 3-540-93523-1 Springer-Verlag, Berlin · Heidelberg · New York · Tokyo
ISBN 0-387-93523-1 Springer-Verlag, New York · Heidelberg · Berlin · Tokyo

Typesetting, printing, and bookbinding: Universitätsdruckerei H. Stürtz AG, Würzburg

Preface

The present volume is a continuation of Series B on the mononuclear organoiron compounds. It covers the literature completely to the end of 1983 and includes many references to the literature up to mid-1984. An empirical formula index and a ligand formula index for the volumes "Organoiron Compounds" B8, B9, and B10 provide ready access to the ca. 2100 compounds covered.

This volume continues the description of mononuclear organoiron derivatives by treating compounds of the type $^4LFe(CO)_3$ where ^4L includes trimethylenemethane and its derivatives, σ,π-allyl- and σ,σ,π-bound ligands. The last part of this volume describes compounds of the type $^4LFe^2L$ and $^4L(^2L)Fe(CO)_n(^2D)_{2-n}$, compounds with one ^4L and one ^3L ligand, and compounds with two ^4L ligands. In the symbol ^mL_n, n is the number of organic ligands L in the compound, and m is the number of Fe-C bonds formed from an L ligand. The symbol ^2D designates a Lewis base donor with two electrons such as PR_3 as already explained in the prefaces to "Kobalt-Organische Verbindungen" 1, New Suppl. Ser., Vol. 5, 1973, and "Nickel-Organische Verbindungen" 1, New Suppl. Ser., Vol. 16, 1975.

Series B so far comprises volumes B1 to B12, and a survey of this series has been given in the preface to B7 (1981).

Much of the data, particularly in tables, is given in abbreviated form without dimensions; for explanation see p. VIII. Additional remarks, if necessary, are given in the heading of the tables.

Frankfurt
November 1985

Adolf Slawisch

Remarks on Abbreviations and Dimensions

Many compounds in this volume are presented in tables in which numerous abbreviations are used, and the dimensions are omitted for the sake of conciseness. This necessitates the following clarification:

Temperatures are given in °C, otherwise K stands for Kelvin. Abbreviations used with temperatures are m.p. for melting point, b.p. for boiling point, dec. for decomposition, and subl. for sublimation.

NMR represents **nuclear magnetic resonance.** Chemical shifts are given as δ values in ppm; reference substances and signs for δ are shown in the scheme below:

		increasing field	→
		$\delta=0$ for	
^{1}H	+	$Si(CH_3)_4$	−
^{11}B	+	$BF_3 \cdot O(C_2H_5)_2$	−
^{13}C	+	$Si(CH_3)_4$	−
^{19}F	−	$CFCl_3$	+
^{31}P	+	H_3PO_4	−

Coupling constants J, in Hz, are given as J(A, B) or as J(1,3) and refer to labelled structural formulas.

Multiplicities of the signals are abbreviated as s, d, t, q (singlet to quartet), quint, sext, sept, oct (quintet to octet), and m (multiplet); terms like dd (double doublet) and t's (triplets) are also used. Assignments referring to labelled structural formulas are given in the form C-4, H-3,5. The numbering deviates in some cases from official nomenclature so that corresponding values for compounds in the same chapter can be more easily compared.

Mössbauer spectra are represented by ^{57}Fe-γ and ^{119}Sn-γ: both the isomer shift δ (vs. $Na_2[Fe(CN)_5NO]$ or $BaSnO_3$ at room temperature) and the quadrupole splitting Δ are given in mm/s; the experimental error has generally been omitted. Other reference substances for δ are indicated after the numerical value, e.g., $\delta=0.23$ (Fe).

Optical spectra are labelled as IR (infrared), R (Raman), and UV (electronic spectrum including the visible region): IR bands and Raman lines are given in cm^{-1}; the assigned bands are usually labelled with the symbols ν for stretching vibration and δ for deformation vibration whereas unlabelled bands belong to CO stretching vibrations. Intensities occur in parentheses either in the common qualitative terms (s, m, w, vs, br, etc.) or as numerical relative intensities; str., def., pol., and dp mean stretching, deformation, polarized, and depolarized, respectively. The CO stretching force constant and CO, CO interaction constant (in mydn/Å) are denoted as k and k′, respectively. The UV absorption maxima, λ_{max}, are given in nm followed by the extinction coefficient ε ($L \cdot cm^{-1} \cdot mol^{-1}$) or log ε in parentheses; sh means shoulder.

Solvents, or the **physical state** of the sample, and the temperature (in °C or K) are given in parentheses immediately after the spectral symbol, e.g., R (solid), ^{13}C NMR (C_6D_6, 50 °C). Common solvents are given by their formula (cyclo-C_6H_{12}=cyclohexane) except THF, which represents tetrahydrofuran.

Figures give only selected parameters. Bond lengths (in Å) or angles are mean values.

Table of Contents

Organoiron Compounds, Part B

Mononuclear Compounds 10

Organoiron Compounds, Part B

Mononuclear Compounds 10

1.4.1.4.2 $^4LFe(CO)_3$ Compounds where ^4L is Trimethylenemethane and Derivatives

General References:

T.-Y. Luh, Trimethylamine N-Oxide — A Versatile Reagent for Organometallic Chemistry, Coord. Chem. Rev. **60** [1984] 255/76.

M. Franck-Neumann, Synthetic Applications of Some Metal Carbonyl Complexes, Pure Appl. Chem. **55** [1983] 1715/32.

T.A. Albright, Rotational Barriers and Conformations in Transition-Metal Complexes, Accounts Chem. Res. **15** [1982] 149/55.

J.C. Green, Gas Phase Photoelectron Spectra of d- and f-Block Organometallic Compounds, Struct. Bonding [Berlin] **43** [1981] 37/112.

A.H. Cowley, UV Photoelectron Spectroscopy in Transition Metal Chemistry, Progr. Inorg. Chem. **26** [1979] 45/160.

C. Furlani, C. Cauletti, He(I) Photoelectron Spectra of d-Metal Compounds, Struct. Bonding [Berlin] **35** [1978] 119/69.

R. Pettit, Role of Cyclobutadieneiron Tricarbonyl in the "Cyclobutadiene Problem", J. Organometal. Chem. **100** [1975] 205/17.

B.E. Mann, ^{13}C NMR Chemical Shifts and Coupling Constants of Organometallic Compounds, Advan. Organometal. Chem. **12** [1974] 135/213.

The compounds known so far have a staggered geometry, see e.g., No. 1, pp. 18/21, or Formula VIII, p. 20. Many complexes exhibit three distinct ^{13}CO resonances at room temperature, a clear indication of a relatively high barrier of rotation around the central C-Fe bond [48].

The compounds in Table 1, pp. 4/15, are prepared by the following methods:

Method I:

a. $H_2C{=}C(CH_2Cl)_2$ reacts with $Fe_2(CO)_9$ in ether at room temperature for 12 to 24 h [1, 14, 33, 51]. The compound $(CH_2)_3CFe(CO)_3$ (No. 1) is purified by fractional distillation [1, 14, 33] or by chromatography on neutral alumina with pentane as eluent [51].

b. $H_2C{=}C(CH_2Cl)_2$ reacts with $Na_2Fe(CO)_4$ in tetrahydrofuran [20, 26, 33].

c. $RR'C{=}C(CH_2Br)_2$ reacts with $Fe_2(CO)_9$ in hexane at 40 to 60 °C for 6 to 10 h. The compounds $RR'CC(CH_2)_2Fe(CO)_3$ are isolated by elution from alumina with pentane or by fractional distillation [25, 33].

Method II: $RR'C{=}C(CH_3)CH_2Cl$ and $Fe_2(CO)_9$ react together without isolation of intermediate $^3LFe(CO)_3Cl$ ($^3L{=}RR'CC(CH_3)CH_2$) under the following conditions [14, 15, 33, 40] (cf. Method IV and V):

a. Room temperature for 3 h and refluxing for 3 h in cyclohexane [33].

b. Mixing of the components and fractional distillation of this mixture to yield $RR'CC(CH_2)_2Fe(CO)_3$ [15, 40].

References on pp. 39/42

c. Excess $Fe_2(CO)_9$ at 40 °C in hexane solution, subsequently refluxed until CO development ceases. Fractional distillation gives $RR'C{=}C(CH_3)_2$ (19% yield), $RR'CHC(CH_3){=}CH_2$ (14%), and the product $RR'CC(CH_2)_2Fe(CO)_3$ [15, 33].

Method III: a. Cyclo-$RR'C_3H_2{=}CH_2$ (Formula I) reacts with $Fe_2(CO)_9$ in C_6H_6 at room temperature for 5 to 20 h yielding $RR'CC(CH_2)_2Fe(CO)_3$ [12, 14, 72, 73, 83]. The solution is filtered and the solvent removed. The residue is purified by column chromatography on activity III alumina (for product separation and identification) or Florisil (for measurement of the ratio of $RR'CC(CH_2)_2Fe(CO)_3$: syn-$RC_4H_5Fe(CO)_3$ ($R{=}C_6H_5$, $R'{=}H$), since they were not separated under these conditions) with pentane/benzene. On alumina, the bottom (largest R_f) band is $RR'CC(CH_2)_2Fe(CO)_3$. The middle band is syn-$RC_4H_5Fe(CO)_3$ and at the top of the column an unidentified compound (or mixture of compounds) is separated for $R{=}C_6H_5$, $R'{=}H$ with ^1H NMR: $\delta(CS_2) = 1.4$ (m, 1.5H), 2.1 (m, 1.0H), 3.7 (m, 1.5H), 7.3 (m, 5.0H) ppm, and IR (CS_2): many bands below 1500; 1680, 1725, 1980, 2050, and 2800 to 3100 cm^{-1} (for mass spectrum see [83]) [73, 83]. For a qualitative molecular orbital picture concerning this reaction with cyclo-$RC_3H_2{=}CH_2$ (see Formula I with $R'{=}H$, $R{=}CH{=}CH_2$ [72] or C_6H_5 [72, 73, 83]) see [72, 73, 83]. Cyclo-$RR'C_3H_2{=}CH_2$ ($R{=}CH{=}CH_2$, $R'{=}H$) reacts at 35 to 40 °C for 2 h. The solvent is distilled off. Shortway distillation of the residue gives No. 6 [45].

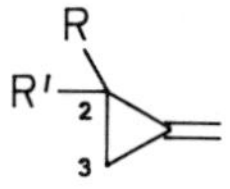

R R' 2 3 — CH_3O_2C CH_3O_2C — CO_2CH_3 CO_2CH_3 — O O $(CO)_3Fe$

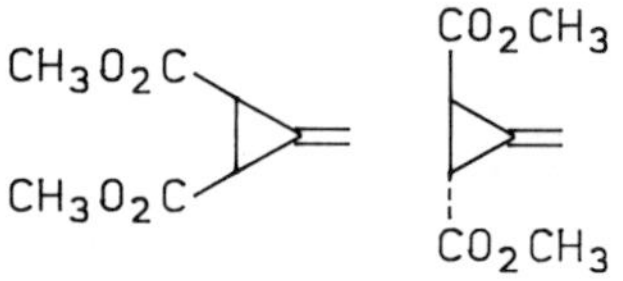

cis trans

I II III

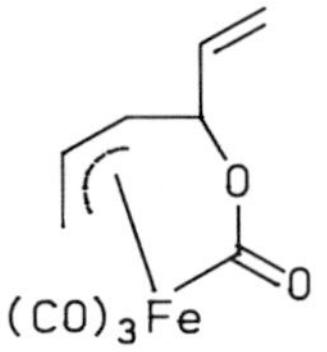

b. Cis- and trans-$(CH_3O_2C)_2C_3H_2{=}CH_2$ (Formula II) readily undergo reactions with excess $Fe_2(CO)_9$ at room temperature in benzene in less than 5 h. The reaction mixtures are filtered, the filtrates evaporated, and the residues washed with pentane. The cis isomer gives No. 36, the trans isomer No. 37. For other isomers see "Further information" for Nos. 36 and 37 [36].

c. $(BrCH_2)_2C{=}CHCH(CH_3)O_2CCH_3$ is stirred at 50 °C with $Fe_2(CO)_9$ in hexane for 6 h. Filtration and evaporation of the filtrate leaves a dark residue. Chromatography on silica gel with benzene gives a red band with high R_f which on standing appeared to decompose to give the second component which elutes as a yellow band at $R_f = 0.5$ containing the compound. No satisfactory C, H, Fe analysis is obtained [68].

Method IV: π-Allyl complexes of the type $RR'CC(CH_3)CH_2Fe(CO)_3Cl$ are disproportionated [15, 25, 33] (isobutene or its derivatives, $FeCl_2$ and CO are byproducts [15, 33]).

a. The π-allyl complex is decomposed at room temperature in hydrocarbon solvents [15].

b. The dry, solid π-allyl complex is heated to 100 °C [15].

c. The π-allyl complex is heated in a solvent to 100 °C [15].

d. The π-allyl complex is refluxed in cyclohexane for 3 h [33].

References on pp. 39/42

Method V: The π-allyl complex $BrCH_2C(CH_2)_2Fe(CO)_3Br$ is decomposed [15, 25, 33]:

a. The π-allyl complex is refluxed in cyclohexane for 30 min yielding $CH_2{=}C(CH_2Br)_2$ (47% yield), $BrC(CH_2Br)_3$ (6%), $FeBr_2$, and $(CH_2)_3CFe(CO)_3$ [33].

b. The π-allyl complex (1.1 mmol) reacts with $Fe_2(CO)_9$ (1.4 mmol) in ether at room temperature for 24 h [33].

Method VI: Al_2O_3 is suspended in an ether solution of σ, π-allyl complex $C_7H_8O_2Fe(CO)_3$ (Formula III). Addition of NH_3 gas at room temperature yields $(CH_2)_3CFe(CO)_3$ [53].

Method VII: a. A mixture of 1- and 3-chloro-2-chloromethylpropene reacts with $Fe_2(CO)_9$ in ethyl ether at room temperature for 24 h. The reaction mixture is filtered. Distillation of the solution yields $Fe(CO)_5$, $(CH_2)_3CFe(CO)_3$, and $ClCHC(CH_2)_2Fe(CO)_3$. The latter is chromatographed on neutral alumina. Elution with pentane gives a yellow oil, which crystallized from pentane at -78 °C [33].

b. An excess of $Fe_2(CO)_9$ reacts with a mixture of 1- and 3-chloro-2,3-dimethylpropene under conditions similar to those in Method IIc. The products are $CH_3CHC(CH_2)_2Fe(CO)_3$ and $^4LFe(CO)_3$ with ^4L = isoprene (mole ratio 1:3, overall yield 36%) [15].

Method VIII: a. $(CH_2)_3CFe(CO)_3$ (No. 1) is mixed with $RCOCl/AlCl_3$ at 5 °C in CS_2 (for $R{=}C_6H_5$) or CCl_4 (for $R{=}CH_3$) and refluxed for 2 h. The upper layer is discarded, and the lower dark red layer hydrolyzed on ice and extracted with ether. Chromatography on neutral alumina with hexane/C_6H_6 as eluent yields the product as a yellow oil and $RCOCH{=}C(CH_3)_2$ [33].

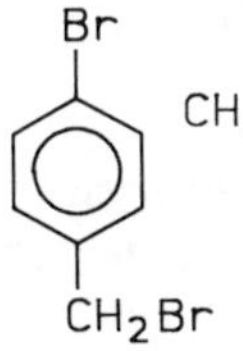

IV 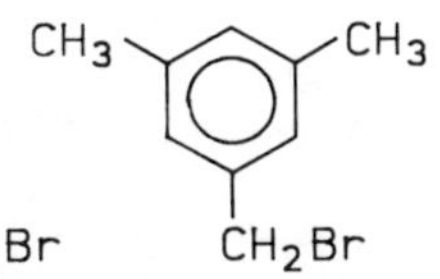 V 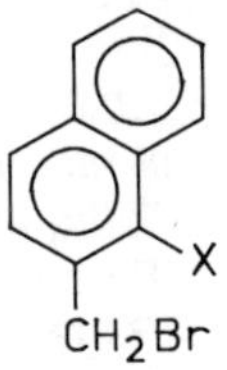 VI 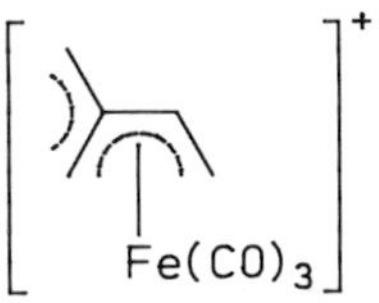 VII

b. Modification of Method VIIIa by use of CH_2Cl_2 as solvent. An aqueous NH_3 quench gives the pure product [68].

Method IX: a. The compounds No. 40, 41, 43, and 44 are prepared by reaction of IV, V, or VI (X = H or Br), respectively, with $Fe_2(CO)_9$ in hexane at 45 °C for 3 to 4 h. The reaction mixture is filtered and the solvent removed. Chromatographic work-up (alumina, petroleum ether/benzene (9:1) as eluent) yields the products [4, 22, 27, 31, 42, 43].

b. For the preparation of No. 42 the procedure IXa is employed except that a 1:1 mixture of 1- and 2-bromo-methylnaphthalene is used in the reaction [21 to 23, 27, 31].

References on pp. 39/42

Table 1
Compounds of the Type $^{4}LFe(CO)_3$ ($^{4}L = C(CH_2)_3$ and Derivatives).
Further information on numbers preceded by an asterisk is given at the end of the table, pp. 16/39.
For abbreviations and dimensions see p. VIII.

No.	^{4}L ligand	method of preparation (yield in %), properties and remarks	Ref.
*1		I a (30 to 70), I b (32), II a, II b (14 to 20), III a, IV a to d (up to 39), V a (47), V b (ca. 100), VI (50), VII a m.p. 28 to 29°, m.p. 28.4 to 29.6°, m.p. 29°, m.p. 32 to 33°, m.p. 32.5 to 33.0°, b.p. 53 to 54°/16 Torr, b.p. 53 to 55°/16 Torr, b.p. 58 to 59°/30 Torr, pale yellow prisms (from pentane at −78°) ^{1}H NMR: 2.00 (s) or 2.01 (s) ^{13}C NMR (CD_2Cl_2, at 245 K): 212.2 (CO; $^{1}J(^{13}C, ^{57}Fe) = 28.5 \pm 0.1$) ^{13}C NMR ($CH_2Cl_2$): 54.9 ($CH_2$), 106.1 (central C), 211.9 (CO) ^{13}C NMR (CS_2): 52.0 (t, CH_2; J(C, H) = 158 ± 10), 105.0 (central C), 211.6 (CO) ^{57}Fe NMR (C_6H_6 at 300 K): −76.7 ± 0.8 (rel. to neat $Fe(CO)_5$ ^{57}Fe-γ (neat, 78 K): δ = 0.265 ± 0.06 ^{57}Fe-γ (neat, 294 K): Δ = 0.618 ± 0.06 ^{57}Fe-γ (frozen solution in $C_2H_5OH/(CH_3)_2CHOH$/ether (16:42:42)): δ = 0.278 ± 0.06, Δ = 0.606 ± 0.06 IR (solid): 1979, 2059 (CO) IR (liquid film): ca. 1980, ca. 2062 (CO) IR (gas): 2004, 2010, 2015, 2070, 2074, 2079 (CO) IR (hexane): 1994, 2064 (CO) IR (cyclohexane): 1993, 2059 (CO) IR (cyclohexane): 1995, 2063 (CO) IR (CCl_4 or hexane): 1966, 1995, 2061 (CO) IR (?): 1998, 2064 (CO) mass spectrum: $[C_4H_6Fe(CO)_n]^+$ (n = 0 to 3)	[1 to 3, 12, 14, 15, 20, 25, 26, 30, 33, 37, 38, 40, 47, 48, 51, 53, 68, 82]
2	H–4–H′; Cl–2(H); 3(H′)–H	VII a (low) melts at room temp., b.p. 78 to 89°/15 Torr, yellow oil (from pentane at −78°) ^{1}H NMR (CS_2): 1.67 (H-3′; J(H-3′, 4) = 4.3), 1.85 (H-3), 2.39 (H-4′), 3.08 (H-4), 4.20 (H-2; J(H-2, 4′) = 3.0) IR (hexane): 1967, 2001, 2071 (CO)	[25, 33]
*3	H–4–H′; CH_3–2(H); 3(H′)–H	VII b (9), see also No. 14, p. 25 ^{1}H NMR (CCl_4): 1.63 (H-3′; J(H-3′, 4′) = 4.5), 1.77 (H-3), 2.11 (H-4′), 2.51 (H-4), 3.14 (H-2; J(H-2, 4′) = 2.5) IR (hexane): 1979, 2058 (CO)	[15, 33]

Table 1 [continued]

No.	^{4}L ligand	method of preparation (yield in %), properties and remarks	Ref.
*4	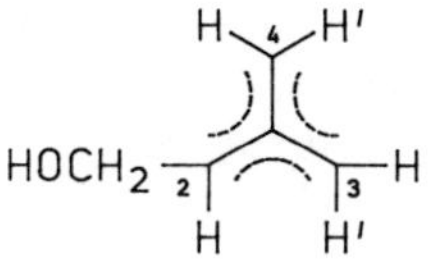 HOCH$_2$ (structure)	see No. 14, p. 25 m.p. 45 to 46°, red crystals ^{1}H NMR ($CDCl_3$): 1.82 (d, H-3'; J(H-3',4') = 4.5), 1.89 (s, H-3), 2.00 (s, OH), 2.20 (d, H-4'; J(H-2,4') = 2), 2.54 (d, H-4; J(H-3',4) = 4.5), 2.9 to 3.5 (m, H-2), 3.5 to 3.9 (m, CH_2O) IR (film): 1000, 1470; 1980, 2075 (CO); 3350 (OH) mass spectrum: $[M-nCO]^+$ (n = 0 to 3), $[M-OH]^+$, $[M-3CO-H_2O]^+$, $[M-Fe(CO)_3]^+$, $[Fe(CO)_n]^+$ (n = 1, 2, 3), $[Fe]^+$	[71]
5	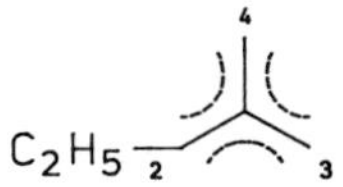 C_2H_5 (structure)	see No. 9, p. 24 ^{13}C NMR (CH_2Cl_2): 17.0 (CH_3), 23.4 (CH_2), 39.9 (C-4), 50.8 (C-3), 82.7 (C-2), 103.2 (C-1) (the CO signals are too weak to locate accurately)	[48]
6	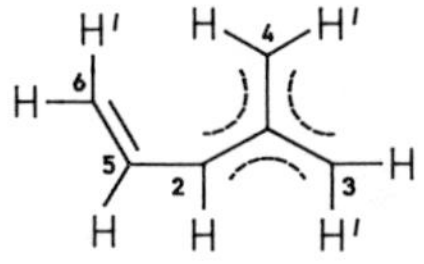 (structure)	III a (43) b.p. ca. 92°/2.5 Torr, green oil ^{1}H NMR: 1.79 (s, H-3), 1.80 (d, H-3'; J(H-3',4) = 4.3), 2.18 (m, H-4'), 2.66 (d, H-4), 3.67 (dd, H-2; J(H-2,5) = 9.7, J(H-2,4') = 2.1), 4.83 to 6.03 (m, H-5,6,6') ^{13}C NMR: 51.9 (C-3 or 4), 52.5 (C-3 or 4), 80.2 (C-2), 105.2 (C-1), 118.7 (C-6), 137.6 (C-5), 214.8 (CO) IR: 1984, 2048 (CO) mass spectrum: $[M]^+$	[34, 45, 72]
*7	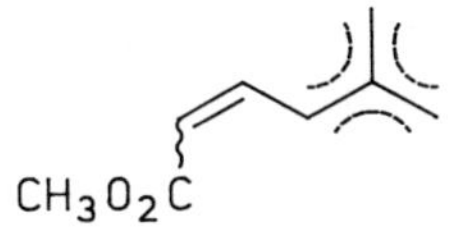 CH_3O_2C (structure)	see No. 8, p. 23	[102]
*8	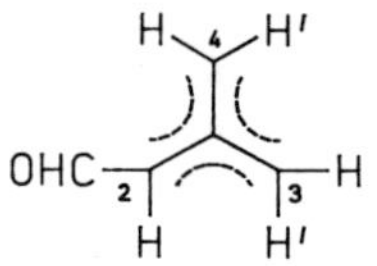 OHC (structure)	see No. 4, p. 23 m.p. 75°, subl. 25 to 30°/0.001 Torr, waxy yellow crystals (from n-pentane at −20°) ^{1}H NMR ($CDCl_3$): 2.21 (s, H-3), 2.38 (d, H-3'), 2.47 (d, H-4'), 3.05 to 3.25 (m, 2H, analyzed as 3.15 (dd, H-2; J(H-2,4') = 2, J = 7.5) and 3.17 (d, H-4; J(H-3',4) = 5)), 8.96 (d, CHO; J = 7.5) IR (CCl_4): 1150; 1670 (C=O); 2020, 2100 (CO) mass spectrum: $[M-nCO]^+$ (n = 0 to 4), $[Fe(CO)_n]^+$ (n = 0 to 2)	[68]

References on pp. 39/42

Table 1 [continued]

No.	^{4}L ligand	method of preparation (yield in %), properties and remarks	Ref.
*9	[structure: $CH_3C(O)$–C-2(H)–C-3(H,H')–C-4(H,H')]	VIIIa, VIIIb (ca. 25) b.p. 44 to 45°/0.05 Torr, b.p. 55 to 59°/0.05 Torr (recrystallized from ether/pentane (1:10) at −78°) ^{1}H NMR ($CDCl_3$): 1.90 (s, H-3), 1.98 (s, CH_3), 1.98 (d, H-3′), 2.41 (d, H-4′), 2.99 (d, H-2; J(H-2,4′) = 2), 3.82 (d, H-4; J(H-3′,4) = 4) ^{1}H NMR (CCl_4): 2.07 (H-3), 2.12 (H-4′), 2.49 (H-3′), 3.16 (H-4; J(H-3′,4) = 2.0), 3.88 (H-2; J(H-2,4′) = 4.0) ^{13}C NMR (CCl_4): 29.9 (CH_3; J = 123), 53.3 (C-4; J = 162), 58.4 (C-3; J = 162), 68.8 (C-2; J = 157), 106.8 (C-1), 200.2 (C=O); 208.8, 209.7, 209.8 (CO) ^{13}C NMR (CH_2Cl_2): 30.4 (CH_3), 54.3 (C-4), 58.4 (C-3), 69.4 (C-2), 107.3 (C-1), 202.5 (C=O); 209.8, 210.3 (CO) ^{13}C NMR ($CH_3C_6H_5$): 29.9 (CH_3), 53.7 (C-4), 58.6 (C-3), 69.4 (C-2), 107.1 (C-1), 201.0 (C=O); 209.7, 210.3, 210.4 (CO) IR (hexane): 1971, 2003, 2069 (CO) IR (CCl_4): 1685 (C=O); 1990, 2070 (CO)	[33, 48, 68]
*10	[structure: $CH_3CH(OH)$–C-2(H)–C-3(H,H')–C-4(H,H')] ψ-endo	see Nos. 8 and 9, pp. 23/4 ^{1}H NMR (CCl_4): 1.63 (H-3′), 1.73 (H-3), 2.09 (H-4′), 2.44 (H-4; J(H-3′,4) = 4.0), 2.93 (H-2; J(H-2,4′) = 2.0) ^{13}C NMR (CH_2Cl_2): 27.5 (CH_3; J = 123), 51.4 (C-4; J = 172), 52.9 (C-3; J = 172), 66.7 (COH; J = 138), 85.9 (C-2; J = 147), 102.6 (C-1); 211.0, 211.4 (CO) ^{13}C NMR (1,4-dioxane): 28.0 (CH_3), 51.2 (C-4), 53.1 (C-3), 66.2 (COH), 87.3 (C-2), 102.9 (C-1); 211.3, 212.2, 212.3 (CO) IR (hexane): 1963, 1992, 2062	[33, 48, 68]
*11	[structure: $CH_3CH(OCH_3)$(C-5)–C-2(H)–C-3(H,H')–C-4(H,H')] ψ-endo	^{1}H NMR (CCl_4): 1.25 (m, CH_3), 1.57 (d, H-3′), 1.65 (s, H-3), 2.00 (br,d, H-4′; J(H-2,4′) ≈ 2), 2.35 (d, H-4; J(H-3′,4) = 4), 2.85 (m, H-2,5), 3.14 (s, OCH_3) IR (CCl_4): 1100 (COC); 1970, 2060 (CO); 2830 (CH_3O) mass spectrum: $[M-nCO]^+$ (n = 0 to 3)	[68]

References on pp. 39/42

Table 1 [continued]

No.	^4L ligand	method of preparation (yield in %), properties and remarks	Ref.
*12	ψ-endo	IIIc (57), see also No. 10, p. 25 b.p. 60 to 65°/0.005 Torr, yellow ^1H NMR ($CDCl_3$): 1.40 (d, CH_3), 1.75 (H-3′), 1.80 (s, H-3), 1.98 (s, CH_3CO_2), 2.15 (d, H-4′), 2.55 (d, H-4; J(H-3′,4) = 4), 2.85 (dd, H-2; J(H-2,4′) = 2.5), 4.40 (dq, CH; J(CH, H-2) = 10, J(CH_3, CH) = 6) ^{13}C NMR (CH_2Cl_2): 21.1 (CH_3-acetate), 24.5 (CH_3), 51.7 (C-4), 70.0 (CH), 79.9 (C-2), 103.6 (C-1), 169.9 (C=O); 211.3, 212.3 (CO) (C-3 is hidden by solvent) ^{13}C NMR ($CH_3C_6H_5$): 24.3 (CH_3-acetate), 51.2 (C-4), 52.7 (C-3), 69.4 (CH), 80.0 (C-2), 103.3 (C-1), 171.5 (C=O); 213.4, 214.1, 214.3 (CO) (CH_3 is hidden by solvent) IR (neat): 1240, 1375, 1460, 1740; 1990, 2075 (CO) mass spectrum: $[M-nCO]^+$ (n = 1 to 3), $[C_7H_9O_2Fe]^+$, $[C_6H_8Fe]^+$ and other fragments	[48, 68]
13		see No. 14, p. 25 ^1H NMR (CD_2Cl_2): 2.16 (H-3), 2.22 (H-4′), 2.61 (H-3′; J(H-3′,4) = 2.0), 2.92 (H-4), 4.00 (H-2; J(H-2,4′) = 3.5)	[33]
*14		Ic b.p. 34 to 37°/0.04 Torr, pale yellow ^1H NMR (CCl_4): 1.97 (H-3′), 2.02 (H-4′), 2.42 (H-3′; J(H-3′,4) = 2.3), 2.87 (H-4), 3.91 (H-2; J(H-2,4′) = 4.2) IR (hexane): 1978, 2009, 2074	[25, 33]
*15		see No. 7, p. 23	[102]
16		see No. 7, p. 23	[102]
*17		Ic, IIc (32), IIIa (40) m.p. 62 to 64°, m.p. 63 to 64°, m.p. 63.5 to 64.5°, b.p. 109 to 115°/0.5 Torr, b.p. 110 to 114°/ 0.5 Torr, plates (from 95% C_2H_5OH)	[12, 15, 19, 25, 33, 72, 73, 83]

References on pp. 39/42

Table 1 [continued]

No.	^{4}L ligand	method of preparation (yield in %), properties and remarks	Ref.
		^{1}H NMR (CCl_4): 1.84 (s, H-3,3′), 2.30 (d, H-4′), 2.87 (d, H-4; J(H-3′,4) = 5.4), 4.28 (d, H-2; J(H-2,4′) = 3.3), 7.11 (s, C_6H_5) ^{1}H NMR (benzene-d_6): 1.4 (2H, 2 peaks, area ratio 3.4:1), 1.9 (d, 1H; J = 2.5), 2.6 (d, 1H; J = 4.5), 3.9 (d, 1H; J = 2.5), 6.9 to 7.1 (m, 5H) ^{1}H NMR (CS_2): 1.8 (doublet with a singlet in the middle, 2H; J = 4.5 for the doublet), 2.3 (d, 1H; J = 2.5), 2.8 (d, 1H; J = 4.5), 4.2 (d, 1H; J = 2.5), 7.1 (s, 5H) IR (hexane): 1962, 1993, 2061 (CO) IR (CCl_4): 1966, 1996, 2058 (CO) IR (CCl_4): 2000, 2074 (CO) mass spectrum: $[M-nCO]^+$ (n = 0 to 3) and other fragments	
18	H 4 H′; O; C$_6$H$_5$ 2 H; 3 H, H′	VIIIa m.p. 54 to 56°, crystals (from ether/pentane (1:10) at −78°) ^{1}H NMR (CCl_4): 2.00 (H-3), 2.17 (H-4′), 2.53 (H-3′; J(H-3′,4) = 1.9), 3.73 (H-4), 4.10 (H-2; J(H-2,4′) = 4.1) IR (hexane): 1970, 2003, 2070 (CO)	[33]
*19	[H 4 H′; CH$_3$ 2 O; 3 H, H′]$^-$	see "Further information", pp. 27/8	[65, 67, 74 to 77]
*20	H 4 H′; CH$_3$ 2 HO; 3 H, H′	orange oil ^{1}H NMR ($CDCl_3$ at −10°): 1.21 (t, CH_3; J = 7.5), 1.51 (s, H-4′; J(H-4,4′) < 1.5), 1.69 (d, H-4), 1.91 (s, H-3), 2.00 (q, CH_2; J = 7.5), 2.81 (d, H-3′; J(H-3′,4) = 4.5), 4.75 (br,s, OH) IR (pentane): 1975, 1985, 2050 (CO); 3590 (OH)	[65, 67, 74 to 76]
*21	H 4 H′; CH$_3$ 2; 3 H, H′; OSi(CH$_3$)$_3$	yellow oil ^{1}H NMR ($CDCl_3$): 0.24 (s, $Si(CH_3)_3$), 1.20 (t, CH_3; J = 7.5), 1.54 (d, H-4′), 1.72 (dd, H-4; J(H-4,4′) = 1.5), 1.91 (s, H-3), 1.93 (q, CH_2), 2.64 (d, H-3′; J(H-3′,4) = 4.5) ^{13}C NMR ($CDCl_3$ at −20°): 14.67 (CH_3), 28.97 (CH_2), 41.28 (C-4; J(C,H) = 161), 47.25 (C-3; J(C,H) = 160), 93.43 (C-1), 131.48 (C-2); 212.40, 212.55, 212.69 (CO) IR (pentane): 1980, 1990, 2055 or 2060 (CO)	[65, 67, 76, 77]

Table 1 [continued]

No.	^{4}L ligand	method of preparation (yield in %), properties and remarks	Ref.
*22	$[n\text{-}C_3H_7\text{-}C(O)\text{-}C_3H_4]^-$	not isolated	[74]
*23	$n\text{-}C_3H_7\text{-}C(HO)\text{-}C_3H_4$	not isolated	[74]
24	$[n\text{-}C_4H_9\text{-}C(O)\text{-}C_3H_4]^-$	see No. 22, p. 28 not isolated	[74]
25	$n\text{-}C_4H_9\text{-}C(HO)\text{-}C_3H_4$	see No. 23, p. 28 not isolated	[74]
26	$[n\text{-}C_5H_{11}\text{-}C(O)\text{-}C_3H_4]^-$	see No. 22, p. 28 not isolated	[74]
27	$n\text{-}C_5H_{11}\text{-}C(HO)\text{-}C_3H_4$	see No. 23, p. 28 not isolated	[74]
28	$[n\text{-}C_6H_{13}\text{-}C(O)\text{-}C_3H_4]^-$	see No. 22, p. 28 not isolated	[74]
29	$n\text{-}C_6H_{13}\text{-}C(HO)\text{-}C_3H_4$	see No. 23, p. 28 not isolated	[74]

References on pp. 39/42

Table 1 [continued]

No.	^{4}L ligand	method of preparation (yield in %), properties and remarks	Ref.
30	$[n\text{-}C_7H_{15}\text{–}C(O)\ldots]^-$	see No. 22, p. 28 not isolated	[74]
31	$n\text{-}C_7H_{15}$, HO	see No. 23, p. 28 not isolated	[74]
*32	H, H′ (C-4); CH_3 (C-2); CH_3; H, H′ (C-3)	Ic pale yellow ^{1}H NMR (CCl_4): 1.72 (H-3, 4′), 2.23 (H-3′, 4) IR (hexane): 1953, 1985, 2056 (CO)	[25, 33]
*33	H, H′ (C-4); $C_2H_5O_2C$ (C-2); $C_2H_5O_2C$; H, H′ (C-3)	Ic m.p. 27 to 29°, b.p. 69 to 70°/0.006 Torr, yellow crystals (from ether/pentane (1:10) at −78°) ^{1}H NMR (CCl_4): 2.10 (H-3, 4′), 3.18 (H-3′, 4) IR (hexane): 1978, 2011, 2075 (CO)	[25, 33]
34	H, H′ (C-4); CH_3 (C-2); C_6H_5; H, H′ (C-3)	III a (60) oil ^{1}H NMR (CS_2): 1.85 (H-4′), 1.91 (H-3), 2.11 (H-4), 2.41 (H-3′; J(H-3′, 4) = 4.2) IR (CCl_4): 1960, 1993, 2010, 2054, 2087 (CO) mass spectrum: $[M-nCO]^+$ (n = 0 to 3) and other fragments	[12, 33]
*35	H, H′ (C-4); C_6H_5 (C-2); C_6H_5; H, H′ (C-3)	III a (50) m.p. 157 to 158° IR (CCl_4): 1965, 1995, 2057 (CO)	[12, 83, 84]
*36	CH_3O_2C, CO_2CH_3 (E,E)-isomer	III b (60) m.p. 94 to 96° IR ($CHCl_3$): 1020, 1135, 1190, 1295, 1345, 1435, 1725; 2000, 2020, 2095 (CO); 2940, 3000	[36]
*37	CH_3O_2C, CO_2CH_3 racemic (E,Z)-isomers (see Formula XLI, p. 29)	III b (78) b.p. 90 to 100°/0.5 Torr, oil IR ($CHCl_3$): 1155, 1180, 1300, 1425, 1720; 2000, 2020, 2090 (CO); 2930, 2990	[36]

Table 1 [continued]

No.	${}^4L$ ligand	method of preparation (yield in %), properties and remarks	Ref.
*38	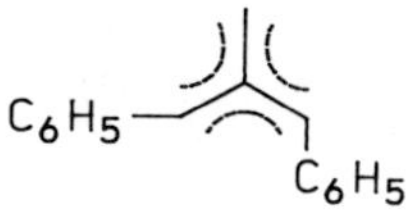 C_6H_5 … C_6H_5	see "Further information", p. 30	[84]
*39		yellow liquid mass spectrum: $[M]^+$	[98, 101]
*40	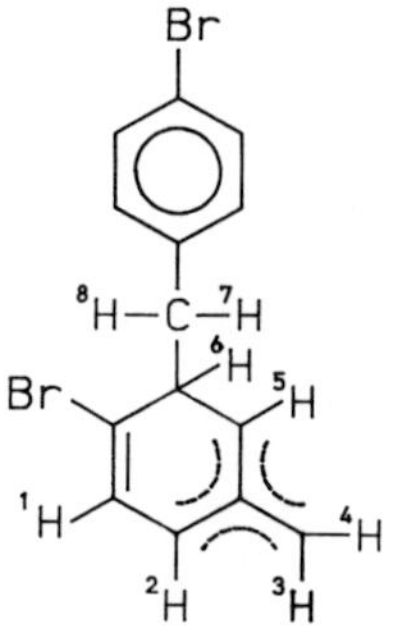 Br; $^8H-C-^7H$; ^6H; ^5H; Br; ^1H; ^4H; ^2H; ^3H	IXa (4) m.p. 109°, pale yellow crystals (from hexane) ^1H NMR (CS_2): 1.52 (H-3), 1.57 (H-4), 2.80 (H-6), 2.94 (H-7; J(H-6,7) = 3.5), 3.06 (H-2), 3.18 (H-8; J(H-7,8) = 12.2, J(H-6,8) = 7.5), 3.55 (H-5; J(H-5,6) = J(H-2,5) = 2.5), 6.09 (H-1; J(H-1,2) = 6.0, J(H-1,6) = −1) IR: 1990, 2054 (CO); 2944, 2965 (CH_2) Raman: 1590 (CC of benzene), 1600 (C=C) mass spectrum: $[M]^+$	[42]
41	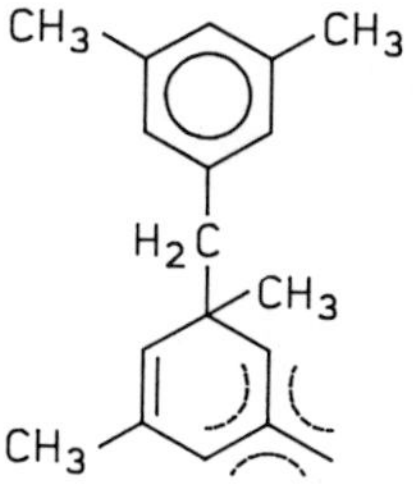 CH_3, CH_3; H_2C; CH_3; CH_3	IXa (ca. 1) m.p. 48°	[42]
*42	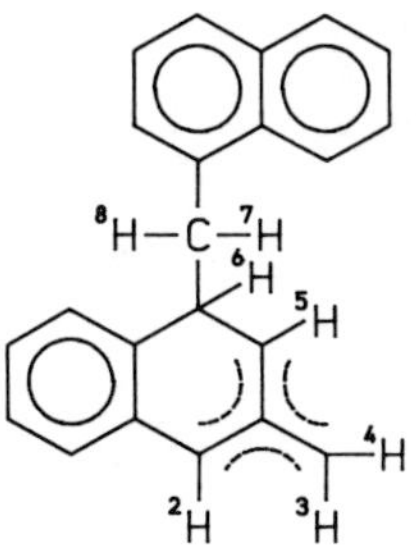 $^8H-C-^7H$; ^6H; ^5H; ^4H; ^2H; ^3H	IXb (ca. 10) m.p. 101°, m.p. 101 to 102°, m.p. 102°, yellow prismatic crystals (from hexane or C_2H_5OH) IR ($CHCl_3$): 1988, 2060 (CO); 2928, 2953 (CH_2) mass spectrum: $[M-nCO]^+$ (n = 0 to 3), $[C_{11}H_9Fe]^+$, $[C_{22}H_{18}]^+$, $[C_{21}H_{16}]^+$, $[C_{20}H_{14}]^+$, $[C_{11}H_9]^+$, $[CO]^+$	[21 to 23, 27, 31]

References on pp. 39/42

Table 1 [continued]

No.	^4L ligand	method of preparation (yield in %), properties and remarks	Ref.
*43	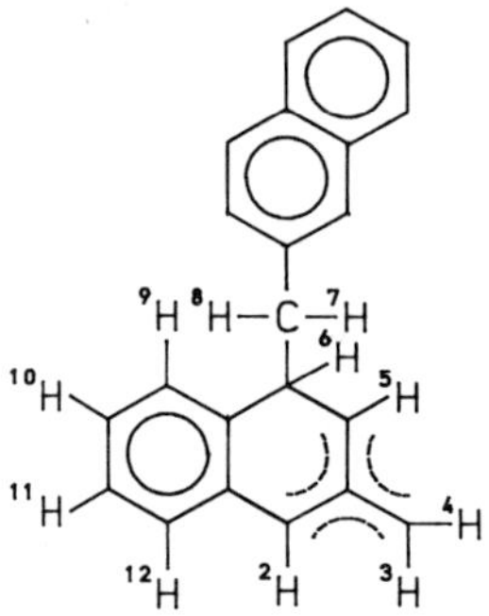	IXa (ca. 10 to 14) m.p. 141°, m.p. 141.5°, subl. p. 115°/10^{-7} Torr, light yellow amorphous powder (from hexane or C_2H_5OH) IR ($CHCl_3$): 1988, 2055 (CO) IR (hexachlorobutadiene): 2920, 2935 (CH_2) or 2906, 2926 (CH_2) IR (KBr): 1978, 1990, 2063 (CO); 2912, 2928 (CH_2) mass spectrum: $[M-nCO]^+$ (n=0 to 3), $[C_{11}H_9Fe]^+$, $[C_{22}H_{18}]^+$, $[C_{21}H_{16}]^+$, $[C_{20}H_{14}]^+$, $[C_{11}H_9]^+$, $[CO]^+$	[4, 22, 27, 31]
*44	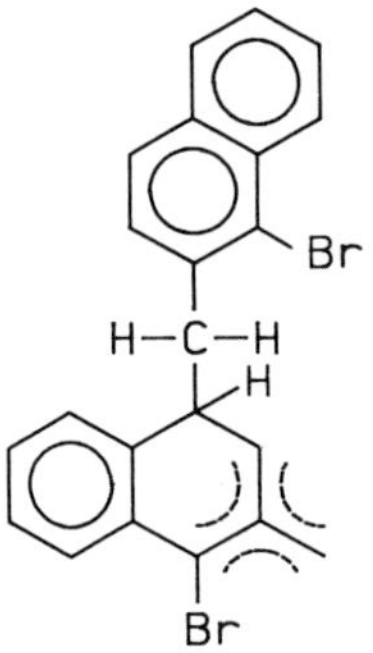	IXa (ca. 10) m.p. 174°, m.p. 174 to 175°, yellow needles IR ($CHCl_3$): 2000, 2063 (CO); 2320 (CH_2)	[27, 31, 43]
*45	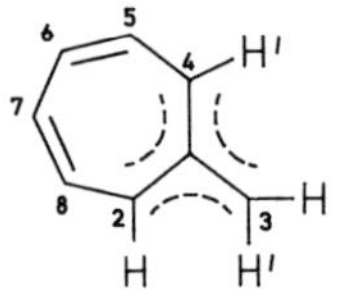	m.p. 37 to 41°, m.p. 38 to 41°, red crystals ^{1}H NMR (cyclohexane (?)): 1.36 (s, H-3,3′), 3.62 (d, H-2,4′; J(H-2,8)=6.8), 5.6 to 6.1 (m, H-5 to 8) ^{1}H NMR (CS_2, −60 to +170°): 1.40 (s, H-3,3′), 3.70 (dd, H-2,4′; J(H-2,8)=6, J(H-2,7)=1), 5.87 (m, H-5 to 8) IR (CH_2Cl_2): 798, 879, 918, 933, 961, 1158, 1322, 1410, 1456, 1484, 1552, 1622, 1699, 1778, 1790, 1855; 1980, 2049 (CO); 2959 IR (cyclohexane): 1973, 1982, 2050 (CO) UV (CH_2Cl_2): $\lambda_{max}(\varepsilon)$=272 (25600), 355 (2100) nm (long tail in the visible region)	[24, 32, 35]
46	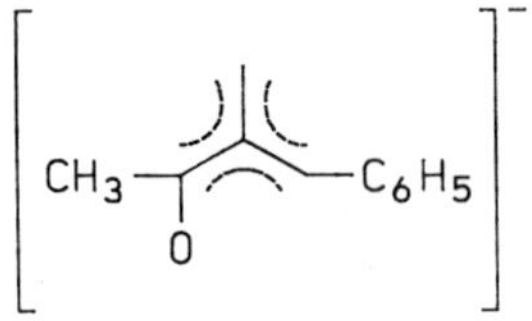 $[CH_3C(O)C(CH_2)CHC_6H_5]^-$	see No. 22, p. 28 not isolated	[74]

Table 1 [continued]

No.	⁴L ligand	method of preparation (yield in %), properties and remarks	Ref.
47	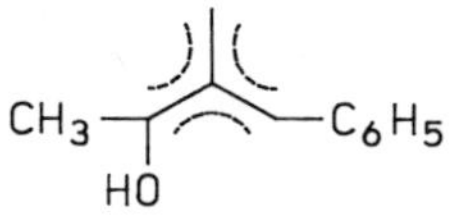 CH_3–C(HO)–C–C_6H_5 allyl	see No. 23, p. 28 not isolated	[74]
48	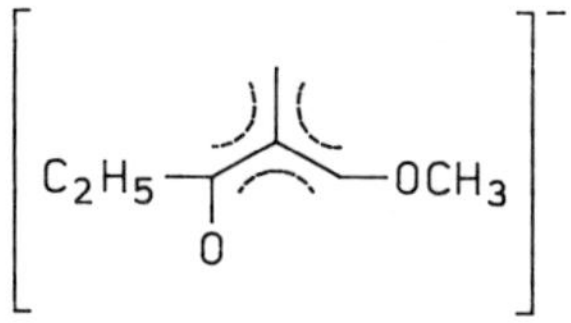 [C_2H_5–C(O)–C–OCH_3]$^-$	see No. 22, p. 28 not isolated	[74]
49	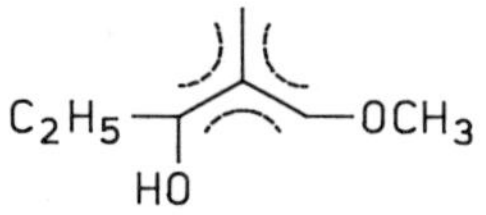 C_2H_5–C(HO)–C–OCH_3	see No. 23, p. 28 not isolated	[74]
*50	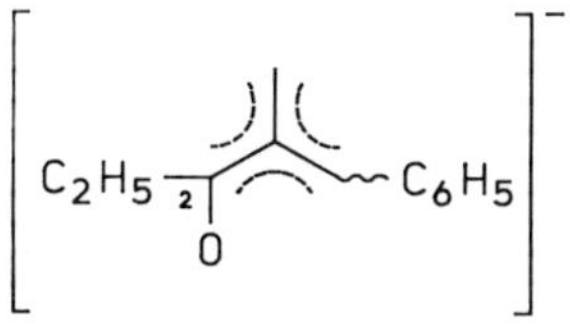 [C_2H_5–C²(O)–C~C_6H_5]$^-$	–	[74]
*51	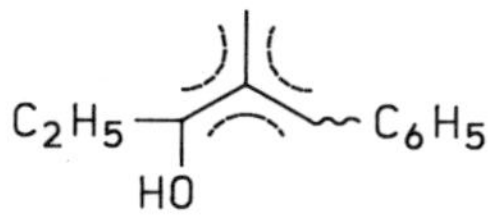 C_2H_5–C(HO)–C~C_6H_5	IR (hexane or pentane): 1970, 1975, 2057 (CO)	[74, 76]
*52	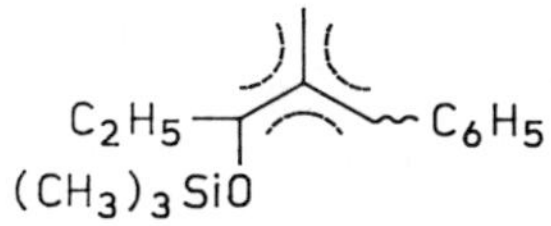 C_2H_5–C($(CH_3)_3SiO$)–C~C_6H_5	see "Further information", p. 36	[76]
*53	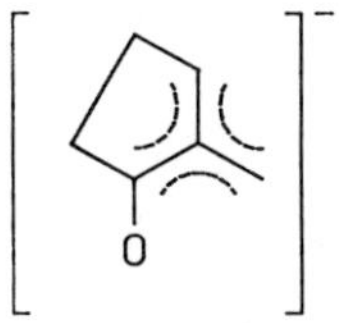 [cyclopentenyl–O]$^-$	IR (tetrahydrofuran): 1600 (C=O); 1880, 1970 (CO)	[64, 67, 77]
*54	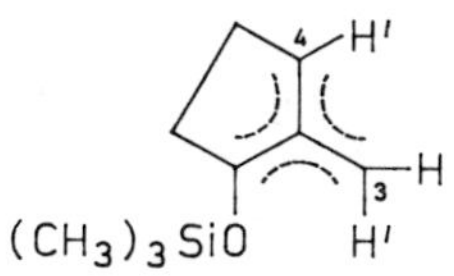 $(CH_3)_3SiO$–cyclopentenyl (H-3, H′, H-4′)	m.p. ≅40° (dec.), subl. p. 21°/10^{-4} Torr, red-orange very air-sensitive solid ¹H NMR (benzene-d_6): 0.1 (s, CH_3), 2.1 (m, H-3, CH_2), 2.7 (m, H-3′, 4′) ¹H NMR ($CDCl_3$): 0.1 (s, CH_3), 2.2 (s, H-3), 2.4 (m, CH_2), 2.7 (s, H-3′), 3.2 (d, H-4′) IR (pentane): 1970, 1980, 2050 (CO)	[64, 76]

References on pp. 39/42

Table 1 [continued]

No.	^{4}L ligand	method of preparation (yield in %), properties and remarks	Ref.
*55	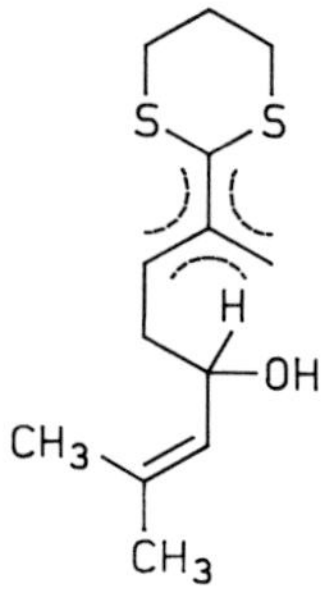 S S H OH CH_3 CH_3	–	[102]
*56	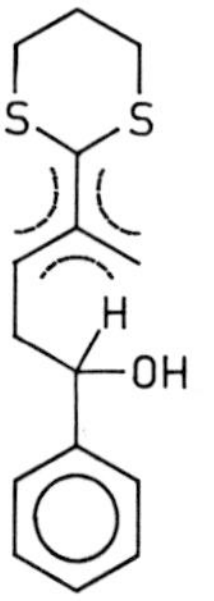 S S H OH	–	[102]
*57	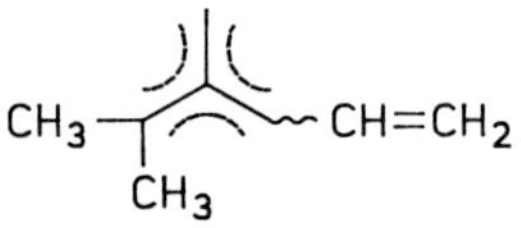 CH_3 CH=CH_2 CH_3	see "Further information", pp. 37/8	[39, 83]
*58	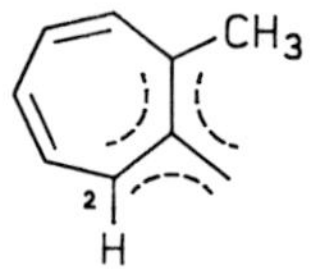 CH_3 2 H	reddish liquid ^{1}H NMR ($CDCl_3$): 1.46 (s, 1H), 1.66 (CH_3), 1.89 (s, 1H), 3.31 (d, H-2; J=7.6), 5.8 to 6.2 (m, 4H) IR (neat): 1970, 2030 UV (C_2H_5OH): λ_{max}(log ε)=268 (4.34), 342 (sh, 3.31), 468 (sh, 2.18) nm mass spectrum: $[M-nCO]^+$ (n=0 to 3)	[88]
*59	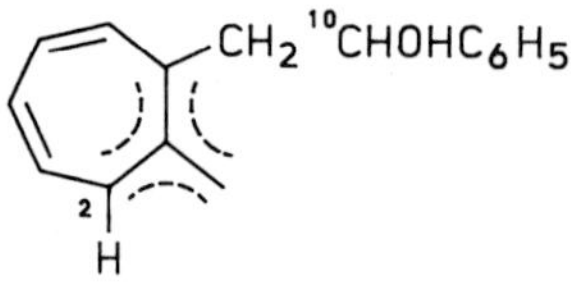 CH_2 ^{10}CHOHC_6H_5 2 H	reddish viscous liquid ^{1}H NMR ($CDCl_3$): 1.20 (s, 1H), 1.47 (s, 1H), 1.84 (dd, 1H; J=6.5, J=13.5), 2.12 (br,s, OH), 2.65 (dd, 1H; J=6.5, J=13.5), 3.24 (d, H-2; J=7.5), 4.86 (t, H-10; J=6.5), 5.8 to 6.4 (m, 4H), 7.25 (br,s, 5H) IR (neat): 1960, 2030; 3300 (OH) UV (C_2H_5OH): λ_{max}(log ε)=269 (4.28), 344 (sh, 3.32), 474 (sh, 2.29) nm mass spectrum: $[M-nCO]^+$ (n=1 to 3)	[88]

References on pp. 39/42

Table 1 [continued]

No.	^{4}L ligand	method of preparation (yield in %), properties and remarks	Ref.
*60	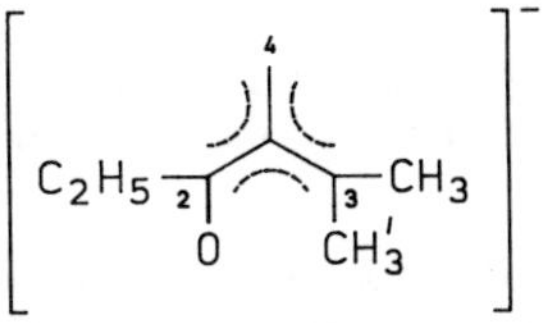 $[$4; C_2H_5 2, 3 CH_3; O, $CH_3'$$]^-$	see "Further information", p. 38	[74 to 77]
*61	H 4 H'; C_2H_5 2, 3 CH_3; HO, CH_3'	^{1}H NMR ($CDCl_3$ at −10°): 1.24 (t, CH_3), 1.47 (d, H-4'; J(H-4,4') = 2), 1.57 (s, CH_3-3), 1.80 (d, H-4), 2.10 (s, CH_3'-3), 2.87 (m, CH_2), 4.38 (br,s, OH) IR (pentane or hexane): 1965, 1975, 2040 (CO)	[74 to 77]
*62	H 4 H'; C_2H_5 2, 3 CH_3; $(CH_3)_3SiO$, CH_3'	^{1}H NMR ($CDCl_3$): 0.26 (s, $Si(CH_3)_3$), 1.23 (t, CH_3; J = 7.5), 1.43 (d, H-4'; J(H-4,4') = 2.5), 1.61 (s, CH_3-3), 1.74 (d, H-4), 2.00 (s, CH_3'-3), 2.1 (m, CH_2; J = 7.5) ^{13}C NMR ($CDCl_3$ at −20°): 16.55 (CH_3), 26.41 (CH_3-3), 26.52 (CH_3'-3), 32.49 (CH_2), 37.74 (C-4), 85.90 (C-3), 88.51 (C-1), 126.67 (C-2), 211.78, 212.68, 213.57 (CO) IR (pentane or hexane): 1967, 1975, 2042 (CO)	[76, 77]
*63	O, OC_2H_5; O=; CH_3	m.p. 78°, yellow crystals ^{1}H NMR ($CDCl_3$): 1.34 (t, CH_3; J = 7), 1.41 (s, CH_3), 3.10 (d, 1H; J = 2), 3.23 (d, 1H; J = 2), 3.63 (m, 1H), 4.20 (m, 1H) IR (CCl_4): 1780, 1990, 2000, 2060	[95]
*64	O, OC_2H_5; O=; CH_3	m.p. 58°, yellow crystals ^{1}H NMR ($CDCl_3$): 1.14 (t, CH_3; J = 7), 1.34 (t, CH_3; J = 7), 1.00 to 2.14 (m, 2H), 3.04 (d, 1H; J = 2), 3.20 (d, 1H; J = 2), 3.66 (qd, 1H; J = 7, J = 9), 4.24 (qd, 1H; J = 7, J = 9) IR (CCl_4): 1775 (C=O); 1990, 2000, 2060 (CO)	[94]
*65	$[$n-C_4H_9; C_2H_5, CH_3; O, $CH_3$$]^-$	–	[74]
*66	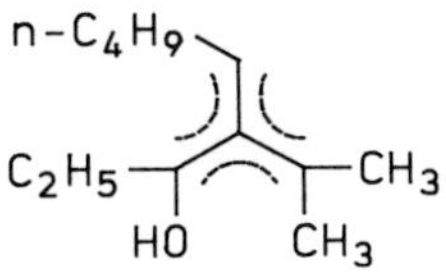 n-C_4H_9; C_2H_5, CH_3; HO, CH_3	–	[74]

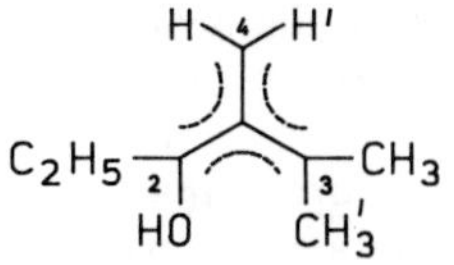

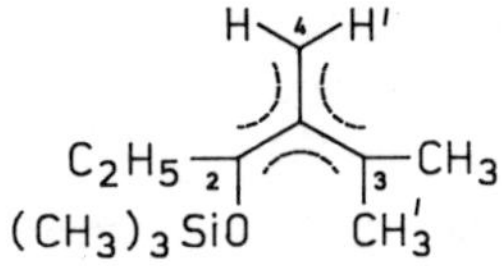

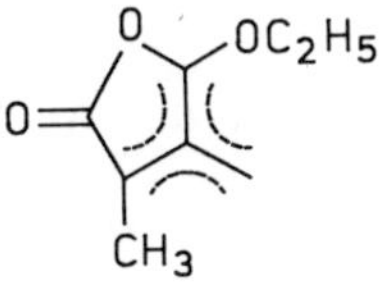

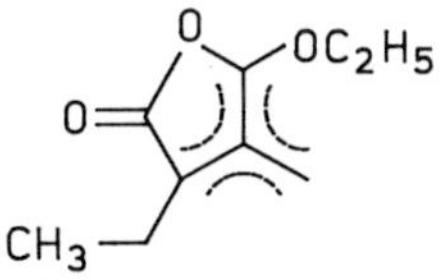

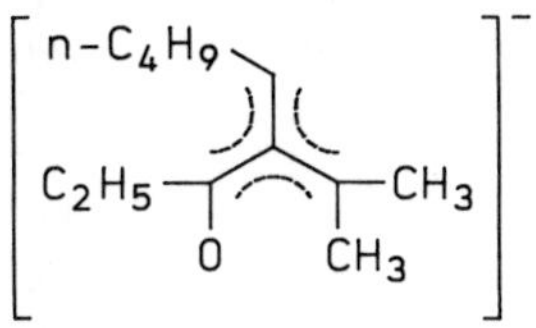

References on pp. 39/42

*Further information:

$(CH_2)_3CFe(CO)_3$ (Table **1**, No. **1**). **Preparation:** Method IIIa (p. 2) gave $^4LFe(CO)_3$ (^4L = buta-1,3-diene) in ca. 2% yield rather than $(CH_2)_3CFe(CO)_3$ [12]. The perdeuterated compound $(CD_2)_3CFe(CO)_3$ is obtained by Method II (p. 1) with $CD_2{=}C(CD_3)CD_2Cl$ and $Fe_2(CO)_9$ in ether [51]. For $(CH_2)_3CFe(^{12}CO)_2{}^{13}CO$ see exchange reactions with ^{13}CO, p. 21.

Physical Properties. $(CH_2)_3CFe(CO)_3$ sublimes readily at room temperature [25, 51]. Its refractive index is $n_D^{30} = 1.5879$ [1]. Since an attempt to record its microwave spectrum was unsuccessful, evidently the dipole moment must be small [9].

From the ^{13}C satellites in its ^1H NMR spectrum, each resonance appears to be a doublet with a splitting of 5 Hz, J(C, H) = 162 ± 1 Hz. No broadening of the proton signal was observed at −60 °C [1]. The ^1H NMR spectrum of $(CH_2)_3CFe(^{12}CO)_3$ in N-(p-methoxybenzylidene)-p-n-butylaniline at 15 °C is consistent with a staggered conformation. The plane containing all six protons intersects the symmetry axis between the Fe and C atoms. The spectrum observed for $(CH_2)_3CFe(^{12}CO)^{13}CO$ in the same solvent resembled a broadened spectrum of $(CH_2)_3CFe(^{12}CO)_3$. This could arise from an intermediate rate of rotation or from insufficient resolution. The attempt to exclude the latter possibility by measuring in a 1:1 mixture of N-(p-methoxybenzylidene)-p-n-butylaniline and N-(p-propoxybenzylidene)-p-n-heptylaniline at temperatures near 0 °C failed. So the nature of the intramolecular motion could not be determined [38].

Some IR active bands at 803, 917, 1025, 1349, 1456, 1478, 1566, 1805, 2475, 2885, 3000, and 3067 cm^{-1} are given in [1] without assignment. The assignments given in Table 2 [51] differ in five assignments from those in [47]. In [47] no values are given for ν_{10} and ν_{29}; ν_{27}, ν_{26}, and ν_8 are assigned to Fe-CO stretching (A_1), in-plane CC_3 deformation (E), and Fe-CO stretching (E) vibrations [40, 47].

Table 2
Observed Fundamental Vibrations of $(CH_2)_3CFe(CO)_3$ (h_6) and $(CD_2)_3CFe(CO)_3$ (d_6) in cm^{-1}. For abbreviations see p. VIII.

C_{3v} species and activity	No.	schematic description	values from [40, 47]: h_6 (liquid)	values from [51]: h_6 (gas)	d_6 (gas)
A_1	1	CH_2 str., sym.	2976	3024	2210
R(p)	2	C≡O str.	2061	2074	2074
IR(‖)	3	CH_2 scissor	1474	1478	1184
	4	CH_2 wagging	989	987	881
	5	C-C str.	918	916	790
	6	CC_3 def.	802	802	707
	7	Fe-C-O def.	604	603?	594?
	8	Fe-CO str.	440	488	482
	9	Fe-C$(CH_2)_3$ str.	372	369	355
	10	$Fe(CO)_3$ def.		130?	128?
A_2	11	CH_2 str., antisym.			
	12	CH_2 twist			
	13	CH_2 rock			
	14	Fe-C-O def.			
	15	torsion			

Table 2 [continued]

C_{3v} species and activity	No.	schematic description	values from [40, 47]: h_6 (liquid)	values from [51]: h_6 (gas)	d_6 (gas)
E	16	CH_2 str., antisym.	3086	3086	2328
R(dp)	17	CH_2 str., sym.	3019	3022	(2215) [a)]
IR(⊥)	18	C≡O str.	1994	2010	2010
	19	CH_2 scissor	1456	1459	1040
	20	C-C str.	1348	1348	1374
	21	CH_2 wagging	1025	1028	844
	22	CH_2 twist	900	900?	
	23	CH_2 rock	815	821	613
	24	Fe-C-O def.	581	582	549
	25	Fe-C-O def.	510	512	507
	26	Fe-CO str.	493	470	461
	27	CC_3 def.	471	438	377
	28	Fe-$C(CH_2)_3$ tilt	351	350	321
	29	$Fe(CO)_3$ def.		104 [b)]	104 [b)]
	30	OC-Fe-$C(CH_2)_3$ def.	95	88	86

a) Estimated from Fermi resonance pair.
b) Liquid value.

The positions of ν(CO) bands in the IR spectra of $(CH_2)_3CFe(CO)_3$ isolated in different matrices are shown in the following table (ν(CO) in cm^{-1}) [62], cf. [66] together with CO stretching force (k) [62, 66] and interaction (k_i) constants in mdyn/Å [62]:

matrix at 12 K	E	A_1	k	k_i
Ar	1998.7, 2000.4, 2003.3	2070.0	16.553	0.385
CH_4	1983.8, 1998.4, 1993.5, 1996.5	2060.0, 2063.7, 2064.4	16.485	0.381
N_2	1990.5, 1992.4, 1996.5, 1998.1, 2000.4	2063.3, 2067.3	16.469	0.387

For weighted mean positions of centers of multiplets in these matrices see [62, 66]. Force fields have been calculated [47, 51, 62, 66, 70]. The most essential valence force constants (in mdyn/Å) are k_{CO}=16.18 [47], 16.43 [70], 16.51 [51]; k_{CH}=5.36 [47], 5.10 [51]; k_{CC}=4.41 [47], 5.47 [51]; k_{Fe^4L}=2.83 [47], 3.70 [51]; and k_{Fe-CO}=3.04 [47], 3.08 [51]; the stretch-stretch interaction constants in mdyn/Å are $k_{CO, CO}$=0.36 [47], 0.28 [51]; $k_{Fe-CO, Fe-CO}$=0.0 [51]; k_{Fe-CO, Fe^4L}=0.0 [51], 0.57 [47]; $k_{Fe-CO, CO}$=0.70 (constrained value) [47], 0.75 [51]. These force constants suggest that a model with bonds between Fe and the three C-C-π orbitals is more suitable than one with a bond along the Fe-central carbon line [47, 51].

The He(I) photoelectron spectra are:

band No.	I		II	III	IV	V
IP in eV [54]	8.00(2) (adiabatic)	8.62(2)	9.26(2)	11.07(6)	12.11(7)	12.57(3)
IP in eV [16]	—	8.32(2)	9.150(36)	10.80(1)	11.070(24)	12.480(29)
assignment [54]	$^2E(13e)$ and	$^2A_1(17a_1)$	$^2E(15_e)$	$^2A_1(16a_1)$	$^2E(14e)$	$^2A_2(2a_2)$

References on pp. 39/42

band No.	VI	VII	VIIa	VIII	IX	X
IP in eV [54]	13.04(4)	13.88(3)	14.28(5)	15.27(5)	17.60(3)	20.78(11)
IP in eV [16]	12.820(21)	—	—	15.230(27)	18.650(21)	—
assignment [54]	(12e)	11e and 15a_1	$1a_2+10e$	9e	8e and 12a_1	7e

Similar values: $I_2=8.63$ (Fe) [78], $I_3=9.25$ (π_1), and $I_4=12.01$ (π_2) eV [63, 76]; I_1 is not resolved [78]. The band reported at 18.65 eV [16] could not be reproduced [54].

In the He(II) photoelectron spectrum bands I and II interchange their relative intensities, IV to VI have very low intensities, III is not detectable, VII and VIII decrease in intensity at the position where III is expected, and IX has a much higher intensity. No distinct bands can be assigned to the 14a_1 and 13a_1 MO's, although there is a fairly high intensity observed before IX in the He(I) spectrum. The 11a_1 MO is not detected and probably lies beyond 24 eV. The position of the low energy photoelectron bands arising from the Fe orbitals are inaccurately predicted by Koopman's theorem [54]. The X-ray photoelectron spectrum shows the following peaks (in eV) and assignments:

290.18	291.47	293.17	539.36	715.14	[81]
290.5	—	293.2	539.0	—	[54]
C1s in CH_2	C1s in $C_{central}$	C1s in CO	O1s	Fe2$p_{3/2}$	[54, 81]

The C1s and O1s binding energies of a wide variety of transition-metal carbonyl compounds in the gas phase are subjected to correlative analysis showing that these energies are linearly related to the k_{CO} force constants. From the relative binding energy shifts it is calculated that in back bonding to CO groups more charge is transferred to the C atoms than to the O atoms [70]. The net electron-withdrawing ability of the $C(CH_2)_3$ group is comparable to that of a butadiene, -S-S-, or CH_3S group [96].

The main distances and angles obtained by gas-phase electron-diffraction investigation of $(CH_2)_3CFe(CO)_3$ at 50 °C are shown in **Fig. 1** [8, 18]. The angle between the C(1)-C(2) bond and the line bisecting the C(2)H_2 angle is 14.4° (±5.1°) [8]. The three Fe-CH_2 distances are essentially equal [8, 18]. Hence the kind of valence tautomerism discussed in [1] can be ruled out [8].

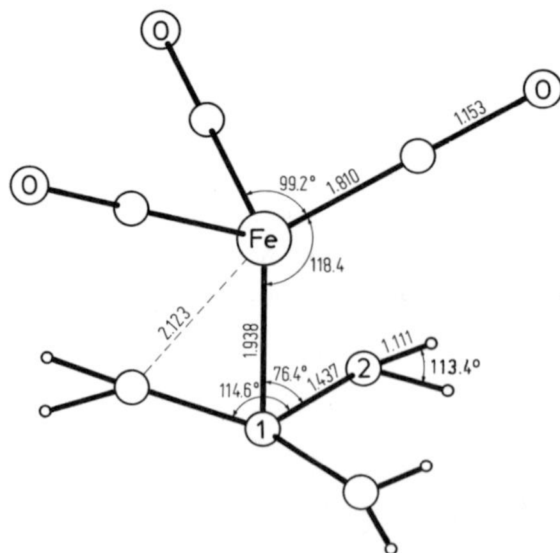

Fig. 1. Molecular structure of $(CH_2)_3CFe(CO)_3$ (No. 1) by gas-phase electron-diffraction (distances in Å) [8, 18].

References on pp. 39/42

The possible coordination polyhedra for metal complexes with π-donor ligands are considered using topology and group theory [13]. The existence of $(CH_2)_3CFe(CO)_3$ can be derived from the elplacarnet tree (**el**emental **pla**nar **car**bon **net**works) [56]. The utility of the effective atomic number rule in predicting the stability, structure, and reactivity is demonstrated by the synthesis of this complex [26]. The conformational preferences [58, 60, 69, 99] and rotational barriers [48, 57 to 61] are analyzed. An orbital interaction diagram for a planar trimethylenemethane and $Fe(CO)_3$ in the preferred staggered geometry is shown in **Fig. 2**.

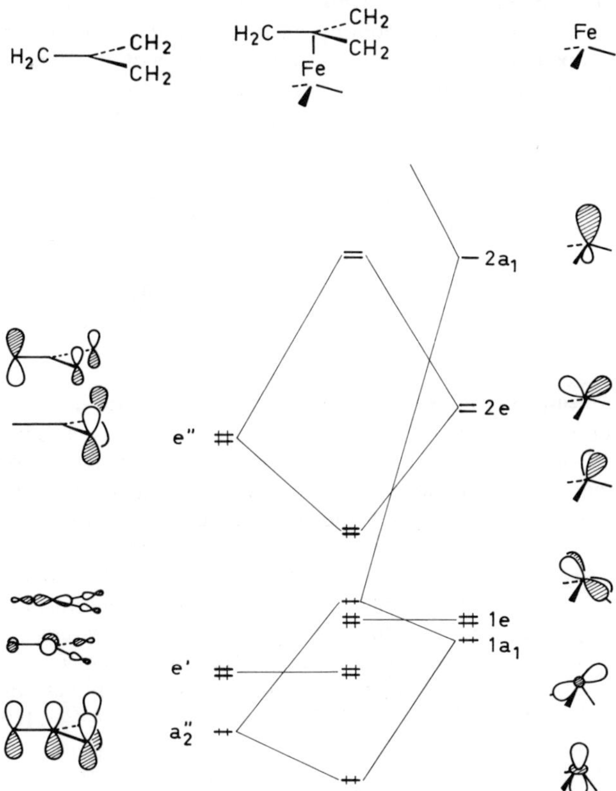

Fig. 2. Orbital interaction diagram for a planar trimethylenemethane and iron tricarbonyl in the staggered geometry [92], cf. [58, 60].

The primary bonding interaction in the complex is that between the 2e set on the $Fe(CO)_3$ fragment and the e″ on $(CH_2)_3C$. However, upon rotation about the Fe–$C(CH_2)_3$ axis by 60 °C into the eclipsed geometry the interaction of these orbitals is decreased because the overlap between them decreases. Therefore the energy of the HOMO in the molecule increases in the eclipsed form and this is the main factor behind the rotation barrier. In the staggered geometry the overlap between the 1e set and e″ is almost zero (0.0116) since that portion of e pointing up toward $C(CH_2)_3$ lies in the nodal region of e″. However, upon rotation to the eclipsed geometry the overlap increases by an order of magnitude (0.1193). The interaction between 1e and e″ is a four-electron repulsive one, the greater the interaction, the less stable the structure. This is then another factor contributing to

References on pp. 39/42

the overall preference for the staggered conformation. Extended Hückel calculations gave a barrier of 87.0 kJ/mol using a planar $C(CH_2)_3$ ligand and carbonyl-iron-carbonyl angles of 90 °C. Considering only the changes in the upper two filled e sets in Fig. 2, p. 19, the staggered geometry is more stable than the eclipsed by 138.9 kJ/mol [58], cf. [49, 57, 60]. According to calculations with an extended CNDO/2 formalism, the staggered conformation is more stable by (only!) 0.696 kJ/mol than the eclipsed one [50]. The trimethylenemethane ligand is puckered in a manner which brings the terminal carbons somewhat closer to the iron and the central carbon away from it (compare Fig. 1, p. 18). There are two reasons for this puckering motion. Firstly, there is a repulsive interaction between a_2'' on trimethylene-methane and $1a_1$. Increasing θ reorients a_2'' so that the methylene orbitals lie in the nodal region of z^2. It also mixes s character into the central carbon of a_2'' from a higher, unoccupied orbital. This mixing occurs in such a manner as to hybridize that component away from the iron (cf. VIII):

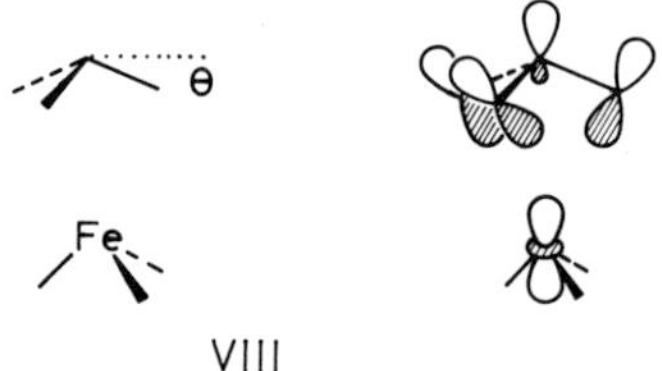

VIII

This factor does not alter the magnitude of the rotational barrier since the orbital is centro-symmetric. The second factor which plays an important role in increasing θ is the mixing between e'' and e' on $C(CH_2)_3$. A detailed analysis shows that e' mixes into e'' so that the orbitals are not only directed more toward the corners of an octahedron and consequently overlap in the staggered conformation more effectively with the 2e set on $Fe(CO)_3$, but they are also hybridized toward the iron. Therefore, as θ increases, the mismatch of overlap between e'' and 2e, and increasing repulsions from 1e in the eclipsed geometry, create a larger barrier. With $\theta = 12°$ a barrier of 87.4 kJ/mol is calculated [58], see also [60, 99]. This "parachute" type electronic structure of the molecule, where Fe is more strongly bonded to the outer three carbon atoms of $C(CH_2)_3$ than to the central one, is also supported by the calculated force field constants (see p. 17) [47, 51] and the net atomic charges obtained from extended CNDO/2 [50] and ab initio SCF MO [54] calculations, cf. [66]:

	atomic charge	
	CNDO/2 [50]	ab initio SCF MO [54]
Fe	+0.67	+1.96
C(carbonyl)	+0.09	+0.11
C(central)	+0.17	−0.24
C(terminal)	−0.30	−0.71
O	−0.05	−0.32
H	−0.01	+0.17

The bond overlap populations obtained from the ab initio SCF MO calculation predict that the iron-hydrocarbon bonding is dominated by the metal 3d and 4p interactions with the 2p orbitals of the outer carbon atoms. A small bonding interaction between the central carbon 2p and the metal 3d and 4s orbitals is noted, although the central carbon-iron overlap population is negative [54].

The OC-Fe-CO bond angles in Ar, CH_4, and N_2 matrices derived by a CNDO/2 formalism are 95.5°, 92.1°, and 94.8°, respectively [66].

A reparametrized semiempirical INDO Hamiltonian is used to calculate the electronic structure of the molecule in the ground state and in the cationic hole-state. Molecular orbital electronegativities for $Fe(CO)_3$ and other transition metal fragments are deduced from this new Hamiltonian by means of the "transition operator method" (TOM). A significant covalent character on the bonding of $Fe(CO)_3$ to $C(CH_2)_3$ is indicated as a result of a synergic bonding interaction due to iron to ligand and ligand to iron charge transfer [93]. Calculated ionization potentials using the "transition operator method" as well as the semiempirical ΔSCF procedure based on the reparametrized Hamiltonian are compared with measured vertical ionization potentials and with ab initio ΔSCF findings derived on the near minimal-basis level [89 to 91, 97]. In the case of strongly localized metal 3d electrons, the validity of Koopman's theorem cannot be assumed due to large electronic rearrangements, which are the sum of relaxation and correlation. The ΔSCF and TOM procedure only account for the relaxation contribution of the whole electronic rearrangement. The reorganization energies for $(CH_2)_3CFe(CO)_3$ are therefore determined by means of the many-body perturbation theory with Green's function formalism and the reparametrized Hamiltonian [89, 97, 100].

Chemical Behavior. Irradiation in an N_2 matrix at 20 K produces $(CH_2)_3CFe(CO)_2N_2$ [41, 46, 62, 66]. Photolysis in an Ar or CH_4 matrix at 12 K produces $(CH_2)_3CFe(CO)_2$ [62]. A vigorously stirred solution of $(CH_2)_3CFe(^{12}CO)_3$ in n-hexane exchanges the CO groups in the presence of ^{13}CO upon irradiation; the IR spectrum of the reaction mixture showed that approximately 30% $(CH_2)_3CFe(^{12}CO)_2{}^{13}CO$ is formed [38]. Irradiation of the compound in hexane in the presence of PF_3 gives $(CH_2)_3CFe(CO)_{3-n}(PF_3)_n$ (n = 1 to 3) [30]. Irradiation of $(CH_2)_3CFe(CO)_3$ in the presence of $(CH_3)_3CBr$ and isobutylene gives good yields of 1,1-dineopentylethylene and 2,2,4,6,6-pentamethylhept-3-ene [10, 11]. The compound reacts with CF_2=CF_2 in hexane upon irradiation, yielding $C_6H_6F_4Fe(CO)_3$ (see Formula X with R = H) [29].

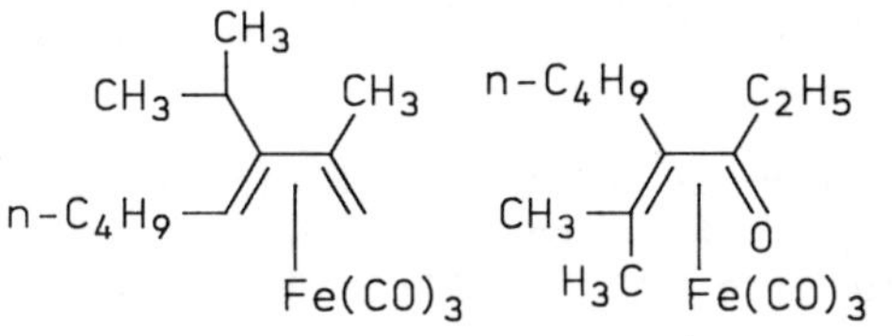

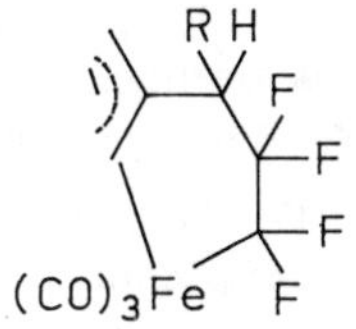

The photolysis of $(CH_2)_3CFe(CO)_3$ in n-pentane through quartz gives ca. 20 products. Some are (yield in %): XI (6), XII (6), XIII (3), XIV (20), XV (3); XVI, XVII, and XVIII (ca. 1).

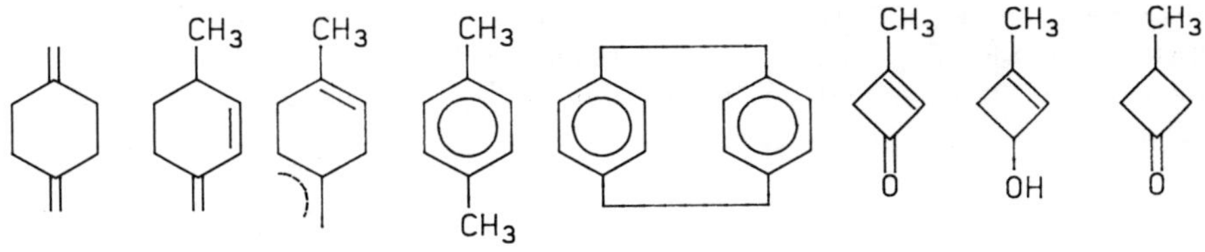

The presence of O_2 is not essential for the formation of XIV although exposure of the reaction mixture to air increases the yield of XIV at the expense of the dienes. The photolysis of $(CH_2)_3CFe(CO)_3$ in cyclopentene as potential trapping agent for the ^4L gives ca. 16 products.

References on pp. 39/42

The formation of XI to XIV is completely suppressed. Products (yield in %): XVI (29), XVII (11), XVIII (3), XIX (5), XX (4, exo and endo), XXI (?, 18), and XXII (1).

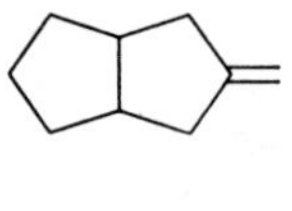

XIX 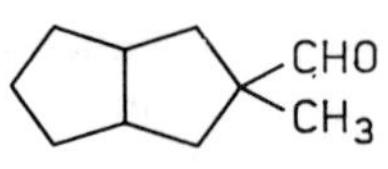 XX 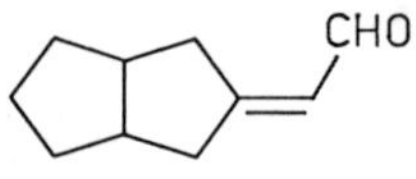 XXI 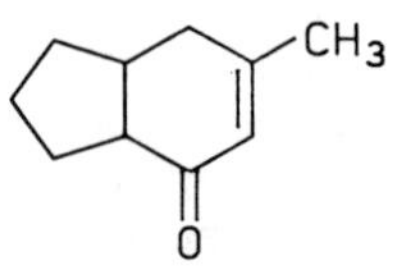 XXII

In the photolysis of $(CH_2)_3CFe(CO)_3$, through pyrex in cyclopentadiene as a potential trapping agent for the ^4L, ca. 16 products are formed. None of the C_8 compounds XI to XIV has been identified. The compound XXIII (23%) has been isolated. A bicyclic hydrocarbon C_9H_{16} (11%) is tentatively formulated as XXIV; XXV and XXVI (ca. 1 to 2%).

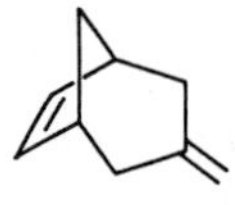

XXIII 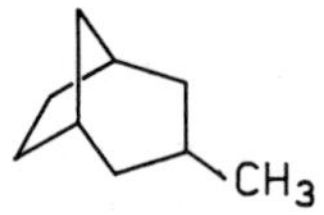 XXIV 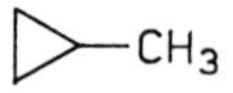 XXV 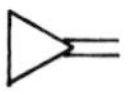 XXVI 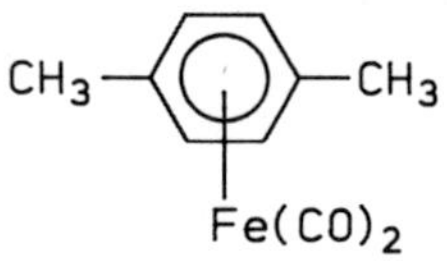 XXVII

XXV and XXVI are products in all three photolyses. Photolysis in cyclopentadiene produces some Fe complexes such as $C_5H_5Fe(CO)_2H$ (32%) and ferrocene (10%). Fe complexes are also formed in n-pentane, e.g.: XXVII (6%), XXVIII (?, 8%), and $^4LFe(CO)_3$ (^4L = isoprene, 5%). Photolyses in cyclopentene and cyclopentadiene solvent produce dimers. The carbonylated products XVI, XVII, XVIII, XX, XXI, XXII provide no evidence of free $C(CH_2)_3$ in the reactions since they all must be formed within the ligand sphere of the Fe. In contrast, such products as XI, XIV, and XXVI strongly suggest that free $C(CH_2)_3$ is involved in the above reactions particularly since the formation of dimers and p-xylene is suppressed by cyclopentene and cyclopentadiene [6].

$(CH_2)_3CFe(CO)_3$ is slowly oxidized by air, $FeCl_3$, and $[NH_4]_2[Ce(NO_3)_6]$ [25]. Oxidation by the latter in the presence of $(NC)_2C{=}C(CN)_2$ gives $C_{10}H_6N_4$ (Formula XXIX) [20]. The reaction with Br_2 in CCl_4 gives on cooling to 0 °C $^3LFe(CO)_3Br$ ($^3L = BrCH_2C(CH_2)_2$) after 2 h [15, 33].

The compound reacts with 96% H_2SO_4 yielding a yellow solution of $[^3LFe(CO)_3]^+$ ($^3L = CH_3C(CH_2)_2$) [1, 33]. Addition of $HOSO_2F$ in liquid SO_2 at −65 °C results in the formation of $^3LFe(CO)_3OSO_2F$ [68]. $(CH_2)_3CFe(CO)_3$ is inert towards $HgCl_2$ or $POCl_3$/N-methyl-formanilide [33]. It gives only low yields on Friedel-Crafts acetylation [14, 25], compare Method VIII (p. 3). Reaction with $CH_3OCCl_2H/TiCl_4$ in CH_2Cl_2 gives $CH_3C(CH_2)_2Fe(CO)_3Cl$ as major product and $OHCCHC(CH_2)_2Fe(CO)_3$ (No. 8) in minor amounts. Use of $AlCl_3$ and $(C_2H_5)_2OBF_3$ catalysts give even poorer results, and an $HC(OC_2H_5)_3/(C_2H_5)_2OBF_3$ reagent does not react with $(CH_2)_3CFe(CO)_3$ [71].

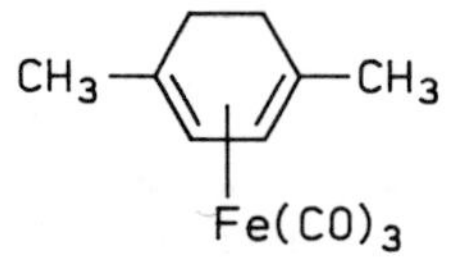

XXVIII 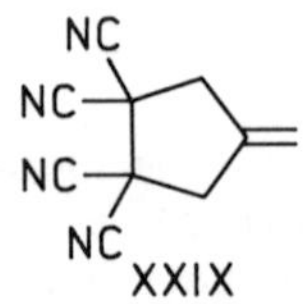 XXIX 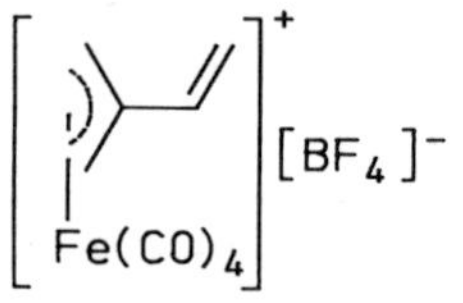 XXX

$CH_3CHC(CH_2)_2Fe(CO)_3$ (Table **1**, No. **3**) is also formed together with $^4LFe(CO)_3$ ($^4L=$ isoprene) on refluxing a solution of $^3LFe(CO)_3Br$ ($^3L=1,2$-dimethylallyl) in cyclohexane for several hours (total yield 36%) [33]. The compound is also obtained by refluxing $^4LFe(CO)_3$ ($^4L=CH_3O_2CCHC(CH_2)_2$) (cf. No. 14, p. 25).

Protonation [14, 25, 33] of the 1:2 mixture of No. 3 and $^4LFe(CO)_3$ obtained from No. 14 with 86% H_2SO_4 gives a red solution. Dilution with H_2O and extraction with ether yields pure $^4LFe(CO)_3$ ($^4L=$isoprene) [14, 33] or isomeric diene complexes [25]. The protonation occurs initially on the Fe atom [25].

$HOCH_2CHC(CH_2)_2Fe(CO)_3$ (Table **1**, No. **4**) reacts with excess $HOSO_2F$ in liquid SO_2 at $-65\,°C$ to give $[C_5H_7Fe(CO)_3]^+$ (Formula VII, p. 3). The same ion is generated by reaction of No. 4 with $HBF_4/O(COCH_3)_2$ at 0 °C under 1 atm of CO; no isolable products are observed and the IR spectra of crude products give no evidence for formation of $[CH_2{=}CHC_3H_4Fe(CO)_4]BF_4$ (XXX) [71], cf. [44]. Oxidation with active MnO_2 in C_6H_6 at room temperature for 5 h followed by filtration and evaporation leaves a crude product which is purified by chromatography on silica gel with CH_2Cl_2 to give No. 8 (74% yield) [68].

Fe(CO)3 CO2CH3 a — Fe(CO)3 CO2CH3 b — Fe(CO)3 CO2CH3 CH3 CH3 N=N

XXXI XXXII

$CH_3O_2CCH{=}CHCHC(CH_2)_2Fe(CO)_3$ (Table **1**, No. **7**). Reaction of XXXIa with $(CH_3)_2CN_2$ gives XXXII in 65% yield (No. 16) which upon heating gives a 4:1 mixture of isomers of XXXIII (No. 15) in 80% yield. Reaction of the isomer XXXIb with $(CH_3)_2CN_2$ gives also only one diastereoisomer of XXXII (No. 16) in 45% yield, which upon heating gives XXXIII (No. 15) stereospecifically in 92% yield [102].

Fe(CO)3 CO2CH3 H3C CH3 — CH3 O O P O CH3 CH2CO2CH3

XXXIII XXXIV

$OHCCHC(CH_2)_2Fe(CO)_3$ (Table **1**, No. **8**). Reaction of $(CH_2)_3CFe(CO)_3$ (No. 1) with $CH_3OCCl_2H/TiCl_4$ gives 10% of impure No. 8, but the major product is π-$CH_3C(CH_2)_2Fe(CO)_3Cl$. Use of $AlCl_3$ and $(C_2H_5)_2OBF_3$ catalysts gives even poorer results [71]. Treatment with CH_3MgI in ether produces ψ-endo-$CH_3(HO)CHCHC(CH_2)_2Fe(CO)_3$ (No. 10) in addition to two unidentified products [68]. Reaction with NaH and XXXIV gives No. 7 (96% yield) as a 1:1 mixture of isomers XXXIa and b separable by chromatography on SiO_2 [102].

$$\left[H_3C \quad Fe(CO)_3 \right]^+ \qquad FO_2SO{-}Fe{-}CO\ (OC\ CO),\ CH_3$$

a b

XXXV

$CH_3COCHC(CH_2)_2Fe(CO)_3$ (Table **1**, No. **9**). The deuterium labelled compound $CD_3COCHC(CH_2)_2Fe(CO)_3$ is prepared by H-D exchange of No. 9 in $NaOCH_3/CH_3OD$. The ^1H NMR spectrum of No. 9 in $CDCl_3$ (see Table 1, p. 6) is assigned on the basis of the deuterated compound [68]. The compound exhibits three distinct ^{13}CO resonances at room temperature, a clear indication of a relatively high barrier of rotation. The signals undergo a reversible broadening and merge into a single sharp signal on warming. The spectra have been taken at approximately five degree intervals from 32 to 88 °C with a two degree interval at the first coalescence point. The two downfield metal carbonyl peaks coalesce at 59 °C, and this peak coalesces with the upfield resonance at 71 °C; $\Delta G^{\ddagger}_c = 18 \pm 1$ kcal/mol. Cooling again reproduces the original spectrum [48]. Reaction with $NaBH_4$ [33, 48, 68] (or $NaBD_4$ [48, 68]) in CH_3OH at 0 °C yields quantitatively reasonably pure [68] No. 10 [33, 48, 68] or ψ-endo-$CH_3(DO)CDCHC(CH_2)_2Fe(CO)_3$ (No. 10) [48, 68]. Pure No. 10 is obtained using dry-column chromatography on silica gel (ca. 50% yield) [68]. Compound No. 5 is obtained with $LiAlH_4/AlCl_3$ [48].

ψ-endo-$CH_3(RO)CHCHC(CH_2)_2Fe(CO)_3$ (Table **1**, Nos. **10** to **12** with R = H, CH_3, CH_3CO). Compound No. 10 is also obtained from the cation XXXV a by addition of H_2O at low temperatures. The solution is neutralized with $NaHCO_3$ and the organic products are extracted with CCl_4 (40% yield of impure No. 10). The deuterium labelled derivatives, ψ-endo-$CD_3HOCHCDC(CH_2)_2Fe(CO)_3$ and ψ-endo-$CD_3(DO)CDCDC(CH_2)_2Fe(CO)_3$, are also reported [68]. For other preparations of No. 10 and its derivative ψ-endo-$CH_3(DO)CDCHC(CH_2)_2Fe(CO)_3$ see No. 9. Compound No. 11 is prepared by methanolysis of the cation $[CH_3C_5H_5Fe(CO)_3]^+$ (Formula XXXV a with CH_3 in anti position). The solution of the cation in liquid SO_2 at −78 °C is quenched by addition of CH_3OH at −78 °C. After 10 min at −78 °C the solution is warmed to room temperature, and H_2O is added. Organic products are extracted with CCl_4, the combined extracts are washed with H_2O and dried (molecular sieve Type 5A), and the volume is reduced (the product is 95% of the ψ-endo diastereoisomer). Quenching a mixture of XXXV a and b in liquid SO_2 at −78 °C with CH_3OH and work up as above gives a mixture (yield 48%) which consists to 76% of No. 11, to 12% of $^4LFe(CO)_3$ (^4L = syn-$CH_3CH{=}CHC(CH_2OCH_3){=}CH_2$), and to 12% of another unidentified methyl ether [68]. For the preparation of No. 12 see the chemical behavior of No. 14. The ^1H NMR spectrum of No. 12 in $CDCl_3$ is assigned on the basis of the deuterated derivatives of No. 10 [68]. Successive addition of $Eu(fod)_3$ (fod = 1,1,1,2,2,3,3-heptafluoro-7,7-dimethyl-octan-4,6-dione) to the ^1H NMR sample of No. 11 leads to resolution of the H-2, 5 multiplet into a downfield H-5 multiplet partly obscured by the CH_3O signal and an upfield H-2 doublet of doublets, J(H-2, 5) = 10 Hz, J(H-2, 4′) = 2 Hz, and to simplification of the CH_3 signal to a doublet, J(H-5, CH_3) = 5.5 Hz [68]. The ^{13}C NMR spectrum of ψ-endo-$CH_3DOCDCHC(CH_2)_2Fe(CO)_3$ in CH_2Cl_2 shows chemical shifts at δ = 27.3 (CH_3), 51.6 (C-4), 53.1 (C-3), 85.9 (C-2), 102.8 (C-1); 211.2, 211.5, 211.6 (CO) ppm [48]. The compounds No. 10 and 12 exhibit three distinct ^{13}CO resonances at room temperature. The signals undergo a reversible broadening and merge into a single sharp signal on warming. For No. 10, spectra are run in 1,4-dioxane at 27, 47, 57, 77 °C, and again 27 °C; at the coalescence temperature at 57 °C $\Delta G^{\ddagger}_c = 17 \pm 2$ kcal/mol for the $Fe(CO)_3$ rotation. The peak sharpens considerably at 77 °C. The rotational barrier is electronic in origin [48]. Compound No. 10 (R=H) reacts with $O(COCH_3)_2$/pyridine to give No. 12 (R=$COCH_3$) [48, 68]. The reaction mixture is stirred at room temperature for 5 h and poured into H_2O. No. 12 is isolated by ether extraction and purified by chromatography on silica gel [48]. $^4LFe(CO)_3$ ($^4L = CH_2$=$C(CH_2OH)CH$=$CHCH_3$) is formed in the reaction of No. 10 with strong acids [25]. No. 10 gives a red solution in 96% H_2SO_4. Addition of ice, extraction with ether, and drying the organic layer and evaporation of the solvent left a yellow oil. This oil in CCl_4 shows bands in the IR spectrum at 1978, 1990, 2048, and 2062 cm^{-1} and chemical shifts in the ^1H NMR spectrum at $\delta = 0.11$ (d; J=8), 0.7 to 2.5 (m, with impurities), 4.1 to 4.4 (m, with impurities), 5.29 (d; J=8) ppm. It consists of a 4:1 mixture of $^4LFe(CO)_3$ ($^4L = CH_2$=$C(CH_2OH)CH$=$CHCH_3$) and No. 10 (deduced from the IR bands at 2048 and 2062 cm^{-1}). In 70% $HClO_4$ a red precipitate is formed from No. 10 which rapidly turns gummy on exposure to air. The IR spectrum of the precipitate in Nujol shows bands at 1984, 2058, 2105, and 2146 cm^{-1} [33]. A mixture of XXXVa and b is obtained from No. 10 and $HOSO_2F$ in liquid SO_2 at −78 °C and warming to −25 °C for 10 min. Only the cation XXXVa is formed in the reaction of No. 10 with excess $HOSO_2F/SbF_5$ (1:1) in liquid SO_2 at −78 °C. No. 12 (R=O_2CCH_3) hydrolyzes with NaOH in CH_3OH/H_2O (1:1) at room temperature for 2 h. This mixture is poured into H_2O, and the product is isolated by ether extraction. Drying ($MgSO_4$) and evaporation leaves No. 10 [68].

$CH_3O_2CCHC(CH_2)_2Fe(CO)_3$ (Table **1**, No. **14**) is volatile and slowly oxidized by air or oxidizing agents such as $[NH_4]_2[Ce(NO_3)_6]$ or $FeCl_3$. It yields No. 13 with strong bases [25] and gives a mixture of $CH_3CHC(CH_2)_2Fe(CO)_3$ (No. 3) and $^4LFe(CO)_3$ (^4L=isoprene) in a 1:2 molar ratio by reduction with $LiAlH_4/AlCl_3$ in refluxing ether after 3 h (chromatography on alumina, pentane as eluent) [25, 33]. Reduction with $LiAlH_4$ in ether at −60 to +65 °C gives No. 4. After 30 min cold $CH_3CO_2C_2H_5$ is added. Slow warming to −20 °C is followed by dropwise addition of 10% aqueous NH_4Cl. The reaction mixture is filtered, the ether layer separated, and the aqueous layer is further extracted with ether. Drying with $MgSO_4$ and evaporation of the combined extracts leaves a dark residue which is purified by chromatography. Elution with CH_2Cl_2 gives the product as a red band (yield 79%) preceded by smaller green and orange bands. Cooling at 4 °C causes crystallization of No. 4 [68, 71].

$CH_3O_2C(CH_3)_2C_3H_2CHC(CH_2)_2Fe(CO)_3$ (Table **1**, No. **15**) is irradiated in CH_3CO_2H to give a mixture of XXXVI a and b in high yield (~85%) [102].

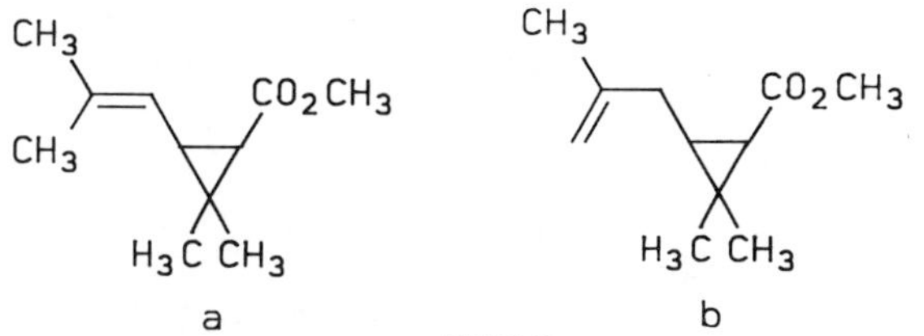

a b

XXXVI XXXVII

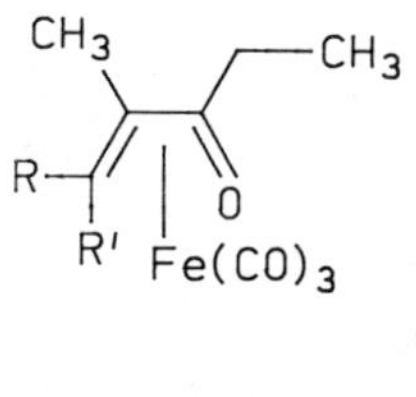

$C_6H_5CHC(CH_2)_2Fe(CO)_3$ (Table **1**, No. **17**) is also prepared from cyclo-$C_6H_5C_4H_5$ (see Formula I, p. 2, with R′=H), $ON(CH_3)_3$ and $Fe(CO)_5$ [73, 83], or $Fe_2(CO)_9$ in benzene [83] according to Method IIIa, p. 2 [83]. The compound No. 17 is also formed in the reaction of cyclo-$C_6H_5C_4H_5$ with $^3LFe(CO)_3$ (^3L=benzylideneacetone) in benzene. The reaction conditions together with the product ratio No. 17: $^4LFe(CO)_3$ (^4L=syn-$C_6H_5C_4H_5$) are shown in the follow-

References on pp. 39/42

ing table [83]:

reagent	reaction time in h	No. 17: syn-$C_6H_5C_4H_5Fe(CO)_3$
$Fe_2(CO)_9$ (Method III a)	5	$>$10:1
$Fe_2(CO)_9$ (Method III a)	20	1.3:1
$Fe(CO)_5/ON(CH_3)_3$	20	$>$20:1
$Fe_2(CO)_9/ON(CH_3)_3$	20	12:1
$^3LFe(CO)_3$	20	$>$10:1
$^3LFe(CO)_3$	96	1.25:1

To clarify the reaction mechanism, cyclo-2-$C_6H_5C_4H_4D$ (see Formula I, p. 2, deuterated in position 3, trans and cis with respect to $R=C_6H_5$), is subjected to these reactions. I, deuterated in trans position 3 gives $C_6H_5CHC(CH_2)CDHFe(CO)_3$, deuterated in position H-3′ (see Table 1, pp. 7/8) and the cis deuterated I gives $C_6H_5CHC(CH_2)CHDFe(CO)_3$ deuterated in position H-3 (for a figure of the ^1H NMR spectra of these deuterated isomers and No. 17 see [73, 83]). These reactions with the labelled I clearly show that the reaction mechanism, discussed in [12, 34, 39], cannot be correct. Comparison of the ^1H NMR spectra reveals that the ring-opening occurs in a disrotatory sense and the reaction is stereospecific, a result that rules out the previously proposed zwitterionic mechanism [12, 34, 39] but which is consistent with the frontier molecular orbital predictions [72] for a pericyclic process [73, 83].

The complex No. 17 crystallizes in the centrosymmetric monoclinic space group $P2_1/a-C^5_{2h}$ with $a=16.322\pm0.012$, $b=6.632\pm0.005$, $c=12.542\pm0.008$ Å, $\beta=117.18°\pm0.05°$, and Z=4 molecules for the unit cell. The observed density is 1.33 ± 0.05 g/cm^3; the calculated 1.289 g/cm^3 [7, 17]. The main distances and angles are given in **Fig. 3**. The angles Fe-C(1)-C(2), Fe-C(1)-C(3), Fe-C(1)-C(4) are $76.1°\pm0.6°$, $77.1°\pm0.6°$, $78.4°\pm0.5°$ [7, 17]. The Fe atom is π-bonded to the four C atoms C(1) to C(4) [17], and the ^4L ligand is stabilized by the $Fe(CO)_3$ moiety [5]. The carbon atom skeleton of the trimethylenemethane is significantly nonplanar, C(1) being displaced away from the Fe atom by 0.315 Å relative to the plane defined by C(2), C(3), and C(4). The plane of the C atom adjacent to the phenyl ring (i.e., C(4), H(4), and C(5)) makes an angle of 34° 7′ with the phenyl group. The p_π-p_π overlap integral for the C(4)-C(5) bond is thus reduced to approximately 83% of the value it would take if C(4), H(4), and C(5) were coplanar with the phenyl group [17].

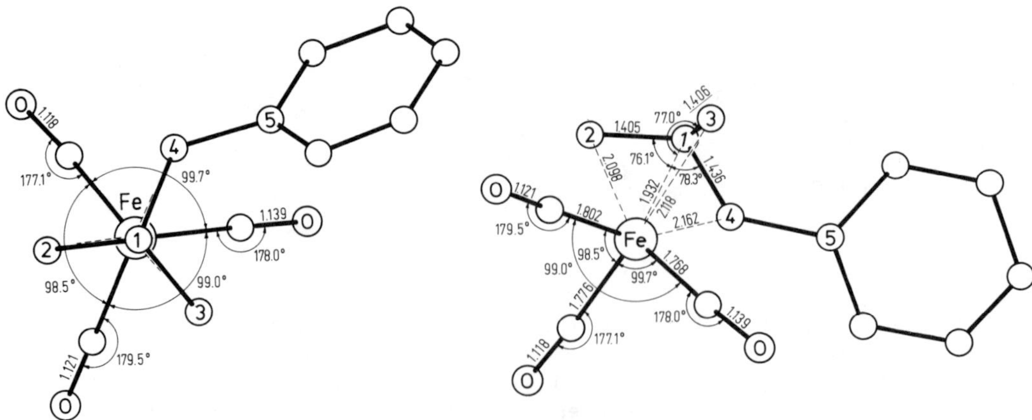

Fig. 3. Molecular structure of $C_6H_5CHC(CH_2)_2Fe(CO)_3$ (No.17) [17].

References on pp. 39/42

The mass spectrum shows fragments which resemble those of cyclo-$C_6H_5C_4H_5$ (see Formula I, p. 2, with R=H, R′=C_6H_5) [12]. The volatile compound is only slowly oxidized by air or oxidizing agents such as $[NH_4]_2[Ce(NO_3)_6]$ or $FeCl_3$ [25]. Irradiation of a mixture of $C_6H_5CHC(CH_2)_2Fe(CO)_3$ and CF_2=CF_2 in hexane gives $C_{12}H_{10}F_4Fe(CO)_3$ (see Formula X, p. 21, with R=C_6H_5) [29].

$[C_2H_5(OC)C(CH_2)_2Fe(CO)_3]^-$ (Table **1**, No. **19**) is obtained as the Na^+ [65, 67, 74 to 77] and isolated as the $[N(P(C_6H_5)_3)_2]^+$ salt [65, 77]. 4.5×10^{-2} mol of gaseous allene is dissolved at −5 °C in tetrahydrofuran which contains 1.2×10^{-2} mol $Na_2Fe(CO)_4 \cdot 2O(CH_2CH_2)_2O$ and 1.0×10^{-2} mol of C_2H_5Br is added dropwise to this solution [76, 77], cf. [65, 67, 74, 75]. The reaction mixture is filtered by Celite [76, 77]. This solution of $Na[C_2H_5COC(CH_2)_2Fe(CO)_3]$ may be used [76, 77] or the solvent may be evaporated in vacuum and the oily residue can be redissolved in CH_2Cl_2 [77]. $Na[C_2H_5COC(CH_2)_2Fe(CO)_3]$ is also obtained by the reaction of $C_2H_5(HO)CC(CH_2)_2Fe(CO)_3$ (No. 20) with Na or NaOH in tetrahydrofuran [65]. $[N(P(C_6H_5)_3)_2][C_2H_5COC(CH_2)_2Fe(CO)_3]$ is prepared by the reaction of $Na[C_2H_5COC(CH_2)_2$-$Fe(CO)_3]$ with $[N(P(C_6H_5)_3)_2]Cl$ in CH_2Cl_2 [77], cf. [65]. After 12 h the reaction mixture is filtered through Celite. Addition of cold ether results in crystals, which are purified by CH_2Cl_2/ether (1:1.5) at −20 °C (yield 85%) [77]. $[N(P(C_6H_5)_3)_2][C_2H_5COC(CH_2)_2Fe(CO)_3]$ forms yellow crystals, m.p. 114 to 115 °C [65, 77]. The ^1H NMR spectrum of the $[N(P(C_6H_5)_3)_2]^+$ salt is given in the following table (δ in ppm, J in Hz, for assignment see Table 1, p. 8):

t in °C	solvent	H-3	H-4′	H-3′	H-4	CH_2	CH_3	Ref.
+25	CD_2Cl_2	0.9(s)	0.9(s)	2.16(s)	2.16(s)	—	—	[65]
−5	CD_2Cl_2	0.9(s)	0.9(s)	2.16(s) J(H-3′,4)=0	2.16(s)	2.35(q) J=7	1.15(t) J=7	[77]
−90	HCF_2Cl	1.06(s) or 0.94(d)	0.94(br,s) or 1.06(d)	2.37(d) or 1.94(d) J(H-3′,4)=4	1.94(d) or 2.37(d)	2.35(q) J=7	1.15(t) J=7	[65, 77]

The ^1H NMR spectrum at low temperatures shows a restricted rotation about the C(1)-C(2) bond. At temperatures >−30 °C the chemical shifts for H-3′,4 and H-3,4′ coalesce; the free activation energy for the rotation at −30 °C is $\Delta G_c^{\ddagger} = 11 \pm 0.5$ kcal/mol [65]. For H-3′,4 the coalescence temperature is $T_c = -60 \pm 2$ °C, for H-3,4′ $T_c = -40 \pm 2$ °C [77].

The ^{13}C NMR spectrum of $[N(P(C_6H_5)_3)_2][C_2H_5COC(CH_2)_2Fe(CO)_3]$ in CD_2Cl_2 at −20 °C shows chemical shifts at δ=12.2 (CH_3), 28.7 (CH_2), 35.16 (C-3,4), 76.60 (C-1), 201.4 (C-2), and 221.40 (CO) ppm [76]. The IR bands for both salts in different solvents are shown in the following table (ν(CO) and ν(=C=O) in cm^{-1}):

cation	solvent	ν(=C=O)	ν(CO)	Ref.
Na^+	tetrahydrofuran	1560	1890, 1980	[65, 77]
Na^+	N-methylpyrrolidone	1600	1875, 1965	[65]
Na^+	C_2H_5OH	—	1900, 1990	[65]
$[N(P(C_6H_5)_3)_2]^+$	tetrahydrofuran	1605	1875, 1965	[65, 77]

Several mesomeric structures are discussed for the anion, see e.g., [65, 77]. By monitoring the change in ν(CO) and ν(=C=O) in the IR spectrum, ion pairing phenomena as a function of the solvent and/or the counter ion are observed [77].

References on pp. 39/42

Reaction of the Na^+ salt [65, 67, 75 to 77] or the $[N(P(C_6H_5)_3)_2]^+$ salt [65] with CH_3COOH gives $C_2H_5(HO)CC(CH_2)_2Fe(CO)_3$ (No. 20) [65, 67, 75 to 77]. Reaction of the $Na[C_2H_5CO\text{-}C(CH_2)_2Fe(CO)_3]$ with $ClSi(CH_3)_3$ results in the formation of $C_2H_5((CH_3)_3SiO)CC(CH_2)_2\text{-}Fe(CO)_3$ (No. 21) [65, 67, 76, 77].

$C_2H_5(HO)CC(CH_2)_2Fe(CO)_3$ (Table **1**, No. **20**) is prepared by acidification of the anion $[C_2H_5COC(CH_2)_2Fe(CO)_3]^-$ (No. 19) [65, 67, 74 to 76] starting from the $[N(P(C_6H_5)_3)_2]^+$ salt [65] or from the Na^+ salt [65, 67, 75, 76]. The solution of the Na^+ salt is reacted with CH_3CO_2H at room temperature. The resulting violet solution is filtered through Celite and the deep red filtrate is concentrated. The oily residue is extracted with pentane, and the pentane solution is filtered several times. This solution contains the complex $C_2H_5(HO)CC(CH_2)_2Fe(CO)_3$ (No. 20) which can be stored at −20 °C or redissolved in $CDCl_3$ (yield 85%) [76]. The complex isomerizes with great readiness to XXXVII (see p. 25) with R=R′=H [65, 67]. A solution of No. 20 in pentane, which is concentrated and chromatographed on silica gel (eluent pentane (50%)/CH_2Cl_2 (50%)) gives XXXVII with R=R′=H in 90% yield, which makes the separation of No. 20 difficult [75]. Also, distillation at 50 °C/10^{-4} Torr gives XXXVII (R=R′=H) in 80% yield [75, 76]. In tetrahydrofuran at room temperature the isomerization is complete after 3 d and at 60 °C within 1 h [75]. Reaction with Na or NaOH gives the anion $[C_2H_5COC(CH_2)_2Fe(CO)_3]^-$ (No. 19) [65, 76], and reaction with $[N(P(C_6H_5)_3)_2]Cl$ gives its salt [65]. Reaction with $ClSi(CH_3)_3$ gives $C_2H_5((CH_3)_3SiO)C\text{-}C(CH_2)_2Fe(CO)_3$ (No. 21) [65, 67].

$C_2H_5((CH_3)_3SiO)CC(CH_2)_2Fe(CO)_3$ (Table **1**, No. **21**) is prepared from $[C_2H_5COC(CH_2)_2\text{-}Fe(CO)_3]^-$ (No. 19) and $ClSi(CH_3)_3$ [65, 67, 76] at room temperature in tetrahydrofuran [76]. After filtration on Celite, the filtrate is evaporated in vacuum, and the residue is extracted with pentane. The pentane solution is concentrated and chromatographed on silica gel with pentane (80%)/CH_2Cl_2 (20%) (yield 90%) [76]. The complex is also obtained by the reaction of $C_2H_5(HO)CC(CH_2)_2Fe(CO)_3$ (No. 20) with $ClSi(CH_3)_3$ [65, 67]. Protonation of the complex with CF_3COOH in CH_2Cl_2 at room temperature gives XXXVI with R=R′=H [65, 67, 76] in 70% yield [76].

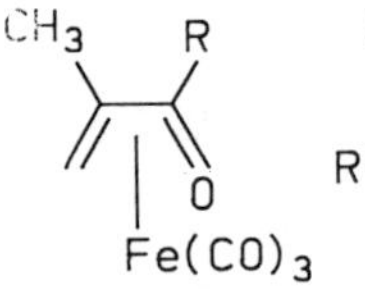

XXXVIII 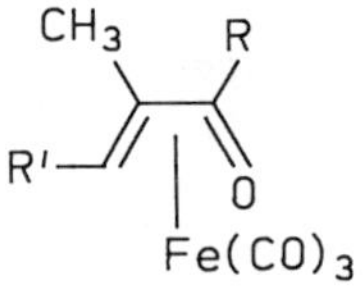 XXXIX

$[n\text{-}C_3H_7COC(CH_2)_2Fe(CO)_3]^-$ and **$n\text{-}C_3H_7(HO)CC(CH_2)_2Fe(CO)_3$** (Table **1**, Nos. **22** and **23**) are discussed as intermediates in the reaction of $n\text{-}C_3H_7X$, $Na_2Fe(CO)_4 \cdot 1.5\ O(CH_2CH_2)_2O$, and $CH_2{=}C{=}CH_2$ in tetrahydrofuran yielding XXXVIII with $R{=}n\text{-}C_3H_7$ upon acidification with CH_3CO_2H. Similar reaction of RX ($R{=}n\text{-}C_mH_{2m+1}$ (m=4 to 7), X=halogen, $OSO_2C_6H_4CH_3\text{-}4$) results in the formation of XXXVIII with $R{=}n\text{-}C_mH_{2m+1}$ with compounds No. 24 to 31 as intermediates, which is indicated by IR absorption at 1560 (C=O), 1885 to 1890, and 1965 to 1980 (CO) [55, 74]. Compounds No. 46 and 47 are discussed as intermediates in a similar reaction of CH_3X, $Na_2Fe(CO)_4 \cdot 1.5\ O(CH_2CH_2)_2O$, and $C_6H_5CH{=}C{=}CH_2$ yielding XXXIX with $R{=}CH_3$, $R'{=}C_6H_5$. Similar reaction with C_2H_5X and $CH_3OCH{=}C{=}CH_2$ gives XXXIX with $R{=}C_2H_5$, and $R'{=}OCH_3$, and compounds No. 48 and 49 as intermediates [74].

$R_2CC(CH_2)_2Fe(CO)_3$ (Table **1**, Nos. **32** and **33** with $R{=}CH_3$ and $CO_2C_2H_5$) are volatile and only slowly oxidized by air or oxidizing agents such as $[NH_4]_2[Ce(NO_3)_6]$ or $FeCl_3$ [25].

Protonation of No. 32 in 86% H_2SO_4 followed by dilution with H_2O yields $^4LFe(CO)_3$ (^4L = 2,3-dimethylbuta-1,3-diene) in 95% yield [33].

$(C_6H_5)_2CC(CH_2)_2Fe(CO)_3$ (Table **1**, No. **35**). Reaction of cyclo-$(C_6H_5)_2C_3H(D){=}CH_2$ (see Formula I, p. 2, with R = R' = C_6H_5, monodeuterated at C-3) with $Fe_2(CO)_9$ as described in Method IIIa, p. 2, gives $(C_6H_5)_2CC(CH_2)CHDFe(CO)_3$ monodeuterated in position H-4' (see Table 1, p. 10) and $(C_6H_5)_2CC(CH_2)CDHFe(CO)_3$ monodeuterated in position H-3' (see Table 1, p. 10) in a ratio (1.02 ± 0.02) : 1 [84]. Compound No. 35 is also formed in the reaction of cyclo-$(C_6H_5)_2C_3H_2{=}CH_2Fe(CO)_4$ with $Fe_2(CO)_9$ or $ON(CH_3)_3$ at 20 °C in benzene [83].

Fe(CO)3
CH3O2C CO2CH3
E,E
CH3O2C CO2CH3
Z,Z
isomers
XL

Fe(CO)3
CH3O2C CO2CH3
E,Z
Fe(CO)3
CH3O2C CO2CH3
E,Z
isomers
XLI

$(CH_3O_2CCH)_2CCH_2Fe(CO)_3$ (Table **1**, Nos. **36** and **37**). From the four possible isomers XL and XLI only the (E,E)-isomer XL (No. 36) is observed in the reaction of cis-Feist's methyl ester (see Formula II, p. 2) with $Fe_2(CO)_9$ by Method IIIb, p. 2. Trans-Feist's methyl ester (see Formula II) yields a racemic mixture of (E,Z)-isomers XLI in a ratio of 1 (No. 37) under similar conditions. Attempts were made in vain to alter the ratio of 1 of the enantiomers by changing the reaction medium to optically active isopentyl alcohol and by adding d-camphor to the benzene reaction medium. No. 37 can be easily purified by preparative thin layer chromatography. No. 36 gives the correct osmometric molecular weight in ethyl acetate.

The ^1H NMR spectra of (E,E)-XL (No. 36) and racemic (E,Z)-XLI (No. 37) in different solvents are given in the following table:

complex	solvent	^1H NMR spectra, δ in ppm		
		CH_2	CH	CH_3
(E,E)-XL	$CDCl_3$	2.15 (s)	2.65 (s)	3.7 (s)
(No. 36)	benzene-d_6	1.7 (s)	2.2 (s)	3.4 (s)
	acetonitrile-d_3	2.2 (s)	2.62 (s)	3.65 (s)
	pyridine-d_5	2.5 (s)	2.73 (s)	3.67 (s)
(E,Z)-XLI (No. 37)	$CDCl_3$	2.1 to 2.4 (AB, dd, $J_A = J_B = 3.0$ Hz, $J_{AB} = 9.0$ Hz)	2.75 (apparent singlet)	3.71 (s) 3.74 (s)
	benzene-d_6	2.0 (apparent singlet)	2.8 (apparent singlet)	3.32 (s) 3.37 (s)

References on pp. 39/42

Table [continued]

complex	solvent	^1H NMR spectra, δ in ppm CH$_2$	CH	CH$_3$
	acetonitrile-d_3	2.15 to 2.45 (AB, dd, $J_A=J_B=3.0$ Hz, $J_{AB}=9.0$ Hz)	2.75 (apparent singlet)	3.67 (s) 3.71 (s)
	pyridine-d_5	2.25 to 2.45 (AB, dd, $J_A=J_B=3.0$ Hz, $J_{AB}=6.0$ Hz)	2.95 (apparent singlet)	3.63 (s) 3.68 (s)
	CCl_4	2.0 to 2.35 (AB, dd, $J_A=J_B=3.0$ Hz, $J_{AB}=9.0$ Hz)	2.70 (s)	3.66 (s) 3.70 (s)

The ^1H NMR spectrum of (E, E)-XL (No. 36) is completely solvent independent. In contrast, the ^1H NMR spectrum of racemic (E, Z)-XLI (No. 37) is solvent-dependent and exhibits two multiplets and an apparent singlet in the ratio of 3:1:1. The ^1H NMR spectrum of No. 37 is examined as a function of added chiral shift reagent A (tris[3-(2,2,2-trifluoro-1-hydroxyethyl)-d-camphorato]europium (III)). The data obtained are:

chiral reagent A (see text)	^1H NMR spectrum in CCl_4, δ in ppm
none	δ=2.13 to 2.34 (AB pattern, dd, 1H, CH_2, $J_A=J_B=3.0$ Hz, $J_{AB}=9.0$ Hz), 2.73 (s, 1H, CH), 3.73 (s, 1.5H, CH_3), 3.76 (s, 1.5H, CH_3)
3 drops	δ=2.08 to 2.33 (br, m), 2.83 (br, s), 3.70 (s), 3.73 (s)
8 drops	δ=2.23 to 2.53 (br, s), 3.00 to 3.42 (br, m), 3.81 (apparent s), 3.97 (s), 4.02 (s); area of δ=4.02+3.97 equals area of δ=3.81
12 drops	δ=2.23 to 2.53 (br, s), 3.26 to 3.67 (br), 3.87 (br, s), 4.10 (s), 4.15 (s); area of δ=4.15 equals area of δ=4.10; area of δ=4.15+4.10 equals area of δ=3.87

Under the assumption that the ester carbonyl is complexed with A, the hydrogens nearest to the ester carbonyl exhibit the greatest downfield shift (Δδ=0.53 ppm) while the hydrogens furthest from the center of complexation exhibit the least downfield shift (Δδ=0.19 ppm). Thus in the case of (E, Z)-XLI, the two CH_3 singlets at δ=4.10 and 4.15 ppm may be assigned to a CH_3CO_2 group outside the CC_3 system, and the CH_3 singlet at δ=3.87 ppm may be assigned to a CH_3CO_2 group inside the CC_3 system. The equal areas suggest that the ratio of racemic (E, Z)-XLI is exactly one. The remaining signals are too broad and ill-defined to assign multiplicities or areas. The mass spectra of both complexes show mainly $[(CH_3O_2CCH)_2CCH_2Fe(CO)_n]^+$ with n=0, 1, 2, 3 and also some unexpected ions of greater mass which probably originate from dimer-like species [36].

$(C_6H_5CH)_2CCH_2Fe(CO)_3$ (Table **1**, No. **38**) is prepared from trans-2,3-diphenylmethylenecyclopropane XLII (R=H) and $Fe_2(CO)_9$. Similar reaction of the monodeuterated XLII (R=D) gives XLIIIb and a in the ratio 1.22±0.03. The apparent isotope effect and the reaction mechanism is discussed [84].

References on pp. 39/42

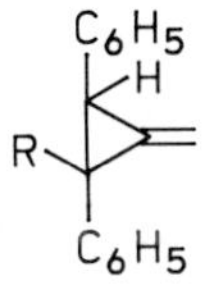

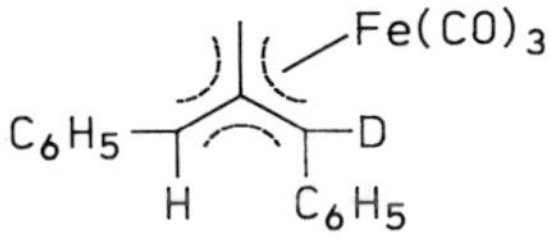

XLII XLIII a b XLIV

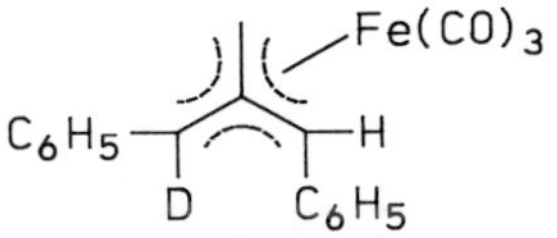

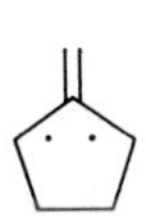

$C_5H_6CH_2Fe(CO)_3$ (Table **1**, No. **39**) is prepared by heating cyclopent-1-enylmethyl chloride and $Fe_2(CO)_9$ at reflux in C_6H_6 for 12 h followed by gas chromatography (3% yield). Oxidation with $ON(CH_3)_3$ in C_6H_6 at 60 °C gives a 14% yield of a mixture of hydrocarbon products, at least ten of which are, empirically, dimers of XLIV. When the oxidation of No. 39 is carried out in the presence of a large excess of diethylfumarate as trapping agent, two fused cycloadducts, XLV a and b and the trans bridged adduct XLVI are observed in combined yields ranging from 9 to 19%. The yield of CO_2 is 0.22 mol/mol of No. 39. Control experiments established that the product ratios are kinetically determined and do not change during prolonged reaction times, but depend on the concentration of the trapping agent. The result suggest that the intermediate XLIV is liberated in its triplet state [98, 101].

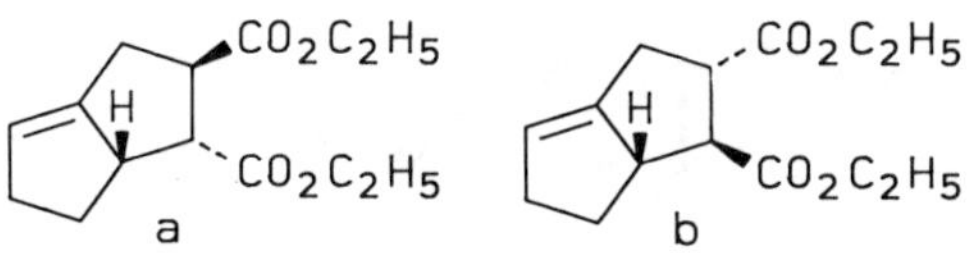

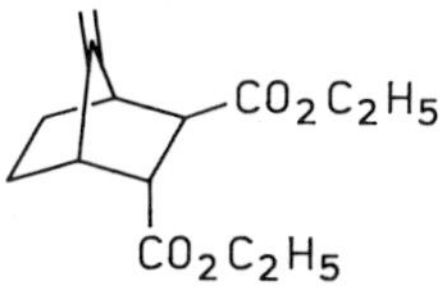

XLV a b XLVI

$C_{14}H_{12}Br_2Fe(CO)_3$ (Table **1**, No. **40**). The byproducts from Method IXa, p. 3, are $(p\text{-}BrC_6H_4CH_2)_2CO$ (33% yield), $FeBr_2$, $Fe(CO)_5$, and CO [42]. The compound crystallizes in the monoclinic space group B2/b-C^6_{2h} with a = 31.539(9), b = 12.523(3), c = 9.417(2) Å, γ = 109.65(15)°, and Z = 8 molecules in the unit cell. The observed and calculated densities are 1.80 and 1.819 g/cm^3, respectively. The main bond distances and angles are given in **Fig. 4**, p. 32 [52].

$C_{22}H_{18}Fe(CO)_3$ (Table **1**, No. **42**) shows in the ^{1}H NMR spectra (at 60, 100, and 220 MHz; $(CH_3)_3SiSi(CH_3)_3$ as reference) in CS_2 chemical shifts at δ = 1.57 (s, CH_2, H-3 or H-4), 1.7 (s, CH_2, H-4 or H-3), 3.60 (complex m, N), 6.92 (s, aromatic H), 7.52 (complex m, aromatic H) ppm and in C_6H_6 (at 100 MHz) shifts at δ = 1.24 (s, CH_2, H-3 or H-4), 1.302 (s, CH_2, H-4 or H-3) and an ABCDE type pattern of N with shifts at δ = 3.305 to 3.512 (complex m, H-6, 7), 3.136 (four lines H-8), 3.555 (t, H-5), and 3.614 (d, H-2) ppm. The multiplet at 7.52 ppm resolves at 220 MHz in CS_2 into areas of ratio 4:1:1:1, representing chemical shifts in the naphthyl group with δ = 7.16 to 7.50(m), 7.62(d), 7.73(d), and 8.05(d) ppm (numbering see Table 1, No. 42, p. 11) [27].

Compound No. 42 crystallizes in the monoclinic space group $P2_1/c$-C^5_{2h} with a = 8.36 ± 0.03, b = 17.63 ± 0.02, c = 13.73 ± 0.02 Å, β = 95° ± 1°, and Z = 8 molecules per unit cell. The measured density is 1.40 g/cm^3; the calculated is 1.4 g/cm^3. **Fig. 5**, p. 32, gives the principal bond angles and distances. The distances from the Fe atom to the C atoms of the trimethylenemethane ligand are Fe-C(1) = 1.95, Fe-C(2) = 2.10, Fe-C(3) = 2.11, and Fe-C(4) = 2.11 Å.

The bond angles C(2)-C(1)-C(3) = 116°, C(3)-C(1)-C(4) = 108°, and C(4)-C(1)-C(2) = 119° indicate that the ligand is nonplanar. The central atom C(1) is displaced by 0.26 Å away from the Fe atom in respect to the plane passing through C(2), C(3), and C(4) [21, 23].

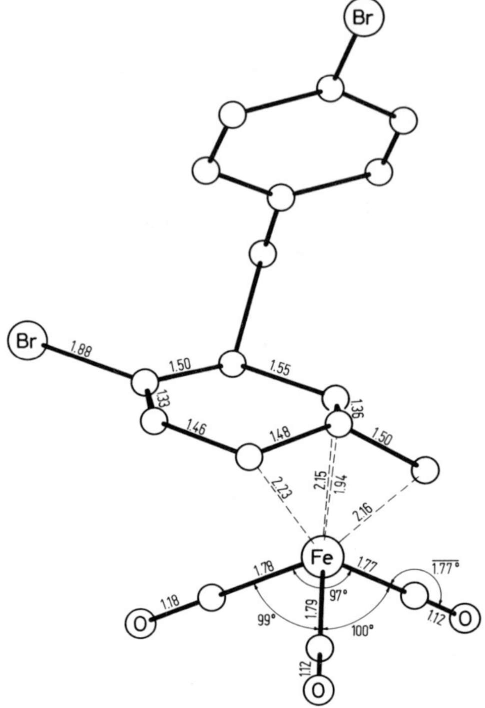

Fig. 4. Molecular structure of $C_{14}H_{12}Br_2Fe(CO)_3$ (No. 40) [52].

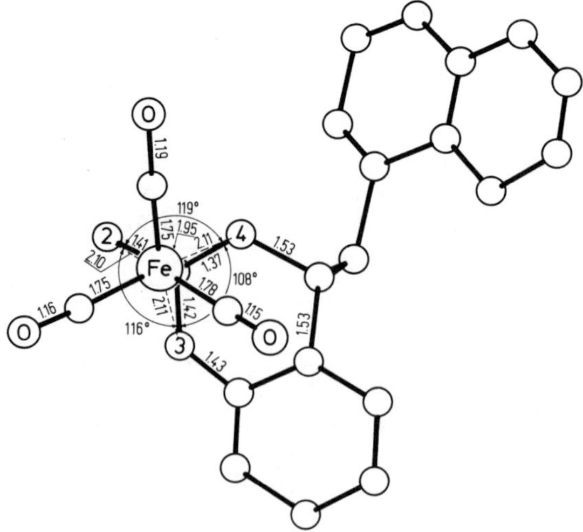

Fig. 5. Molecular structure of $C_{22}H_{18}Fe(CO)_3$ (No. 42). The molecule is projected on the plane of the benzene ring [21, 23].

References on pp. 39/42

$C_{22}H_{18}Fe(CO)_3$ (Table **1**, No. **43**), erroneously written earlier as $^3L(^1L)Fe(CO)_3$ ($^3L = \pi$-allyl bonded 2-$CH_2C_{10}H_7$, $^1L = \sigma$-bonded 2-$CH_2C_{10}H_7$) [4], has the structure given in Table 1, p. 12 [21, 23, 27].

The monodeuterated compound (H-6 substituted by D) obtained similarly by Method IXa, p. 3, forms light yellow crystals, m.p. 141 °C, and shows bands at 1988(CO), 2062(CO), 2930(CH_2), and 2945(CH_2) cm^{-1} in the IR spectrum in $CHCl_3$ [27].

$C_{22}H_{18}Fe(CO)_3$ is diamagnetic. The ^1H NMR spectrum ($(CH_3)_3SiSi(CH_3)_3$ as reference) at 60 MHz in CS_2 at 20 °C shows chemical shifts at $\delta = \sim 1.6$ (H-3, 4), 2.9 to 3.8 (complex m), and 6.7 to 7.7 (m, aromatic H) ppm [4]. In CS_2 solution at 100 MHz the peak at ~1.6 ppm splits into two singlets, $\delta = 1.595$ (H-3 or H-4) and 1.639 (H-3 or H-4) ppm. The complex multiplet is resolved to give $\delta = 3.091$ (four lines, H-8) [4], 3.361 (group of lines, H-6, 7) [4], 3.635 (t, H-5) [4, 27], and 3.816 (d, H-2; J = 2.5 Hz) ppm [4, 27]. Instead of the multiplet at 6.7 to 7.7 ppm, peaks at $\delta = 6.77$ (H-9 to H-12, $\Delta\nu_{1/2} \approx 2.5$ Hz), 7.26 (m, 4 naphthyl-H), and 7.56 (m, 3 naphthyl-H) ppm are observed [4]. In benzene solution at 100 MHz chemical shifts at $\delta = 3.35$ (d, H-2) and 3.42 (t, H-5) ppm are observed; at 60 MHz there is only the complex multiplet $\delta = 2.9$ to 3.8 ppm [27]. The singlets at $\delta = 1.595$ and 1.639 ppm for H-3 and H-4 collapse at 97.5 °C [4, 27]. Similar changes and spectra were observed in solutions such as CCl_4, $CDCl_3$, and CS_2/C_6H_6 (4:1) [4].

Polarographic reduction at an Hg-dropping electrode (0.1 N $[(C_2H_5)_4N]ClO_4$, $c = 1 \times 10^{-3}$ mol/L) occurs at $E_{1/2} = -1.6$ V relative to a saturated calomel electrode. No oxidation is observed up to -0.5 V [4, 27]. The compound is stable to air and is unchanged up to 150 °C under 10 Torr. $C_{10}H_7CH_2CH_2C_{10}H_7$ is produced by irradiation [4].

$C_{22}H_{16}Br_2Fe(CO)_3$ (Table **1**, No. **44**) shows complex multiplets, $\delta = 2.8$ to 4.1 and 6.9 to 8.9 ppm, and in the high field region two singlets in the ^1H NMR spectrum in CS_2 ($(CH_3)_3$-$SiSi(CH_3)_3$ as internal standard) [27]. X-ray diffraction shows that the crystals belong to the monoclinic space group $P2_1/n-C^5_{2h}$ with a = 19.597(8), b = 8.232(6), c = 13.770(7) Å, $\beta = 89.38(5)°$, and Z = 4 molecules. The measured and calculated densities are 1.70 and 1.74 g/cm³, respectively. The principal bond distances and angles are given in **Fig. 6**. The Fe-C distances to the outer C atoms of the trimethylenemethane ligand are effectively equal with 2.09, 2.10, and 2.10 Å; the distance to the central C atom is 1.81 Å. The trimethylenemethane fragment is nonplanar since the central C atom is out of the plane of the other three by 0.22 Å in the direction towards the Fe atom. The angles about the center C atom total only 352°. The facial angle between the two mean planes of the naphthalene nuclei is 32° [43].

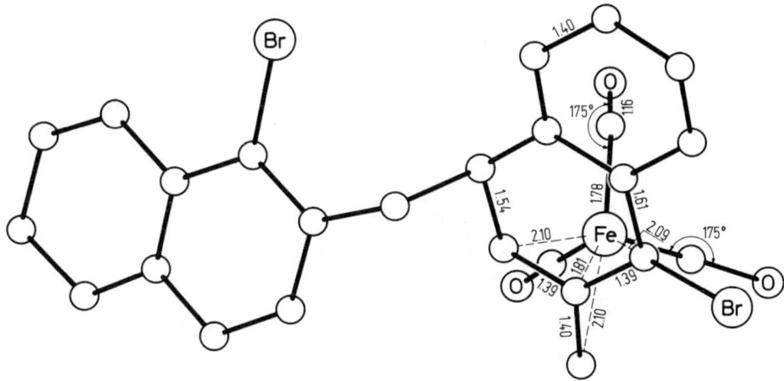

Fig. 6. Molecular structure of $C_{22}H_{16}Br_2Fe(CO)_3$ (No. 44) [43].

References on pp. 39/42

CH_3 C_6H_5 O $Fe(CO)_3$ — XLVII

$(CO)_3Fe$ $Fe(CO)_3$ — XLVIII

$Fe(CO)_3$ — XLIX

$[\ \ Fe(CO)_3]^+$ — L

$\mathbf{C_8H_8Fe(CO)_3}$ (Table **1**, No. **45**) can be prepared by the reaction of 7-(hydroxymethyl)cycloheptatriene with a large excess of $Fe_2(CO)_9$ in ether [24, 35, 87]. The mixture is stirred for 17 h at room temperature, then refluxed for 4 h, and again stirred for 2 d at room temperature. After filtration through Kieselgur, the solvent is evaporated, and the residue is distilled at reduced pressure. The products are a phenylsubstituted substance, a red liquid (boiling point 84 to 115 °C/0.2 mm Hg), and an orange liquid which may be a $^4LFe(CO)_3$ (^4L=7-(hydroxymethyl)cycloheptatriene) complex. The red liquid [48] gives, on standing, red crystals of No. 45 (yield 25% [24], 48% [35]) [48, 80]. No. 45 can also be obtained by the reaction of XLVII with 7-(hydroxymethyl)cycloheptatriene (mole ratio 1:1) in boiling C_6H_6 (reaction time: 6 h, yield: 27%) [32]. It is also obtained by sublimation of $C_8H_8(Fe(CO)_3)_2$ (see Formula XLVIII) at 80 °C/0.1 mm Hg [35]. $C_8H_8Fe(CO)_3$ itself sublimes at 40 °C/0.1 mm Hg [24, 35].

The compound crystallizes in the centrosymmetric monoclinic space group $P2_1/c-C^5_{2h}$ with the unit cell a=6.119(2), b=6.979(2), c=24.954(7) Å, β=90.46(2)°, and Z=4. The observed density is 1.514±0.010, the calculated 1.521 g/cm³. The main distances and angles are given in **Fig. 7**. The heptafulvene and $Fe(CO)_3$ moieties interact via a trimethylenemethane-Fe linkage. Within the seven-membered carbocyclic ring, the dihedral angle between the planes defined by C(1) through C(6) and by C(6), C(7), and C(1) is 29.88°. Angles within the trimethylenemethane fragment are C(6), C(7), C(1)=119.9(2)°, C(8), C(7), C(1)=114.9(3)°, and C(8), C(7), C(6)=113.4(3)°. The molecule or at least the heptafulvene moiety appears to undergo a rigid-body libration about an axis perpendicular to the seven-membered ring which intersects the plane of the ring at a point lying about one-third of the way along from C(7) to the midpoint of C(3)-C(4). The carbon atoms exhibiting the greatest amount of thermal motion are C(3) with B_{av}=10.5 and C(4) with 10.2 Å [80], cf. [79]. The compound is predicted to be stabilized by a localization energy approximation of the Hückel type [85].

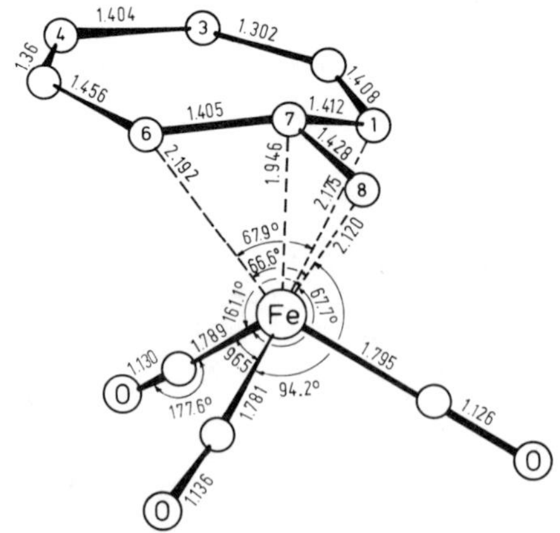

Fig. 7. The molecular structure of $C_8H_8Fe(CO)_3$ (No. 45) [80].

References on pp. 39/42

The compound is more stable than its isomer XLIX [35] which is the sole product observed in the reaction of L with $N(C_2H_5)_3$ in CH_2Cl_2 [28], cf. [86]. Thermolysis in xylene at 150 °C for 7 h gives a most likely dimeric product which is separated by filtration through Kieselgur and chromatography on neutral alumina (eluent: C_6H_6) [35].

The mass spectrum of No. 45 [24, 35] at 7.5 eV shows peaks for $[C_8H_8Fe(CO)_n]^+$ (n=0 to 3), $[C_6H_nFe]^+$ (n=6, 7), $[C_2H_4Fe(CO)_3]^+$ (?), $[C_6H_nCO]^+$ (n=4 to 6), $[C_8H_n]^+$ (n=6 to 8, 9(?)), $[C_nH_n]^+$ (n=6, 7), $[FeH]^+$, and $[Fe]^+$; at 75 eV the same fragments are detected together with $[C_7H_5]^+$, $[C_6H_5]^+$, $[FeC]^+$, $[C_3O_2]^+$, $[C_5H_5]^+$, and $[C_5H_3]^+$ [35].

If $C_8H_8Fe(CO)_3$ is treated with H_2/10% Pd/C in CH_3OH for 5 h under a pressure of 1.4 kp/cm^2, 90% of the starting material is recovered. It reacts with 48% HBF_4/H_2O in CH_3NO_2 with gas evolution. The organic layer turns black, but no ^{1}H NMR signal is observed [35]. It reacts with excess $Fe_2(CO)_9$ yielding 77% of $C_8H_8(Fe(CO)_3)_2$ (see Formula XLVIII) [35], see also [24, 87]. $C_8H_8Fe(CO)_3$ does not react with $(NC)_2C{=}C(CN)_2$ in C_6H_6 at room temperature [24, 35], and 70% of the starting complex is recovered if a mixture of $C_8H_8Fe(CO)_3$ and $(NC)_2C{=}C(CN)_2$ is refluxed in C_6H_6/acetone (volume ratio 55:10) for 8 h [35]. The thermolysis of $C_8H_8Fe(CO)_3$ (0.6 mmol) in the presence of $(NC)_2C{=}C(CN)_2$ (0.25 mmol) in refluxing $(n\text{-}C_4H_9)_2O$ for 10 h yields a complex mixture of products: a black-brown solid by filtration of the mother liquor (insoluble in organic solvents, IR bands in KBr at 1107, 1280, 1387, 1453, 1592, 1988, 2062, 2114, 2217, 2865, 2941, and 3300 cm^{-1}), a yellow-orange liquid by chromatography of the filtrate on neutral alumina with hexane as eluent (IR bands for the liquid film at 706, 742, 868, 1435, 1445, 1456, 1499, 1972, 2045, 2833, 2890, and 2976 cm^{-1}), an orange waxy oil by elution with ethyl acetate (IR bands for the film at 682, 701, 759, 872, 912, 1034, 1072, 1109, 1174, 1236, 1366, 1435, 1484, 1621, 1664, 1701, 1730 (CO?), 1835, 1957, 2024, 2817, 2882, 2890, and 3367 cm^{-1}), and a red-brown liquid by elution with CH_3OH (IR bands as film at 703, 1263, 1385, 1456, 1724, 1742 (CO?), 1992, 2066, 2941, and 3425 cm^{-1}).

By attempted photolysis of $C_8H_8Fe(CO)_3$ in the presence of $CH_3O_2CC{\equiv}CCO_2CH_3$ (mole ratio 1:1) in $(n\text{-}C_4H_9)_2O$ at room temperature for 1 h, 85% of the starting complex is recovered [35, 87]. Thermolysis in xylene at 150 °C in the presence of $CH_3O_2CC{\equiv}CCO_2CH_3$ (mole ratio 1:2.3) yields most likely a dimeric compound (cf. in refluxing xylene) [35] and a product which gives, after dehydrogenation with air in the presence of Pd/C, 1,2-dicarbomethoxyazulene with a 16% yield [24, 35].

$[C_2H_5CO(C_6H_5CH)CCH_2Fe(CO)_3]^-$ (Table **1**, No. **50**) is prepared as the Na^+ salt [74 to 76] similar to $[C_2H_5COC(CH_2)_2Fe(CO)_3]^-$ (No. 19) using $C_6H_5CH{=}C{=}CH_2$ instead of allene. The complex is also obtained by the reaction of $C_2H_5(HO)C(C_6H_5CH)CCH_2Fe(CO)_3$ (No. 51) with Na or NaOH in tetrahydrofuran [76]. The phenyl group is placed arbitrarily; it also very probably replaces each of the three trimethylenemethane H (compare Formula LI). For No. 50 the activation energy for rotation about the C(1)–C(2) bond is low [75, 76]. Acidification of the anion with CH_3CO_2H results in the formation of the isomers of $C_2H_5(HO)$-$C(C_6H_5CH)CCH_2Fe(CO)_3$ (No. 51) [75, 76], and reaction with $ClSi(CH_3)_3$ gives the isomers of $C_2H_5((CH_3)_3SiO)C(C_6H_5CH)CCH_2Fe(CO)_3$ (No. 52) [76].

LI

References on pp. 39/42

$C_2H_5(HO)C(C_6H_5CH)CCH_2Fe(CO)_3$ (Table **1**, No. **51**) is prepared similarly to No. 20 by acidification of the anion $[C_2H_5CO(C_6H_5CH)CCH_2Fe(CO)_3]^-$ (No. 50) with CH_3CO_2H at room temperature. The phenyl group is placed arbitrarily (see Table 1, p. 13), but it is very probable that all four isomers LI with R=H exist in solution [76]. Isomerization of the complex [75, 76] in tetrahydrofuran at 50 to 60 °C gives LIIa (yield 65%) and b (15%); LIIc could not be observed. Possibly, other isomers of LII are additionally formed [75]. Reaction of the compound No. 51 with Na or NaOH in tetrahydrofuran gives the anion No. 50 [76].

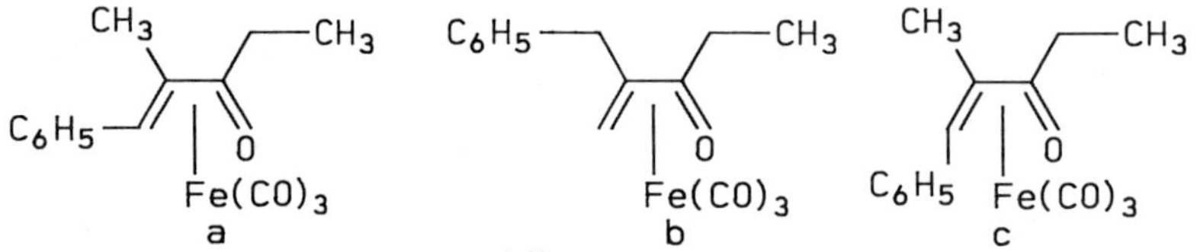

LII LIII

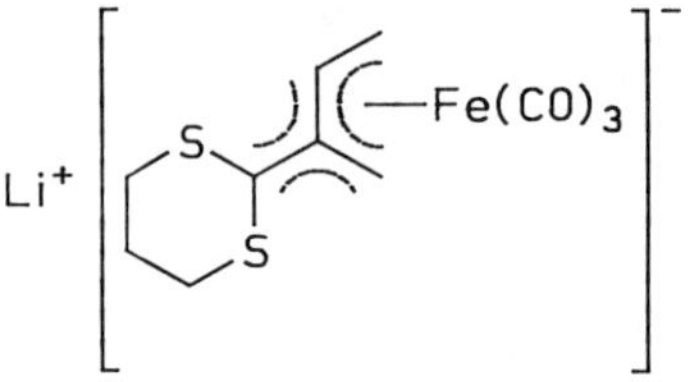

$C_2H_5((CH_3)_3SiO)C(C_6H_5CH)CCH_2Fe(CO)_3$ (Table **1**, No. **52**) is prepared from the anion $[C_2H_5CO(C_6H_5CH)CCH_2Fe(CO)_3]^-$ (No. 50) by the same method as $C_2H_5((CH_3)_3SiO)C$-$C(CH_2)_2Fe(CO)_3$ (No. 21) in 85% yield. The ^1H NMR spectrum shows the compound to be a mixture of isomers LIa to d (R=$(CH_3)_3Si$) in the ratio 3:4:6:7. A second thin layer chromatography with pentane (80%)/CH_2Cl_2 (20%) permits to separation of LIa and b (R=$(CH_3)_3Si$). The ^1H NMR spectra of the isomers in $CDCl_3$ are shown in the following table (for numbering see Formula LI, δ in ppm, J in Hz):

complex LI (R= $Si(CH_3)_3$)	H-3′	H-3	H-4′	H-4	CH_2	CH_3	$Si(CH_3)_3$	C_6H_5
a	—	4.14(s)	1.17(d) J(H-4,4′) =1.5	1.52(d)	1.16(m)	1.30(t)	−0.14(s)	7.20(m)
b	2.44(s)	1.62(s)	3.72(s)	—	1.84(m)	1.00(t)	0.27(s)	7.20(m)
c	5.10(d) J(H-3′,4) =3	—	2.41(d) J(H-4,4′) =1.5	2.08(dd)	2.00(m)	1.28(t)	0.25(s)	7.20(m)
d	3.04(d) J(H-3′,4) =3	2.84(s)	—	4.15(d)	2.00(m)	1.33(t)	0.25(s)	7.2(m)

The IR spectra of LIa and b (R=$Si(CH_3)_3$) in hexane or pentane are shown in the following table:

LI (R=$Si(CH_3)_3$)	ν(CO) in cm^{-1}
a	1970, 1980, 2050
b	1977, 1987, 2055

Acidification of $C_2H_5((CH_3)_3SiO)C(C_6H_5CH)CCH_2Fe(CO)_3$ with CF_3COOH gives LII [76].

References on pp. 39/42

[$OC_6H_7Fe(CO)_3$]$^-$ (Table **1**, No. **53**) is prepared from $Na_2Fe(CO)_4 \cdot 1.5\ O(CH_2CH_2)_2O$ and $Br(CH_2)_2CH{=}C{=}CH_2$ [64, 67, 77] in tetrahydrofuran as the Na^+ salt [64]. Acidification of the anion with CH_3CO_2H [64, 67, 77] gives 2-methylcyclopent-2-en-1-one in 30 to 40% yield dependent on the reaction conditions [64]. Similar yields of 2-methylcyclopent-2-en-1-one are obtained by attempted alkylation of the anion with CH_3I, CD_3I, $CH_2{=}CHCH_2Cl$ or $C_6H_5CH_2Cl$ at room temperature for 6 d [64]. Reaction with $ClSi(CH_3)_3$ gives $(CH_3)_3SiOC_6H_7Fe(CO)_3$ (No. 54) [64, 76].

$(CH_3)_3SiOC_6H_7Fe(CO)_3$ (Table **1**, No. **54**) is prepared by the reaction of the anion [OC_6H_7-$Fe(CO)_3$]$^-$ (No. 53) with $ClSi(CH_3)_3$ [64, 76] in tetrahydrofuran at room temperature [64]. The solvent is evaporated, and the red oily residue is extracted with pentane. After evaporation of the pentane an oil is obtained which crystallizes partially at −80 °C and decomposes when chromatographed on alumina. Purification is by sublimation (yield 30%) [64]. Reaction with CH_3COOH [64, 76] in tetrahydrofuran gives quantitatively 2-methylcyclopent-2-en-1-one [64].

$R(HO)CHCH_2CH(C_3H_6S_2C)CCH_2Fe(CO)_3$ (Table **1**, Nos. **55** and **56** with $R{=}CH{=}C(CH_3)_2$, C_6H_5) are prepared from LIII and RCHO. A 1:1 mixture of diastereoisomers of No. 55 is obtained with $(CH_3)_2CH{=}CHCHO$, and a 4:1 mixture of diastereoisomers of No. 56 is obtained with C_6H_5CHO [102].

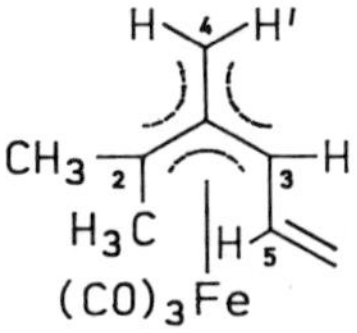

LIV 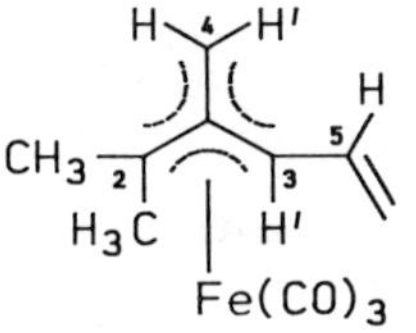 LV LVI LVII

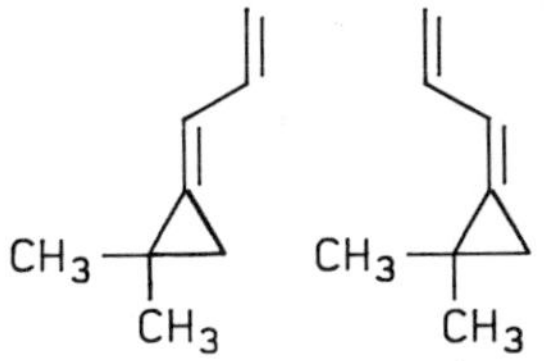

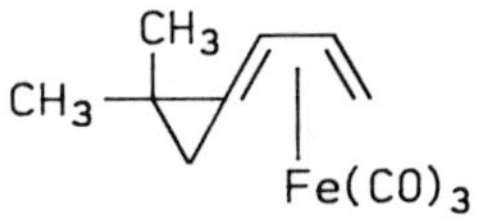

$(CH_3)_2CC(CH_2)C_3H_4Fe(CO)_3$ (Table **1**, No. **57**) exists in two isomers, LIV and LV. Isomer LV (yield ca. 19%) and traces of XL are obtained by the reaction of isomers LVIa and b with $Fe_2(CO)_9$ in boiling benzene for 50 min; the reaction also takes place at lower temperatures, but the additional reaction time and resulting decomposition give lower yields. Bulb to bulb distillation gives $C_8H_{12}Fe(CO)_3$ complexes in 32% yield. Chromatography on silica gel and elution with CCl_4 yield LVII (ca. 13%) and LV (yield ca. 19%). The analogous reaction of pure isomer LVIa with $Fe_2(CO)_9$ produced LVII and LV in an overall yield of 46% in a ratio of 67:33. Traces of isomer LIV and a 21% yield of isomer LV are obtained by the analogous reaction with isomer LVIb. The molecular weight of isomer LV is confirmed by mass spectrometry. The ^1H NMR spectrum of isomer LV shows chemical shifts $\delta=1.53$ (s, CH_3), 1.60 (s, CH_3), 2.46 (d, H-4), 2.53 (s, H-4′), 4.11 (dd, H-3′), and 4.75 to 6.00 (m, vinyl-H) ppm with coupling constants J(H-3′,4) = 2.43 and J(H-3′,5) = 9.43 Hz. Isomer LIV shows chemical shifts at $\delta=1.46$ (s, CH_3), 1.48 (s, CH_3), 2.20 (s, H-4), 2.24 (s, H-4′), 3.47 (d, H-3),

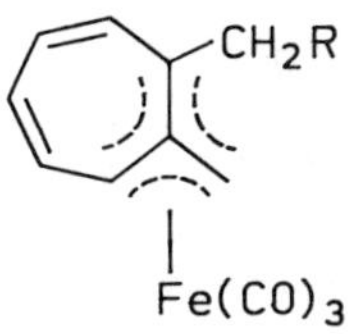

LVIII 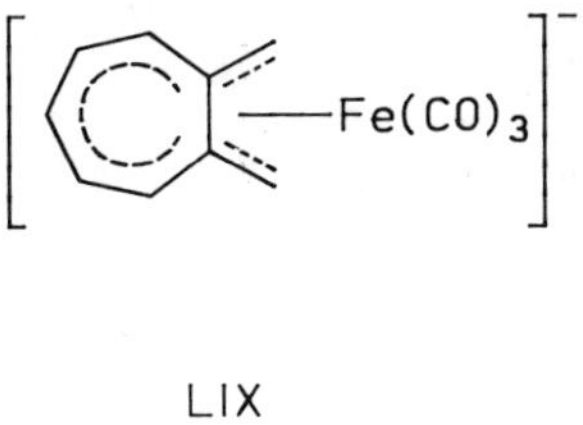 LIX 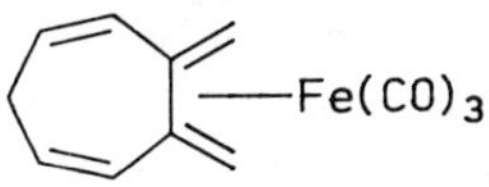 LX

References on pp. 39/42

and 4.55 to 5.80 (m, vinyl-H) ppm with J(H-3,5) = 9.4 Hz (see Formulas LIV and LV for assignments). LV has IR absorptions at 1610 (C=C); 1975 and 2045 cm^{-1}. Refluxing in C_6H_6 or CCl_4 decomposes isomer LV; there is no isomerization [39].

$RCH_2C_7H_5CH_2Fe(CO)_3$ (Table **1**, Nos. **58** and **59** with R = H, $CHOHC_6H_5$, compare LVIII). Addition of a molar equivalent of LiC_4H_9-n in hexane to a solution of LX in tetrahydrofuran at −35 °C forms an unstable anion LIX as deep reddish solution. Quenching of the anion with H_2O and C_6H_5CHO affords No. 58 (see Formula LVIII with R = H) in 28% yield and No. 59 (see Formula LVIII with R = $CHOHC_6H_5$) in 5% yield, respectively. Quenching of the anion with D_2O instead of H_2O gives LVIII with R = D [88].

$[C_2H_5CO((CH_3)_2C)CCH_2Fe(CO)_3]^-$ (Table **1**, No. **60**) is obtained as the Na^+ salt [74 to 77] and isolated as the $[N(P(C_6H_5)_3)_2]^+$ salt [77]. The Na^+ salt is prepared in the same manner as described for $Na[C_2H_5COC(CH_2)_2Fe(CO)_3]$ (No. 19) using $(CH_3)_2C{=}C{=}CH_2$ instead of allene (yield 85%) [76, 77]. $Na[C_2H_5CO((CH_3)_2C)CH_2Fe(CO)_3]$ is also obtained by the reaction of $C_2H_5(HO)C((CH_3)_2C)CCH_2Fe(CO)_3$ (No. 61) with Na or NaOH in tetrahydrofuran [76]. $[N(P(C_6H_5)_3)_2][C_2H_5CO((CH_3)_2C)CCH_2Fe(CO)_3]$ is prepared as described for $[N(P(C_6H_5)_3)_2]$-$[C_2H_5COC(CH_2)_2Fe(CO)_3]$ (No. 19) from the Na^+ salt of No. 60 with $[N(P(C_6H_5)_3)_2]Cl$ in 80% yield. The $[N(P(C_6H_5)_3)_2]^+$ salt forms orange crystals, m.p. 132 to 134 °C [77]. The ^{13}C NMR spectrum in CD_2Cl_2 at −20 °C shows chemical shifts (for assignment see Table 1, p. 15) at δ = 12.60 (CH_3), 22.40 (CH_3'-3), 29.7 (CH_2), 30.52 (CH_3-3), 33.76 (C-4), 65.84 (C-3), 77.40 (C-1), 205.16 (C-2), and 222.02 (CO) ppm [76]. The IR bands for both salts in different solvents are shown in the following table (ν(CO) and ν(C=O) in cm^{-1}) [77]:

cation	solvent	ν(C=O)	ν(CO)
Na^+	tetrahydrofuran	1570	1880, 1970
	N-methylpyrrolidone	1600	1865, 1950
	$OS(CH_3)_2$	1600	1865, 1950
	C_2H_5OH	hidden	1900, 1980
$[N(P(C_6H_5)_3)_2]^+$	tetrahydrofuran	1605	1865, 1955
	N-methylpyrrolidone	1600	1860, 1950
	$OS(CH_3)_2$	1600	1860, 1950
	C_2H_5OH	hidden	1900, 1980

Several mesomeric structures are discussed for the anion, and by monitoring the change in ν(CO) and ν(C=O) in the IR spectrum, ion pairing phenomena as function of the solvent and/or the counter ion are observed [77]. Reaction of the Na^+ salt with CH_3COOH gives $C_2H_5(HO)C((CH_3)_2C)CCH_2Fe(CO)_3$ (No. 61) [75 to 77], and reaction with $ClSi(CH_3)_3$ results in the formation of $C_2H_5((CH_3)_3SiO)C((CH_3)_2C)CCH_2Fe(CO)_3$ (No. 62) [76, 77].

$C_2H_5(HO)C((CH_3)_2C)CCH_2Fe(CO)_3$ (Table **1**, No. **61**) is prepared by acidification of the anion $[C_2H_5CO((CH_3)_2C)CCH_2Fe(CO)_3]^-$ (No. 60) with CH_3COOH [75 to 77] at room temperature as described for $C_2H_5(HO)CC(CH_2)_2Fe(CO)_3$ (No. 20) [76]. No. 61 isomerizes readily to XXXVII (see p. 25) with R = R′ = H and its isomer, chiefly by warming under reduced pressure or by chromatography on SiO_2. A mixture of products is obtained. Thus, isomerization in tetrahydrofuran at 50 to 60 °C gives XXXVII (30% yield), LXI (26%), LXII (9%), and LXIII (10%) [75]. Reaction with Na or NaOH in tetrahydrofuran gives $Na[C_2H_5CO((CH_3)_2C)$-$CCH_2Fe(CO)_3]$ (No. 60) [76].

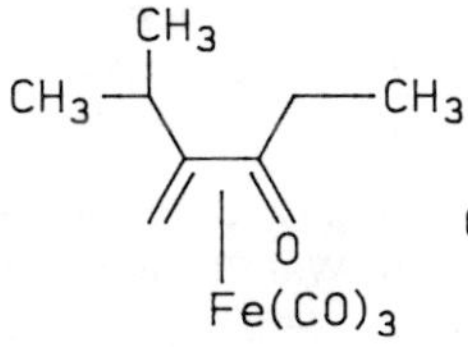

LXI 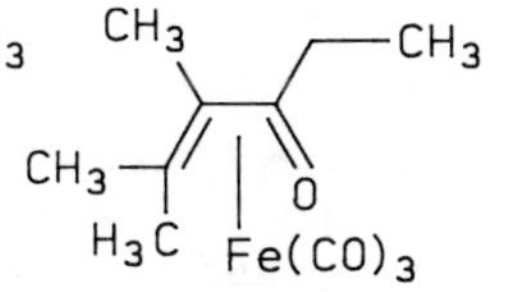 LXII 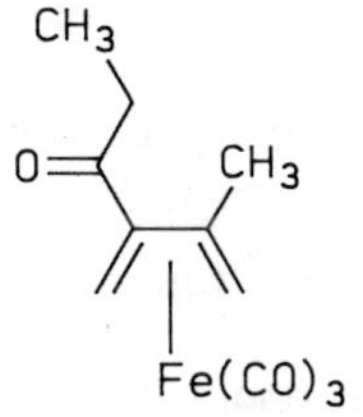 LXIII 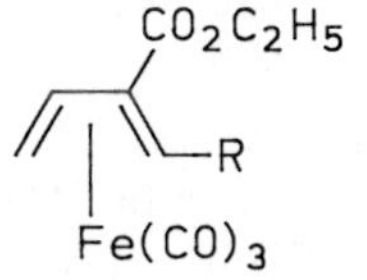 LXIV

$C_2H_5((CH_3)_3SiO)C((CH_3)_2C)CCH_2Fe(CO)_3$ (Table **1**, No. **62**) is prepared from $[C_2H_5CO((CH_3)_2C)CCH_2Fe(CO)_3]^-$ (No. 60) and $ClSi(CH_3)_3$ [76, 77] at room temperature in tetrahydrofuran as described for $C_2H_5((CH_3)_3SiO)CC(CH_2)_2Fe(CO)_3$ (No. 21) [76].

$R(C_2H_5O)C_3O_2CCH_2Fe(CO)_3$ (Table **1**, Nos. **63** and **64** with $R = CH_3$, C_2H_5) are prepared by quickly heating a solution of $R(CH_3O_2C)C{=}C{=}CH_2$ and $Fe_2(CO)_9$ in C_6H_6 at 70 °C for 5 min. Reaction with $CH_3(CH_3O_2C)C{=}C{=}CH_2$ affords a mixture of complexes No. 63 (60% yield) and LXIV (R = H) in 25% yield [94, 95]. Similar reaction with $C_2H_5(CH_3O_2C)C{=}C{=}CH_2$ gives No. 64 in 60% yield [94]. Compound No. 63 crystallizes in the monoclinic space group $P2_1/n\text{-}C^5_{2h}$ with a = 12.824(6), b = 6.941(4), c = 13.983(7) Å, β = 100.27(4)°; Z = 4 molecules per unit cell. $D_{calc} = 1.59$ g/cm³. The main bond distances and angles are shown in **Fig. 8** [95].

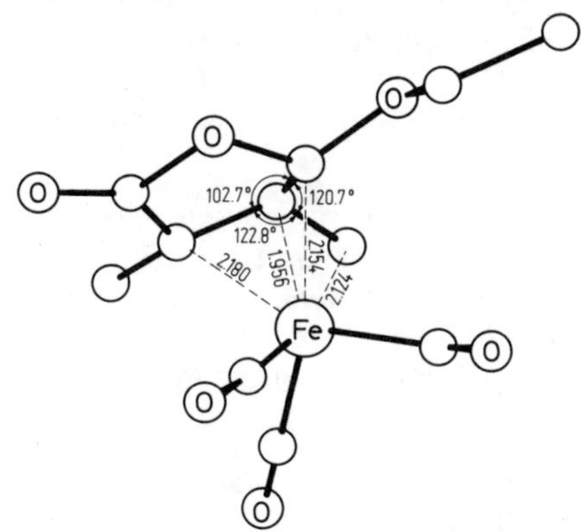

Fig. 8. Molecular structure of $CH_3(C_2H_5O)C_3O_2CCH_2Fe(CO)_3$ (No. 63) [95].

The complex No. 63, when heated in C_6H_6 (80 °C) for 2 h without a Lewis acid, either in the presence or absence of $Fe_2(CO)_9$ remains unchanged. Similar reaction in the presence of $(C_2H_5)_2OBF_3$ gives LXIV (R = H) in 90% yield. Analogous reaction of No. 64 gives LXIV ($R = CH_3$) in 60% yield [94].

$[C_2H_5CO((CH_3)_2C)CCHC_4H_9\text{-}n\ Fe(CO)_3]^-$ and $C_2H_5(HO)C((CH_3)_2C)CCHC_4H_9\text{-}n\ Fe(CO)_3$ (Table **1**, Nos. **65** and **66**) are discussed as intermediates in the reaction of C_2H_5X (X = halogen or $OSO_2C_6H_4CH_3$-4), $Na_2Fe(CO)_4 \cdot 1.5\ O(CH_2CH_2)_2O$ and $(CH_3)_2C{=}C{=}CHC_4H_9$-n in tetrahydrofuran yielding IXa and b, p. 21 [74].

References:

[1] G.F. Emerson, K. Ehrlich, W.P. Giering, P.C. Lauterbur (J. Am. Chem. Soc. **88** [1966] 3172/3). — [2] R. Grubbs, R. Breslow, R. Herber, S.J. Lippard (J. Am. Chem. Soc. **89** [1967]

6864/70). — [3] R.H. Herber (Symp. Faraday Soc. No. 1 [1967/68] 86/96). — [4] A.N. Nesmeyanov, I.I. Kritskaya, G.P. Zol'nikova, Yu.A. Ustynyuk, G.M. Babakhina, A.M. Vainbert (Dokl. Akad. Nauk SSSR **182** [1968] 1091/4; Dokl. Chem. Proc. Acad. Sci. USSR **178/183** [1968] 903/6). — [5] G.F. Emerson (personal communication from [7]).

[6] A.C. Day, J.T. Powell (Chem. Commun. **1968** 1241/3). — [7] M.R. Churchill, K. Gold (Chem. Commun. **1968** 693/4). — [8] A. Almenningen, A. Haaland, K. Wahl (Chem. Commun. **1968** 1027/8). — [9] H. Møllendahl (private communication from [18]). — [10] J.C. Hogan (Diss. Boston College 1969; Diss. Abstr. Intern. B **30** [1969/70] 4012).

[11] J.C. Hogan (Diss. Boston College 1969; Diss. Abstr. Intern. B **30** [1969/70] 4012 from E. Koerner von Gustorf, Fortschr. Chem. Forsch. **13** [1969/70] 366/450). — [12] R. Noyori, T. Nishimura, H. Takaya (Chem. Commun. **1969** 89). — [13] R.B. King (J. Am. Chem. Soc. **91** [1969] 7217/23). — [14] K. Ehrlich, G.F. Emerson (4th Intern. Conf. Organometal. Chem., Bristol 1969, Abstr. No. J5). — [15] K. Ehrlich, G.F. Emerson (Chem. Commun. **1969** 59/60).

[16] M.J.S. Dewar, S.D. Worley (J. Chem. Phys. **51** [1969] 1672/3). — [17] M.R. Churchill, K. Gold (Inorg. Chem. **8** [1969] 401/7). — [18] A. Almenningen, A. Haaland, K. Wahl (Acta Chem. Scand. **23** [1969] 1145/50). — [19] G.F. Emerson, K. Ehrlich (unpublished results from [17]). — [20] J.S. Ward, R. Pettit (Chem. Commun. **1970** 1419/20).

[21] A.N. Nesmeyanov, I.S. Astakhova, G.P. Zol'nikova, I.I. Kritskaya, Yu.T. Struchkov (Chem. Commun. **1970** 85). — [22] I.I. Kritskaya, G.P. Zol'nikova, Yu.T. Struchkov, Yu.A. Ustynyuk, A.N. Nesmeyanov (Proc. 13th Intern. Conf. Coord. Chem., Cracow-Zakopane 1970, Vol. 2, pp. 327/8). — [23] I.S. Astakhova, Yu.T. Struchkov (Zh. Strukt. Khim. **11** [1970] 472/8; J. Struct. Chem. [USSR] **11** [1970] 432/7). — [24] D.J. Ehntholt, R.C. Kerber (Chem. Commun. **1970** 1451/2). — [25] K.C. Ehrlich (Diss. State Univ. New York 1970; Diss. Abstr. Intern. B **31** [1970] 115).

[26] J.S. Ward (Diss. Univ. Texas 1970; Diss. Abstr. Intern. B **31** [1971] 6525). — [27] I.I. Kritskaya, G.P. Zol'nikova, I.F. Leshcheva, Yu.A. Ustynyuk, A.N. Nesmeyanov (J. Organometal. Chem. **30** [1971] 103/14). — [28] G.T. Rodeheaver, G.C. Farrant, D.F. Hunt (J. Organometal. Chem. **30** [1971] C22/C24). — [29] A. Bond, M. Green, B. Lewis, S.F.W. Lowrie (Chem. Commun. **1971** 1230/1). — [30] R.J. Clark, M.R. Abraham, M.A. Busch (J. Organometal. Chem. **35** [1972] C33/C36).

[31] A.N. Nesmeyanov, I.I. Kritskaya (Dokl. Akad. Nauk SSSR **202** [1972] 1079/82; Dokl. Chem. Proc. Acad. Sci. USSR **202/207** [1972] 139/42). — [32] J.A.S. Howell, B.F.G. Johnson, P.L. Josty, J. Lewis (J. Organometal. Chem. **39** [1972] 329/33). — [33] K. Ehrlich, G.F. Emerson (J. Am. Chem. Soc. **94** [1972] 2464/70). — [34] W.E. Billups, L.P. Lin, O.A. Gansow (Angew. Chem. **84** [1972] 684). — [35] R.C. Kerber, D.J. Ehntholt (J. Am. Chem. Soc. **95** [1973] 2927/34).

[36] I.S. Krull (J. Organometal. Chem. **57** [1973] 363/72). — [37] D.H. Finseth (Diss. Univ. Pittsburg 1973; Diss. Abstr. Intern. B **34** [1974] 4896). — [38] A.D. Buckingham, A.J. Rest, J.P. Yesinowski (Mol. Phys. **25** [1973] 1457/60). — [39] W.E. Billups, L.P. Lin, B.A. Baker (J. Organometal. Chem. **61** [1973] C55/C58). — [40] D.C. Andrews, G. Davidson (J. Organometal. Chem. **43** [1973] 393/9).

[41] A.J. Rest (unpublished results from I. Fischler, Diss. Ruhr-Univ. Bochum 1974). — [42] A.N. Nesmeyanov, G.P. Zol'nikova, I.F. Leshcheva, I.I. Kritskaya (Izv. Akad. Nauk SSSR Ser. Khim. **1974** 2388; Bull. Acad. Sci. USSR Div. Chem. Sci. **1974** 2306). — [43] V.S. Kuz'min, G.P. Zol'nikova, Yu.T. Struchkov, I.I. Kritskaya (Zh. Strukt. Khim. **15** [1974] 162/4; J. Struct. Chem. [USSR] **15** [1974] 153/5). — [44] B.R. Bonazza, C.P. Lillya (J. Am. Chem. Soc. **96** [1974] 2298/300). — [45] L.P. Lin (Diss. Rice Univ. 1975; Diss. Abstr. Intern. B **36** [1975] 1714).

[46] A. Rest (private communication from I. Fischler, K. Hildenbrand, E. Koerner von Gustorf, Angew. Chem. **87** [1975] 35/7). – [47] D.C. Andrews, G. Davidson, D.A. Duce (J. Organometal. Chem. **97** [1975] 95/104). – [48] E.S. Magyar, C.P. Lillya (J. Organometal. Chem. **116** [1976] 99/102). – [49] R. Hoffmann (private communication from [48]). – [50] J.-M. Savariault, J.-F. Labarre (Inorg. Chim. Acta **19** [1976] L53/L54).

[51] D.H. Finseth, C. Sourisseau, F.A. Miller (J. Phys. Chem. **80** [1976] 1248/61). – [52] G.G. Aleksandrov, G.P. Zol'nikova, I.I. Kritskaya, Yu.T. Struchkov (Koord. Khim. **2** [1976] 272/7; Soviet J. Coord. Chem. **2** [1976] 206/9). – [53] Y. Becker, A. Eisenstadt, Y. Shvo (Tetrahedron **32** [1976] 2123/6). – [54] J.A. Connor, L.M.R. Derrick, I.H. Hillier, M.F. Guest, D.R. Lloyd (Mol. Phys. **31** [1976] 23/32). – [55] A. Guinot, P. Cadiot, J.L. Roustau (J. Organometal. Chem. **128** [1977] C35/C38).

[56] R.B. King (Israel J. Chem. **15** [1976/77] 181/8). – [57] R. Hoffmann (8th Intern. Conf. Organometal. Chem., Kyoto 1977, pp. 1/2). – [58] T.A. Albright, P. Hofmann, R. Hoffmann (J. Am. Chem. Soc. **99** [1977] 7546/57). – [59] T.A. Albright, R. Hoffmann (unpublished results from [60]). – [60] R. Hoffmann, T.A. Albright, D.L. Thorn (Pure Appl. Chem. **52** [1978] 1/9).

[61] C.P. Lillya (private communication from [60]). – [62] A.J. Rest, J.R. Sodean, P.J. Taylor (J. Chem. Soc. Dalton Trans. **1978** 651/6). – [63] S.D. Worley, T.R. Webb, D.H. Gibson, T.-S. Ong (J. Organometal. Chem. **168** [1979] C16/C20). – [64] J.Y. Mérour, J.L. Roustan, C. Charrier, J. Benaim, J. Collin, P. Cadiot (J. Organometal. Chem. **168** [1979] 337/49). – [65] A. Guinot, P. Cadiot, J.L. Roustan (J. Organometal. Chem. **166** [1979] 379/83).

[66] N.J. Fitzpatrick, A.J. Rest, D.J. Taylor (J. Chem. Soc. Dalton Trans. **1979** 352/4). – [67] J. Collin, A. Guinot, J.L. Roustan, P. Cadiot (9th Intern. Conf. Organometal. Chem., Dijon 1979, Abstr. No. P53W). – [68] B.R. Bonazza, C.P. Lillya, E.S. Magyar, G. Scholes (J. Am. Chem. Soc. **101** [1979] 4100/6). – [69] T.A. Albright (J. Organometal. Chem. **198** [1980] 159/68). – [70] S.C. Avanzino, A.A. Bakke, H.-W. Chen, C.J. Donahue, W.L. Jolly, T.H. Lee, A.J. Ricco (Inorg. Chem. **19** [1980] 1931/6).

[71] P.A. Dobosh, C.P. Lillya, E.S. Magyar, G. Scholes (Inorg. Chem. **19** [1980] 228/32). – [72] A.R. Pinhas, B.K. Carpenter (J. Chem. Soc. Chem. Commun. **1980** 15/7). – [73] A.R. Pinhas, B.K. Carpenter (J. Chem. Soc. Chem. Commun. **1980** 17/9). – [74] J.L. Roustan, A. Guinot, P. Cadiot (J. Organometal. Chem. **194** [1980] 367/78). – [75] J.L. Roustan, A. Guinot, P. Cadiot (J. Organometal. Chem. **194** [1980] 357/65).

[76] J.L. Roustan, A. Guinot, P. Cadiot (J. Organometal. Chem. **194** [1980] 191/202). – [77] J.L. Roustan, A. Guinot, P. Cadiot, A. Forgues (J. Organometal. Chem. **194** [1980] 179/90). – [78] S.D. Worley, T.R. Webb, D.H. Gibson, T.-S. Ong (J. Electron Spectrosc. Relat. Phenomena **18** [1980] 189/98). – [79] M.R. Churchill (Inorg. Chem. **12** [1973] 1213/4). – [80] M.R. Churchill, B.G. DeBoer (Inorg. Chem. **12** [1973] 525/31).

[81] J.W. Koepke, W.L. Jolly, G.M. Bancroft, P.Å. Malmquist, K. Siegbahn (Inorg. Chem. **16** [1977] 2659/61). – [82] T. Jenny, W. von Philipsborn, J. Kronenbitter, A. Schwenk (J. Organometal. Chem. **205** [1981] 211/22). – [83] A.R. Pinhas, A.G. Samuelson, R. Risemberg, E.V. Arnold, J. Clardy, B.K. Carpenter (J. Am. Chem. Soc. **103** [1981] 1668/75). – [84] A.G. Samuelson, B.K. Carpenter (J. Chem. Soc. Chem. Commun. **1981** 354/6). – [85] B.J. Nicholson (J. Am. Chem. Soc. **88** [1966] 5156/65).

[86] G.C. Farrant, G.T. Rodeheaver, D.F. Hunt (5th Intern. Conf. Organometal. Chem., Moscow 1971, Vol. 1, Abstr. No. 20). – [87] D.J. Ehntholt (Diss. State Univ. New York 1971; Diss. Abstr. Intern. B **32** [1972] 4486). – [88] M. Oda, N. Morita, T. Asao (Chem. Letters **1981** 397/8). – [89] M.C. Böhm, R. Gleiter (J. Comput. Chem. **1** [1980] 407/16). – [90] M.C. Böhm, R. Gleiter (Theor. Chim. Acta **59** [1981] 153/79).

[91] M.C. Böhm (Z. Naturforsch. **36a** [1981] 1205/12). — [92] R. Hoffmann (Angew. Chem. **94** [1982] 725/808). — [93] M.C. Böhm, P.C. Schmidt, K.D. Sen (J. Mol. Struct. **87** [1982] 43/52). — [94] F. Brion, D. Martina (Tetrahedron Letters **23** [1982] 861/4). — [95] D. Martina, F. Brion, A. De Cian (Tetrahedron Letters **23** [1982] 857/60).

[96] D.B. Beach, J.L. Hoskins, W.L. Jolly, S.P. Smit, S.F. Xiang (J. Electron Spectrosc. Relat. Phenomena **28** [1983] 299/302). — [97] M.C. Böhm (J. Mol. Struct. **92** [1983] 73/92). — [98] J.A. Mondo, J.A. Berson (J. Am. Chem. Soc. **105** [1983] 3340/1). — [99] T.A. Albright, P. Hofmann, R. Hoffmann, C.P. Lillya, P.A. Dobosh (J. Am. Chem. Soc. **105** [1983] 3396/411). — [100] M.C. Böhm (J. Chem. Phys. **78** [1983] 7044/64).

[101] J.A. Mondo (Diss. Yale Univ. 1982; Diss. Abstr. Intern. B **44** [1983] 499). — [102] M. Franck-Neumann, M.P. Heitz, D. Martina (unpublished results from [103]). — [103] M. Franck-Neumann (Pure Appl. Chem. **55** [1983] 1715/32).

1.4.1.4.3 $^4LFe(CO)_3$ Compounds where ^4L is a σ,π Allyl Bound Ligand

General References:

T.-Y. Luh, Trimethylamine N-Oxide: A Versatile Reagent for Organometallic Chemistry, Coord. Chem. Rev. **60** [1984] 255/76.

I. Omae, Organometallic Intramolecular-Coordination Compounds Containing a π-Allyl Donor Ligand, Coord. Chem. Rev. **53** [1984] 261/91.

M. Franck-Neumann, Synthetic Applications of Some Metal Carbonyl Complexes, Pure Appl. Chem. **55** [1983] 1715/32.

J.A.S. Howell, P.M. Burkinshaw, Ligand Substitution Reactions at Low-Valent Four- Five- and Six-Coordinate Transition-Metal Centers, Chem. Rev. **83** [1983] 557/99.

A.J. Pearson, Natural Products Synthesis Using Organoiron Complexes, Pure Appl. Chem. **55** [1983] 1767/79.

M.I. Rybinskaya, On the α-Carbenium Centre Stabilization in Olefin Iron Carbonyl Complexes: Realization of Olefin and Allyl Structures, Pure Appl. Chem. **54** [1982] 145/59.

A.N. Nesmeyanov, Chemistry of Sigma Pi Complexes of Iron, Manganese, and Rhenium: Selected Works 1969/1979, Nauka, Moscow 1980, pp. 1/562 [russ.]; C.A. **94** [1981] No. 47465.

P.L. Pauson, Nucleophilic Addition to Transition Metal Complexes, J. Organometal. Chem. **200** [1980] 207/21.

B.L. Booth, Complexes Containing Metal-Carbon σ-Bonds, Organometal. Chem. **7** [1978] 238/81.

R. Fields, Per- and Poly-Fluorinated Aliphatic Derivatives of the Transition Elements, Fluorocarbon Relat. Chem. **3** [1976] 308/55.

J.M. Kelly, Photochemistry of Inorganic and Organometallic Compounds, Photochem. **7** [1976] 151/210.

E. Weissberger, P. Laszlo, Stereospecific Cyclic Ketone Formation with Iron(O): Anatomy of an Interligand Reaction, Accounts Chem. Res. **9** [1976] 209/17.

B.L. Booth, Complexes Containing Metal-Carbon σ-Bonds, MTP [Med. Tech. Publ. Co.] Inter. Rev. Sci. Inorg. Chem. Ser. Two **6** [1975] 137/88; C.A. **85** [1976] No. 21511.

M. Cooke, η^3-Allylic Complexes, Organometal. Chem. **4** [1975] 336/52.

R. Fields, Per- and Poly-Fluorinated Aliphatic Derivatives of the Transition Elements, Fluorocarbon Relat. Chem. **2** [1974] 290/349.

I.I. Kritskaya, Allyl Derivatives of Metals and Related Compounds, Metody Elementoorg. Khim. Tipy Metalloorg. Soedin. Perekhodnykh Metal. **2** [1975] 734/908; C.A. **84** [1976] No. 31165.

L.A. Paquette, The Renaissance in Cyclooctatetraene Chemistry, Tetrahedron **31** [1975] 2855/83.

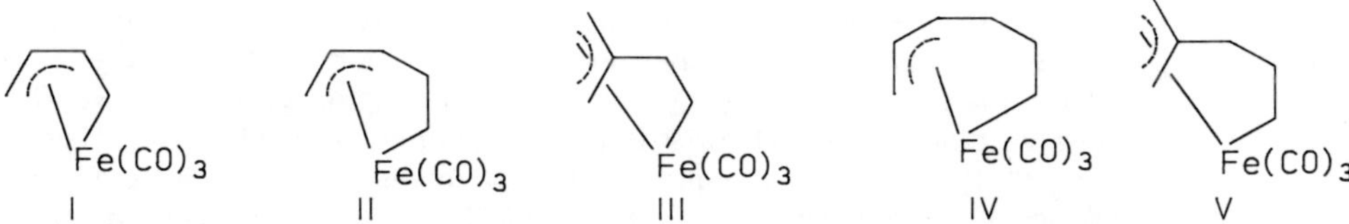

This Section 1.4.1.4.3 is divided into two subsections. The first (1.4.1.4.3.1) describes compounds with acyclic ligands of the Type I to V (substituents and heteroatoms omitted for clarity) in the sequence shown. The second subsection (1.4.1.4.3.2) describes compounds with cyclic ligands as shown in VI and VII, and at the end of the table, some compounds with a particular structure. The cyclic compounds of Type VI are arranged according to the increasing size of the ring as shown in VIII to XIV. The second part of Table 4 in 1.4.1.4.3.2, pp. 135/53, describes compounds with formally bicyclic ligands regardless of possibly condensed ring systems. The principal arrangement is shown in XV to XIX. Further subdivision is given by the size of the ring designated as "curves" in XV to XIX. The last four compounds in Table 4, pp. 153/4, have a special structure (e.g., No. 147, tricyclic), or the structure is unknown (e.g., Nos. 149, 150).

1.4.1.4.3.1 ^{4}L is an Acyclic Ligand

The compounds described in this section can be generally prepared by the following methods:

Method I: $Fe(CO)_5$ and an organic substrate are photolyzed in an inert solvent with a high or medium pressure mercury lamp through Pyrex or Duran glass.

a. A solution of $Fe(CO)_5$ and $(C_6H_5)_2C{=}C{=}O$ in ether is irradiated and strongly stirred. After 6.5 h the reaction mixture is filtered to separate $Fe_2(CO)_9$, and the filtrate is evaporated in vacuum. The oily residue is chromatographed on SiO_2 (Merck 7734; activity II to III). Elution with n-pentane gives $(CO)_4Fe(\mu\text{-}C{=}C(C_6H_5)_2)Fe(CO)_4$ in 1 to 3% yield (see 2.1.4.2.1 in "Fe-Organische Verbindungen" C 1, 1979, p. 226). The column is washed with n-pentane/ C_6H_6 (2:1), and further elution with C_6H_6 gives No. 10 by evaporation and recrystallization from n-pentane at −50 °C [63, 71, 84]. Similar reaction with $(4\text{-}CH_3OC_6H_4)_2C{=}C{=}O$ followed by chromatography on SiO_2 with petroleum ether gives a small amount of $Fe_3(CO)_{12}$. Subsequent elution with petroleum ether/C_6H_6 (1:1 to 1:2) gives No. 11 which is purified by chromatography with C_6H_6/ether (1:1) and recrystallized from pentane/ether (1:1) [71]. Irradiation of $Fe(CO)_5$ and I in ether for 4 h at −50 °C gives II and No. 103 in the ratio 10:1 (total yield 89%). Chromatography on neutral silica (Woelm) at −20 °C with pentane gives II (80%), and further elution with ether gives pure No. 103 [12], cf. [41].

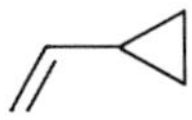

Fe(CO)4

I 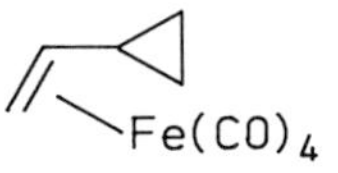 II 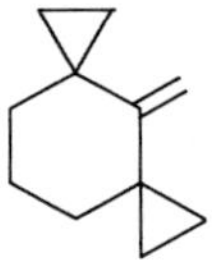 III 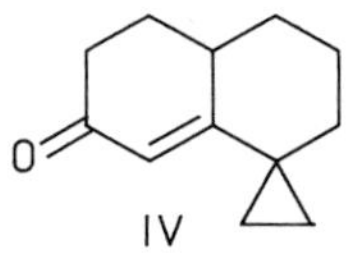 IV

b. Irradiation of $Fe(CO)_5$ and III in light petroleum ether (b.p. 40 to 60 °C) gives No. 54 as the major product and IV in minor amounts [16].

c. Irradiation of $Fe(CO)_5$ and (Z)-$ClCH_2CH{=}CHCH_2OH$ in CH_2Cl_2 for 48 h gives No. 75 [2], cf. [3].

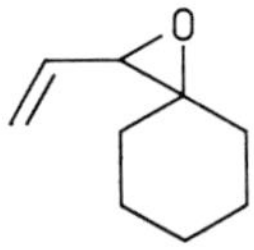

CH3 H3C

V 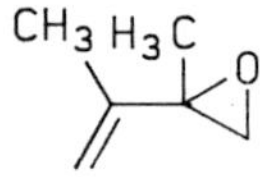 VI 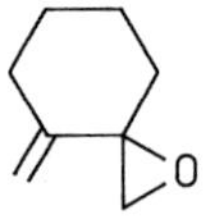 VII 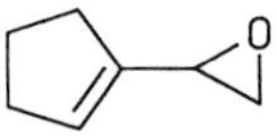 VIII 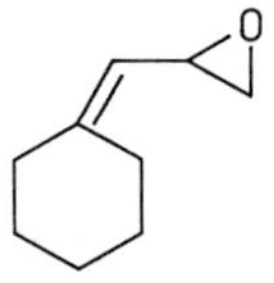 IX

d. $Fe(CO)_5$ and the organic substrates are irradiated in benzene solution at room temperature for 2 to 6 h. Removal of the volatiles under vacuum, taking care not to heat the mixture above 10 °C, gives the crude product. The product is stirred with ether, filtered through Celite, and triturated with petroleum ether to give the pure compound [69]. Thus, photolysis of $Fe(CO)_5$ and V to IX gives No. 84 and Nos. 89 to 92, respectively [39, 55, 69]. Similar irradiation of $Fe(CO)_5$ with X or XI (R = H or CH_3) for 2 h gives Nos. 74, 75, and 79, respectively [13]. Irradiation of $Fe(CO)_5$ with XII (R = CH_3) for 5 h gives No. 81 [55]. Similarly, $Fe(CO)_5$ and a mixture of XI and XIII (R = CH_3) are photolyzed until the development of CO ceases. The volatiles are distilled at 30 °C/ 10 Torr, and $Fe(CO)_5$ is added to the distillate and irradiation is continued. This procedure is repeated once. The combined distillation residues are recrystallized from CH_2Cl_2/pentane to give No. 79 (main product) and very small amounts of No. 80 (total yield 80%). No. 81 can be enriched in the mother liquor by recrystallization with CS_2 [55].

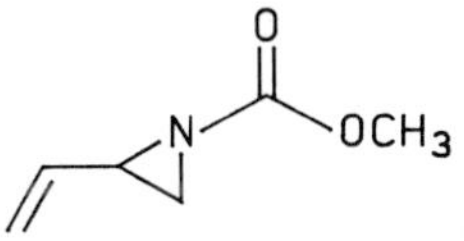

X 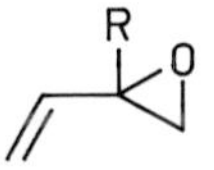 XI 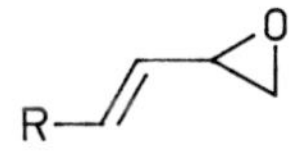 XII 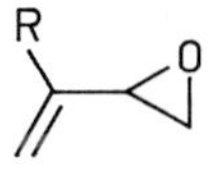 XIII

Method II: $Fe_2(CO)_9$ reacts thermally with an organic substrate in an inert solvent. The products are generally purified by chromatography.

a. Equimolar amounts of $Fe_2(CO)_9$ and the appropriate cyclopropene XIV are stirred in pentane at room temperature for 3 d (for Nos. 3 and 13) or in tetrahydrofuran at 0 °C for 24 h (for Nos. 9 and 31). The solvent and generated $Fe(CO)_5$ are removed in vacuum. The residue is chromatographed on Kieselgel 60. Elution with hexane gives $Fe_3(CO)_{12}$, and elution with hexane/toluene (5:1) gives the complexes. The complexes No. 9 and 31 are crystallized from ether/hexane, and Nos. 3, 13 from pentane. Thus XIV, with $R=R'=C_6H_5$, $R''=R'''=H$ gives No. 3, XIV ($R=R'=R''=C_6H_5$, $R'''=H$) gives No. 9, XIV ($R=R'=R''=R'''=CH_3$) gives No. 13, and XIV ($R=R'=R''=C_6H_5$, $R'''=CH_3$) gives No. 31 [47]. Similar reaction with XIV ($R=t\text{-}C_4H_9$, $R'=R''=H$, $R'''=CO_2CH_3$) affords No. 4 as the sole reaction product. But XIV ($R=C_6H_5$, $R'=CH_3$, $R''=H$, $R'''=CO_2CH_3$) gives an equimolar mixture of Nos. 7 and 8 [80, 81]. Similarly, $Fe_2(CO)_9$ is refluxed with XIV ($R=R'=R''=C_6H_5$, $R'''=OCH_3$) in pentane. The reaction mixture is filtered, and the filtrate is evaporated. The residue is dissolved in small amounts of toluene and chromatographed on SiO_2 (Merck 60). Hexane elutes some $Fe_3(CO)_{12}$, and toluene/hexane (3:1) gives No. 29 and XV ($R=CH_3$). Similar reaction of XIV ($R=R'=R''=C_6H_5$, $R'''=OC_2H_5$) gives No. 30 and XV ($R=C_2H_5$) together with XVI [82].

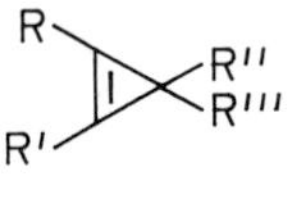

XIV 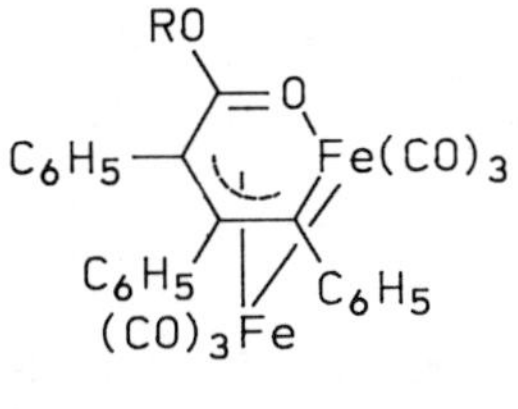 XV 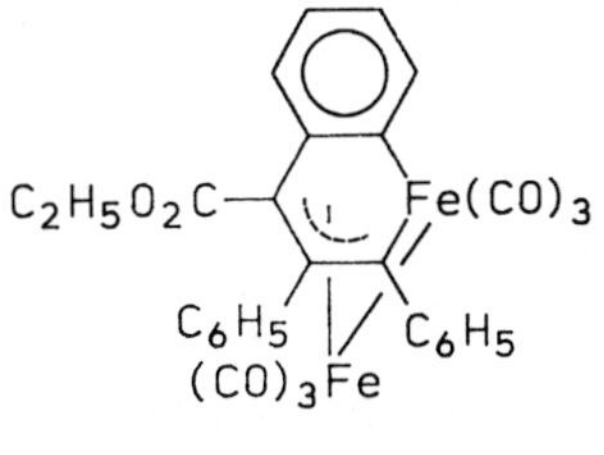 XVI

b. $Fe_2(CO)_9$ reacts with (Z)- and (E)-2,3-dimethylbut-2-en-1,4-diol in petroleum ether at 40 °C to give $Fe_3(CO)_{12}$ as the main product and, after chromatographic work-up, No. 89. Analogous reaction with $HOCH_2CH{=}CHCH_2OH$ gives No. 75, and $CH_2{=}CH(CH_2OH)_2$ gives No. 119 [1]. Analogous reaction with (Z)-$ClCH_2CH{=}CHCH_2OH$ gives No. 75 [1, 4].

c. $Fe_2(CO)_9$ reacts with the appropriate cyclopropene XIV in CH_2Cl_2 or C_6H_6 at 25 °C for 18 h. The reaction mixture is filtered, and the filtrate is chromatographed on silica gel (Merck 60). The details are gathered in the following table (see p. 46) [61]:

Similarly, $Fe_2(CO)_9$ reacts with (Z)-$HOCH_2CH{=}CHCH_2NHC_6H_5$ (see Formula L, p. 85) in C_6H_6 for 45 min. Filtration and evaporation of the filtrate gives an

References on pp. 106/8

R's in organic substrate XIV ($R''=R'''=CH_3$) R	R'	solvent	eluents	reaction products No.
H	CO_2CH_3	C_6H_6	cyclohexane/10 to 30% $CH_3CO_2C_2H_5$	12
CH_3	CO_2CH_3	CH_2Cl_2	hexane+hexane/2% ether	23 and 14 (1:1)
n-C_3H_7	CO_2CH_3	CH_2Cl_2	hexane	24 and 15 (2:3)
i-C_3H_7	$COCH_3$	CH_2Cl_2	CH_2Cl_2/5% ether	17
i-C_3H_7	CO_2CH_3	CH_2Cl_2	CH_2Cl_2	18
t-C_4H_9	$COCH_3$	CH_2Cl_2	CH_2Cl_2/2% ether	19
cyclo-C_6H_{11}	$COCH_3$	CH_2Cl_2	CH_2Cl_2/2.5% ether	20
$CH=C(CH_3)_2$	CO_2CH_3	C_6H_6	C_6H_6	25 and 21 (12:1)
$CH=C(CH_3)_2$	$COCH_3$	CH_2Cl_2	CH_2Cl_2/2% ether	22
CO_2CH_3	CO_2CH_3	C_6H_6	C_6H_6/10% ether	26
C_6H_5	CO_2CH_3	C_6H_6	hexane	27 and 28 (2:1)

oily residue which is dissolved in CH_2Cl_2, washed with 5% aqueous HCl, H_2O, and dried ($MgSO_4$). Removal of the solvent gives No. 75 after recrystallization from hexane/CH_2Cl_2. Neutralization of the acidic extracts gives aniline (56% yield) [11], cf. [43]. Similar reaction with XVII in C_6H_6 at 46 °C for 1 h followed by work-up as above and chromatography on neutral alumina (activity III) with C_6H_6/CH_2Cl_2 gives No. 80. Analogous reaction with (Z)-$HOCH_2CH=CHCH_2NHCH_3$ in C_6H_6 at 45 °C for 50 min followed by purification of the residue with active carbon gives No. 57 [11].

XVII XVIII XIX XX

d. Excess $Fe_2(CO)_9$ reacts with XVIII ($R=C_6H_5$) in C_6H_6 containing H_2O (XVIII: $H_2O=1:1$) at 44 °C for 0.5 h. After filtration, the C_6H_6 is removed and the residue is chromatographed on basic alumina (activity III). Washing the column with hexane followed by hexane/CH_2Cl_2 (9:1) elutes a very labile material which decomposes immediately. Hexane/CH_2Cl_2 (1:1) elutes No. 71 [11, 31]. Similar reaction with XVIII ($R=C_6H_5$) at 42 to 44 °C for 20 to 30 min, followed by filtration, and evaporation of the filtrate gives a brown semi-solid which is recrystallized from hexane. Removal of aniline from the mixture by extraction with 5% HCl facilitates the purification of No. 75. Similar reaction with XVIII (R=H) gives a red oil which with chromatography on basic alumina gives No. 56 on elution with CH_2Cl_2 [11].

Method III: $^2LFe(CO)_4$ complexes (^2L=alkene) react in an inert solvent at −78 °C with BF_3 or $(C_2H_5)_2OBF_3$ followed by RNH_2.

a. BF_3 is passed at −78 °C through a solution of $^2LFe(CO)_4$ (^2L=(E)-$C_6H_5COCH=CHCOC_6H_5$) in toluene until the reaction mixture acquires a yellow-orange color. Cyclo-$C_6H_{11}NH_2$ is added at −78 °C and the mixture is stirred for 30 min and then shaken for 30 min with $[NH_4]BF_4$. The mixture is filtered, and the precipitate is dissolved in $CHCl_3/H_2O$. After 1 to 1.5 h, the $CHCl_3$ layer is separated and the solvent evaporated to give No. 48 [72]. Similar reaction of $^2LFe(CO)_4$ ($^2L = C_6H_5CH{=}CHCOR$ with $R = C_4H_9$-t, C_6H_5) with BF_3 and NH_3 gives Nos. 122 and 123 [90]. Similar reaction of $^2LFe(CO)_4$ ($^2L = C_6H_5CH{=}CHCOC_6H_5$) with BF_3 and cyclo-$C_6H_{11}NH_2$ in CH_2Cl_2 gives No. 50 [45].

b. $(C_2H_5)_2OBF_3$ is added to $^2LFe(CO)_3$ ($^2L = C_6H_5CH{=}CHCOCH_3$) in CH_2Cl_2 at −78 °C followed by CH_3NH_2 in CH_2Cl_2. The reaction mixture is evaporated, and the residue is washed with pentane to give No. 36. Analogous reaction with i-$C_3H_7NH_2$ gives No. 43. Similarly, $^2LFe(CO)_4$ in toluene treated with $(C_2H_5)_2OBF_3$, and then with cyclo-$C_6H_{11}NH_2$, gives No. 49. No. 47 is analogously obtained from $^2LFe(CO)_4$ with $(C_2H_5)_2OBF_3$ and $C_6H_5CH_2NH_2$ [53]. Similar reactions give Nos. 38 [52, 54], 39, 40, 41 [78, 90], 42 [78], 45 [54, 90], 46 [78, 90]. Analogous reaction of $^2LFe(CO)_4$ ($^2L = (E)$-$C_6H_5COCH{=}CHCOC_6H_5$) in CH_2Cl_2 with $(C_2H_5)_2OBF_3$ at −78 °C followed by i-$C_3H_7NH_2$ gives No. 44. Similar reaction with CH_3NH_2 gives No. 35 [72]. Analogous reaction of BF_3 adducts of $^2LFe(CO)_4$ ($^2L = C_6H_5CH{=}CHCOR$) with primary amines $R'NH_2$ gives the compounds No. 37, 51, 124 to 136 [90].

Method IV: $^4LFe(CO)_3$ (^4L = diolefin or trimethylenemethane or its derivative) is photolyzed with a 250 W Hanovia lamp in n-hexane in the presence of alkenes or alkynes.

a. $^4LFe(CO)_3$ (^4L = butadiene) and excess $CF_2{=}CF_2$ in n-hexane are irradiated in a Carius tube for 24 h. The solvent is removed and the residue is chromatographed on alumina. Elution with C_6H_6/n-hexane (1:9) gives No. 107 [5, 24]. Analogous reaction with $CF_3CF{=}CF_2$, but elution with hexane gives No. 108 [19]. Similar reaction with a slight excess of perfluoronorbornadiene for 48 h gives No. 109 in 20% yield. A repeat of this reaction, using a 2:1 and a 4:1 ratio $^4LFe(CO)_3$ to perfluoronorbornadiene, and irradiation for 8 d gives No. 109 in 34 and 36% yield, respectively [75]. Similar irradiation of $^4LFe(CO)_3$ (^4L = isoprene) with $CF_2{=}CF_2$ followed by recrystallization gives No. 111. The formation of No. 111 occurs approximately 5 times as readily as does reaction to give No. 110 [24]. Similar reaction with $CF_3CF{=}CF_2$ gives No. 112 [5, 19] and with perfluoronorbornadiene (3 d irradiation) gives No. 114 [75]. Similar irradiation of syn-$CH_3CH{=}CHCH{=}CH_2Fe(CO)_3$ with $CF_2{=}CF_2$ for 65 h followed by recrystallization gives No. 113 [24]. Similar irradiation of $^4LFe(CO)_4$ (^4L = 2,3-dimethylbutadiene) with $CF_2{=}CFCl$ [19], $CF_2{=}CF_2$ [24], or $CF_3CF{=}CF_2$ for 40 h [19] gives after recrystallization the compounds No. 115 [19], 116 [24], and 117 [19]. Similar irradiation of $^4LFe(CO)_3$ ($^4L = C(CH_2)_3$ or $C(CH_2)_2CHC_6H_5$, see Formula CXVII, p. 104 with R = H or C_6H_5) with $CF_2{=}CF_2$ gives Nos. 120 and 121 [5].

b. $^4LFe(CO)_3$ (^4L = butadiene) and $CF_3C{\equiv}CH$ are irradiated in hexane in a Carius tube for 7 d. The solvent is removed and the residue is chromatographed on alumina. Elution with hexane gives first unchanged $^4LFe(CO)_3$ followed by No. 94 [40]. Similar irradiation with $CF_3C{\equiv}CCF_3$ [20, 23, 25] in hexane [23] for 24 h in a Carius tube, followed by work-up as before but eluting with C_6H_6/hexane (1:9), gives No. 97 [40]. Analogous irradiation of $^4LFe(CO)_3$ (^4L = isoprene) with $CF_3C{\equiv}CH$ gives a 2.5:1 mixture of isomers No. 95 and 96 in 31% yield [40]. Similar irradiation of $^4LFe(CO)_3$ (^4L = 2,3-dimethylbutadiene) and $CF_3C{\equiv}CH$ in hexane in a Carius tube for 6 d gives No. 98. Similar reaction

References on pp. 106/8

with $CF_3C{\equiv}CCF_3$ for 8 d, work-up as above but elution with CH_2Cl_2/hexane (1:1), gives No. 99 [40].

Some of the compounds, No. 2 to 32 of the general type XIX, are first formulated as shown in XX. But most of them are later shown to be σ,π-allyl complexes of type XIX, see, e.g., [6, 47, 51, 61], compare 1.4.1.4.1.1.1.10 and 1.4.1.4.1.1.1.15 in "Organoiron Compounds" B6, 1981, pp. 358 and 417. For the arrangement of the compounds in Table 3 see also 1.4.1.4.3 (p. 43).

Table 3
Compounds of the Type $^4LFe(CO)_3$ where ^4L is an Open-Chained σ,π-Allyl System.
Further information on numbers preceded by an asterisk is given at the end of the table, pp. 76/106.
For abbreviations and dimensions see p. VIII.

No.	compound	method of preparation (yield in %), properties and remarks	Ref.
^{4}L is a 1σ, 2-4π allyl ligand:			
*1	$[H^x{-}Fe(CO)_3$ (allyl C-1 to C-4)$]^+$	^{1}H NMR: $J(H\text{-}1, H_x) = 22$ ^{13}C NMR (excess $HOSO_2F$/liq. SO_2 at −80 and −20°): −2.76 (td, C-1; $J(^{13}C\text{-}1, H^x) = 73.7$, J(C, H) = 146.5), 53.06 (t, C-4; J(C,H) = 165.5), 80.92 (d, C-2; J(C,H) = 1, J(C,H) = 184), 100.29 (d, C-3; J(C,H) = 173.9); 196.04, 199.23, 203.11 (CO) ^{13}C NMR (excess CF_3CO_2H/HBF_4, at 0°): −1.9 (C-1), 54.0 (t, C-4; J(C,H) = 161), 81.7 (d, C-2; J(C,H) = 179), 101.9 (d, C-3; J(C,H) = 176); 196.5, 199.7, 203.0 (CO) IR (CF_3CO_2H/HBF_4): 2087, 2136 (CO)	[30, 35, 36]
*2	CH_3O_2C, OCH_3, =O, $Fe(CO)_3$	yellow crystals	[46, 50, 51]
*3	C_6H_5, C_6H_5, H-4, H′, =O, $Fe(CO)_3$	IIa (60) m.p. 88 to 89°, yellow crystals monomeric in $CHCl_3$ (by osmometry) ^{1}H NMR ($CDCl_3$): 1.74 (d, H′-4; J = 1.8), 3.20 (d, H-4), 7.1 to 7.4 (m, C_6H_5) IR (hexane): 1783 (C=O); 1998, 2012, 2068 (CO)	[47]
4	$C(CH_3)_3$, CH_3O_2C, =O, $Fe(CO)_3$	IIa	[80, 81]

Table 3 [continued]

No.	compound	method of preparation (yield in %), properties and remarks	Ref.
*5	CH_3O_2C OCH_3 CH_3O_2C =O $Fe(CO)_3$	—	[51]
6	CH_3O_2C CO_2CH_3 CH_3O =O $Fe(CO)_3$	see No. 5, p. 77	[51]
7	C_6H_5 CH_3 CH_3O_2C =O $Fe(CO)_3$	IIa	[80, 81]
8	CH_3 C_6H_5 CH_3O_2C =O $Fe(CO)_3$	IIa	[80, 81]
*9	C_6H_5 C_6H_5 C_6H_5 =O $Fe(CO)_3$	IIa (89) m.p. 142/4° (dec.), yellow crystals monomeric in $CHCl_3$ (by osmometry) ^{1}H NMR ($CDCl_3$): 3.28 (s, H), 7.1 to 7.5 (m, C_6H_5) IR (hexane): 1783 (C=O); 1995, 2006, 2062 (CO)	[47]
*10	7 8 C_6H_5 6 3 2 1 =O 5 4 $Fe(CO)_3$	Ia (38) m.p. 63°, ruby red, air-stable crystals ^{1}H NMR ($CDCl_3$): 4.43 (d, H; ^3J(H, H) = 5.6), 7.30 to 7.65 (m, C_6H_5, C_6H_4) ^{13}C NMR ($CDCl_3$): 50.99 (C-2), 71.28 (C-4), 113.14 (C-3), 124.01, 127.76, 128.40, 128.85, 130.32, 130.86 (C-1 in C_6H_5), 132.69, 136.62, 208.09 (CO), 230.66 (C-1) IR (KBr): 1723, 1746 (C=O); 1971, 1994, 2056 (CO) IR (C_6H_{14}): 1785 (C=O); 1995, 2006, 2065 (CO) mass spectrum: $[M-nCO]^+$ (n = 0 to 4), $[C(C_6H_5)_2]^+$	[63, 71]
*11	$C_6H_4OCH_3$-4 CH_3O =O $Fe(CO)_3$	Ia (17) bright red, air-stable crystals IR (CCl_4): 1770 (C=O); 1990, 2050 (CO)	[71]

References on pp. 106/8

Table 3 [continued]

No.	compound	method of preparation (yield in %), properties and remarks	Ref.
*12	CO_2CH_3 CH_3 =O CH_3 $Fe(CO)_3$	IIc (63) m.p. 84°, yellow or orange-yellow crystals ^1H NMR (benzene-d_6): 0.90 (s, CH_3), 1.2 (s, CH_3), 3.27 (s, CO_2CH_3), 5.85 (s, CH) ^1H NMR ($CDCl_3$): 1.18 (s, CH_3), 1.83 (s, CH_3), 3.70 (s, CO_2CH_3), 6.14 (s, CH) IR (CCl_4): 1710 (CO_2CH_3), 1780 (C=O); 2000, 2065 (CO)	[61, 64]
13	CH_3 CH_3 CH_3 =O CH_3 $Fe(CO)_3$	IIa (41) m.p. 61 to 63° (from pentane), yellow crystals monomeric in $CHCl_3$ (by osmometry) ^1H NMR ($CDCl_3$): 1.14 (s, CH_3), 1.76 (s, CH_3), 1.88 (s, CH_3), 2.72 (s, CH_3) IR (hexane): 1774 (C=O); 1988, 1993, 2057 (CO)	[47]
*14	CH_3O_2C CH_3 (2) CH_3 =O CH_3 $Fe(CO)_3$	IIc (42) m.p. 39 to 40°, m.p. 40°, pale yellow crystals ^1H NMR ($CDCl_3$): 1.23 (s, CH_3), 1.84 (s, CH_3), 1.92 (s, CH_3-2), 3.94 (s, CO_2CH_3) IR (CCl_4): 1725 (CO_2CH_3), 1770 (C=O); 1995, 2060 (CO)	[61, 64]
*15	CH_3O_2C C_3H_7-n CH_3 =O CH_3 $Fe(CO)_3$	IIc (46) yellow oil ^1H NMR ($CDCl_3$): 0.97 (t, CH_3 in n-C_3H_7), 1.25 (s, CH_3), 1.68 (m, CH_2), 1.90 (s, CH_3), 2.12 (t, CH_2), 3.83 (s, CO_2CH_3) IR (CCl_4): 1730, 1740 (CO_2CH_3); 1770 (C=O); 1995, 2060 (CO)	[61, 64]
16	$CH_3(OH)CH$ $CH(CH_3)_2$ CH_3 =O CH_3 $Fe(CO)_3$	see No. 17, p. 80 m.p. 90° (from CH_2Cl_2/pentane at −78°), yellow crystals ^1H NMR ($CDCl_3$): 1.0 (d, CH_3; J=7), 1.16 (s, CH_3), 1.33 (d, CH_3; J=7), 1.66 (d, CH_3; J=7), 1.92 (s, CH_3), 2.16 (d, CH; J=3), 3.40 (sept, CH; J=7), 5.12 (q, CH; J=3, J=7) IR (CCl_4): 1740, 1765 (C=O); 1980, 1990, 2050 (CO); 3200, 3600 (OH)	[61]
*17	CH_3OC $CH(CH_3)_2$ CH_3 (4) =O CH_3 $Fe(CO)_3$	IIc (57) yellow oil ^1H NMR ($CDCl_3$): 1.10 (d, CH_3), 1.30 (s, CH_3-4), 1.32 (d, CH_3), 1.83 (s, CH_3-4), 2.30 (sept, CH), 2.54 (s, $COCH_3$) IR (CCl_4): 1710 ($COCH_3$); 1745, 1770 (C=O); 1995, 2060 (CO)	[61]

References on pp. 106/8

Table 3 [continued]

No.	compound	method of preparation (yield in %), properties and remarks	Ref.
*18	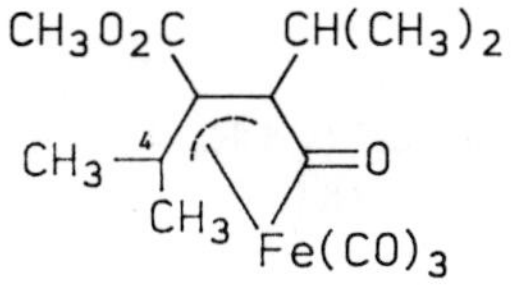 CH_3O_2C, $CH(CH_3)_2$, CH_3-4, CH_3, =O, $Fe(CO)_3$	IIc (81) yellow oil ^{1}H NMR ($CDCl_3$): 1.13 (d, CH_3), 1.25 (s, CH_3-4), 1.30 (d, CH_3), 1.88 (s, CH_3-4), 2.41 (m, CH), 3.93 (s, CO_2CH_3) IR (CCl_4): 1730, 1740 (CO_2CH_3); 1765 (C=O); 1995, 2060 (CO)	[61]
*19	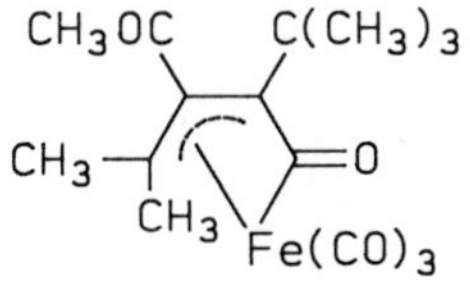 CH_3OC, $C(CH_3)_3$, CH_3, CH_3, =O, $Fe(CO)_3$	IIc (65) yellow oil ^{1}H NMR ($CDCl_3$): 1.22 (s, C_4H_9), 1.27 (s, CH_3), 1.76 (s, CH_3), 2.59 (s, $COCH_3$) IR (CCl_4): 1710 ($COCH_3$); 1750 (C=O); 1990, 2060 (CO)	[61]
20	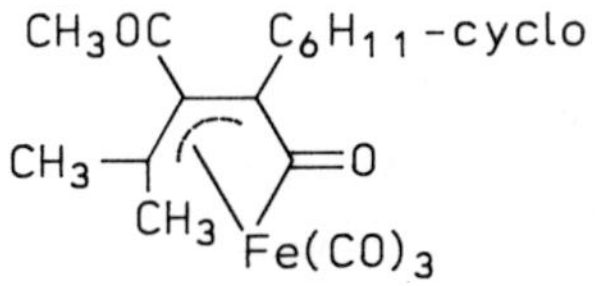 CH_3OC, C_6H_{11}-cyclo, CH_3, CH_3, =O, $Fe(CO)_3$	IIc (40) yellow oil ^{1}H NMR ($CDCl_3$): 1.29 (s, CH_3), 1.42 to 1.90 (m, 10H), 1.81 (s, CH_3), 2.15 (m, CH), 2.52 (s, $COCH_3$) IR (CCl_4): 1710 ($COCH_3$); 1760 (C=O); 1995, 2060 (CO)	[61]
*21	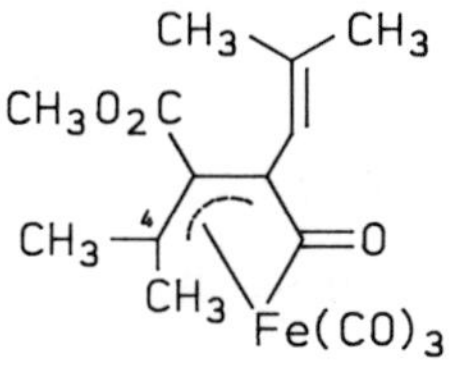 CH_3, CH_3, CH_3O_2C, CH_3-4, CH_3, =O, $Fe(CO)_3$	IIc (5) yellow oil ^{1}H NMR (benzene-d_6): 1.46 (s, CH_3-4), 1.67 (s, CH_3-4), 1.80 (s, CH_3), 1.85 (s, CH_3), 3.40 (CO_2CH_3), 5.70 (m, CH)	[61]
22	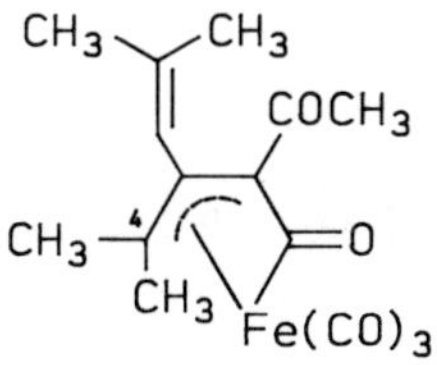 CH_3, CH_3, $COCH_3$, CH_3-4, CH_3, =O, $Fe(CO)_3$	IIc (34) yellow oil ^{1}H NMR ($CDCl_3$): 1.36 (s, CH_3-4), 1.81 (s, CH_3), 1.87 (s, CH_3-4), 1.96 (s, CH_3), 2.30 (s, $COCH_3$), 6.06 (m, CH) IR (CCl_4): 1680 ($COCH_3$); 1760, 1785 (C=O); 1995, 2060 (CO)	[61]
*23	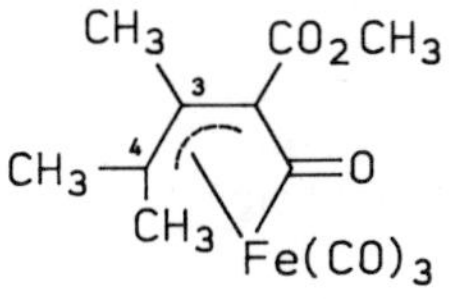 CH_3-3, CO_2CH_3, CH_3-4, CH_3, =O, $Fe(CO)_3$	IIc (40) m.p. 48 to 49°, m.p. 49°, orange-yellow crystals ^{1}H NMR ($CDCl_3$): 1.28 (s, CH_3-4), 1.92 (s, CH_3-4), 2.51 (s, CH_3-3), 3.80 (s, CO_2CH_3) IR (CCl_4): 1715 (CO_2CH_3); 1780 (C=O); 1995, 2060 (CO)	[61, 64]

References on pp. 106/8

Table 3 [continued]

No.	compound	method of preparation (yield in %), properties and remarks	Ref.
*24	n-C_3H_7, CO_2CH_3, CH_3-4, CH_3, =O, $Fe(CO)_3$	IIc (35) m.p. 59°, yellow crystals ^{1}H NMR ($CDCl_3$): 1.11 (t, CH_3), 1.27 (s, CH_3-4), 1.85 (q, CH_2), 1.93 (s, CH_3-4), 2.78 (t, CCH_2), 3.76 (s, CO_2CH_3) IR (CCl_4): 1715 (CO_2CH_3); 1780 (C=O); 1995, 2060 (CO)	[61]
*25	CH_3, CH_3, CO_2CH_3, CH_3-4, CH_3, =O, $Fe(CO)_3$	IIc (60) m.p. 86°, yellow crystals ^{1}H NMR (benzene-d_6): 1.10 (s, CH_3-4), 1.42 (s, CH_3), 1.52 (s, CH_3-4), 1.62 (s, CH_3), 3.32 (s, CO_2CH_3), 6.06 (m, CH) IR (CCl_4): 1715 (CO_2CH_3); 1770 (C=O); 2000, 2060 (CO)	[61]
*26	CH_3O_2C-3, CO_2CH_3-2, CH_3, CH_3, =O, $Fe(CO)_3$	IIc (62) m.p. 127°, yellow crystals ^{1}H NMR (benzene-d_6): 0.94 (s, CH_3), 1.47 (s, CH_3), 3.18 (s, CO_2CH_3-2), 3.48 (s, CO_2CH_3-3) IR (CCl_4): 1715 (CO_2CH_3); 1770 (C=O); 2010, 2075 (CO)	[61]
*27	C_6H_5, CO_2CH_3, CH_3, CH_3, =O, $Fe(CO)_3$	IIc (49) m.p. 118°, yellow crystals ^{1}H NMR ($CDCl_3$): 1.51 (s, CH_3), 1.89 (s, CH_3), 3.62 (s, CO_2CH_3), 7.46 (s, C_6H_5) IR (CCl_4): 1715 (CO_2CH_3); 1755, 1775 (C=O); 2000, 2060 (CO)	[61]
*28	CH_3O_2C, C_6H_5, CH_3, CH_3, =O, $Fe(CO)_3$	IIc (24) yellow oil ^{1}H NMR ($CDCl_3$): 1.40 (s, CH_3), 1.99 (s, CH_3), 3.70 (s, CO_2CH_3), 7.36 (s, C_6H_5) IR (CCl_4): 1730 (CO_2CH_3); 1760 (C=O); 1995, 2060 (CO)	[61]
29	C_6H_5-3, C_6H_5-2, C_6H_5-4, CH_3O_2C, =O (1), $Fe(CO)_3$	IIa (13) m.p. 133 to 135° (dec.), bright yellow prisms (from toluene/hexane) ^{1}H NMR ($CDCl_3$): 3.84 (s, CH_3), 7.1 (m, C_6H_5) ^{13}C NMR ($CDCl_3$): 58.5 (CH_3), 69.3 (C-4), 84.3 (C-2), 111.4 (C-3), 125.1 to 135.0 (9 signals, C_6H_5), 169.4 (CO_2) IR (CH_2Cl_2): 1731 (C=O); 2002, 2064 (CO) mass spectrum: $[M]^+$	[82]

References on pp. 106/8

Table 3 [continued]

No.	compound	method of preparation (yield in %), properties and remarks	Ref.
30	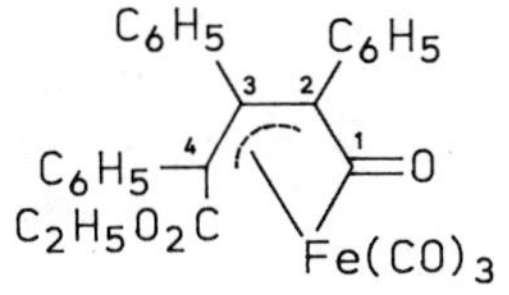 C_6H_5, C_6H_5, 3, 2, 4, 1, C_6H_5, $C_2H_5O_2C$, =O, $Fe(CO)_3$	II a (8) m.p. 129 to 130° (dec.), bright yellow prisms (from toluene/hexane) ^1H NMR ($CDCl_3$): 1.32 (t, CH_3; J=7.8), 3.80, 4.41 (m, CH_2), 7.1 (m, C_6H_5) ^{13}C NMR ($CDCl_3$): 14.7 (CH_3), 68.1 (CH_2), 83.7 (C-2), 111.4 (C-3), 124.5 to 135.0 (9 signals, C_6H_5), 169.5 (CO_2) IR (CH_2Cl_2): 1730 (C=O); 1985, 2000, 2062 (CO) mass spectrum: $[M]^+$	[82]
*31	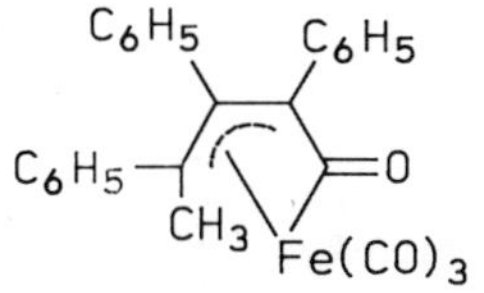 C_6H_5, C_6H_5, C_6H_5, CH_3, =O, $Fe(CO)_3$	II a (91) m.p. 120 to 122° (dec.), yellow air-stable crystals monomeric in $CHCl_3$ (by osmometry) ^1H NMR ($CDCl_3$): 2.31 (s, CH_3), 7.1 to 7.6 (m, C_6H_5) IR (hexane): 1781 (C=O); 1993, 2000, 2060	[47]
32	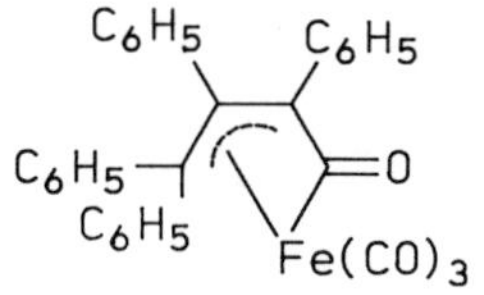 C_6H_5, C_6H_5, C_6H_5, C_6H_5, =O, $Fe(CO)_3$	cannot be isolated by Method II a	[47]
33	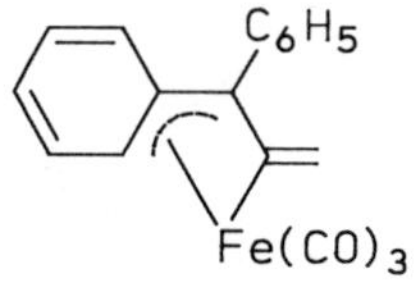 C_6H_5, $Fe(CO)_3$	see No. 10, p. 79	[84]
34	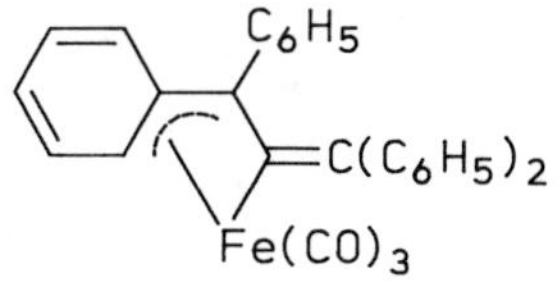 C_6H_5, $=C(C_6H_5)_2$, $Fe(CO)_3$	see No. 10, p. 79 yellow, viscous, rather air-stable oil IR (KBr): 1940, 1985, 1990, 2035 (CO)	[71]

^{4}L is a 1σ, 3-5π allyl ligand (continued at No. 122, p. 73):

No.	compound	method of preparation (yield in %), properties and remarks	Ref.
*35	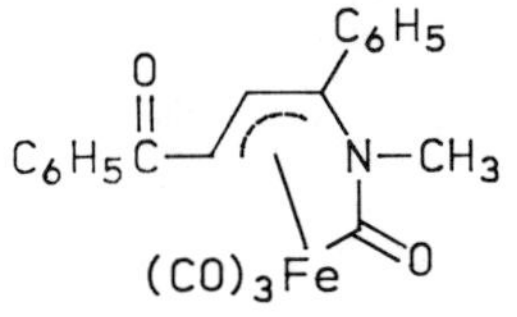 C_6H_5, O, C_6H_5C, $N-CH_3$, $(CO)_3Fe$, O	III b (25) dec. 108° (from CH_3OH), red-orange, air-stable crystals IR: 2010, 2030, 2080 (CO)	[72, 78]

References on pp. 106/8

Table 3 [continued]

No.	compound	method of preparation (yield in %), properties and remarks	Ref.
*36	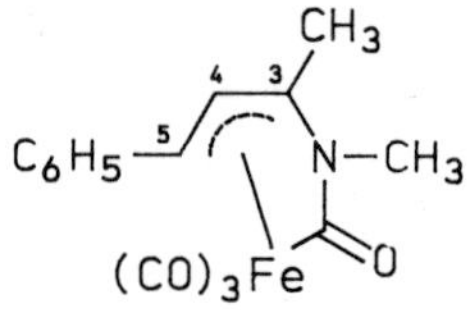 CH_3 4 3 C_6H_5 5 N–CH_3 $(CO)_3Fe$ O	IIIb (30) m.p. 79 to 80° (dec., from CH_3OH) ^{13}C NMR (CH_2Cl_2): 61.62 (C-5), 87.74 (C-4), 100.02 (C-3), 184.69 (at −70°, C=O); 201.91, 209.98, 210.16 (at −70°, CO) ^{13}C NMR (CH_2Cl_2): 62.66 (C-5), 88.07 (C-4), 103.62 (C-3) IR (KBr): 1650 (C=O) IR (hexane): 1991, 2009, 2068	[52 to 54, 73, 74, 78, 90]
*37	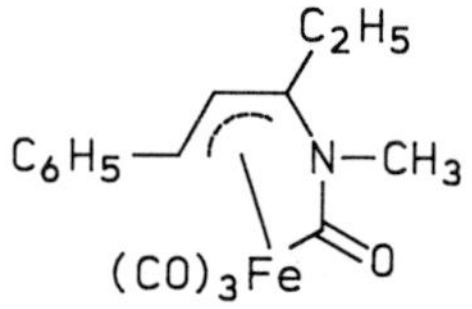 C_2H_5 C_6H_5 N–CH_3 $(CO)_3Fe$ O	IIIb (47) dec. p. 82 to 83° IR (hexane): 1994, 2010, 2065 (CO)	[78, 90]
*38	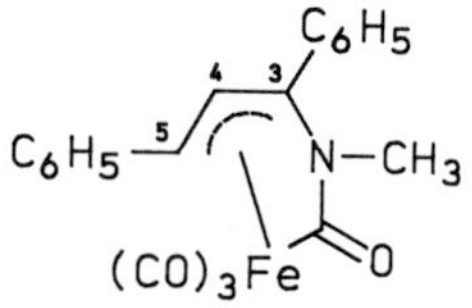 C_6H_5 4 3 C_6H_5 5 N–CH_3 $(CO)_3Fe$ O	IIIb (70) m.p. 135° (dec.), yellow, air-stable crystals (from CH_3OH) ^1H NMR ($CDCl_3$): 2.70 (CH_3), 4.96 (H-5), 6.63 (H-4; J=9.7), 7.0 to 8.0 (C_6H_5) ^{13}C NMR (CH_2Cl_2): 64.48 (C-5), 85.27 (C-4), 114.8 (C-3) IR (KBr): 1650 (C=O) IR (hexane): 1997, 2012, 2068 (CO)	[45, 52 to 54, 73, 74, 78]
*39	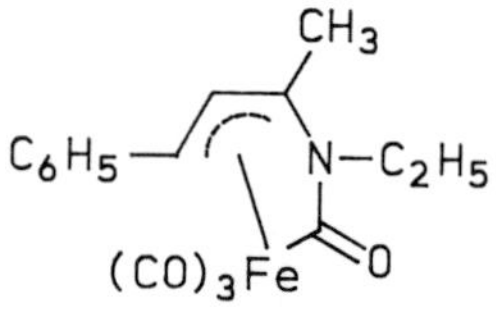 CH_3 C_6H_5 N–C_2H_5 $(CO)_3Fe$ O	IIIb (43) dec. p. 76 to 77° IR (hexane): 1992, 2010, 2065 (CO)	[78, 90]
*40	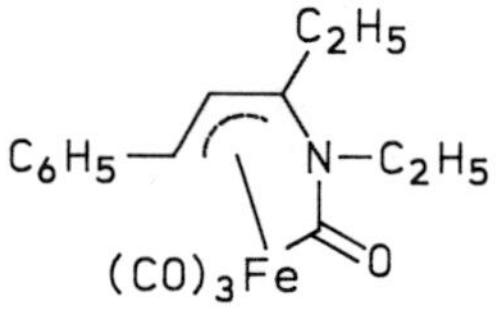 C_2H_5 C_6H_5 N–C_2H_5 $(CO)_3Fe$ O	IIIb (45) dec. p. 85 to 86° IR (hexane): 1994, 2010, 2065 (CO)	[78, 90]
*41	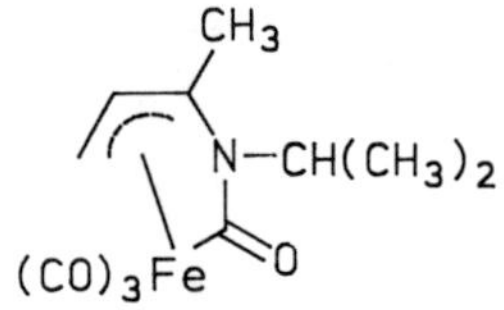 CH_3 N–$CH(CH_3)_2$ $(CO)_3Fe$ O	IIIb (49)	[78, 90]
*42	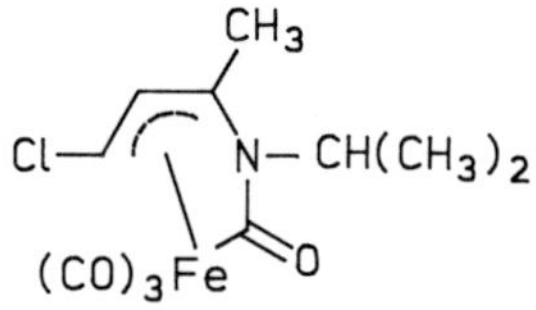 CH_3 Cl N–$CH(CH_3)_2$ $(CO)_3Fe$ O	IIIb	[78]

References on pp. 106/8

Table 3 [continued]

No.	compound	method of preparation (yield in %), properties and remarks	Ref.
*43	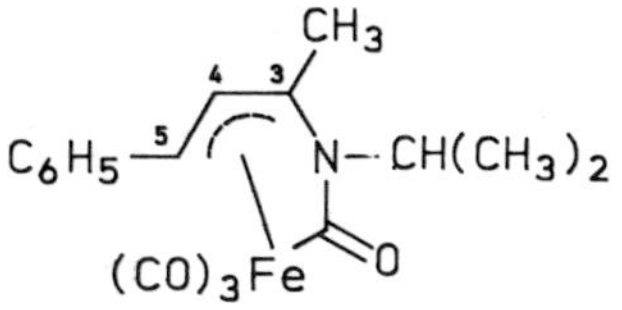 CH_3 (4, 3, 5) C_6H_5 N–$CH(CH_3)_2$ $(CO)_3Fe$ O	III b (33) m.p. 81 to 82° (dec., from CH_3OH) ^{13}C NMR (CH_2Cl_2): 63.05 (C-5), 90.34 (C-4), 95.15 (C-3); 184.00 (at −70°, C=O); 201.91, 210.29, 210.42 (at −70°, CO) ^{13}C NMR (CH_2Cl_2): 63.77 (C-5), 89.69 (C-4), 95.15 (C-3) ^{13}C NMR (CH_2Cl_2): 63.77 (C-5), 84.69 (C-4), 95.15 (C-3) IR (hexane): 1993, 2008, 2068 (CO), (overlaps with bands of another tautomer)	[52 to 54, 73, 74, 78, 90]
*44	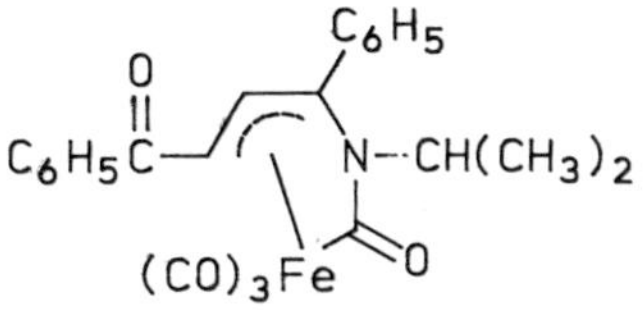 C_6H_5 O C_6H_5C N–$CH(CH_3)_2$ $(CO)_3Fe$ O	III b (27) dec. p. 109° (from CH_3OH), red-orange, air-stable crystals IR: 2010, 2030, 2080 (CO)	[72, 78]
*45	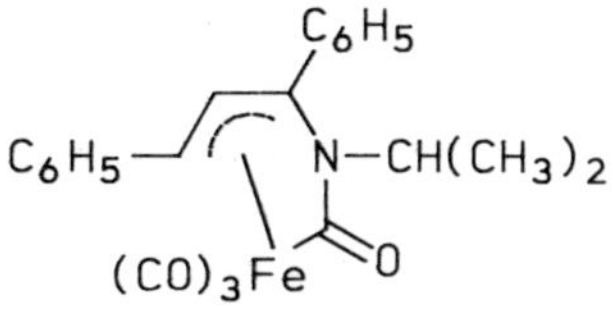 C_6H_5 C_6H_5 N–$CH(CH_3)_2$ $(CO)_3Fe$ O	III b (43)	[52, 54, 73, 78, 90]
*46	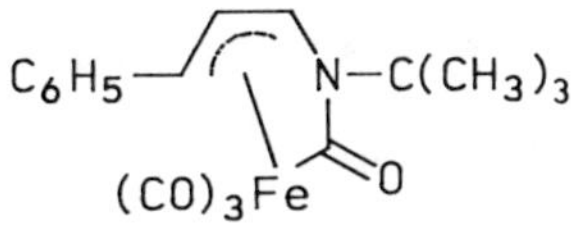 C_6H_5 N–$C(CH_3)_3$ $(CO)_3Fe$ O	III b (50) light yellow solid IR (KBr): 2000, 2070 (CO) IR (hexane): 2070 (CO), (the remaining bands overlap with bands of another tautomer)	[78, 90]
*47	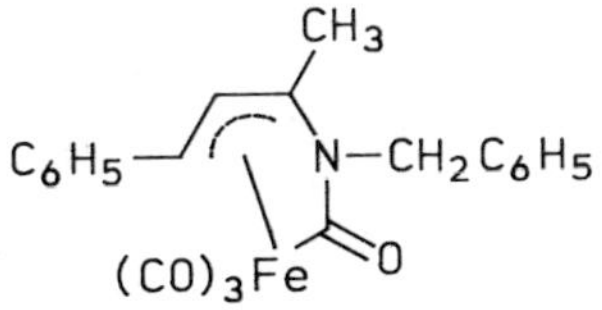 CH_3 C_6H_5 N–$CH_2C_6H_5$ $(CO)_3Fe$ O	III b (40) m.p. 94 to 95° (dec., from CH_3OH)	[52 to 54, 73]
*48	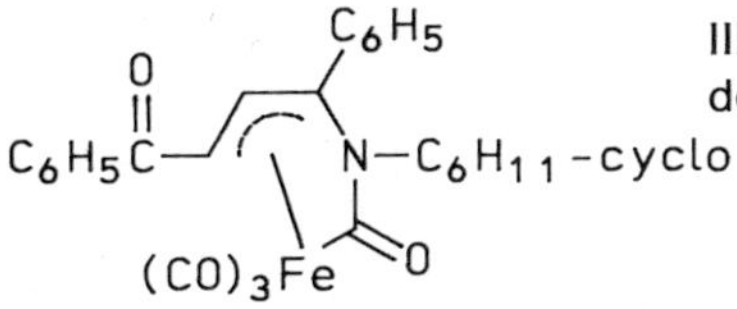 C_6H_5 O C_6H_5C N–C_6H_{11}-cyclo $(CO)_3Fe$ O	III a (20) dec. p. 115° (from CH_3OH), orange-red, air-stable crystals	[72, 78]
*49	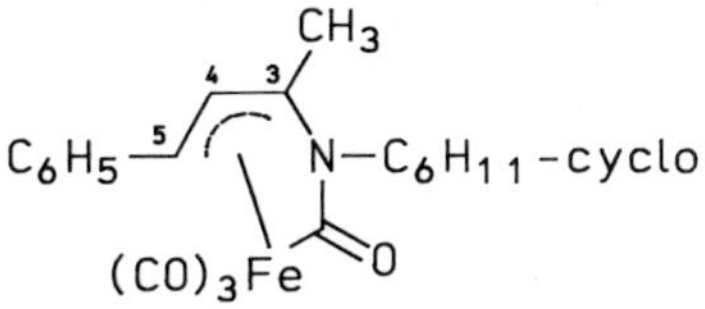 CH_3 (4, 3, 5) C_6H_5 N–C_6H_{11}-cyclo $(CO)_3Fe$ O	III b (30) m.p. 88 to 89° (dec., from CH_3OH) ^{13}C NMR (CH_2Cl_2): 63.31 (C-5), 90.34 (C-4), 94.83 (C-3), 184.49 (at −70°, C=O); 201.97, 210.09, 210.35 (at −70°, CO)	[52 to 54, 73, 74, 78, 90]

References on pp. 106/8

Table 3 [continued]

No.	compound	method of preparation (yield in %), properties and remarks	Ref.
		^{13}C NMR (CH_2Cl_2): 64.09 (C-5), 89.82 (C-4), 98.46 (C-3) IR (KBr): 1650 (C=O) IR (hexane): 1992, 2008, 2068 (CO), (overlaps with bands of another tautomer)	
*50	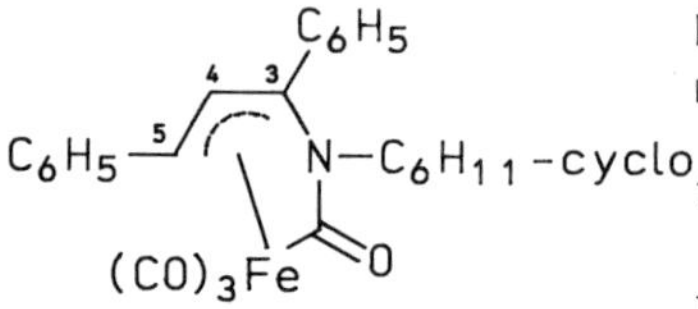 C_6H_5; C_6H_5; N$-C_6H_{11}$-cyclo; $(CO)_3Fe$; O; 3; 4; 5	III a (50) m.p. 95° (dec., from CH_3OH), yellow, air-stable crystals ^{1}H NMR ($CDCl_3$): 4.81 (H-5), 6.16 (H-4; J = 9.7), 7.0 to 8.0 (C_6H_5) ^{13}C NMR (CH_2Cl_2): 65.52 (C-5), 87.03 (C-4), 99.24 (C-3) IR (KBr): 1650 (C=O) IR (hexane): 1993, 2006, 2065 (CO)	[45, 52 to 54, 73, 74, 78]
51	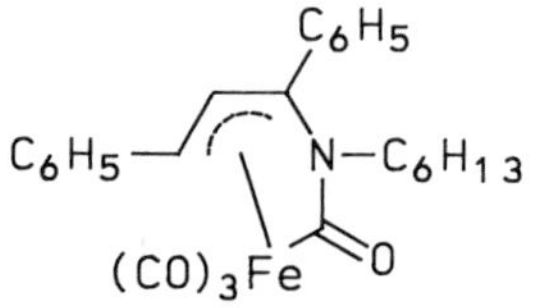 C_6H_5; C_6H_5; N$-C_6H_{13}$; $(CO)_3Fe$; O	III b (38) dec. p. 95 to 96°	[90]
*52	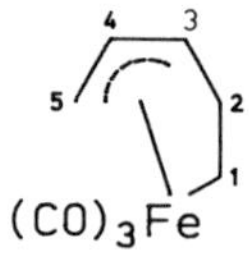 $(CO)_3Fe$; 1; 2; 3; 4; 5	see No. 103, p. 101 yellow oil ^{1}H NMR (CS_2 at $-20°$): 2.2 to 2.6 (m, H-1, 2, 5 anti), 3.50 (d, H-5 syn), 3.7 (m, H-3), 3.9 (m, H-4) IR (hexane): 1989, 1994, 2053 (CO) mass spectrum: $[M]^+$	[8, 12]
*53	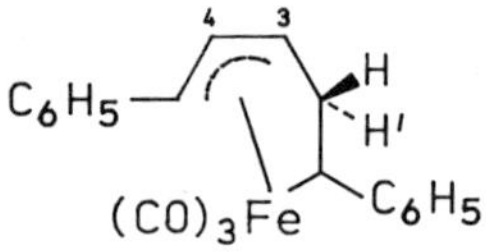 C_6H_5; H; H′; C_6H_5; $(CO)_3Fe$; 3; 4	air-sensitive, thermally unstable yellow oil ^{1}H NMR (benzene-d_6): 1.1 (t, H-1; J(H-1, 2) ~ J(H-1, 2′) = 10), 2.65 (H-2; J(H-2, 2′) = 14), 2.9 (H-2′; J(H-2′, 3) = 10), 3.6 (m, H-3; J(H-2, 3) = 3), 4.2 (d, H-5; J(H-4, 5) = 12), 4.6 (dd, H-4; J(H-3, 4) = 7), 7.1 (m, C_6H_5) IR (hexane): 1981, 1988, 2058	[32]
54	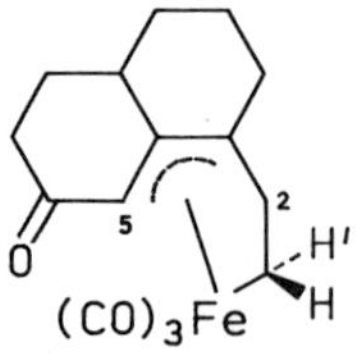 O; 5; 2; H′; H; $(CO)_3Fe$	I b yellow ^{1}H NMR (CCl_4): −0.26 (dd, H-1; J = 8, J(^{13}C-1, H-1) = 120), 0.36 (dd, H-1′; J = 8), 1.36 (m, H-2), 3.58 (s, H-5) IR (light petroleum ether): 1678 (C=O); 1975, 1998, 2056 (CO) mass spectrum: $[M-nCO]^+$ (n = 0 to 3), $[M-3CO\text{-}Fe]^+$, $[M-4CO\text{-}Fe]^+$	[16]

References on pp. 106/8

Table 3 [continued]

No.	compound	method of preparation (yield in %), properties and remarks	Ref.
*55	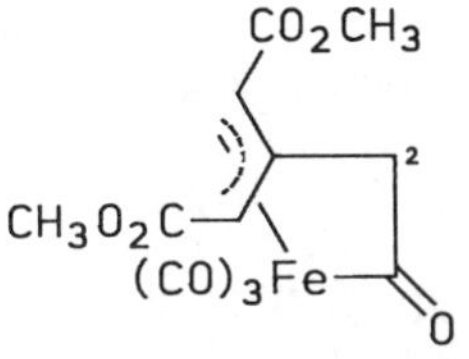	unstable oil (changes color on standing under N_2 in the refrigerator) ^{1}H NMR (benzene-d_6): 3.30 (s, CH_3), 3.45 (s, CH_3), 4.10 (H-2; ABX: $J_{AB}=9$, $J_{AX}=0$, $J_{BX}=2$; $\Delta\nu=39$ Hz), 4.6 (d, CH; J=2), 5.45 (d, CH; J=2) IR (CS_2): 1670, 1709, 1739 (CO_2CH_3); 2024, 2070, 2123 (CO)	[17]

^{4}L is a 1σ, 4-6π allyl ligand (continued at No. 137, p. 75):

No.	compound	method of preparation (yield in %), properties and remarks	Ref.
56	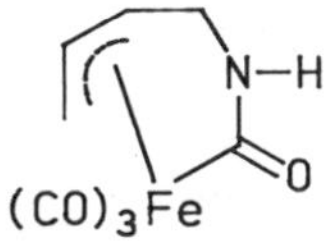	IId m.p. 106 to 107° (from ether/hexane) crystals ^{1}H NMR ($CDCl_3$): 2.96 (d, 1H; J=12), 3.38 (m, 2H), 3.65 (d, 1H; J=8), 4.65 (m, 1H), 4.91 (m, 1H), 5.72 (m, NH) IR ($CHCl_3$): 1595 (C=O); 3200, 3465 (NH) IR (hexane): 2000, 2019, 2070 (CO) mass spectrum: $[M]^+$	[11, 43]
57	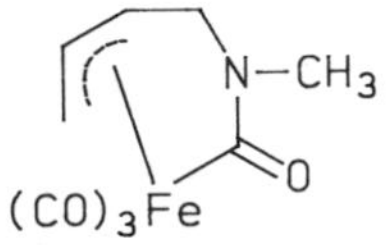	IIc (70), see No. 75, p. 87 m.p. 99 to 100° (dec.), crystals (from CH_2Cl_2/hexane) ^{1}H NMR ($CDCl_3$): 2.46 (s, CH_3), 2.85 (d, 1H; J=12.5), 3.38 (m, 2H), 3.65 (dd, 1H; J=1, J=8), 4.50 (m, 1H), 4.93 (m, 1H) IR ($CHCl_3$): 1580 (C=O) IR (hexane): 1998, 2008, 2070 (CO) mass spectrum: $[M]^+$	[11, 43]
58	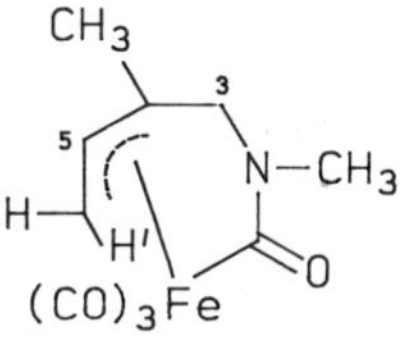	see No. 59, p. 84 m.p. 100 to 101° (from pentane at −20°), nearly colorless crystals ^{1}H NMR ($CDCl_3$): 2.62 (NCH_3), 2.64 (H-6′), 3.42 (H-3), 3.47 (H-6; J(H-6,6′)=2.0), 4.84 (H-5; J(H-5,6)=8.0, J(H-5,6′)=13) ^{13}C NMR ($CDCl_3$): 28.14 (CH_3-4), 30.9 (NCH_3), 54.1 (C-6), 55.3 (C-3), 85.3 (C-4), 94.4 (C-5); 201.7, 204.7, 207.9, 210.9 (CO) IR (KBr): 540, 565, 575, 605, 665, 720, 800, 900, 950, 960, 970, 1060, 1160, 1190, 1225, 1255, 1300, 1330, 1385, 1405, 1430, 1455, 1475, 1580, 1970, 2000, 2075, 2880 IR (hexane): 1987, 2016, 2076 (CO) mass spectrum: $[M]^+$ and fragments given	[55]

References on pp. 106/8

Table 3 [continued]

No.	compound	method of preparation (yield in %), properties and remarks	Ref.
*59	H, H′, CH_3, N–CH_3, H, H′, O, $(CO)_3Fe$	see No. 79, p. 89 m.p. 76 to 77° (from CH_2Cl_2/pentane), pale yellow crystals ^{1}H NMR ($CDCl_3$): 2.08 (CH_3-5), 2.62 (NCH_3), 2.77 (H-6′), 3.24 (H-3′; J(H-3′,3) = −12, J(H-3′,4) = 2.0), 3.47 (H-3; J(H-3,4) = 6.0), 3.58 (H-6; J(H-6,6′) = 2.0), 4.37 (H-4) ^{13}C NMR ($CDCl_3$): 28.9 (CH_3-5), 30.9 (NCH_3), 50.4 (C-6), 59.2 (C-3), 69.0 (C-4), 111.2 (C-5); 201.6, 204.1, 207.6, 210.6 (CO) IR (KBr): 580, 600, 610, 660, 715, 800, 880, 918, 925, 950, 990, 1035, 1060, 1100, 1200, 1240, 1315, 1355, 1385, 1405, 1430, 1440, 1480, 1580, 1975, 2060, 2900 IR (hexane): 1990, 2017, 2077 (CO) mass spectrum: $[M]^+$ and fragments given	[55]
*60	CH_3, N–CH_3, CH_3, O, $(CO)_3Fe$	see "Further information", pp. 84/5	[55]
*61	H, H′, N–$CH_2C_6H_5$, H, H′, O, $(CO)_3Fe$	see No. 75, p. 87 m.p. 76 to 78°, white crystals (from petroleum ether) ^{1}H NMR ($CDCl_3$): 2.88 (dd, H-6′; J = 13, J = 2), 3.16 (dd, H-3; J = 13, J = 2), 3.29 (dd, H-3′; J = 13, J = 6), 3.73 (ddd, H-6; J = 8, J = 2), 4.14 (d, 1H; J = 15), 4.26 (d, 1H; J = 15), 4.40 (m, H-4), 4.84 (ddd, H-5; J = 13, J = 8, J = 8), 7.04 to 7.35 (m, 5H) IR (Nujol): 1165, 1405, 1570, 1585, 2010, 2070, 2850, 2980, 3080	[59, 62, 88]
*62	H, H′, CH_3, N–$CH_2C_6H_5$, H, H′, O, $(CO)_3Fe$	see No. 79, pp. 89/90 m.p. 70 to 72°, white crystals (from petroleum ether) ^{1}H NMR ($CDCl_3$): 1.94 (s, CH_3), 2.70 (d, H-6′; J = 2), 3.14 (dd, H-3; J = 13, J = 2), 3.33 (dd, H-3′; J = 13, J = 6), 3.56 (dd, H-6; J = 2, J = 2), 4.24 (m, N-CH_2, H-4), 7.06 to 7.34 (m, C_6H_5) IR (Nujol): 1165, 1405, 1570, 1585, 2010, 2070, 2850, 2980, 3080	[62, 88]

References on pp. 106/8

Table 3 [continued]

No.	compound	method of preparation (yield in %), properties and remarks	Ref.
*63	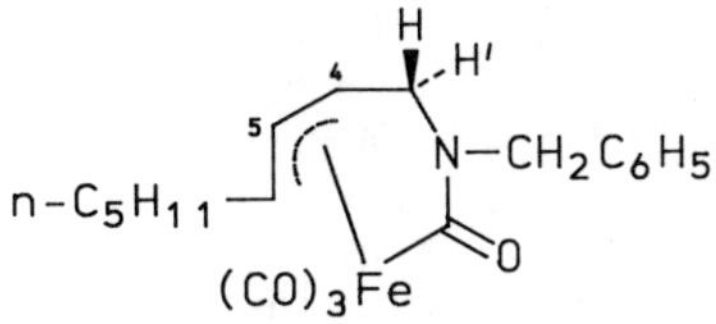	see No. 77, p. 89 m.p. 80 to 86° (from petroleum ether), oil which solidifies to give crystals ^{1}H NMR ($CDCl_3$): 0.91 (t, CH_3; J=6), 1.35 (m, 4H), 1.41 to 1.64 (m, 2H), 1.72 (m, 1H), 2.27 (m, 1H), 3.09 (dd, H-3; J=13, J=2), 3.26 (dd, H-3′; J=13, J=6), 3.78 (ddd, H-6; J=12, J=9, J=5), 4.15 (d, 1H; J=14), 4.17 (ddd, H-4; J=8, J=6, J=2), 4.31 (d, 1H; J=14), 4.64 (dd, H-5; J=12, J=8), 7.13 (d, 2H; J=6), 7.25 (m, 3H) IR (Nujol): 1590, 1980, 1995, 2030, 2850, 2900	[59, 62, 88]
*64	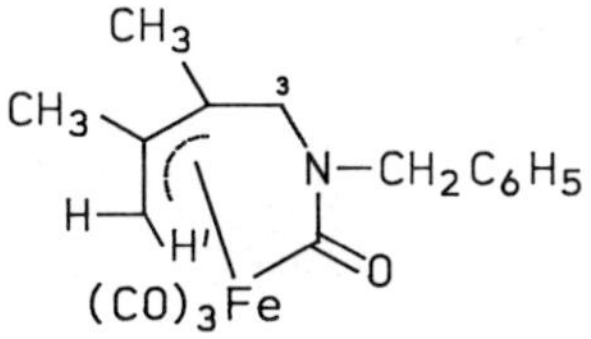	see No. 89, p. 98 m.p. 98 to 100°, yellow crystals (from petroleum ether) ^{1}H NMR ($CDCl_3$): 1.89 (s, CH_3), 2.08 (s, CH_3), 2.44 (d, H-6′; J=2), 3.14 (s, H-3), 3.42 (d, H-6; J=2), 4.28 (s, 2H), 7.08 to 7.33 (m, 5H) IR ($CHCl_3$): 705, 1185, 1440, 1570, 1990, 2025, 2900	[59, 62, 88]
*65	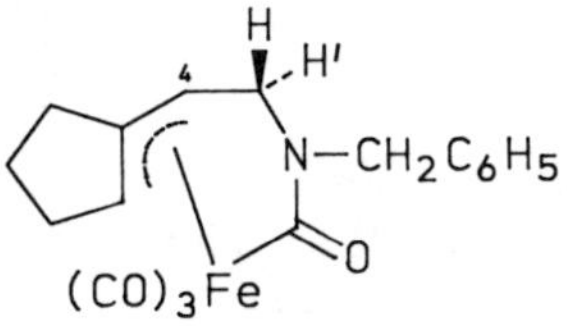	see Nos. 85, p. 93 and 91, p. 98 m.p. 95 to 98°, m.p. 98 to 100° (dec.), white crystals (from petroleum ether) ^{1}H NMR ($CDCl_3$): 1.60 to 2.14 (m, 4H), 2.26 to 2.60 (m, 2H), 3.27 (m, 2H), 4.06 to 4.30 (m, 4H), 7.09 to 7.39 (m, 5H) ^{1}H NMR ($CDCl_3$): 1.37 to 1.58 (m, 1H), 1.82 to 1.96 (m, 1H), 2.02 (dd, 1H; J=8, J=15), 2.37 to 2.48 (m, 2H), 2.75 to 2.92 (m, 1H), 3.21 (d, H-3′; J=13), 3.37 (dd, H-3′; J=6, J=13), 4.13 (d, 1H; J=15), 4.22 (br,s, H-6), 4.27 (d, H-4; J=6), 4.35 (d, 1H; J=15), 7.07 to 7.16 (m, 2H), 7.18 to 7.34 (m, 3H) IR: 680, 1580, 1980, 1990, 2020, 2900 IR (mull): 660, 1171, 1490, 1590, 2000, 2070, 2860, 2960, 3060, 3080	[59, 62, 88]
*66	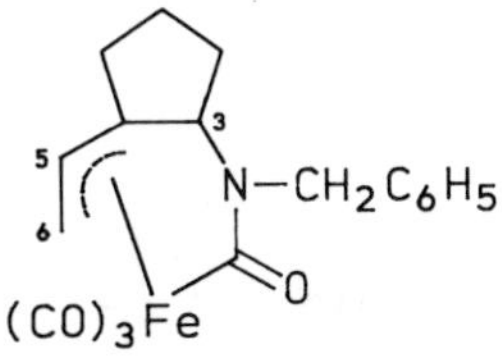	see No. 91, p. 98 m.p. 95 to 98°, white crystals (from petroleum ether) ^{1}H NMR ($CDCl_3$): 1.52 to 2.65 (m, 6H), 3.32 to 3.44 (m, 3H), 3.68 (ABq, 1H; J=14), 4.88 (ABq, 1H), 5.07 (dd, 1H; J=8, J=12), 7.04 to 7.17 (m, 2H)	[59, 62, 88]

References on pp. 106/8

Table 3 [continued]

No.	compound	method of preparation (yield in %), properties and remarks	Ref.
		^{1}H NMR ($CDCl_3$): 1.52 to 2.65 (m, 7H), 3.31 (dd, H-3; J=6.7, J=8), 3.39 (dd, H-6; J=1.8, J=8.0), 3.68 (d, 1H; J=14), 4.88 (d, 1H; J=14), 5.07 (dd, H-5; J=13.4, J=8), 7.04 to 7.17 (m, 2H), 7.17 to 7.34 (m, 3H) IR: 665, 700, 1170, 1590, 1960, 2010, 2070, 2860, 2960, 3020 IR (mull): 660, 1170, 1390, 1400, 1490, 1590, 2000, 2070, 2860, 2960, 3020, 3060, 3080	
*67	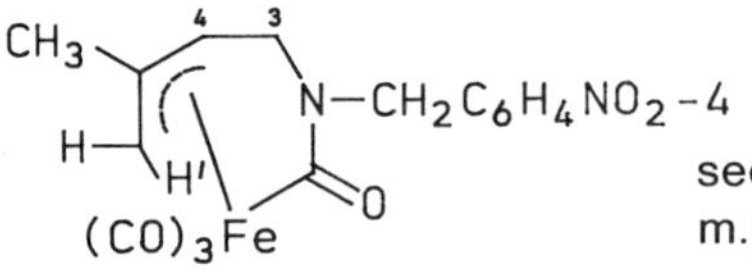 CH_3, 4, 3, N–$CH_2C_6H_4NO_2$-4, H, H′, O, $(CO)_3Fe$	see No. 79, pp. 89/90 m.p. 118 to 123°, yellow crystals (from petroleum ether) ^{1}H NMR ($CDCl_3$): 2.03 (s, CH_3), 2.66 (br,s, H-6′), 3.18 to 3.34 (m, H-3), 3.51 to 3.68 (m, H-6), 4.20 (d, 1H; J=15), 4.19 to 4.39 (m, H-4), 4.39 (d, 1H; J=15), 7.20 to 7.42 (m, 2H), 8.10 to 8.25 (m, 2H) IR ($CHCl_3$): 750, 1215, 1355, 1525, 1580, 2000, 2025, 2095, 2870, 2950, 3000, 3040	[62, 88]
*68	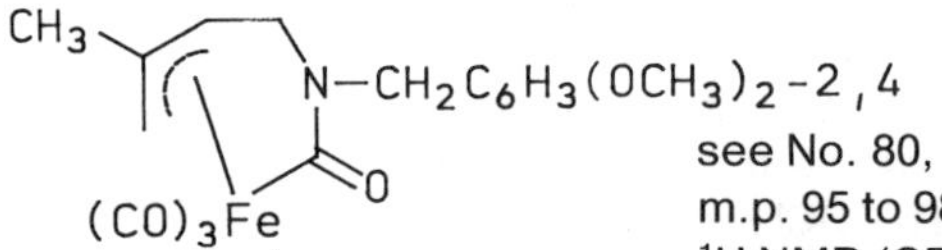 CH_3, N–$CH_2C_6H_3(OCH_3)_2$-2,4, O, $(CO)_3Fe$	see No. 80, pp. 89/90 m.p. 95 to 98° (from petroleum ether) ^{1}H NMR ($CDCl_3$): 1.98 (s, CH_3), 2.73 (br,s, 1H), 3.10 to 3.31 (m, 2H), 3.56 (m, 1H), 3.80 (s, 6H), 4.16 (m, 3H), 6.32 to 6.50 (m, 2H), 6.94 to 7.06 (m, 1H) IR: 740, 910, 1035, 1210, 1505, 1585, 1610, 1960, 2000, 2060, 2945, 2955, 3000	[62]
*69	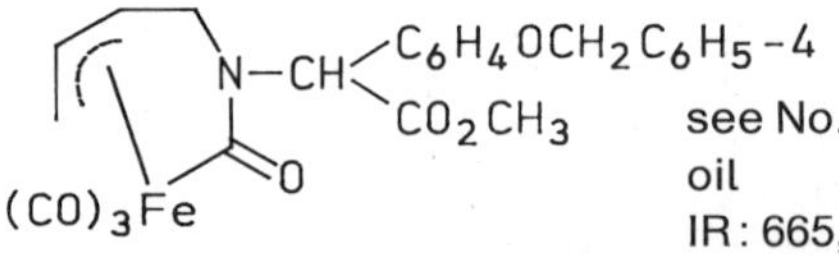 N–CH($C_6H_4OCH_2C_6H_5$-4)(CO_2CH_3), O, $(CO)_3Fe$	see No. 75, p. 87 oil IR: 665, 690, 1160, 1240, 1510, 1590, 1610, 1740, 2000, 2010, 2090, 2870, 2950, 3030, 3060	[59, 62]
*70	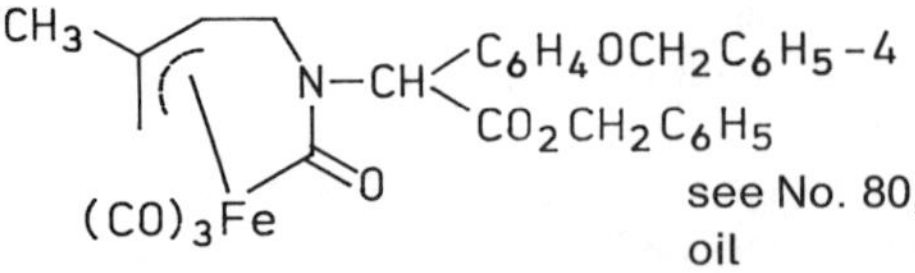 CH_3, N–CH($C_6H_4OCH_2C_6H_5$-4)($CO_2CH_2C_6H_5$), O, $(CO)_3Fe$	see No. 80, pp. 89/90 oil IR: 1170, 1250, 1510, 1588, 1610, 1740, 1980, 2020, 2070, 2930, 2960, 3040, 3070, 3090	[62]

Table 3 [continued]

No.	compound	method of preparation (yield in %), properties and remarks	Ref.
*71	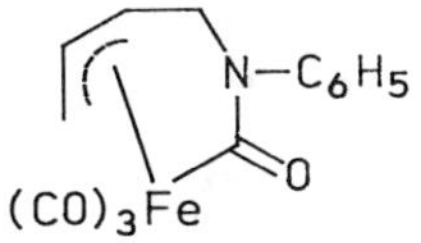 $N-C_6H_5$, O, $(CO)_3Fe$	IId (35), see also No. 75, p. 87 m.p. 104 to 105° (dec.), prisms (from CH_2Cl_2/hexane) IR ($CHCl_3$): 1500 (C_6H_5), 1580 (C=O), 1600 (C_6H_5); 2000, 2020, 2080 (CO) mass spectrum: $[M-nCO]^+$ (n=0 to 4)	[11, 31]
72	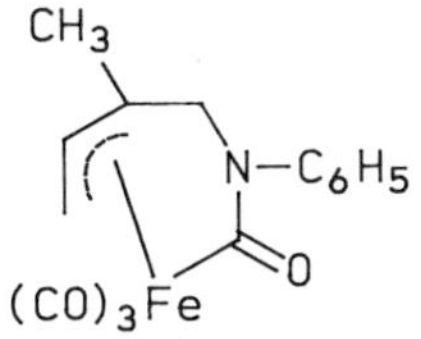 CH_3, $N-C_6H_5$, O, $(CO)_3Fe$	see No. 80, p. 89 m.p. 126 to 127° (dec.), crystals (from CH_2Cl_2/pentane) ^1H NMR ($CDCl_3$): 2.02 (s, CH_3), 2.85 (dd, 1H; J=2, J=14), 3.62 (dd, 1H; J=8, J=2), 3.85 (ABq, 2H; J=14), 4.85 (dd, 1H; J=8, J=14), 7.15 (m, C_6H_5) IR ($CHCl_3$): 1500 (C_6H_5), 1580 (C=O), 1600 (C_6H_5) IR (hexane): 2000, 2018, 2070 (CO) mass spectrum: $[M]^+$	[11]
73	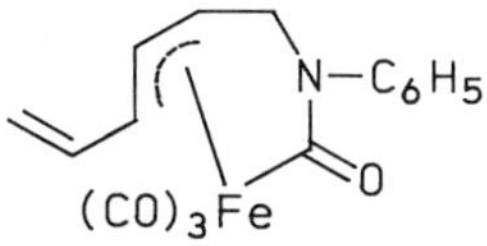 $N-C_6H_5$, O, $(CO)_3Fe$	see No. 78, p. 89 m.p. 110° (dec.), yellow needles ^1H NMR ($CDCl_3$): 3.68 (dd, 1H; J=13.0, J=1.5), 3.94 (dd, 1H; J=6.0), 4.26 (ddd, 1H; J=7.5), 4.58 (dd, 1H; J=9.0), 4.90 (dd, 1H; J=12.5), 5.23 (dd, 1H; J=9.0, J=1.5), 5.53 (dd, 1H; J=17.0), 5.98 (ddd, 1H) IR ($CHCl_3$): 1500 (C_6H_5), 1575 (C=O), 1600 (C_6H_5) IR (hexane): 1995, 2010, 2062 (CO) mass spectrum: $[M-nCO]^+$ (n=0 to 4)	[31]
*74	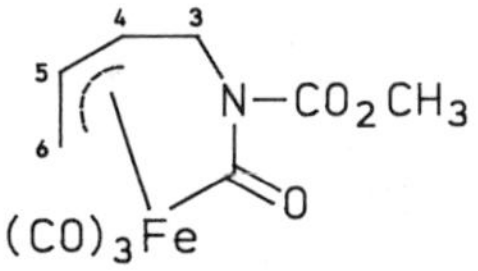 4, 3, 5, 6, $N-CO_2CH_3$, O, $(CO)_3Fe$	Id (71) m.p. 94 to 95° (dec.), pale yellow crystals (from benzene/hexane) ^1H NMR: 2.95 (H-6 anti), 3.71 (CH_3), 3.80 (H-6 syn), 3.88 (H-3), 4.51 (H-4), 4.93 (H-5) IR (hexane): 2007, 2023, 2080 (CO) IR (CH_2Cl_2): 1659, 1712 (CO_2CH_3); 1757 (C=O) mass spectrum: $[M]^+$	[13]
*75	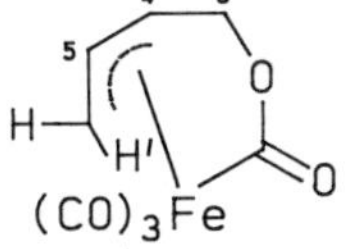 4, 3, 5, O, H, H′, O, $(CO)_3Fe$	Ic (5), Id (90), IIb (80), IIc (78), IId (25 to 40) m.p. 112 to 114° (dec., from hexane or hexane/CH_2Cl_2), m.p. 115 to 116° (dec.), m.p. 118.2 to 118.6° (dec.), m.p. 120°, pale yellow prisms (from ether/CH_2Cl_2 at −80°) or needles (from C_6H_6/petroleum ether) ^1H NMR ($CDCl_3$ at −10 °C): 2.98 (H-6′), 3.73 (H-6; J(H-5,6)=7.0), 4.07 (H-3; J(H-3,3′) = −11.5), 4.94 (H-4,5; J(H-5,6′)=12.5)	[1, 2, 3, 11, 13, 31, 55]

References on pp. 106/8

Table 3 [continued]

No.	compound	method of preparation (yield in %), properties and remarks	Ref.
		^{13}C NMR ($CDCl_3$): 57.65 (C-6), 64.67 (C-3), 76.88 (C-4), 93.26 (C-5); 203.42, 205.69, 206.73, 208.68 (CO) IR (KBr): 660, 750, 810, 865, 950, 982, 1000, 1015, 1065, 1170, 1250, 1260, 1390, 1460, 1480; 1652 (C=O); 2900, 2950, 2970, 3050, 3080, 3280 IR ($CHCl_3$): 1665 (C=O) IR (hexane): 2002, 2022, 2081 (CO) IR (hexane): 1996, 2011, 2082 (CO) IR (heptane): 2014, 2031, 2086 (CO) mass spectrum: $[M-nCO]^+$ (n = 0 to 4)	
*76	CH_3, O, O, $(CO)_3Fe$	see "Further information", pp. 87/8	[55]
*77	C_5H_{11}-n, O, O, $(CO)_3Fe$	see "Further information", pp. 88/9	[59, 69]
*78	O, O, $(CO)_3Fe$	see No. 71, p. 86 m.p. 57 to 59°, yellow needles (from CH_2Cl_2/petroleum ether) ^1H NMR ($CDCl_3$): 3.04 (dd, 1H; J = 13.0), 3.68 (dd, 1H; J = 2.0, J = 8.0), 4.62 (m, 2H), 4.88 (ddd, 1H; J = 8.0), 5.18 (d, 1H; J = 11.0), 5.28 (d, 1H; J = 17.0), 5.86 (ddd, 1H; J = 5.0) IR ($CHCl_3$): 1650 (C=O) IR (hexane): 2010, 2030, 2080 (CO) mass spectrum: $[M-nCO]^+$ (n = 0 to 3), $[M-CO_2]^+$	[31]
*79	CH_3, H, H′, 5, H, H′, O, O, $(CO)_3Fe$	Id (80 to 89) m.p. 101° ^1H NMR ($CDCl_3$): 2.04 (CH_3-4), 2.76 (H-6′; J(H-6,6′) = 2.0), 3.55 (H-6; J(H-5,6) = 8.0), 3.92 (H-3; J(H-3,3′) = −12.0), 4.17 (H-3′), 4.98 (H-5; J(H-5,6′) = 13.0) ^{13}C NMR ($CDCl_3$): 25.70 (CH_3-4), 53.27 (C-6), 68.74 (C-3), 93.79 (C-4,5); 203.25, 205.97, 206.29, 207.91 (CO) IR (KBr): 1656 IR (hexane): 1992, 2005, 2078 (CO)	[13, 55]

References on pp. 106/8

Table 3 [continued]

No.	compound	method of preparation (yield in %), properties and remarks	Ref.
*80	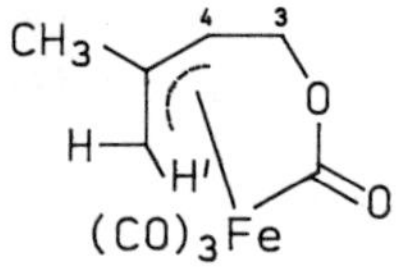	I d (small amounts), II c viscous oil ^{1}H NMR ($CDCl_3$): 1.95 (CH_3-5), 2.74 (H-6'; J(H-6,6')=2.0), 3.47 (H-6), 3.90 (H-3,3'), 4.63 (H-4) ^{13}C NMR ($CDCl_3$): 28.14 (CH_3-5), 57.84 (C-6), 65.12 (C-3), 76.88 (C-4), 111.46 (C-5); 203.74, 206.60, 207.12 (CO) IR ($CHCl_3$): 1660 (C=O) IR (hexane): 2020, 2038, 2070 (CO)	[11, 55]
*81	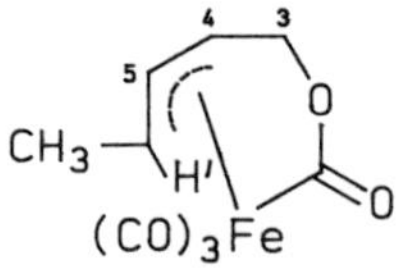	I d (89) m.p. 97 to 98° (dec.), pale yellow crystals (from $CHCl_3$/pentane) ^{1}H NMR ($CDCl_3$): 1.76 (CH_3-6; J(CH_3-6, H-6')=6.0), 3.90 (H-3,3',6'; J(H-5,6')=11.5), 4.60 (H-4; J(H-4,5)=8.0), 4.77 (H-5) ^{13}C NMR ($CDCl_3$): 21.22 (CH_3-6), 64.99 (C-3), 70.77 (C-4), 78.90 (C-6), 94.24 (C-3); 204.26, 206.73, 207.45, 208.68 (CO) IR (KBr): 545, 570, 600, 615, 655, 970, 1050, 1240, 1370, 1380, 1435, 1450, 1470, 1650, 2000 IR (hexane): 2007, 2021, 2081 mass spectrum: $[M]^+$ and fragments given	[55]
*82	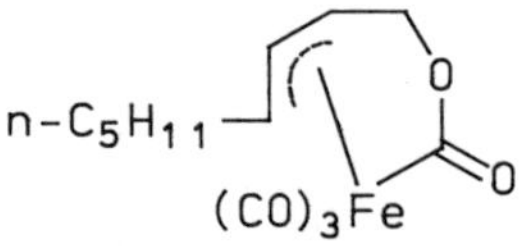	–	[76]
*83	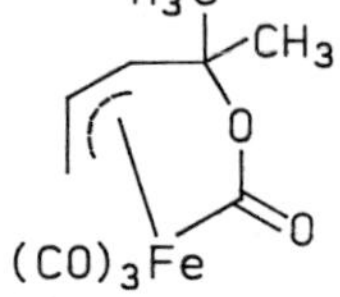	see "Further information", pp. 90/1	[55]
*84	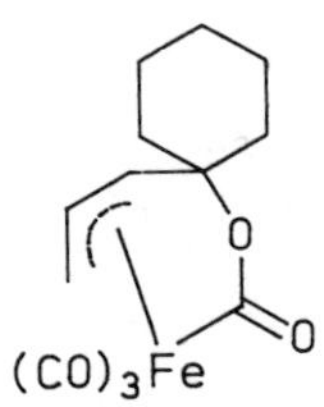	I d (49) m.p. 96° (dec.) ^{1}H NMR ($CDCl_3$): 1.18 to 2.13 (m, 10H), 3.0 (d, 1H; J=13), 3.51 (d, 1H; J=9), 3.90 (d, 1H; J=8), 4.58 to 5.14 (AM_2X, 1H; J=8, J=9, J=13) IR: 875, 960, 985, 1165, 1440, 1450, 1670, 1990, 2050, 2870, 2925	[69]

References on pp. 106/8

Table 3 [continued]

No.	compound	method of preparation (yield in %), properties and remarks	Ref.
*85	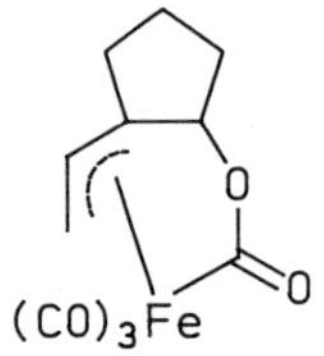	see "Further information", pp. 91/3	[39, 56, 69]
*86	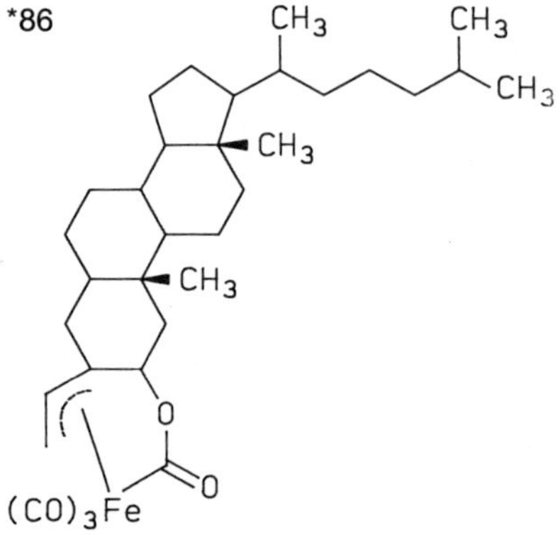	see "Further information", pp. 93/4	[69]
*87	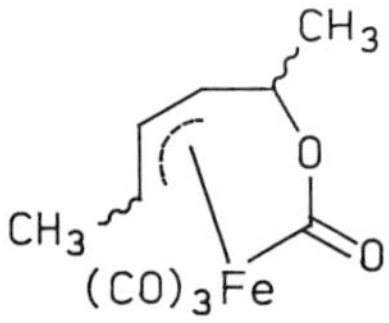	see "Further information", pp. 94/6	[1, 27, 55]
*88	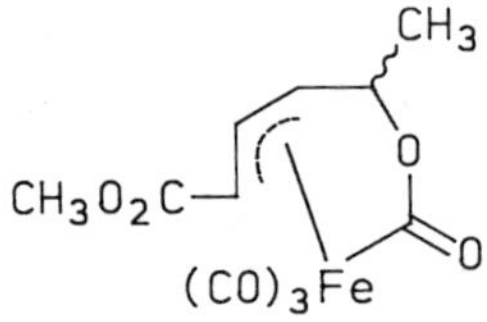	see "Further information", pp. 96/7	[55]
*89	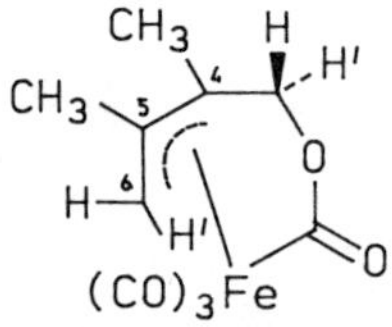	Id (52 to 65), IIb (ca. 1) m.p. 89 to 90°, m.p. 91 to 92° (dec.), m.p. 104°, yellow needles, yellow lustrous crystals (from CH_2Cl_2/pentane) ^1H NMR ($CDCl_3$): 2.00 (CH_3-4), 2.15 (CH_3-5), 2.58 (H-6'; J(H-6, 6') = 2.0), 3.45 (H-6), 3.89 (H-3; J(H-3, 3') = −11.5), 4.04 (H-3') ^{13}C NMR ($CDCl_3$): 23.46, 24.24 (CH_3); 54.29 (C-6), 70.58 (C-3), 90.34 (C-4), 109.96 (C-5); 204.65, 206.40, 207.34, 208.35 (CO)	[1, 39, 55, 69]

References on pp. 106/8

Table 3 [continued]

No.	compound	method of preparation (yield in %), properties and remarks	Ref.
		IR (KBr): 560, 590, 610, 660, 900, 950, 985, 1035, 1060, 1260, 1330, 1370, 1390, 1440, 1460, 1655, 2000 IR (KBr): 685, 784, 819, 900, 950, 970, 990, 1014, 1039, 1060, 1170, 1215, 1322, 1376, 1395, 1442, 1460, 1660, 2003, 2023, 2080, 2882, 2952, 2984, 3000, 3010, 3082, 3290 IR (hexane): 1990, 2022, 2080 (CO) mass spectrum: $[M-CO]^+$ and further fragments	
*90	$(CO)_3Fe$	Id (29) m.p. 83 to 84° (dec.) ^1H NMR ($CDCl_3$): 1.33 to 2.5 (m, 8H), 2.53 (d, 1H; J=3), 3.15 (d, 1H; J=3), 3.77 (ABq, 1H; J=11.5), 4.13 (ABq, 1H; J=11.5) IR: 660, 950, 995, 1045, 1380, 1450, 1465, 1650, 1990, 2050, 2900	[69]
*91	$(CO)_3Fe$	Id (72) m.p. 88.5 to 89.5° (dec.) ^1H NMR ($CDCl_3$): 1.12 to 3.24 (m, 6H), 3.94 to 4.48 (m, 3H), 4.73 (d, 1H; J=4) IR: 1670, 1990, 2100, 2970	[69]
*92	$(CO)_3Fe$	Id (46) m.p. 51 to 53° (dec.) ^1H NMR ($CDCl_3$): 1.24 to 2.76 (m, 10H), 3.86 to 4.70 (m, 3H), 4.98 (m, 1H) IR: 1100, 1660, 2015, 2075, 2855, 2938	[69]
*93	H_3C, CH_3, CH_3, CH_3, $(CO)_3Fe$	see "Further information", pp. 98/100	[55]
94	$(CO)_3Fe$, H, H′, CF_3	IVb (11) m.p. 44°, pale yellow crystals (from hexane at −78°) ^1H NMR ($CDCl_3$): 2.60 (m, H-3; J(H-3, CF_3) =4.0), 2.75 (d, H-6′; J(H-5,6′)=12), 3.9 (d, H-6; J(H-5,6)=8.0), 4.9 (m, H-4,5), 6.3 (m, H-2; J(H-2, CF_3)=2.0, J(H-2,3)=4.0) ^{13}C NMR ($CDCl_3$): 34.8 (s, C-3), 64.4 (s, C-6), 80.9 (s, C-4), 96.7 (s, C-5), 145.0 (q, C-2; J(C,F)=10.0)	[40]

References on pp. 106/8

Table 3 [continued]

No.	compound	method of preparation (yield in %), properties and remarks	Ref.
		^{19}F NMR (acetone-d_6): 59.9 (dt, CF_3; J(CF_3, H-2) = 2.0, J(CF_3, H-3) = 4.0) IR (Nujol): 1630 (C=C) IR (hexane): 2007, 2022, 2079 (CO) mass spectrum: $[M-nCO]^+$ (n = 0 to 3)	
95	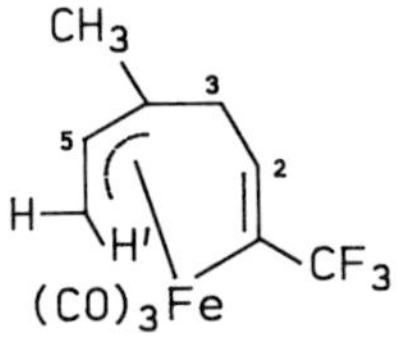	IVb data from a mixture with No. 96 (ratio 1:2.5) m.p. 54°, pale yellow crystals (from hexane at −78°) ^{1}H NMR ($CDCl_3$): 2.10 (s, CH_3), 2.55 (d, H-6'; J(H-5,6') = 14.0), 2.80 (m, H-3), 3.70 (d, H-6; J(H-5,6) = 8.0), 4.75 (m, H-5), 6.30 (m, H-2; J(H-2, CF_3) = 2.0) ^{13}C NMR ($CDCl_3$): 29.7 (s, C-4), 40.2 (s, C-3), 59.2 (s, C-6), 97.1 (s, C-5), 144.5 (q, C-2; J(C,F) = 10.0) ^{19}F NMR (acetone-d_6): 60.2 (m, CF_3; J(CF_3, H-2) = 2.0, J(CF_3, H-3) = 4.0) IR (Nujol): 1630 (C=C) IR (hexane): 2002, 2019, 2076 (CO) mass spectrum: $[M-nCO]^+$ (n = 0 to 3), $[M-CH_2-nCO]^+$ (n = 0 to 2)	[40]
96	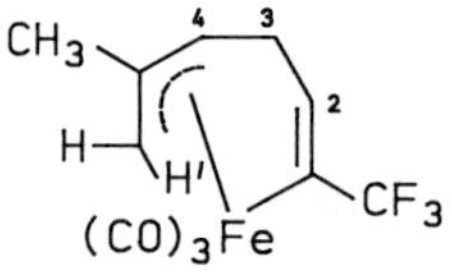	IVb all data from a mixture with No. 95 (ratio 2.5:1) m.p. 54°, pale yellow crystals (from hexane at −78°) ^{1}H NMR ($CDCl_3$): 2.00 (s, CH_3), 2.70 (s, H-6'), 2.75 (m, H-3), 3.70 (s, H-6'), 4.80 (m, H-4; J(H-3,4) = 6.0), 6.25 (m, H-2; J(H-2, CF_3) = 2.0) ^{13}C NMR ($CDCl_3$): 29.2 (s, C-5), 35.9 (s, C-3), 65.3 (s, C-6), 80.2 (s, C-4), 113.7 (s, C-1), 144.5 (q, C-2; J(C,F) = 10.0) ^{19}F NMR (acetone-d_6): 59.8 (m, CF_3; J(CF_3, H-2) = 2.0, J(CF_3, H-3) = 4.0) IR and mass spectrum as for No. 95	[40]
*97	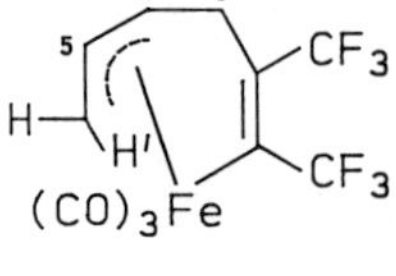	IVb (8) m.p. 38°, pale yellow crystals (from hexane at −40°) ^{1}H NMR ($CDCl_3$): 2.88 (dd, H-6'; J(H-6,6') = 1.5, J(H-5,6') = 13.0), 3.12 (complex m, H-3; J(H-3, CF_3) = 3.0), 4.02 (dd, H-6; J(H-5,6) = 8.0), 4.66 (complex m, H-4; J(H-3,4) = 8.0), 4.86 (ddd, H-5; J(H-4,5) = 8.0)	[20, 23, 25, 40]

Table 3 [continued]

No.	compound	method of preparation (yield in %), properties and remarks	Ref.
		^{19}F NMR (CH_2Cl_2): 54.0 (qt, CF_3-1; J(H-3, CF_3-1) = 3.5, J(CF_3-1,2) = 14.0), 58.5 (q, CF_3-2; J(H-3, CF_3-2) = 3.5) IR (Nujol): 1631 (C=C) IR (hexane): 2021, 2034, 2068 (CO)	
98	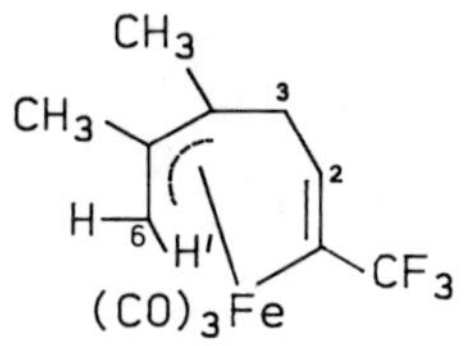	IVb (38) m.p. 62°, pale yellow crystals (from hexane at −78°) ^{1}H NMR ($CDCl_3$): 2.0 (s, CH_3), 2.1 (s, CH_3), 2.55 (s, H-6′), 2.70 (q, H-3; J(H-3, CF_3) = 4.0), 3.70 (s, H-6), 6.35 (dq, H-2; J(H-2,3) = 4.0, J(H-2, CF_3) = 2.0) ^{13}C NMR ($CDCl_3$): 24.2 (s, C-5), 27.3 (s, C-4), 43.6 (s, C-3), 62.1 (s, C-6), 97.0 (s, C-1), 112.0 (s, CF_3), 145.5 (q, C-2; J(C,F) = 10.0) ^{19}F NMR (acetone-d_6): 59.7 (dt, CF_3; J(H-2, CF_3) = 2.0, J(H-3, CF_3) = 4.0) IR (Nujol): 1630 (C=C) IR (hexane): 1987, 2015, 2072 (CO) mass spectrum: $[M-nCO]^+$ (n = 0 to 3)	[40]
*99	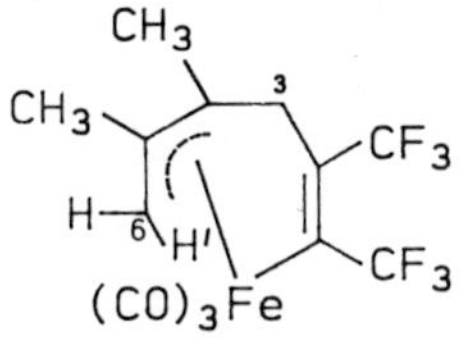	IVb (58) m.p. 45 to 47°, yellow crystals (from hexane at 0°) ^{1}H NMR ($CDCl_3$): 1.98 (s, CH_3), 2.08 (s, CH_3), 2.60 (d, H-6′; J(H-6,6′) = 1.5), 3.11 (q, H-3; J(H-3, CF_3) = 3.5), 3.73 (d, H-6) ^{19}F NMR (CH_2Cl_2): 53.2 (qt, CF_3-1; J(H-3, CF_3) = 3.5), 59.2 (q, CF_3-2; J(CF_3-1,2) = 14.0) IR (Nujol): 1631 (C=C) IR (hexane): 2007, 2025, 2080 (CO) mass spectrum: $[M-nCO]^+$ (n = 1 to 3), $[M-3CO-F]^+$	[23, 40]
*100	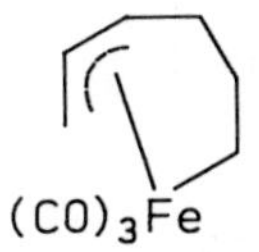	—	[22]
*101	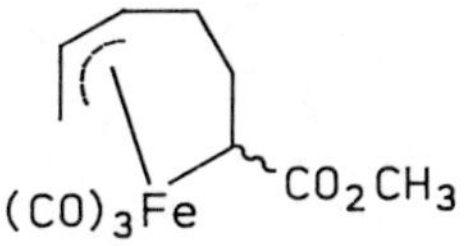	see "Further information", pp. 100/1	[18, 37, 42]

References on pp. 106/8

Table 3 [continued]

No.	compound	method of preparation (yield in %), properties and remarks	Ref.
*102	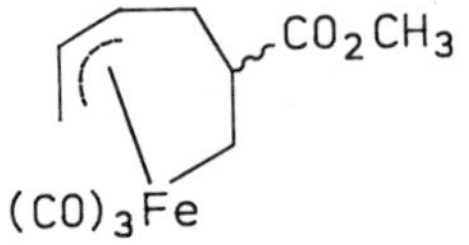 CO_2CH_3; $(CO)_3Fe$	—	[33]
*103	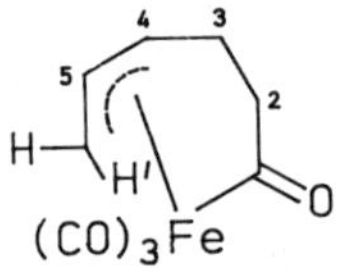 H; H'; O; $(CO)_3Fe$	I a (9), see also No. 52, p. 83 yellow crystals (from pentane at −70°), which gradually decompose on warming ^1H NMR (CS_2 at −20°): 2.1 to 2.5 (m, H-2, 3, 6), 3.25 (d, H-6′), 4.89 (ddd, H-5; J = 8.5, J = 8.5, J = 12), 5.15 (m, H-4) IR (hexane): 2005, 2064 (CO) IR (CH_2Cl_2): 1665 (C=O) mass spectrum: $[M]^+$	[12]
*104	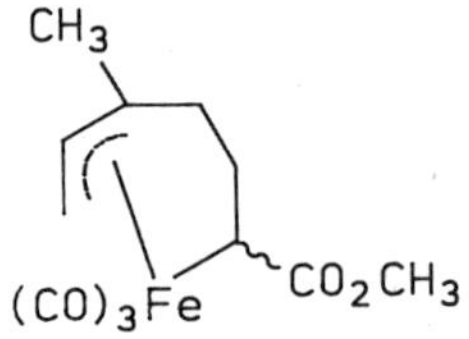 CH_3; $(CO)_3Fe$; CO_2CH_3	see "Further information", pp. 101/2	[37]
*105	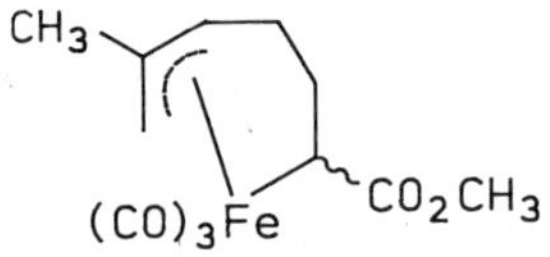 CH_3; $(CO)_3Fe$; CO_2CH_3	see "Further information", p. 102	[37]
*106	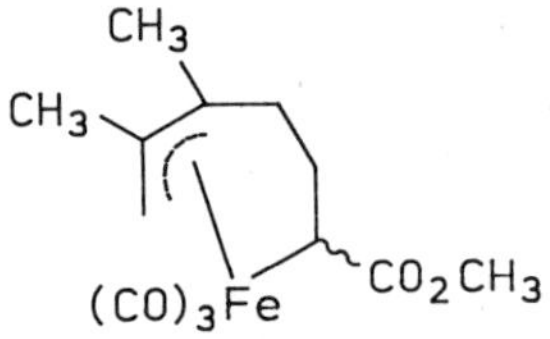 CH_3; CH_3; $(CO)_3Fe$; CO_2CH_3	see "Further information", pp. 102/4	[37, 42, 57, 65, 66]
107	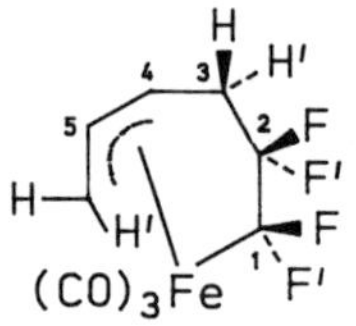 H; H'; F; F'; F; F'; H; H'; $(CO)_3Fe$	IVa (40) m.p. 83°, pale yellow crystals (from n-hexane at 0°) ^1H NMR ($CDCl_3$): 1.40 (m, H-3′; J (H-3, 3′) = 14.0, J (H-3′, 4) = 7.5, J (H-3′, F-2) = 36.0), 2.42 (dddd, H-3; J (H-3, 4) = 7.0, J (H-3, F-2) = 14.0, J (H-3, F-1) = 7.0), 2.68 (ddd, H-6′; J (H-5, 6′) = 13.0, J (H-6, 6′) = 3.0, J (H-6′, F-1) = 3.0), 3.48 (ddd, H-6; J (H-5, 6) = 7.5, J (H-4, 6) = 1.0), 4.68 (dddd, H-4; J (H-4, 5) = 7.5), 5.10 (ddddd, H-5; J (H-5, F-2′) = 3.5, J (H-5, F-1) = 3.5)	[5, 20, 24]

Table 3 [continued]

No.	compound	method of preparation (yield in %), properties and remarks	Ref.
		^{19}F NMR ($CDCl_3$): 76.8 (F-1′; J(F-1, 1′) = 236, J(F-1′, 2) = 14.0, J(F-1′, 2′) = 7.0, J(F-1′, H-3) = 7.0), 87.3 (F-1; J(F-1, 2) = 3.0, J(F-1, 2′) = 3.0, J(F-1, H-5) = 3.5, J(F-1, H-6′) = 3.0), 104.5 (F-2′; J(F-2, 2′) = 222, J(F-1, 2′) = 3.0, J(F-2′, H-5) = 3.5), 108.5 (F-2; J(F-2, H-3) = 14.0, J(F-2, H-3′) = 36.0) IR (n-hexane): 2019, 2035, 2085 (CO) mass spectrum: $[M-nCO]^+$ (n = 0 to 3), $[M-nCO-F]^+$ (n = 1 to 3)	
108	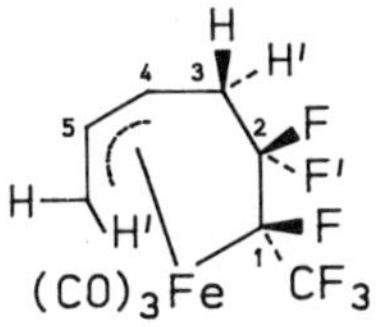	IVa (5) m.p. 75°, yellow crystals (from hexane at −78°) ^{1}H NMR ($CDCl_3$): 1.90 (dm, H-3′; J(H-3, 3′) = 15.0, J(H-3′, 4) = 7.5, J(H-3′, F-2) = 37.0), 2.59 (ddd, H-3; J(H-3, 4) = 7.5, J(H-3, F-2) = 15.0), 3.18 (ddd, H-6′; J(H-6, 6′) = 2.5, J(H-5, 6′) = 13.0, J(H-6′, F-1) = 6.5), 3.62 (dd, H-6; J(H-5, 6) = 8.0), 4.66 (ddd, H-4; J(H-4, 5) = 7.5), 5.25 (m, H-5; J(H-5, F-1) = 3.5, J(H-5, F-2′ = 3.5) ^{19}F NMR ($CDCl_3$): 66.1 (ddd, CF_3; J(CF_3, F-1) = 11.0, J(CF_3, F-2) = 11.0, J(CF_3, F-2′) = 11.0), 98.0 (dm, F-2′; J(F-2, 2′) = 229, J(F-2′, H-5) = 3.5), 116.8 (ddm, F-2; J(CF_3, H-3′) = 37.0, J(CF_3, H-3) = 15.0), 177.5 (m, F-1; J(F-1, H-5) = 3.5, J(F-1, H-6′) = 6.5) IR (hexane): 2021, 2038, 2091 (CO) mass spectrum: $[M-nCO]^+$ (n = 0 to 3), $[M-nCO-F]^+$ (n = 1, 2, 4)	[19]
109	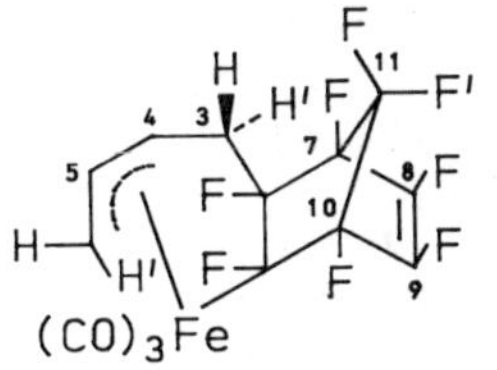	IVa (20 to 36) m.p. 140°, yellow crystals ^{1}H NMR ($CDCl_3$): 2.33 (m, H-3′), 2.47 (m, H-3), 2.96 (m, H-6′), 3.43 (m, H-6), 4.44 (m, H-4), 5.20 (m, H-5) ^{19}F NMR ($CDCl_3$, upfield from CF_3CO_2H): 56 (ABq, F-11, 11′; J(F-11, 11′) = 160), 85 (F-9), 99 (F-8), 105 (F-1), 109 (F-2), 134 (F-10), 149 (F-7) IR ($CDCl_3$): 1758 (CF=CF)	[75]
110	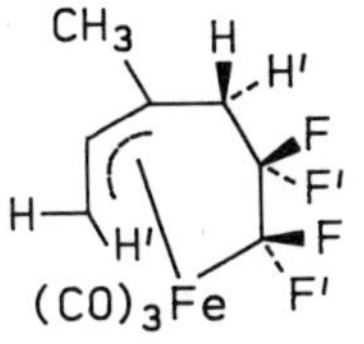	IVa	[24]

References on pp. 106/8

Table 3 [continued]

No.	compound	method of preparation (yield in %), properties and remarks	Ref.
*111	CH_3, H, H′ (C-6), H, H′ (C-3), F, F′ (C-2), F, F′ (C-1), $(CO)_3Fe$	IVa (38) m.p. 83 to 84°, light yellow crystals (from CH_2Cl_2/n-hexane) ^{1}H NMR ($CDCl_3$): 1.05 (m, H-3′; J(H-3,3′) = 14.0, J(H-3′,4) = 8.0, J(H-3′, F-2) = 35.0, J(H-3′, F-2′) = 3.0), 2.05 (s, CH_3), 2.42 (dddd, H-3; J(H-3, F-2) = 14.0, J(H-3,4) = 7.5, J(H-3, F-1′) = 7.0), 2.69 (dd, H-6′; J(H-6′, F-1) = 4.0), 3.34 (dd, H-6; J(H-6,6′) = 2.5), 4.46 (ddd, H-4; J(H-4,6) = 2.5) ^{19}F NMR ($CHCl_3$): 76.6 (F-1′; J(F-1,1′) = 236, J(F-1′,2) = 15.0, J(F-1′, H-3) = 7.0), 88.7 (F-1; J(F-1,2′) = 4.0, J(F-1, H-6′) = 4.0, J(F-1,2) = 3.0), 113.6 (F-2′; J(F-2,2′) = 222, J(F-1′,2′) = 7.0, J(F-2′, H-3′) = 3.0), 116.9 (F-2; J(F-2, H-3′) = 35.0, J(F-2, H-3) = 14.0) IR (n-hexane): 2014, 2032, 2085 (CO) mass spectrum: $[M-nCO]^+$ (n = 0 to 3), $[M-nCO-F]^+$ (n = 1 to 3)	[24]
*112	CH_3, H, H′ (C-6), H, H′ (C-3), F, F′ (C-2), F, CF_3 (C-1), $(CO)_3Fe$	IVa (16) m.p. 97 to 98°, yellow crystals (from CH_2Cl_2/hexane) ^{1}H NMR ($CDCl_3$): 1.80 (m, H-3′; J(H-3,3′) = 14.5, J(H-3′, F-2) = 39.0, J(H-3′, F-2′) = 10.5), 2.09 (s, CH_3), 2.55 (ddd, H-3; J(H-3, 4) = 7.5, J(H-3, F-2) = 14.5), 3.14 (dd, H-6′; J(H-6,6′) = 3.0, J(H-6′, F-1) = 8.0), 3.48 (dd, H-6; J(H-4,6) = 4.0), 3.49 (ddd, H-4; J(H-3′,4) = 7.5) ^{19}F NMR ($CHCl_3$): 65.9 (ddd, CF_3; J(CF_3, F-1) = J(CF_3, F-2) = J(CF_3, F-2′) = 10.5), 101.2 (ddq, F-2′; J(F-2,2′) = 232, J(F-2′, H-3′) = 10.5), 117.0 (dm, F-2; J(F-1,2) = 10.5, J(F-2, H-3′) = 39.0, J(F-2, H-3) = 14.5), 178.3 (m, F-1; J(F-1,2) = 10.5, J(F-1, H-6′) = 8.0) IR (hexane): 2014, 2035, 2088 (CO) mass spectrum: $[M-nCO]^+$ (n = 0 to 3), $[M-nCO-F]^+$ (n = 1 to 3)	[5, 19]
113	CH_3 (C-6), H, H′ (C-3), F, F′ (C-2), F, F′ (C-1), $(CO)_3Fe$	IVa (27) m.p. 55 to 57°, pale yellow crystals (from CH_2Cl_2/n-hexane) ^{1}H NMR ($CDCl_3$): 1.41 (m, H-3′; J(H-3,3′) = 14.0, J(H-3′,4) = 7.0, J(H-3′, F-2) = 36.0), 1.82 (d, CH_3; J(CH_3, H-6) = 6.0), 2.39 (dddd,	[24]

References on pp. 106/8

Table 3 [continued]

No.	compound	method of preparation (yield in %), properties and remarks	Ref.
		H-3; J(H-3,4)=7.0, J(H-3, F-1')=7.0, J(H-3, F-2)=14.0), 3.63 (ddq, H-6; J(H-5, 6)=12.0, J(H-6, F-1)=6.0), 4.41 (ddd, H-4; J(H-4,5)=7.5), 4.94 (dddd, H-5; J(H-5, F-2')=3.5, J(H-5, F-1)=3.5) ^{19}F NMR ($CHCl_3$): 76.3 (F-1'; J(F-1,1')=238, J(F-1',2)=14.0, J(F-1',2')=7.0, J(F-1', H-3)=7.0), 87.1 (F-1; J(F-1,2)=3.0, J(F-1, H-6)=6.0, J(F-1, H-5)=3.5), 110.6 (F-2'; J(F-2,2')=222, J(F-2', H-5)=3.5), 116.9 (F-2; J(F-2, H-3')=36.0, J(F-2, H-3)=14.0) IR (n-hexane): 2013, 2031, 2086 (CO) mass spectrum: $[M-nCO]^+$ (n=0 to 3), $[M-nCO-F]^+$ (n=1 to 3)	
114	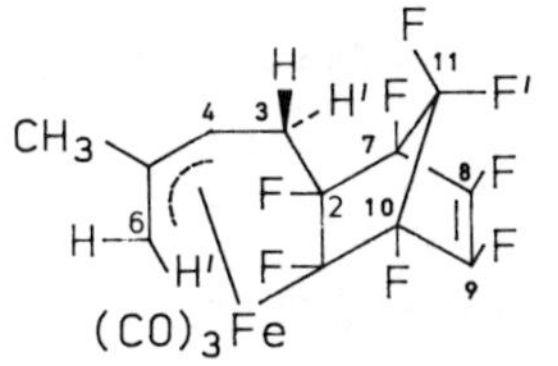	IVa m.p. 166 to 167°, yellow crystals ^{1}H NMR ($CDCl_3$): 2.08 (br,s, CH_3, H-3'), 2.50 (m, H-3), 3.00 (m, H-6'), 4.03 (m, H-6), 4.43 (m, H-4) ^{19}F NMR ($CDCl_3$, upfield from CF_3CO_2H): 55 (F-11, 11'; J(F-11, 11')=160), 84 (F-9), 97 (F-8), 104 (F-1), 109 (F-2), 134 (F-10), 149 (F-7) IR ($CDCl_3$): 1750 (CF=CF)	[75]
115	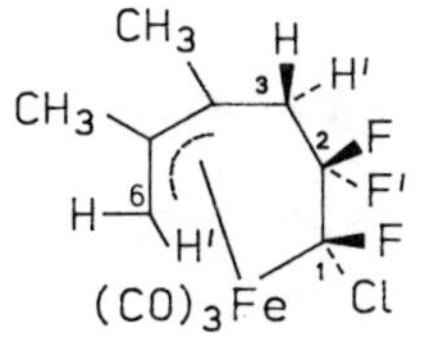	IVa (17) m.p. 98 to 100°, pale yellow crystals (from CH_2Cl_2/hexane at 0°) ^{1}H NMR ($CDCl_3$): 1.75 (ddd, H-3'; J(H-3,3')=14.0, J(H-3', F-2')=J(H-3', F-1)=4.0, J(H-3', F-2)=35.0), 2.02 (s, CH_3), 2.07 (s, CH_3), 2.36 (ddd, H-3; J(H-3, F-2)=15.0, J(H-3, F-2')=1.0), 2.51 (dd, H-6'; J(H-6', F-1)=5.0), 3.22 (d, H-6; J(H-6,6')=3.0) ^{19}F NMR ($CHCl_3$): 104.2 (m, F-1; J(F-1, H-3) =11.0, J(F-1, H-3')=20.0, J(F-1, H-6') =5.0, J(F-1, H-3')=4.0), 104.4 (dm, F-2'; J(F-2,2')=224, J(F-1,2')=10.0, J(F-2', H-3')=4.0, J(F-2', H-3)=1.0), 108.8 (dddd, F-2; J(F-1,2)=11.0, J(F-2, H-3')=35.0, J(F-2, H-3)=15.0) IR (hexane): 2008, 2031, 2083 (CO) mass spectrum: $[M-nCO]^+$ (n=0 to 3), $[M-nCO-HF]^+$ (n=0 to 3), $[M-3CO-HF-HCl]^+$	[19]

References on pp. 106/8

Table 3 [continued]

No.	compound	method of preparation (yield in %), properties and remarks	Ref.
116	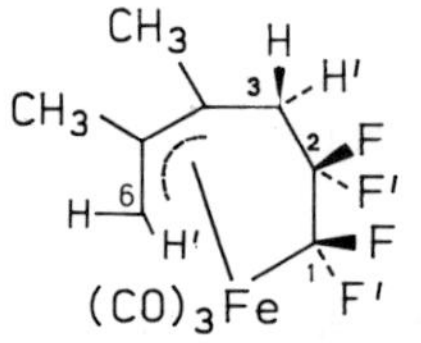	IVa (38) m.p. 116 to 118°, yellow needles (from CH_2Cl_2/n-hexane) ^1H NMR ($CDCl_3$): 1.57 (m, H-3′; J(H-3,3′)=13.0, J(H-3′, F-2)=36.0), 2.01 (s, CH_3-4), 2.09 (s, CH_3-5), 2.23 (ddd, H-3; J(H-3, F-2)=14.0, J(H-3, F-1′)=7.0), 2.39 (dd, H-6′; J(H-6,6′)=3.0, J(H-6′, F-1)=4.0), 3.17 (d, H-6) ^{19}F NMR ($CHCl_3$): 77.6 (F-1′; J(F-1,1′)=236, J(F-1′,2)=14.5, J(F-1′, H-3)=7.0), 89.7 (F-1; J(F-1,2)=3.0, J(F-1, H-6′)=4.0), 110.5 (F-2′; J(F-2,2′)=222, J(F-1′,2′)=7.0), 117.9 (F-2; J(F-2, H-3′)=36.0, J(F-2, H-3)=14.0) IR (n-hexane): 2009, 2028, 2038 (CO) mass spectrum: $[M-nCO]^+$ (n=0 to 3), $[M-nCO-F]^+$ (n=1 to 3)	[24]
117	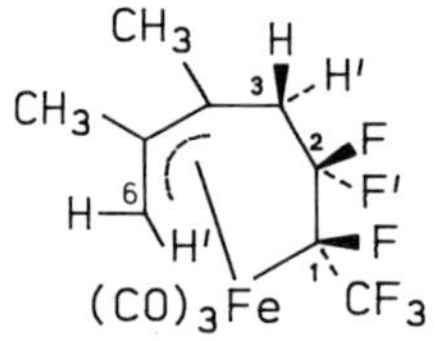	IVa (16) m.p. 85 to 86°, yellow needles (from CH_2Cl_2/hexane) ^1H NMR ($CDCl_3$): 1.91 (dm, H-3′; J(H-3,3′)=15.0, J(H-3′, F-2)=37.0), 1.98 (s, CH_3), 2.13 (s, CH_3), 2.28 (dd, H-3; J(H-3, F-2)=15.0), 2.87 (dd, H-6′; J(H-6′, F-1)=7.0), 3.32 (d, H-6; J(H-6,6′)=3.0) ^{19}F NMR ($CHCl_3$): 66.2 (ddd, CF_3; J(CF_3, F-1)=J(CF_3, F-2)=J(CF_3, F-2′)=10.5), 98.8 (dm, F-2′; J(F-2,2′)=233), 118.6 (dm, F-2; J(F-2, H-3′)=37.0, J(F-2, H-3)=15.0), 179.0 (m, F-1; J(F-1, H-6′)=7.0) IR (hexane): 2009, 2031, 2084 (CO) mass spectrum: $[M-nCO]^+$ (n=0 to 3), $[M-nCO-HF]^+$ (n=1 to 3)	[19]
118	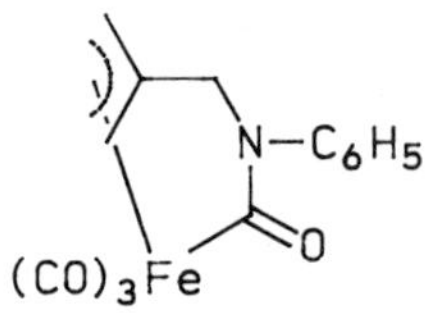	see No. 119, p. 104 m.p. 111 to 112° (dec.), yellow needles (from hexane) ^1H NMR ($CDCl_3$): 2.48 (s, 2H), 3.75 (s, 2H), 4.00 (s, 2H), 7.25 (m, 5H) IR ($CHCl_3$): 1500 (C_6H_5), 1580 (C=O), 1600 (C_6H_5) IR (hexane): 2010, 2075 (CO) mass spectrum: $[M-nCO]^+$ (n=0 to 3)	[31]

Table 3 [continued]

No.	compound	method of preparation (yield in %), properties and remarks	Ref.
*119	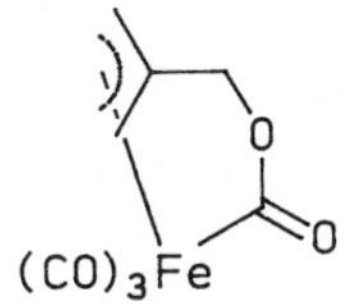	IIb (78) m.p. 143 to 144°, pale yellow needles (from CH_2Cl_2/petroleum ether) ^1H NMR ($CDCl_3$): 2.57 (CH_2), 3.84 (CH_2), 4.10 (OCH_2) IR (KBr): 725, 820, 885, 975, 1000, 1050, 1074, 1240, 1345, 1395, 1400, 1485, 1505, 1652, 2024, 2030, 2084, 2920, 2990, 3021, 3100, 3282	[1]
120	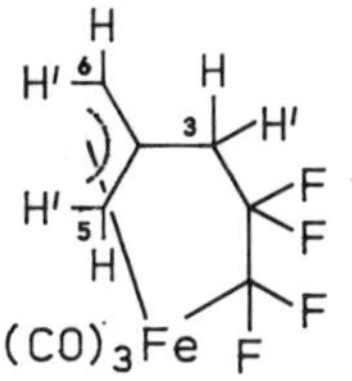	IVa m.p. 73° ^1H NMR: 2.42 (H-5′,6′; J(H-5,5′) = J(H-6,6′) = 2.0), 2.56 (H-3′; J(H′,F) = 3.0), 2.70 (H-3; J(H,F) = 3.0), 3.38 (H-5,6) ^{19}F NMR: 75.8 (m, 2F), 106.2 (m, 2F) IR (hexane): 2026, 2033, 2090 (CO)	[5]
121	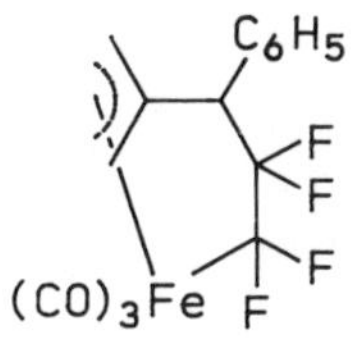	IVa	[5]
^{4}L is a 1σ, 3-5π allyl ligand:			
122	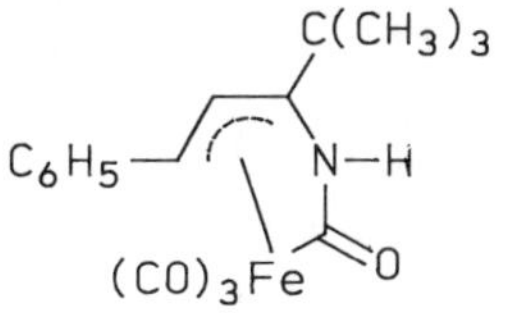	IIIa	[90]
*123	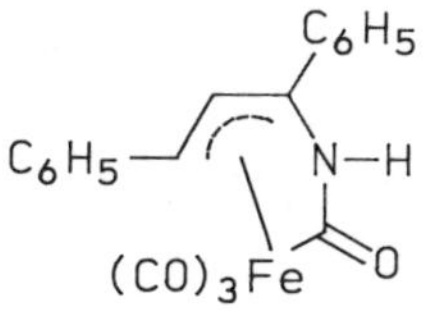	IIIa (47) dec. p. 93 to 94° IR (hexane): 1994, 2009, 2065 (CO)	[90]
*124	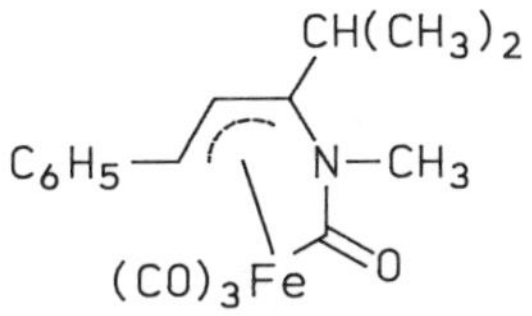	IIIb (53) dec. p. 103 to 104°	[90]

References on pp. 106/8

Table 3 [continued]

No.	compound	method of preparation (yield in %), properties and remarks	Ref.
*125	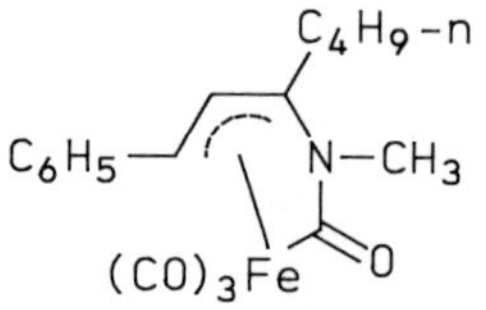 C_4H_9-n, C_6H_5, N–CH_3, $(CO)_3Fe$, O	III b (49) dec. p. 79 to 80° IR (hexane): 1993, 2006, 2065 (CO)	[90]
*126	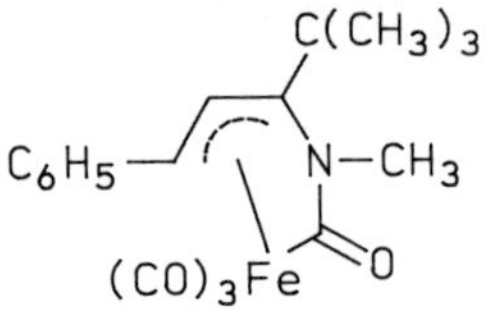 $C(CH_3)_3$, C_6H_5, N–CH_3, $(CO)_3Fe$, O	III b (50) dec. p. 127 to 129°	[90]
127	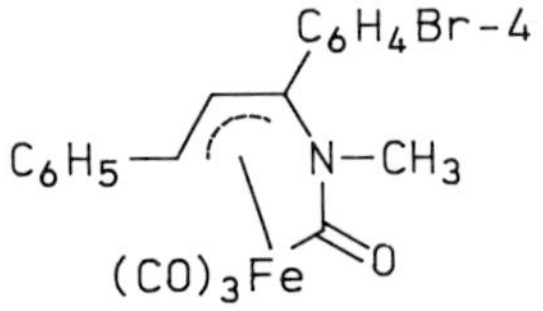 C_6H_4Br-4, C_6H_5, N–CH_3, $(CO)_3Fe$, O	III b (61) dec. p. 119 to 120°	[90]
128	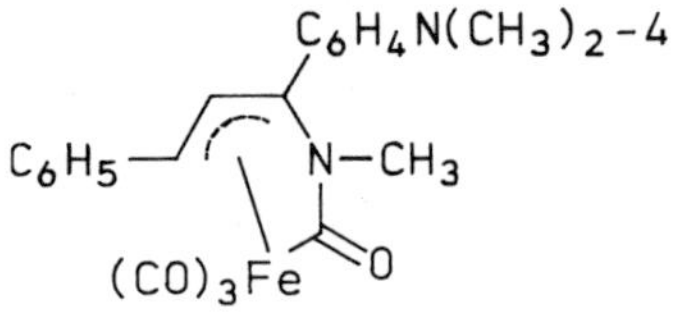 $C_6H_4N(CH_3)_2$-4, C_6H_5, N–CH_3, $(CO)_3Fe$, O	III b (52) dec. p. 100°	[90]
129	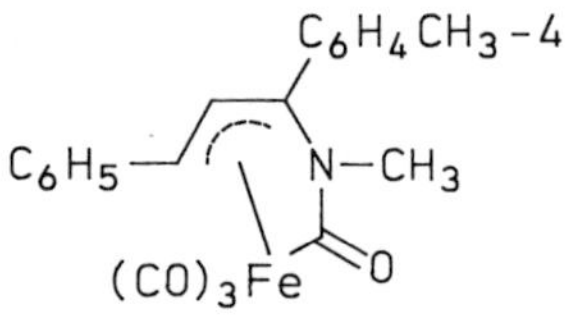 $C_6H_4CH_3$-4, C_6H_5, N–CH_3, $(CO)_3Fe$, O	III b (85) dec. p. 121 to 122°	[90]
*130	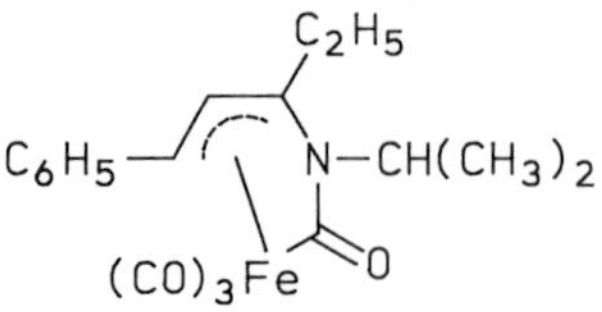 C_2H_5, C_6H_5, N–$CH(CH_3)_2$, $(CO)_3Fe$, O	III b (38) dec. p. 81 to 82°	[90]
*131	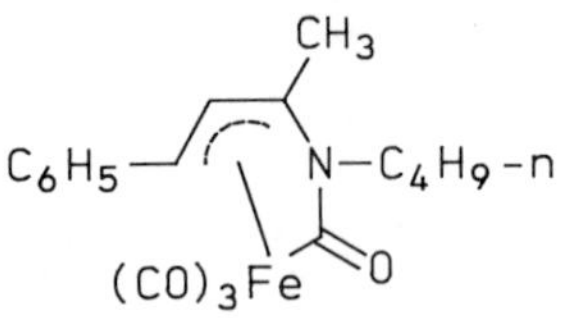 CH_3, C_6H_5, N–C_4H_9-n, $(CO)_3Fe$, O	III b (43) dec. p. 74 to 75°	[90]

References on pp. 106/8

Table 3 [continued]

No.	compound	method of preparation (yield in %), properties and remarks	Ref.
*132	C_6H_5–allyl(CH_3)–N–$CH_2CH(CH_3)_2$, C=O, $(CO)_3Fe$	IIIb	[90]
*133	C_6H_5–allyl(CH_3)–N–$C(CH_3)_3$, C=O, $(CO)_3Fe$	IIIb IR (hexane): 1995, 2011, 2065 (CO)	[90]
*134	C_6H_5–allyl(C_2H_5)–N–C_6H_{11}-cyclo, C=O, $(CO)_3Fe$	IIIb (44) dec. p. 92 to 93° IR (hexane): 1994, 2011, 2068 (CO)	[90]
*135	C_6H_5–allyl(C_4H_9-n)–N–C_6H_{11}-cyclo, C=O, $(CO)_3Fe$	IIIb (45) dec. p. 82 to 83° IR (hexane): 1992, 2006, 2065	[90]
136	C_6H_5–allyl($C_6H_4OCH_3$-4)–N–C_6H_{11}-cyclo, C=O, $(CO)_3Fe$	IIIb (50) dec. p. 106 to 107°	[90]

^{4}L is a 1σ, 4–6π allyl ligand:

No.	compound	method of preparation (yield in %), properties and remarks	Ref.
*137	CH_3, allyl–$CH(CH_2C(OCH_3)_2H)$–N–$CH(CH_3)C_6H_5$, C=O, $(CO)_3Fe$	see No. 140, p. 106	[89]
*138	H, H′, allyl (5, 4)–$CH(C_6H_5)$–N–$CH_2C_6H_5$, C=O, $(CO)_3Fe$	see No. 139, p. 106 m.p. 138 to 140° (dec.), white crystals ^{1}H NMR ($CDCl_3$): 2.97 (d, 1H; J=14), 3.41 (dd, H-6′; J=12.5, J=1), 3.82 (dd, H-6; J=8, J=1), 4.38 (d, H-3; J=6), 4.54 (ddd, H-5; J=12.5, J=8, J=8), 4.68 (dd, H-4; J=8, J=6), 5.12 (d, 1H; J=14), 6.86 to 7.08 (m, 4H), 7.17 to 7.39 (m, 6H) IR ($CHCl_3$): 1582, 2010, 2020, 2078, 2970	[88]

Table 3 [continued]

No.	compound	method of preparation (yield in %), properties and remarks	Ref.
*139	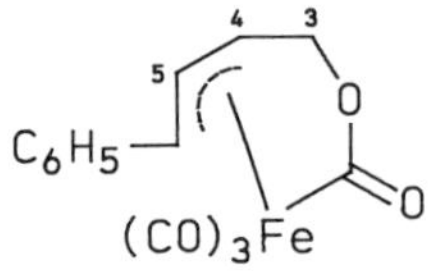	m.p. >120° (dec.), yellow crystals ^{1}H NMR ($CDCl_3$): 4.05 (dd, H-3′; J = 12.5, J = 5.5), 4.17 (dd, H-3; J = 12.5, J = 1.5), 4.78 (ddd, H-4; J = 7.5, J = 5.5, J = 1.5), 4.82 (d, H-6; J = 12.5), 5.62 (dd, H-5; J = 12.5, J = 7.5), 7.28 to 7.42 (m, 5H) IR ($CHCl_3$): 984, 1665, 2030, 2080	[88]
*140	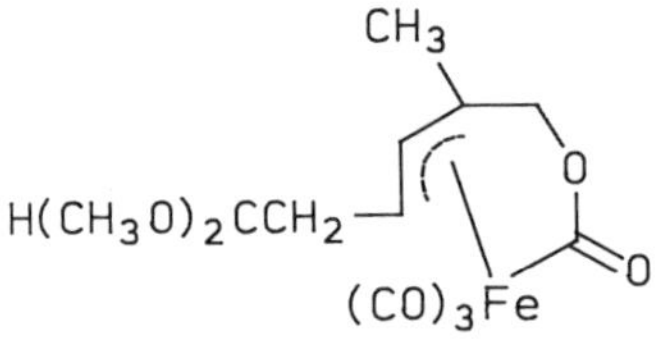	crystalline	[89]

*Further information:

[$C_4H_6Fe(CO)_3H$]$^+$ (Table **3**, No. **1**) is prepared from the reaction of $C_4H_6Fe(CO)_3$ with strong acids (see 1.4.1.4.1.1.1.1 in "Organoiron Compounds" B6, 1981, pp. 150/1) [14, 21, 26, 28, 30, 35, 36, 77]. Thus, reaction of $C_4H_6Fe(CO)_3$ with 1 equivalent HBF_4 [21] or with HBF_4/CF_3CO_2H (ratio $C_4H_6Fe(CO)_3$: acid = ca. 1:15 at 0 °C [30]) [14, 30], cf. [7] or with excess $HOSO_2F$ [14, 30, 36, 77] at −60 °C in liquid SO_2 [14, 19, 77] or with excess $HOSO_2F/SbF_5$ [77] gives No. 1. $C_4H_6Fe(CO)_3$ is monoprotonated in this reaction, and originally other structures of this cation such as XXI [28] or XXII [44] were discussed. But ^{13}C NMR studies show that this cation has the structure shown in Table 3, p. 48 [30, 35, 36]. Reaction of $C_4H_6Fe(CO)_3$ with $DOSO_2F$ in liquid SO_2 at −60 °C gives an ion bearing deuterium on both C-1 and C-4 but not on C-2 and C-3. This observation is in good agreement with a slow intramolecular exchange process as shown in XXIII [36]. The ^{13}C NMR spectrum ($DOSO_2F$/liquid SO_2) of the deuterated No. 1 shows chemical shifts at $\delta = -2.99$ (m, C-1), 52.78 (m, C-4), 80.88 (d, C-2; J(C, H) = 182.9), 100.46 (d, C-3; J(C, H) = 170.3), and 196.16, 199.35, 203.22 (CO) ppm at −60 °C [36].

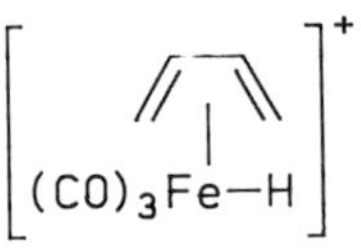

XXI 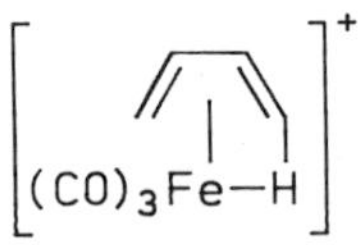 XXII 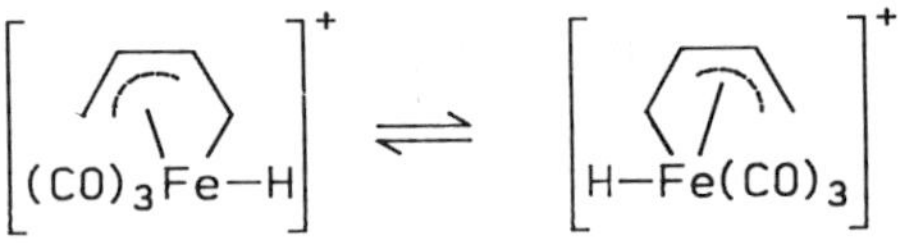 XXIII

$CH_2C(CO_2CH_3)C(OCH_3)C{=}OFe(CO)_3$ (Table **3**, No. **2**) is prepared by treating XXIV (R = H) with CO (1 atm) in CH_2Cl_2 at room temperature for several hours (38%) [46, 50, 51]. Further reaction with CO or ^{2}D = $P(C_6H_5)_3$, $P(C_6H_5)_2CH_3$, or $P(CH_3)_2C_6H_5$ is very slow and no product was isolated [50], but possibly is XXVa formed [51].

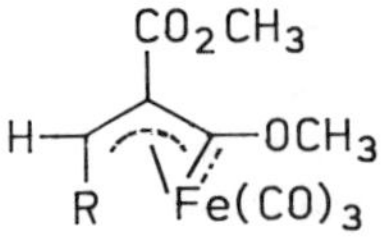

XXIV 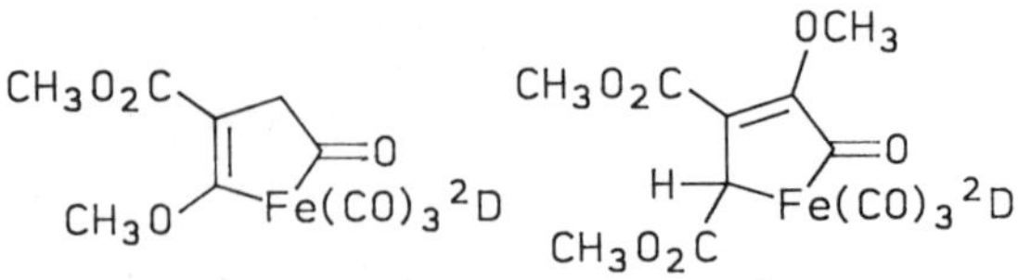 a XXV b

$CH_2C(C_6H_5)C(C_6H_5)C{=}OFe(CO)_3$ (Table **3**, No. **3**) is air-stable in the crystalline state. The solubility in polar organic solvents is good, but the compound is less soluble in nonpolar solvents [47].

XXVI XXVII

$CH_3O_2CHC(CO_2CH_3)C(OCH_3)C{=}OFe(CO)_3$ (Table **3**, No. **5**) is prepared from XXIV ($R=CO_2CH_3$) and CO. Another product of this reaction is an isomer, whose structure (possibly No. 6) is ambiguous. Reaction of this mixture with $^2D{=}CO$, $P(C_6H_5)_2CH_3$, or $P(CH_3)_2C_6H_5$ gives XXVb [51].

$C_6H_5CHC(C_6H_5)C(C_6H_5)C{=}OFe(CO)_3$ (Table **3**, No. **9**). Recrystallization from hexane gives small prisms. The compound crystallizes in the triclinic space group $P\bar{1}-C_i^1, S_2^1$ with the unit cell parameters a = 1241.8(5), b = 989.1(4), c = 1050.8(6) pm; α = 107.99(2)°, β = 104.35(2)°, γ = 112.74(2)°; Z = 2 molecules per unit cell. The calculated density is 1.41 g/cm^3. The main bond distances and angles are shown in **Fig. 9**. The Fe–C(4) bond is relatively short with 191.7(3) pm, thus indicating multiple bond character. The angle C(3)–Fe–C(4) is 76.9(3)°.

The compound is air-stable in the crystalline state, easily soluble in polar organic solvents, but less soluble in nonpolar organic solvents. Reaction with $Fe_2(CO)_9$ in C_6H_6 at 60 °C gives XXVI (R = H) [47].

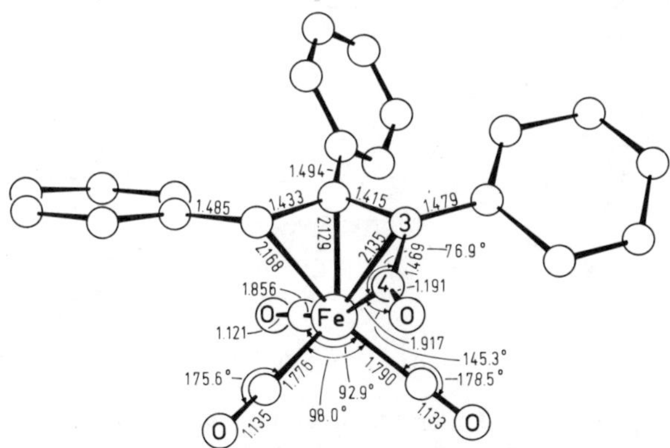

Fig. 9. Molecular structure of $C_6H_5CHC(C_6H_5)C(C_6H_5)C{=}OFe(CO)_3$ (No. 9) [47].

$C_6H_5C(C_6H_5)C{=}OFe(CO)_3$ (Table **3**, No. **10**) is also obtained by irradiation of $Fe(CO)_5$ and $(C_6H_5)_2CN_2$ in ether. After 4 h the reaction mixture is filtered to separate $Fe_2(CO)_9$. The filtrate is evaporated, and the residue is chromatographed. Elution with petroleum ether gives $Fe_3(CO)_{12}$, and with petroleum ether/C_6H_6 (10:1 to 1:1), gives mainly $Fe_3(CO)_9[(C_6H_5)_2CN_2]_2$ (7.5% yield) and $(C_6H_5)_2C{=}C(C_6H_5)_2$. Elution with C_6H_6 finally gives No. 10 in 5% yield [71]. Stirring a benzene solution of No. 10 in a ^{13}CO atmosphere for

References on pp. 106/8

3 d at room temperature gives XXVII (total substitution 17%) which shows additional IR bands in KBr at 1683, 1695, and 1943 cm^{-1} and in C_6H_{14} at 1747 and 1960 cm^{-1} [63, 71, 84].

The compound crystallizes in the triclinic space group $P\bar{1}-C_i^1$ with the unit cell parameters a = 899.6(3), b = 1161.8(5), c = 1485.5(7) pm; α = 106.89(5)°, β = 91.62(5)°, γ = 90.86(8)°; Z = 4 molecules per unit cell. D_{calc} = 1.4940 g/cm^3. The main bond distances and angles are shown in **Fig. 10**.

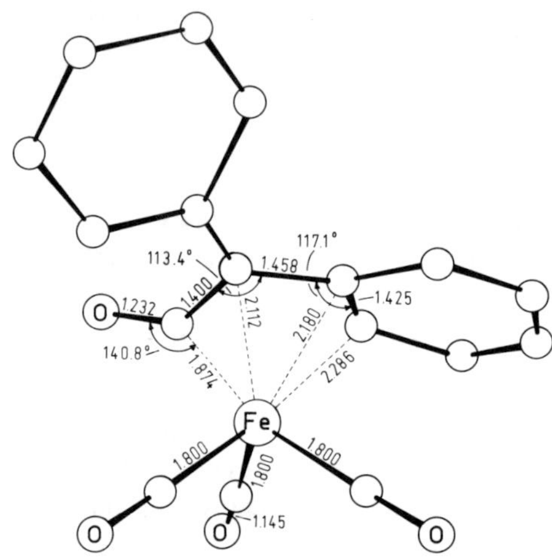

Fig. 10. Molecular structure of $C_6H_5C(C_6H_5)C{=}OFe(CO)_3$ (No. 10) [63].

The compound is very soluble in the usual organic solvents. The solutions decompose rapidly [63, 71]. Heating in C_6H_6 or n-hexane at 75 °C for 3.5 h gives XXVIII (42% yield) [71].

XXVIII XXIX XXX

Oxidation with $ON(CH_3)_3 \cdot 2H_2O$ in boiling acetone gives only organic decomposition products. Reaction with H_2 in boiling n-hexane gives XXVIII and organic products. Reduction with $LiAlH_4$ in tetrahydrofuran at 0 °C gives a pyrophoric product and, with $Li[B(C_2H_5)_3H]$ under analogous conditions, very small amount of oily substances [71]. No apparent reaction is observed with CO; but with ^{13}CO, compound XXVII is obtained (see p. 77) [63, 71]. Reaction with excess $^2D = P(OC_2H_5)_3$ in boiling n-hexane results in the formation of XXIX (90% yield). Similar reaction of XXVII gives the appropriate product XXIX with ^{13}CO groups. Refluxing with $^2D = P(C_6H_5)_3$ in n-hexane for maximal 15 min gives XXIX (84% yield) and $(CO)_3Fe(P(C_6H_5)_3)_2$ [71]; refluxing for 85 min gives only traces of XXIX ($^2D = P(C_6H_5)_3$), but $(CO)_3Fe(P(C_6H_5)_3)_2$ and XXX (90%) [71, 84]. Labelled analogous products are obtained from XXVII [71]. Reaction with SO_2 in boiling n-hexane for ca. 40 min gives S, $Fe_3(CO)_9S_2$, and yellow-green crystals of an unknown main product (m.p. 105 °C, IR (KBr): 1985, 1991, 2030 (νCO) cm^{-1}, $[M]^{+}=504$, air-sensitive crystals, easily soluble in organic solvents, solutions are air-sensitive). Refluxing with CS_2 in petroleum ether for 90 min gives a red, highly fluid, relatively air-stable, unidentified oil (IR (KBr): 1995, 2020, 2040, 2060 cm^{-1}, stable towards humid solvents). Reaction with $C_5H_5V(CO)_4$, thermally or photochemically, in n-hexane gives only XXVIII, and refluxing with $C_5H_5Mn(CO)_2{}^2D$ (^2D = tetrahydrofuran) in n-hexane gives $(C_5H_5Fe(CO)_2)_2$. Heating with $Fe_2(CO)_9$ in C_6H_6 at 75 °C for 4 h gives XXVIII in 77% yield [63, 71, 84]. An attempted exchange reaction of XXVII with No. 11 in CCl_4 at room temperature failed. Refluxing of No. 10 with $(C_5H_5FeNO)_2$ in tetrahydrofuran for 2 h gives $(C_5H_5Fe(CO)_2)_2$ in 18% yield. Refluxing with $Os_3(CO)_{12}$ in C_6H_6 gives only XXVIII [71].

Reaction of No. 10 with CH_3SH in n-hexane at 50 °C for ca. 30 min gives $Fe_2(CO)_6(SCH_3)_2$ (75% yield). Reaction with $LiCH_3$ in ether at room temperature gives a pyrophoric product. No reaction is observed with H_2CN_2 in ether at −180 to +25 °C and refluxing with diazocyclopentadiene in C_6H_6 gives only organic decomposition products. Refluxing with $(4\text{-}RC_6H_4)_2CN_2$ (R = H or CH_3O) in n-hexane for ca. 70 min gives XXVIII and $Fe_3(CO)_9(N_2CC_6H_4R\text{-}4)_2$ (yield for R = H is 6%) [71]. Reaction with $CH_2{=}CH_2$ in boiling n-hexane gives XXXI (24%) and XXVIII (12%) [71, 79, 84]. Refluxing with cyclohexa-1,3-diene in n-hexane gives a yellow, air-sensitive oil (29%) of unknown constitution (IR (CH_2Cl_2): 1980, 2050 (νCO) cm^{-1}, $[M]^{+}$ = 414, soluble in all organic solvents). Refluxing with cyclooctatetraene in petroleum ether for 30 min gives $C_8H_8Fe(CO)_3$. No reaction is observed with $(4\text{-}CH_3OC_6H_4)_2C{=}C{=}O$ in C_6H_6 at room temperature for 4 d. Refluxing with $(CH_3)_2C{=}C{=}CH_2$ in petroleum ether for ca. 2 h gives a yellow, viscous, relatively air-stable oil of unknown constitution (IR (KBr): 1940, 1985, 1990, 2035 (νCO) cm^{-1}) [71]. Reaction with acetylene [84] or with $C_6H_5C{\equiv}CC_6H_5$ [71] in boiling n-hexane for 90 min gives Nos. 33 and 34, respectively. Compound No. 34 is isolated by chromatography on silica gel with petroleum ether/C_6H_6 (10:1 to 10:5) as eluent [71].

XXXI XXXII XXXIII XXXIV

$CH_3OC_6H_4C(C_6H_4OCH_3\text{-}4)C{=}OFe(CO)_3$ (Table **3**, No. **11**) is highly soluble in organic solvents. The solutions are air-sensitive. Attempted exchange reaction with XXVII failed [71].

$(CH_3)_2CCHC(CO_2CH_3)C{=}OFe(CO)_3$ (Table **3**, No. **12**) gives a ca. 1:1 mixture of (Z)- and (E)-$CH_2{=}C(CH_3)CH{=}C(CO_2CH_3)CHOFe(CO)_3$ (XXXII) on irradiation in C_6H_6 (overall yield 65%) [64]. The same photolysis in CH_3OH gives $(CH_3)_2C{=}CHC(CO_2CH_3)_2H$ in quantitative yield after work-up in presence of O_2 [61, 64]. Similar results are obtained with the monodeuterated compound No. 12 (see Formula XXXIII) [64]. Oxidation of No. 12 with $ON(CH_3)_3$ in boiling CH_3OH after ca. 8 h gives $(CH_3)_2C{=}CHC(CO_2CH_3)_2H$ in 55% yield [83].

$(CH_3)_2CC(CO_2CH_3)C(CH_3)C{=}OFe(CO)_3$ (Table **3**, No. **14**) is slightly less stable than No. 23. Refluxing in C_6H_6 for 1 h gives quantitatively XXXIV (R = CH_3) [64]. Decomplexation with $ON(CH_3)_3$ in boiling CH_3OH gives $(CH_3)_2C{=}C(CO_2CH_3)CH(CH_3)CO_2CH_3$ (16%) and XXXIV with R = CH_3 (33%) within 8 h [61, 83].

$(CH_3)_2CC(CO_2CH_3)C(C_3H_7\text{-}n)C{=}OFe(CO)_3$ (Table **3**, No. **15**) is refluxed in C_6H_6 or CH_3OH to give up to 90% yield of XXXIV (R = C_3H_7-n) [64]. Oxidation with $ON(CH_3)_3$ in boiling CH_3OH

References on pp. 106/8

gives a 3:2 mixture of $(CH_3)_2C{=}C(CO_2CH_3)CH(C_3H_7\text{-n})CO_2CH_3$ and XXXIV ($R = C_3H_7$-n) in a total yield of 55% [83].

$(CH_3)_2CC(COCH_3)C(C_3H_7\text{-i})C{=}OFe(CO)_3$ (Table **3**, No. **17**) is refluxed in C_6H_6 or CH_3OH to give XXXV ($R = C_3H_7$-i). At 80 °C up to 90% is achieved within 3 h [64]. Oxidation in boiling CH_3OH with $ON(CH_3)_3$ for 8 h gives $(CH_3)_2C{=}C(CO_2CH_3)C(COCH_3){=}C(CH_3)_2$ (63%) and XXXV ($R = C_3H_7$-i) in 20% yield [83]. Reduction with $NaBH_4$ in CH_3OH at 0 to 20 °C, followed by hydrolysis and drying with $MgSO_4$, gives No. 16 in 98% yield [61].

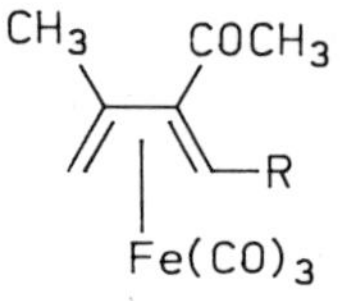

XXXV 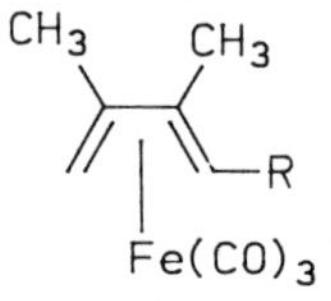 XXXVI 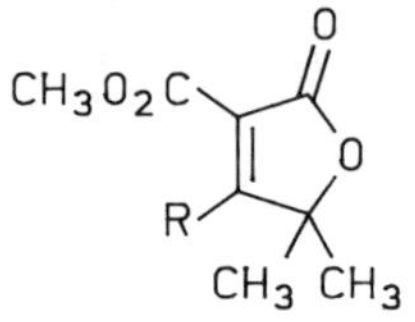 XXXVII 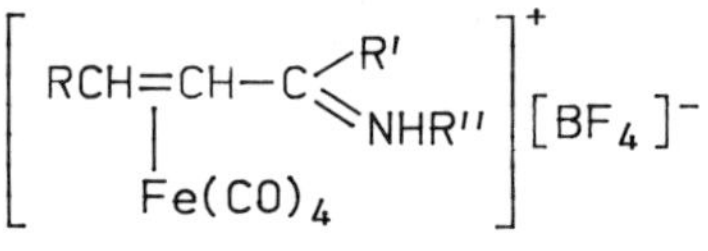 XXXVIII

$(CH_3)_2CC(CO_2CH_3)C(C_3H_7\text{-i})C{=}OFe(CO)_3$ (Table **3**, No. **18**) is refluxed in C_6H_6 or CH_3OH to give XXXIV ($R = C_3H_7$-i). A yield up to 90% is achieved within 3 h at 80 °C [64]. Oxidation in boiling CH_3OH with $ON(CH_3)_3$ for 8 h gives a ca. 1:1 mixture of $(CH_3)_2C{=}C(CO_2CH_3)$-$C(CO_2CH_3){=}C(CH_3)_2$ (29%) and $(CH_3)_2C{=}C(CO_2CH_3)CH(CO_2CH_3)C(CH_3)_2H$ (36%) [83].

$(CH_3)_2CC(COCH_3)C(C_4H_9\text{-t})C{=}OFe(CO)_3$ (Table **3**, No. **19**) is refluxed in C_6H_6 or CH_3OH to give XXXV ($R = C_4H_9$-t); a yield up to 90% is achieved within 3 h at 110 °C [64].

$(CH_3)_2CC(CO_2CH_3)C(CH{=}C(CH_3)_2)C{=}OFe(CO)_3$ (Table **3**, No. **21**) is oxidized in boiling CH_3OH with $ON(CH_3)_3$ for 8 h to give $(CH_3)_2C{=}C(CO_2CH_3)CH(CH{=}C(CH_3)_2)CO_2CH_3$ as the major product in 45% yield [83].

$(CH_3)_2CC(CH_3)C(CO_2CH_3)C{=}OFe(CO)_3$ (Table **3**, No. **23**) is slightly more stable than No. 14. Refluxing in C_6H_6 or CH_3OH gives XXXVI ($R = CO_2CH_3$); a yield up to 90% is achieved at 70 °C within 3 h [64]. Decomplexation with $ON(CH_3)_3$ in boiling CH_3OH gives $(CH_3)_2C{=}C$-$(CH_3)C(CO_2CH_3)_2H$ (56%) within 8 h. Decomplexation with $FeCl_3$ in CH_3OH for 6 h at 40 °C gives $CH_2{=}C(CH_3)C(CH_3){=}C(CO_2CH_3)_2$ (35%) and XXXVII (25%) [83].

$(CH_3)_2CC(C_3H_7\text{-n})C(CO_2CH_3)C{=}OFe(CO)_3$ (Table **3**, No. **24**) gives, on heating in a solvent, a mixture of diolefin complexes in which the syn isomer predominates [64]. Decomplexation with $ON(CH_3)_3$ in boiling CH_3OH gives $(CH_3)_2C{=}C(C_3H_7\text{-n})C(CO_2CH_3)_2H$ (87%) within 8 h [83].

$(CH_3)_2CC(CH{=}C(CH_3)_2)C(CO_2CH_3)C{=}OFe(CO)_3$ (Table **3**, No. **25**). Decomplexation with $ON(CH_3)_3$ in boiling CH_3OH gives $(CH_3)_2C{=}C(CH{=}C(CH_3)_2)C(CO_2CH_3)_2H$ (53%) within 8 h. Decomplexation with $FeCl_3$ in CH_3OH for 2 h at 20 °C gives predominantly $(CH_3)_2C{=}C(CH{=}C$-$(CH_3)_2)C(CO_2CH_3)_2H$ (40%) and, after 2 h at 55 °C, XXXVII ($R = CH{=}C(CH_3)_2$) in 90% yield [83].

$(CH_3)_2CC(CO_2CH_3)C(CO_2CH_3)C{=}OFe(CO)_3$ (Table **3**, No. **26**). Decomplexation with $ON(CH_3)_3$ in boiling CH_3OH gives $(CH_3)_2C{=}C(CO_2CH_3)C(CO_2CH_3)_2H$ (35%) within 8 h. Similar decomplexation with $ON(CH_3)_3$ at 40 °C in $CH_2{=}CHOC_2H_5$ or cyclopentadiene gives $(CH_3)_2C{=}C$-$(CO_2CH_3)CH_2CO_2CH_3$ as the major product and $(CH_3)_2C{=}C(CO_2CH_3)C(CO_2CH_3){=}CH_2$. The last two diesters are also observed in traces in the CH_3OH decomplexation. Similar decomplexation in C_6H_6 gives the two diesters in the ratio of ca. 1:1. Decomplexation with $FeCl_3$ in CH_3OH at 50 °C for 2 h gives the triester in 75% yield. Similar decomplexation at 65 °C for 4 h gives $(CH_3)_2CClC(CO_2CH_3){=}C(CO_2CH_3)_2$ (35%) and XXXVII ($R = CO_2CH_3$) in 17% yield [49, 83].

$(CH_3)_2CC(C_6H_5)C(CO_2CH_3)C{=}OFe(CO)_3$ (Table **3**, No. **27**) gives, on heating in a solvent, a mixture of diolefin complexes with mainly the syn isomer [64]. Decomplexation with $ON(CH_3)_3$ in boiling CH_3OH gives $(CH_3)_2C{=}C(C_6H_5)C(CO_2CH_3)_2H$ (52%) within 8 h [83].

$(CH_3)_2CC(CO_2CH_3)C(C_6H_5)C{=}OFe(CO)_3$ (Table **3**, No. **28**) gives, on heating in a solvent, XXXIV ($R{=}C_6H_5$); at 70 °C, up to 90% is achieved within 3 h [64]. Decomplexation with $ON(CH_3)_3$ in boiling CH_3OH gives $CH_2{=}C(CH_3)C(CO_2CH_3){=}CHC_6H_5$ and traces of the corresponding complex XXXIV ($R{=}C_6H_5$) [83].

$C_6H_5(CH_3)CC(C_6H_5)C(C_6H_5)C{=}OFe(CO)_3$ (Table **3**, No. **31**) is soluble in polar organic solvents, and less soluble in nonpolar organic solvents. Reaction with $Fe_2(CO)_9$ in C_6H_6 at 60 °C gives XXVI with $R{=}CH_3$ (p. 77) [47].

$RCHCHC(R')N(CH_3)C{=}OFe(CO)_3$ (Table **3**, Nos. **35** to **38** with $R{=}C_6H_5CO$, $R'{=}C_6H_5$; $R{=}C_6H_5$, $R'{=}CH_3$, C_2H_5, C_6H_5), **$C_6H_5CHCHC(R')N(C_2H_5)C{=}OFe(CO)_3$** (Table **3**, Nos. **39**, **40** with $R'{=}CH_3$, C_2H_5), **$RCHCHC(R')N(C_3H_7\text{-}i)C{=}OFe(CO)_3$** (Table **3**, Nos. **41** to **45** with $R{=}H$, $R'{=}CH_3$; $R{=}Cl$, $R'{=}CH_3$; $R{=}C_6H_5$, $R'{=}CH_3$; $R{=}C_6H_5CO$, $R'{=}C_6H_5$; $R{=}R'{=}C_6H_5$), **$C_6H_5CHCHCHN(C_4H_9\text{-}t)C{=}OFe(CO)_3$** (Table 3, No. **46**), **$C_6H_5CHCHC(CH_3)N(CH_2C_6H_5)C{=}O\text{-}Fe(CO)_3$** (Table **3**, No. **47**), and **$RCHCHC(R')N(C_6H_{11}\text{-cyclo})C{=}OFe(CO)_3$** (Table **3**, Nos. **48** to **50** with $R{=}C_6H_5CO$, $R'{=}C_6H_5$; $R{=}C_6H_5$, $R'{=}CH_3$, C_6H_5). The protons of the allyl ligand in the compounds No. 35, 44, 48 ($C_6H_5COCHCHC(C_6H_5)NR''C{=}OFe(CO)_3$ with $R''{=}CH_3$, C_3H_7-i, C_6H_{11}-cyclo) give two doublets in the ^1H NMR spectra with spin-spin coupling constants equal to 9.5 Hz. In the IR spectra two bands, or one broadened band, are observed in the 1600 to 1700 cm^{-1} region [72].

$C_6H_5CHCHC(CH_3)N(CH_2C_6H_5)C{=}OFe(CO)_3$ (No. 47) crystallizes at −120 °C in the monoclinic space group $P2_1/c\text{-}C^5_{2h}$ with the unit cell parameters a = 14.28(2), b = 7.016(9), c = 18.34(3) Å, β = 98.14(15)°; Z = 4 molecules per unit cell. D_{calc} = 1.47 g/cm³. The main bond distances and angles are shown in **Fig. 11**. The plane of the π-allyl moiety C(5)C(6)C(7) forms a dihedral angle of 133° with the mean plane through Fe and C(1)O(1) and C(2)O(2) groups. The C(16) to C(21) phenyl ring is inclined by 19.1°, and the C(7)-C(16) bond by 7.4° to the π-allyl plane [10].

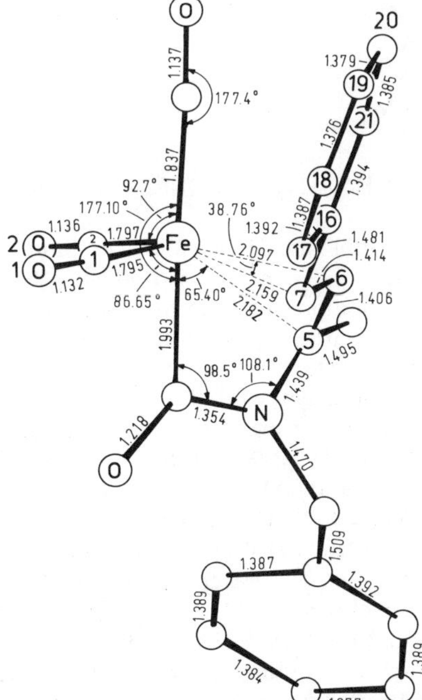

Fig. 11. Molecular structure of $C_6H_5CHCHC(CH_3)N(CH_2C_6H_5)C{=}OFe(CO)_3$ (No. 47) [10].

The compound $C_6H_5CHCHC(C_6H_5)N(C_6H_{11}\text{-cyclo})C{=}OFe(CO)_3$ (No. 50) crystallizes in the monoclinic space group $P2_1/c\text{-}C^5_{2h}$ with the unit cell parameters a = 12.684(2), b = 13.488(2), c = 13.218(2) Å, β = 103.23(1)°; Z = 4 molecules per unit cell. D_{meas} = 1.39 g/cm³, D_{calc} = 1.38 g/cm³. The main bond distances and angles are shown in **Fig. 12**. The plane, defined by carbonyl carbons C(1), C(2), C(3) is inclined at 7.6° to the π-allyl plane. The phenyl ring at C(5) is inclined at 27.6° to the π-allyl plane. For the other phenyl ring at C(7) the corresponding dihedral angle is only 8.2° [45], cf. [74, 78].

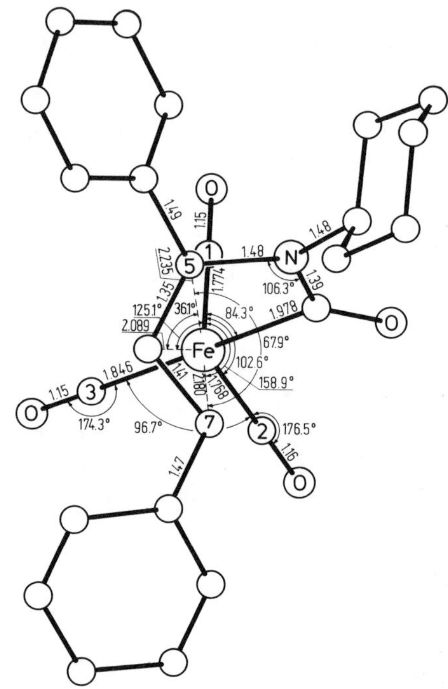

Fig. 12. Molecular structure of $C_6H_5CHCHC(C_6H_5)N(C_6H_{11}\text{-cyclo})C{=}OFe(CO)_3$ (No. 50) [45].

All compounds are readily soluble in $CHCl_3$, C_2H_5OH, and C_6H_6. However, reversible isomerization of all complexes occurs in nonpolar solvents [52 to 54, 73, 74, 78, 90]. The relative amount of XXXIX in the solution depends first of all on the polarity of the solvent. Its concentration is highest in hydrocarbon solvents. In a heptane/$CHCl_3$ mixture, its quantity decreases with increase of $CHCl_3$ content. It is impossible to obtain XXXIX in a pure state since in solution it exists in equilibrium with the original 1σ, 3-5π-allyl complexes XL (Nos. 35 to 49) while on evaporation or freezing of solutions only the XL form is isolated. When R′ = H (No. 46), or when R′ and R″ in XXXIX (R = C_6H_5) are small (Nos. 36, 37, 39, 40), further conversion of XXXIX into XLI and CXVIII is observed, and the IR spectra are obscured by these additional tautomers [53, 74, 78, 90]. The quantitative relations of the tautomeric forms are determined for $C_6H_5CHCHC(CH_3)N(C_6H_{11}\text{-cyclo})C{=}OFe(CO)_3$ (No. 49) by optical density measurements of the solutions at 2068 cm^{-1} for No. 49 and at 2085 cm^{-1} for XXXIX (R = C_6H_5, R′ = CH_3, R″ = C_6H_{11}-cyclo) at 25 °C. It turns out that No. 49, when dissolved in hexane, consists of 30% XL and 70% XXXIX. The addition of CH_2Cl_2 to the solution shifts the equilibrium towards XL, which predominates over XXXIX in pure CH_2Cl_2 or $CHCl_3$ (90% XL and 10% XXXIX) [78]. The thermodynamic characteristics of the equilibrium between No. 50 and its isomer XXXIX (R = R′ = C_6H_5, R″ = C_6H_{11}-cyclo) in hexane and hexane/CH_2Cl_2

References on pp. 106/8

are determined. In hexane: $\Delta H = 11.8$ kJ/mol, $\Delta S = 49.5$ J/mol, $\Delta G^{210} = 1.4$ kJ/mol, $\Delta G^{250} = -0.6$ kJ/mol, $\Delta G^{295} = -28$ kJ/mol; in hexane/CH_2Cl_2 (96:4): $\Delta H = 20.3$ kJ/mol, $\Delta S = 72.8$ J/mol, $\Delta G^{250} = 2.1$ kJ/mol, $\Delta G^{295} = -1.2$ kJ/mol. These thermodynamic parameters of equilibrium are calculated from the temperature dependence of the equilibrium constants K_T, and fit fairly well with the thermodynamic scale of tautomerism and allow a quantitative conclusion on the influence of solvent, temperature, and polarity on the relative stabilities of the tautomeric forms to be reached [87]. The compounds No. 41 and 42 are less stable and, therefore, a quantitative study of the equilibrium is not possible. However, it is shown qualitatively that the appropriate form XXXIX is predominant in these cases [78]. The following table shows the equilibrium constants K_T for the $C_6H_5CHCHC(R')N(R'')C{=}OFe(CO)_3$ complexes [90]:

No.	$C_6H_5CHCHC(R')N(R'')C{=}OFe(CO)_3$ R′	R″	K_T
36	CH_3	CH_3	0.81±0.03
37	C_2H_5	CH_3	0.09±0.01
38	C_6H_5	CH_3	<0.05
39	CH_3	C_2H_5	0.80±0.03
40	C_2H_5	C_2H_5	0.10±0.01
43	CH_3	C_3H_7-i	—
45	C_6H_5	C_3H_7-i	<0.05
46	H	C_4H_9-t	—
49	CH_3	C_6H_{11}-cyclo	2.38±0.10
50	C_6H_5	C_6H_{11}-cyclo	<0.05

Heating complex No. 49 in CH_3OH at 50 °C for 2.5 h gives CXVIII ($R' = CH_3$, $R'' = C_6H_{11}$-cyclo, p. 104) in 50% yield [90]. Reaction of Nos. 35, 36, 38, 43, 44, 48 with aqueous HBF_4 in $CHCl_3$ gives the appropriate XXXVIII (p. 80) which on addition of H_2O regenerates the original complexes XL [53, 72, 74, 78].

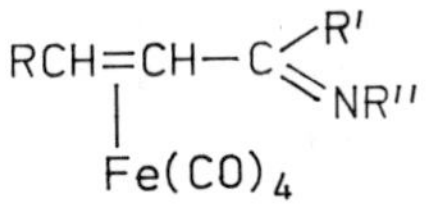

RCH=CH−C(R′)=NR″ / Fe(CO)₄ — XXXIX; R–(allyl)–C(R′)–N–R″, C=O, (CO)₃Fe — XL; RCH=CH−C(R′)=N−CH₃ / Fe(CO)₄ — XLI

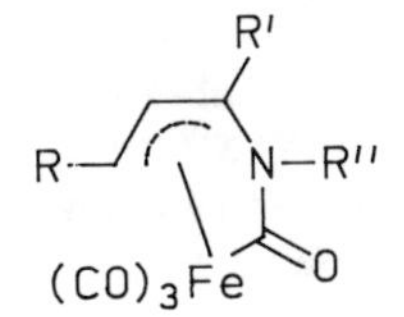

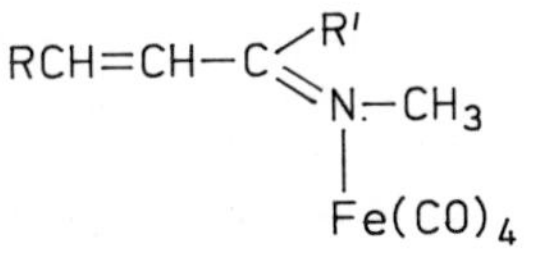

$CH_2CHCHCH_2CH_2Fe(CO)_3$ (Table **3**, No. **52**) reacts with CO in hexane solution at 25 °C to give No. 103 [8, 12] and XLIII (not isolated) [12].

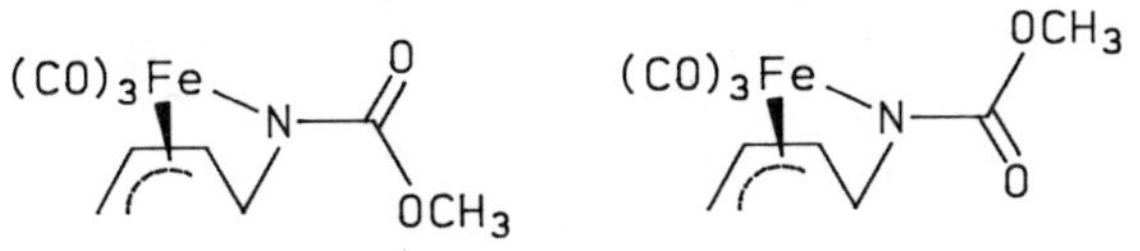

(CO)₃Fe–N–C(=O)OCH₃ — a; (CO)₃Fe–N–C(OCH₃)=O — b; XLII

$C_6H_5CHCHCHCH_2CHC_6H_5Fe(CO)_3$ (Table **3**, No. **53**) is prepared from $[^5LFe(CO)_3]BF_4$ ($^5L = C_6H_5CH(CH)_3CHC_6H_5$) and $Na[NCBH_3]$ in tetrahydrofuran at −22 °C to give XLIV-d_0 and No. 53. The reaction mixture is poured into H_2O, extracted with ether, washed with H_2O and saturated aqueous NaCl, dried ($MgSO_4$), and evaporated. Chromatography with hexane and hexane/C_6H_6 gives XLIV-d_0 and No. 53 (40% yield). Analogous reduction of $[^5LFe(CO)_3]BF_4$ with $Na[NCBD_3]$ gives XLIV-d_0, XLIV-d_1, and XLV. The ^1H NMR spectrum (CS_2) of XLV shows chemical shifts at δ = 1.2 (d, 1H; J = 10 Hz), 3.1 (br, t, 1H; J = 8 Hz), 4.1 (t, 1H; J = 6 Hz), 4.4 (d, 1H; J = 12 Hz), 5.2 (dd, 1H; J = 7 Hz), 7.2 (m, 10H) ppm. Oxidation of No. 53 with $[NH_4]_2[Ce(NO_3)_6]$ in CH_3OH gives (E)-$C_6H_5CH{=}CHCH(OCH_3)CH_2CH(OCH_3)C_6H_5$ (17%) and (E)-$C_6H_5CH(OCH_3)CH{=}CHCH_2CH(OCH_3)C_6H_5$ (17%) along with other unidentified products [32].

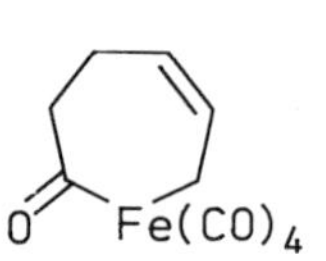

XLIII 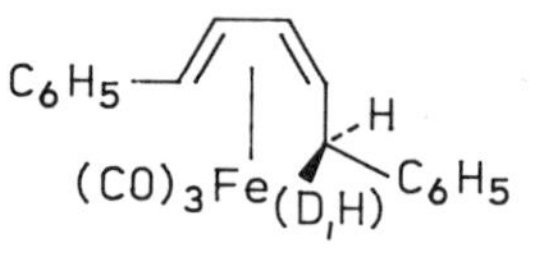 XLIV 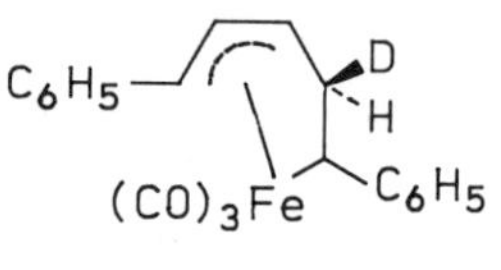 XLV

$(CH_3O_2CCH)_2CCH_2C{=}OFe(CO)_3$ (Table **3**, No. **55**) is prepared by irradiating XLVI in hexane with a high pressure Hg lamp through a pyrex filter for a few minutes. Preparative thick layer chromatography (silica gel PF 254, 5% $O(C_2H_5)_2/H_2O$) gives XLVIIa and b (17%) and No. 55 (34%) [17].

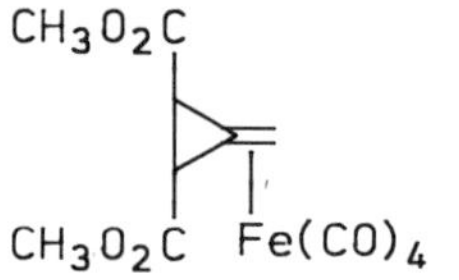

XLVI XLVII (a, b)

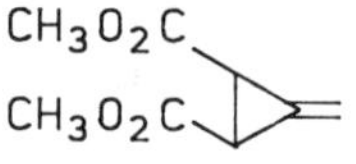

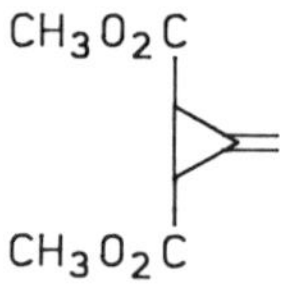

$CH_2C(CH_3)CHCH_2N(CH_3)C{=}OFe(CO)_3$ (Table **3**, No. **59**) reacts with liquid CH_3NH_2 to give $CH_2CHC(CH_3)CH_2N(CH_3)C{=}OFe(CO)_3$ (No. 58) in 90% yield after 1.5 h [55].

$CH_3CHCHCHCH(CH_3)N(CH_3)C{=}OFe(CO)_3$ (Table **3**, No. **60**) exists in two isomeric forms. Isomer XLVIIIa is prepared from isomer LXXXVIIIc, pp. 95/6, of No. 87 with liquid CH_3NH_2 (95% yield) as described for No. 59. Similar reaction of isomer LXXXVIIIb, pp. 95/6, with CH_3NH_2 in ether at 20 °C for 10 min followed by evaporation gives isomer XLVIIIb (95% yield). The physical properties are shown in the following table [55]:

isomers of No. 60	physical properties
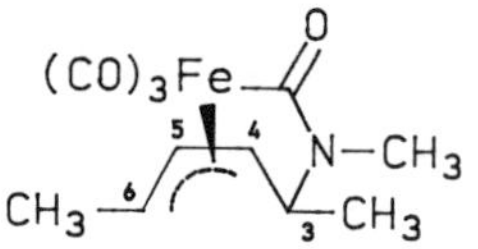 XLVIII a	m.p. 62 to 63° (from pentane), colorless crystals ^1H NMR ($CDCl_3$): 1.27 (CH_3), 2.57 (NCH_3), 3.42 (H-3; J(H-3, CH_3) = 6.0, J(H-3,4) = 2.0), 3.81 (H-6; J(H-6, CH_3) = 6.0), 3.96 (H-4), 4.65 (H-5; J(H-4,5) = 8.0, J(H-5,6) = 12.0) ^{13}C NMR ($CDCl_3$): 20.9, 21.2 (CH_3); 29.3 (NCH_3), 52.24 (C-6), 70.1 (C-4), 79.7 (C-3), 94.0 (C-5); 200.9, 204.6, 208.1, 211.8 (CO) IR (KBr): 570, 600, 610, 660, 825, 1000, 1030, 1065, 1095, 1200, 1305, 1365, 1400, 1450, 1575, 1980, 2060, 2970 IR (hexane): 1989, 2013, 2073 (CO) mass spectrum: $[M]^+$ and fragments given

isomers of No. 60	physical properties
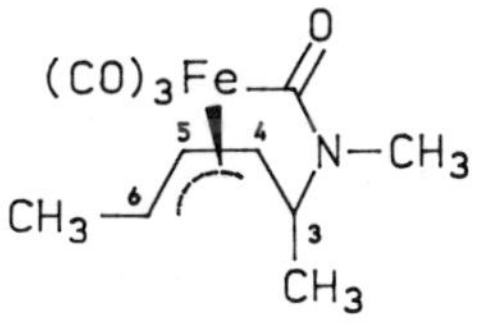 XLVIII b	m.p. 57 to 58° (from pentane at −20°), yellow crystals ^{1}H NMR ($CDCl_3$): 1.26 (CH_3-3), 1.84 (CH_3-6), 2.54 (NCH_3), 3.53 (H-3; J(H-3, CH_3) = 6.5), 3.87 (H-6; J(H-5,6) = 12.0), 4.48 (H-4,5; J(H-3,4) = 6.0) ^{13}C NMR ($CDCl_3$): 21.1, 22.5 (CH_3); 30.6 (NCH_3), 56.1 (C-6), 72.0 (C-4), 79.9 (C-3), 92.3 (C-5); 202.9, 204.6, 208.2, 211.2 (CO) IR (KBr): 575, 600, 610, 660, 750, 815, 860, 930, 980, 1030, 1070, 1160, 1220, 1300, 1320, 1365, 1400, 1450, 1580, 1980, 2040, 2980 IR (hexane): 1990, 2014, 2073 (CO) mass spectrum: $[M]^+$ and fragments given

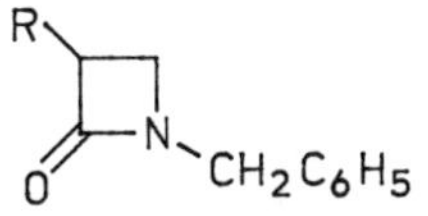

XLIX 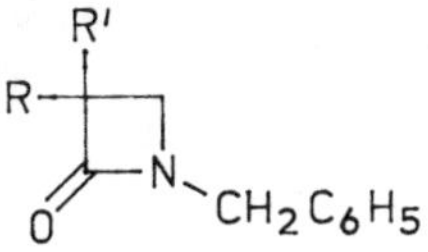 L 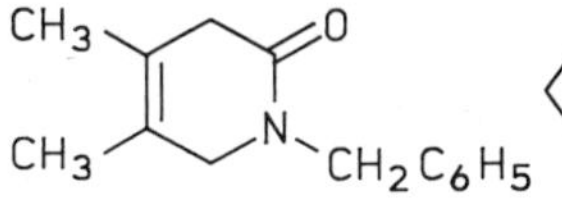 LI 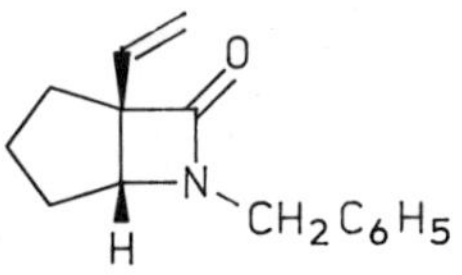 LII

$RCHCR'CR''CHR'''N(CH_2C_6H_5)C{=}OFe(CO)_3$ (Table **3**, Nos. **61** to **66** with R = R′ = R″ = R‴ = H; R = R″ = R‴ = H, R′ = CH_3; R = n-C_5H_{11}, R′ = R″ = R‴ = H; R = R‴ = H, R′ = R″ = CH_3; R″ = R‴ = H, R,R′ = -$(CH_2)_3$-; R = R′ = H, R″,R‴ = -$(CH_2)_3$-), **$CH_2C(CH_3)CHCH_2N(CH_2C_6H_3(R\text{-}2)\text{-}R'\text{-}4)C{=}OFe(CO)_3$** (Table **3**, Nos. **67** and **68** with R = H, R′ = NO_2, and R = R′ = OCH_3), and **$CH_2CRCHCH_2N(CH(C_6H_4OCH_2C_6H_5\text{-}4)CO_2R'$** (Table **3**, Nos. **69** and **70** with R = H, R′ = CH_3, and R = CH_3, R′ = $CH_2C_6H_5$) give β-lactam complexes in excellent yield on oxidation with $[NH_4]_2[Ce(NO_3)_6]$ in C_2H_5OH. Only in one example, No. 64, does oxidation lead to a significant amount of a δ-lactam [59, 62]. The products, reaction conditions, and yields are shown in the following table:

No. of complex	product(s)	reaction conditions	yield in %	Ref.
61	XLIX (R = CH=CH_2)	−30° to room temp.	72 to 75	[59, 62, 88]
62	XLIX (R = $C(CH_3)$=CH_2)	−30° to room temp.	75 to 88	[62, 88]
63	XLIX (R = CH=CHC_5H_{11}-n) (1:1 mixture of (E)-/(Z)-isomers)	−5°	64	[59, 62, 88]
64	L (R = $C(CH_3)$=CH_2, R′ = CH_3) and LI	0° to room temp.	33.5 to 34	[59, 62, 88]
65	L (R = cyclopent-1-en-1-yl, R′ = H)	room temp.	84 to 88	[59, 62, 88]
66	LII	−30° to room temp.	75	[59, 88]
mixture 66/67	LII	−30° to room temp.	75	[62]
67	LIII (R = H, R′ = NO_2)	−30° to room temp.	72 to 74	[62, 88]

References on pp. 106/8

No. of complex	product(s)	reaction conditions	yield in %	Ref.
68	LIII (R = R′ = OCH_3)	−30° to room temp.	13	[62]
69	LIV (R = CH=CH_2, R′ = CH_3)	−30° to room temp.	80 to 81	[59, 62]
70	LIV (R = C(CH_3)=CH_2, R′ = $CH_2C_6H_5$)	−30° to room temp.	60	[62]

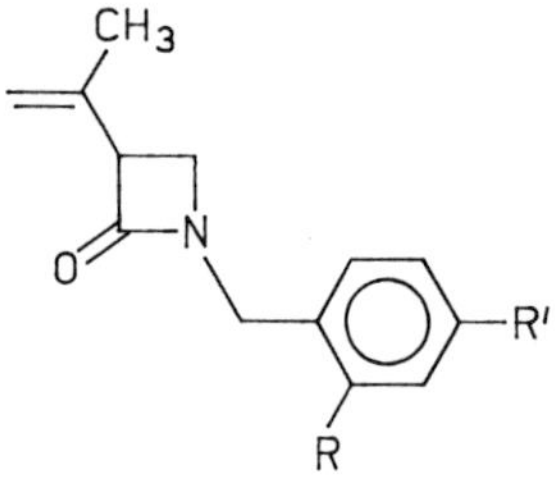

LIII 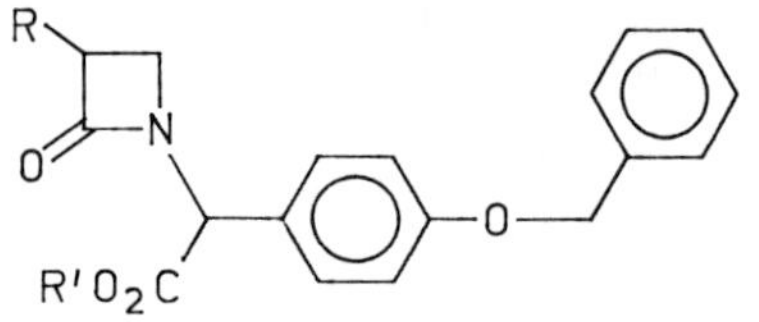 LIV 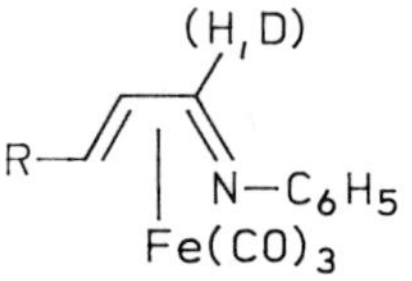 LV 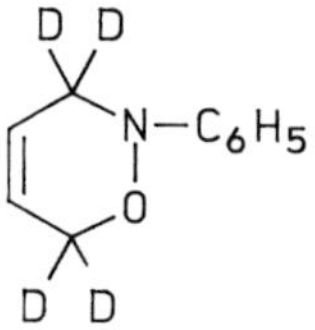 LVI

$CH_2CHCHCH_2N(C_6H_5)C{=}OFe(CO)_3$ (Table **3**, No. **71**). The deuterated complexes $CH_2CDCDCH_2N(C_6H_5)C{=}OFe(CO)_3$ (A), $CD_2CHCHCD_2N(C_6H_5)C{=}OFe(CO)_3$ (B) [11], or $CH_2CHCDCH_2N(C_6H_5)C{=}OFe(CO)_3$ (C) [31] are prepared similarly to the undeuterated complex No. 71 [11, 31]. Thus, analogous to Method IId, starting with LVII gives A, and LVI gives B [11]. For the preparation of C see No. 75 [31]. Compound C melts at 104 to 105 °C (dec.), and its ^1H NMR spectrum is given in a figure in [31]. Its mass spectrum shows the parent ion [31]. Refluxing No. 71 in CH_3OH for 1.5 h gives nondeuterated LV (R = CH_3) in 62% yield. Analogous reaction of compound B gives LV-d_4 (R = CD_3) [11]. The reaction of No. 71 with LVIII (R = CH=CH_2, R′ = C_6H_5) in C_6H_6 at 34 °C for 36 h, followed by washing the reaction mixture with diluted aqueous HCl, saturated aqueous NaCl, evaporation, and chromatography on basic alumina with 20% CH_2Cl_2/petroleum ether as eluent gives a mixture of $^3L = C_6H_5CH{=}CHC(CH_3){=}O$ and $^3LFe(CO)_3$, and elution with 50% CH_2Cl_2/petroleum ether gives No. 78 [31].

$CH_2CHCHCH_2N(CO_2CH_3)C{=}OFe(CO)_3$ (Table **3**, No. **74**) exists in only one form as shown by ^1H NMR spectroscopy in toluene down to −80 °C (cf. No. 51) [13]. Heating for 20 min at 60 °C gives a 2:1 mixture of XLIIa and b and CO in quantitative yield. The reaction is quantitatively reversed at 70 bar CO, 25 °C, in benzene solution after 50 h [13], cf. [43].

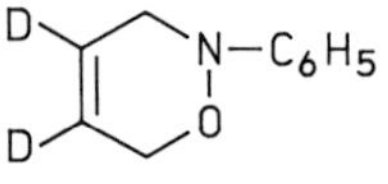

LVII 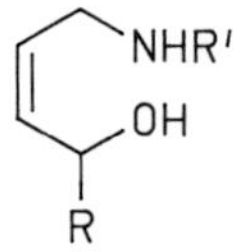 LVIII 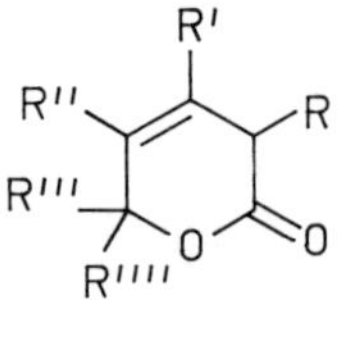 LIX 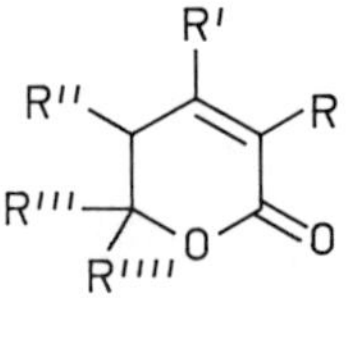 LX 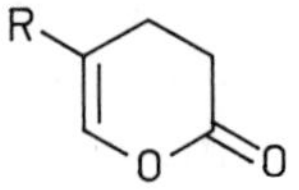 LXI 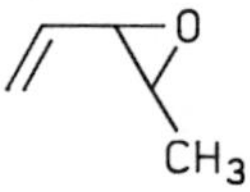 LXII

$CH_2CHCHCH_2OC{=}OFe(CO)_3$ (Table **3**, No. **75**). The monodeuterated isomer $CH_2CDCHCH_2OC{=}OFe(CO)_3$, prepared from (Z)-$HOCH_2CH{=}CDCH_2NHC_6H_5$ as in Method IIc (pp. 45/6), melts at 112 to 114 °C (dec.). Its ^1H NMR spectrum is given in a figure in [31], and its mass spectrum shows the parent ion [31]. $CH_2CDCDCH_2OC{=}OFe(CO)_3$ is prepared from LVII as in Method IId (similar yield), m.p. 112 to 114 °C (dec.). ^1H NMR ($CDCl_3$): δ = 2.94

(s, 1H), 3.67 (s, 1H), 4.00 (d, 2H; Δν≅1 Hz) ppm. $CD_2CHCHCD_2OC{=}OFe(CO)_3$ is prepared from LVI as in Method IId (p. 46), m.p. 112 to 114 °C (dec.). ^1H NMR ($CDCl_3$): 4.75 (ABq, 2H; J=8 Hz) ppm [11].

Compound No. 75 is poorly soluble in nonpolar solvents but easily soluble in polar solvents. The crystals are air-stable, but in solution the compound is air-sensitive [1]. Irradiation in pyridine in the presence of O_2 and two equivalents HNRR′ for 1 h gives (Z)-$HOCH_2CH{=}CHCH_2NRR'$ ($R=R'=C_2H_5$, i-C_3H_7, -$(CH_2)_5$-, -$(CH_2)_2O(CH_2)_2$-). Analogous reaction in pyridine/H_2O (9:1) in the presence of O_2 and two equivalents NaCN gives (Z)-$HOCH_2CH{=}CHCH_2CN$ [48]. Reaction with excess I_2/pyridine at 60 °C gives CO (36.8%) and small amounts of CO_2. Reaction with conc. HCl or $O(COCH_3)_2$/pyridine at 50 °C gives CO, CO_2, and butadiene [1]. CO (300 bar) in C_6H_6 at 75 °C in an autoclave gives LIX (R=R′=R″=R‴=R⁗=H), LX (R=R′=R″=R‴=R⁗=H), and XLI (R=H) in the ratio 3.3:6.1:1.0 (total yield 80%) after 18 h. Analogous reaction (200 bar CO, 50 °C, 16 h) in CH_3OH/saturated aqueous $Ba(OH)_2$ gives LIX (R=R′=R″=R‴=R⁗=H), LX (R=R′=R″=R‴=R⁗=H) in the ratio 3.3:1 (total yield 40%), and (E)-$CH_3OCH_2CH{=}CHCH_2OH$ (45%) [55]. Excess $P(C_6H_5)_3$ in boiling C_6H_6 gives $CH_2CHCHCH_2OC{=}OFe(CO)_2P(C_6H_5)_3$ [1]. Addition of basic alumina (activity III) and aniline to a solution of No. 75 in C_6H_6 gives No. 71. After 2.5 h at room temperature the reaction mixture is filtered and the alumina washed with CH_2Cl_2. Removal of the solvent gives No. 71 (40% yield). Using neutral alumina as a catalyst also yields No. 71 but at a slower rate (ca. 4 h). No reaction occurs when silica gel is used, or when no catalyst is employed. Similar reaction with CH_3NH_2 gives No. 57 (45% yield) [11], cf. [43]. Analogous reaction of $CH_2CDCHCH_2OC{=}OFe(CO)_3$ with aniline gives isomer C of No. 71 [31]. Addition of $C_6H_5CH_2NH_2$ to a solution of No. 75 and $ZnCl_2$ in tetrahydrofuran/ether (1:3) gives No. 61 (81% yield). After 0.5 h at room temperature the mixture is filtered, the solvent removed, and the product is purified by chromatography on Merck-Kieselgel 60 [59, 62] with ether/light petroleum ether (2:1) [88] followed by recrystallization [59, 62, 88]. Similar reaction with 4-$C_6H_5CH_2OC_6H_4CH(CO_2CH_3)NH_2$ in tetrahydrofuran/ether (1:4) gives No. 69 in 80% yield [62]. Refluxing No. 75 in C_6H_6, cyclohexane, or toluene gives CO, CO_2, propene, $Fe(CO)_5$ (49%), and $^4LFe(CO)_3$ (^4L=butadiene) [1].

$CH_2CHCHCH(CH_3)OC{=}OFe(CO)_3$ (Table **3**, No. **76**). Photolysis of $Fe(CO)_5$ and LXII in C_6H_6 at 20 °C for 3 h followed by filtration, evaporation of the filtrate, and recrystallization from CH_2Cl_2/pentane gives a 1:2 mixture of isomers LXIIIa and b of No. 76 (70% yield). The isomers are separated by chromatography on Kieselgel (Merck Type 100) with ether as eluent. The first fraction gives isomer LXIIIa. The physical properties are shown in the following table [55]:

isomers of No. 76	physical properties
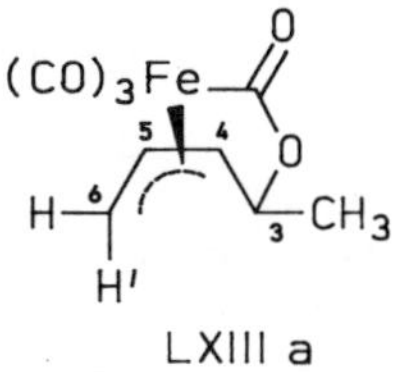 LXIII a	m.p. 81 to 82° (dec.), colorless to pale yellow crystals (from CH_2Cl_2/pentane) ^1H NMR ($CDCl_3$): 1.38 (CH_3-3; J(H-3, CH_3-3)=7.0), 3.01 (H-6′; J(H-6,6′)=2.0), 3.76 (H-6; J(H-5,6)=8.0), 4.32 (H-3; J(H-3,4)=1.5), 4.66 (H-4; J(H-4,5)=8.0), 4.96 (H-5; J(H-5,6′)=13.0) ^{13}C NMR ($CDCl_3$): 23.98 (CH_3-3), 57.84 (C-6), 71.16 (C-3), 82.02 (C-4), 92.16 (C-5); 203.29, 205.95, 206.34, 209.14 (CO) IR (KBr): 545, 570, 605, 610, 640, 660, 840, 950, 985, 1005, 1045, 1080, 1120, 1170, 1305, 1330, 1375, 1440, 1490, 1645, 2000 IR (hexane): 2010, 2027, 2085 (CO) mass spectrum: $[M]^+$ and fragments given

References on pp. 106/8

isomers of No. 76	physical properties
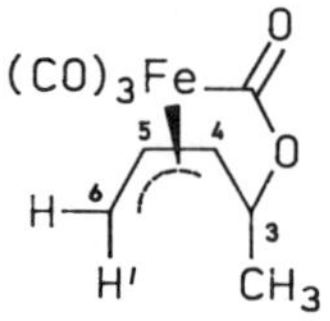 $(CO)_3Fe$ complex, H, H′, CH_3 (structure) LXIII b	m.p. 110 to 111° (dec.), colorless crystals (from CH_2Cl_2/pentane) ^1H NMR ($CDCl_3$): 1.39 (CH_3-3; J(H-3, CH_3-3) = 6.5), 3.13 (H-6′; J(H-6,6′) = 1.0), 3.82 (H-6; J(H-5,6) = 6.0), 4.48 (H-3; J(H-3,4) = 5.0), 4.78 (H-5; J(H-5,6′) = 12.0), 4.90 (H-4) ^{13}C NMR ($CDCl_3$): 21.90 (CH_3-3), 58.62 (C-6), 73.50 (C-3), 82.93 (C-4), 90.92 (C-5); 203.09, 205.76, 206.73, 208.94 IR (KBr): 570, 610, 615, 660, 830, 935, 980, 1025, 1060, 1090, 1170, 1360, 1440, 1450, 1645, 2000 IR (hexane): 2010, 2085 (CO) mass spectrum: $[M]^+$ and fragments given

Isomer LXIIIb in CH_3OH/saturated aqueous $Ba(OH)_2$ at 20 °C gives (E)-$CH_3CH{=}CHCH{=}CH_2Fe(CO)_3$ after 15 h. Reaction with CO (220 bar) in C_6H_6 at 75 °C in an autoclave for 17 h gives LX (R = R′ = R″ = R‴ = H, R⁗ = CH_3) in 65% yield, CO_2 and (E)-$CH_3CH{=}CHCH{=}CH_2$ (ca. 5%). Similar reaction with CO (200 bar) in CH_3OH at 50 °C for 16 h gives LX (R = R′ = R″ = R‴ = H, R⁗ = CH_3) in 15% yield and $CH_3OCH_2CH{=}CHCH(CH_3)OH$ (50%) [55].

$CH_2CHCHCH(C_5H_{11}$-n)OC=OFe(CO)$_3$ (Table **3**, No. **77**) is prepared by irradiation of $Fe(CO)_5$ and LXV in C_6H_6 at room temperature for 2 to 6 h. C_6H_6 and excess $Fe(CO)_5$ are removed under vacuum, taking care not to heat the mixture above 10 °C. Chromatography on Florisil gives the cis conformer LXIVb (18%) and the trans conformer LXIVa (42%) [69], cf. [59]. The physical properties of both are given in the following table [69], cf. [59]:

isomers of No. 77	physical properties
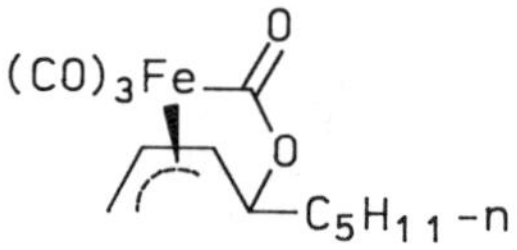 $(CO)_3Fe$ complex, C_5H_{11}-n (structure) LXIV a	m.p. 76 to 77°, crystals ^1H NMR ($CDCl_3$): 0.92 (br,s, 3H), 1.14 to 1.9 (m, 8H), 3.08 (d, 1H; J = 13), 3.78 (d, 1H; J = 8), 4.05 (t, 1H; J = 6.5), 4.6 to 5.0 (m, 2H) IR: 660, 995, 1040, 1070, 1460, 1660, 2010, 2080, 2850
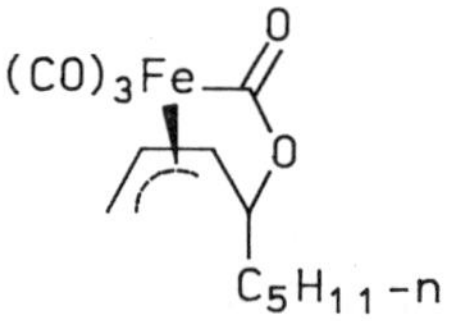 $(CO)_3Fe$ complex, C_5H_{11}-n (structure) LXIV b	m.p. 97° (dec.), crystals ^1H NMR ($CDCl_3$): 0.88 (br,s, 3H), 1.1 to 1.7 (m, 8H), 3.12 (d, 1H; J = 12), 3.7 (d, 1H; J = 8), 4.3 (d, 1H; J = 5), 4.6 to 4.96 (m, 2H) IR: 1455, 1665, 1992, 2050, 2900

Refluxing isomer LXIVb in tetrahydrofuran for 3 h gives LXVIII (R = C_5H_{11}-n) in 39% yield, and LIX (R = R′ = R″ = R‴ = H, R⁗ = C_5H_{11}-n) in 16% yield. Analogous treatment of isomer LXIVa gives LXVIII (R = C_5H_{11}-n) in 40% yield and LIX (R = R′ = R″ = R‴ = H, R⁗ = C_5H_{11}-n) in 12.5% yield [76]. Oxidation of LXIVa with $[NH_4]_2[Ce(NO_3)_6]$ in CH_3CN at −30 °C gives (E)-LXVI (68% yield), and analogous oxidation of LXIVb gives (Z)-LXVI (64%) [69]. Reaction of LXIVa with CO (60 atm) in a bomb in C_6H_6 at 195 °C for 4 h gives a mixture of LIX (R = R′ = R″ = R‴ = H, R⁗ = C_5H_{11}-n) and LX (R = R′ = R″ = R‴ = H, R⁗ = C_5H_{11}-n)

References on pp. 106/8

in a total yield of 65%. Analogous reaction of LXIVb gives LX (R=R′=R″=R‴=H, R⁗= C_5H_{11}-n) in 65% yield [76]. Reaction of LXIVa with $C_6H_5CH_2NH_2/ZnCl_2$ in ether at room temperature for 3 h gives No. 63 in 82% yield [59, 62, 88].

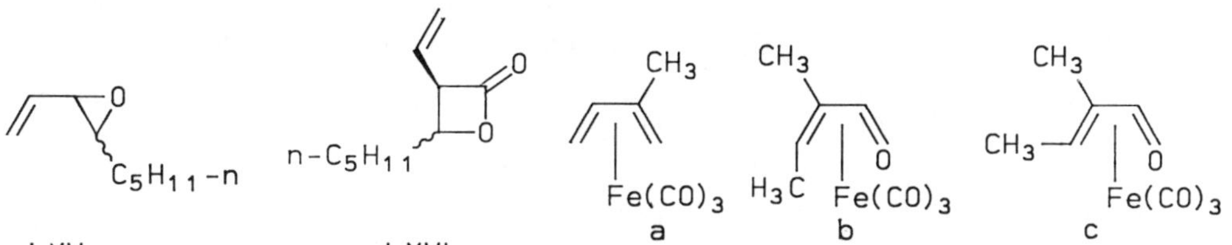

LXV LXVI LXVII (a, b, c)

$CH_2CHCHCH(CH{=}CH_2)OC{=}OFe(CO)_3$ (Table **3**, No. **78**) reacts with $C_6H_5NH_2$ in tetrahydrofuran at room temperature within 36 h. Filtration, evaporation of the filtrate, and chromatography of the residue on basic alumina with 60% CH_2Cl_2/petroleum ether gives No. 73 [31].

$CH_2CHC(CH_3)CH_2OC{=}OFe(CO)_3$, **$CH_2C(CH_3)CHCH_2OC{=}OFe(CO)_3$** (Table **3**, Nos. **79**, **80**) and **$RCHCHCHCH_2OC{=}OFe(CO)_3$** (Table **3**, Nos. **81**, **82** with R=CH_3, n-C_5H_{11}). Heating complex No. 79 in ether in a sealed tube for 4 h gives LXVIIa (22%), LXVIIb and c (combined yield 44%). Similar thermolysis of No. 82 in tetrahydrofuran at 66 °C for 3 h gives LXVIII (R=n-C_5H_{11}) in 27% yield [76]. Reaction of No. 81 with CH_3OH/saturated aqueous $Ba(OH)_2$ at 20 °C for 15 h gives LXVIII (R=CH_3) in 85% yield. Reaction of Nos. 79 and 81 with CO in C_6H_6 or in CH_3OH in an autoclave at 75 °C gives a mixture of products as shown in the following table [55]:

compound No.	reaction conditions	reaction products
79	C_6H_6, CO (300 bar), 17 h	LIX (R=R′=R‴=R⁗=H, R″=CH_3), LX (R=R′=R‴=R⁗=H, R″=CH_3), LXI (R=CH_3), ratio 3.1:2.4:1.0, total yield 99%
	CH_3OH, CO (300 bar), 18 h	LIX (R=R′=R‴=R⁗=H, R″=CH_3), LX (R=R′=R‴=R⁗=H, R″=CH_3), ratio 3.9:1.0, total yield 83%
81	C_6H_6, CO (300 bar), 16 h	LIX (R=CH_3, R′=R″=R‴=R⁗=H), LX (R=CH_3, R′=R″=R‴=R⁗=H), ratio 12.5:1, total yield 67%, and (E)-$CH_3CH{=}CHCH{=}CH_2$, ca. 5% yield
	CH_3OH, CO (220 bar), 16 h	(E)-$HOCH_2CH{=}CHCH(CH_3)OCH_3$, 70% yield

Reaction of compound No. 79 with CH_3NH_2 in ether at 20 °C for 10 min gives No. 59 in 95% yield [55]. Compound No. 80 reacts with aniline in benzene containing basic alumina. After 5 h the mixture is filtered, evaporated, and the residue is chromatographed on basic alumina. Elution with CH_2Cl_2/C_6H_6 gives No. 72 (50% yield) [11]. Similar reactions are observed at room temperature with Nos. 79 and 80 using $RNH_2/ZnCl_2$. The reaction conditions and products are shown in the following table:

References on pp. 106/8

starting complex No.	$RNH_2/ZnCl_2$	reaction conditions	product (yield)	Ref.
79	$C_6H_5NH_2/ZnCl_2$	tetrahydrofuran/ether (1:3), 1 h	No. 62 (95 to 97%)	[62, 88]
	$C_6H_5NH_2/ZnCl_2/HN(CH_2CH_2)_2NH$	tetrahydrofuran/ether (1:3), 0.2 h	No. 62 (100%)	[88]
	$C_6H_5NH_2/ClAl(C_2H_5)_2$	tetrahydrofuran/ether (1:3), 0.5 h	No. 62 (98%)	[88]
79	$4\text{-}O_2NC_6H_4CH_2NH_2/ZnCl_2$	tetrahydrofuran/ether (2:1)	No. 67 (85%)	[62, 88]
80	$2,4\text{-}(CH_3O)_2C_6H_3CH_2NH_2/ZnCl_2$	ether	No. 68 (95%)	[62]
80	$4\text{-}C_6H_5CH_2OC_6H_4\text{-}(C_6H_5CH_2O_2C)CHNH_2/ZnCl_2$	tetrahydrofuran/ether (1:1)	No. 70 (45%)	[62]

The products are chromatographed on Merck-Kieselgel 60 [62] with ether/light petroleum ether (2:1) [88] and recrystallized from petroleum ether [62].

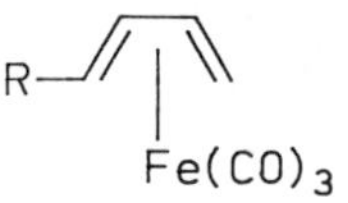

R Fe(CO)$_3$
LXVIII

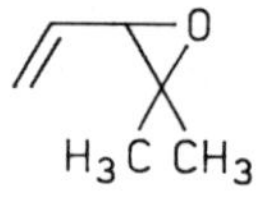

O H$_3$C CH$_3$
LXIX

$CH_2CHCHC(CH_3)_2OC{=}OFe(CO)_3$ (Table **3**, No. **83**). Light induced reaction of LXIX with $Fe(CO)_5$ in C_6H_6 at 20 °C for 6 h gives after recrystallization a 1:1 mixture of isomers LXXa and b (25% yield) not separable by chromatography because trans isomer LXXb is thermodynamically labile. The ratio of isomers is temperature dependent. Analogous reaction of LXIX with $Fe(CO)_5$ at −45 °C for 6 h followed by work-up at −20 °C and recrystallization with cold CH_2Cl_2/pentane gives pure LXXb (46% yield). Warming pure LXXb in C_6H_6 at 60 °C for 1 h gives quantitatively cis isomer LXXa. Pure LXXa is obtained by the following procedure: A solution of isomer LXXb in C_6H_6 is heated at 45 °C for 5 h, concentrated, and excess pentane is added. Cooling to −20 °C for some hours gives pure cis isomer LXXa (79%). The physical properties are shown in the following table [55]:

isomers of No. 83	physical properties
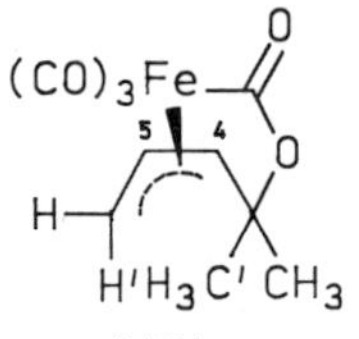 (CO)$_3$Fe O 5 4 O H H' H$_3$C' CH$_3$ LXX a	m.p. 95 to 96°, golden yellow lustrous crystals ^{1}H NMR ($CDCl_3$): 1.40 (CH_3-3), 1.50 (CH_3-3′), 3.14 (H-6′; J(H-5,6′) = 12.0), 3.85 (H-6; J(H-5,6) = 7.0), 4.83 (H-4,5) ^{13}C NMR ($CDCl_3$): 29.51, 31.33 (CH_3); 58.62 (C-6), 78.96 (C-3), 87.87 (C-4), 90.73 (C-5); 202.90, 204.46, 206.34, 209.72 (CO) IR (KBr): 545, 575, 600, 610, 620, 655, 675, 795, 830, 950, 1035, 1110, 1170, 1195, 1225, 1255, 1353, 1375, 1435, 1460, 1650, 2000 IR (hexane): 2010, 2027, 2084 (CO) mass spectrum: $[M]^+$ and fragments given

References on pp. 106/8

isomers of No. 83	physical properties
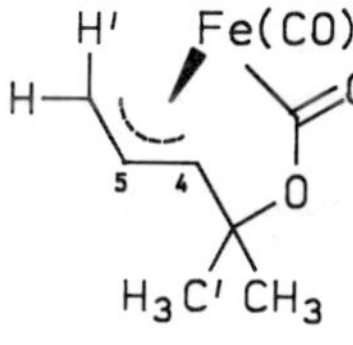 LXX b	m.p. 83 to 85° (dec.), cream-colored needles ^{1}H NMR ($CDCl_3$): 1.55 (CH_3-3, 1.68 (CH_3-3′), 3.06 (H-6′; J(H-6,6′) = 1.0), 3.47 (H-6; J(H-5,6) = 9.0), 3.95 (H-4; J(H-4,5) = 8.0), 4.84 (H-5; J(H-5,6′) = 13.0) ^{13}C NMR ($CDCl_3$): 28.99, 29.31 (CH_3); 62.72 (C-6), 70.51 (C-4), 81.82 (C-3), 91.90 (C-5); 202.77, 203.48, 205.69, 209.72 (CO) IR (KBr): 565, 615, 640, 655, 915, 940, 970, 995, 1025, 1190, 1230, 1300, 1365, 1380, 1450, 1655, (1670), 2000 IR (hexane): 2006, 2018, 2080 (CO) mass spectrum: $[M]^+$ and fragments given

Reaction of the cis isomer LXXa with CF_3CO_2D in acetone-d_6 at −10 to 0 °C gives LXXIa after 1 min. Similar reaction of the trans isomer LXXb at 60 °C gives LXXIb after 30 min. Reaction of the cis isomer with CO (300 bar) in C_6H_6 at 115 °C for 17.5 h in an autoclave gives LIX (R = R′ = R″ = H, R‴ = R⁗ = CH_3) in 25% yield, LX (R = R′ = R″ = H, R‴ = R⁗ = CH_3) in 50% yield (ratio 1:2) and $(CH_3)_2C{=}CHCH{=}CH_2$ (ca. 10%). Similar reaction of the cis isomer in CH_3OH at 100 °C for 20 h gives LIX and LX (R = R′ = R″ = H, R‴ = R⁗ = CH_3) in the ratio 4.0:1.0 (total yield 65%) and (E)-$CH_3OCH_2CH{=}CHC(CH_3)_2OH$ (20% yield). The reaction of the cis isomer LXXa with $P(C_6H_5)_3$ in $CDCl_3$ at 50 °C for 2.5 h gives LXXIIa, whereas the trans isomer at −20 °C gives LXXIIb within 1 min [55].

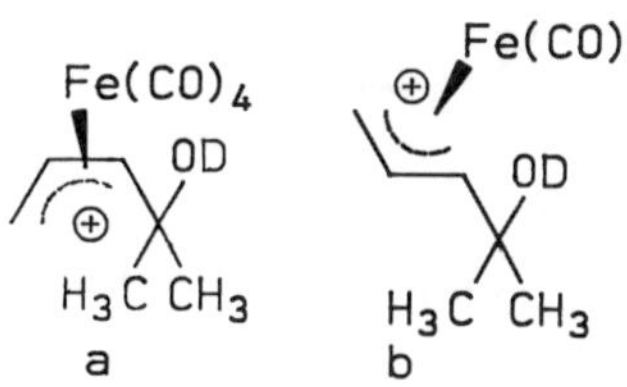

LXXI (a, b) LXXII (a, b)

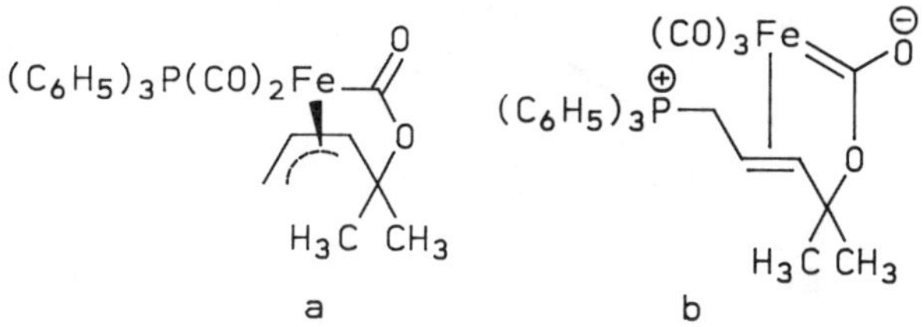

$CH_2CHCHC_6H_{10}OC{=}OFe(CO)_3$ (Table **3**, No. **84**) is oxidized with $[NH_4]_2[Ce(NO_3)_6]$ in acetonitrile at −5 °C to give LXXIII (52%) [69].

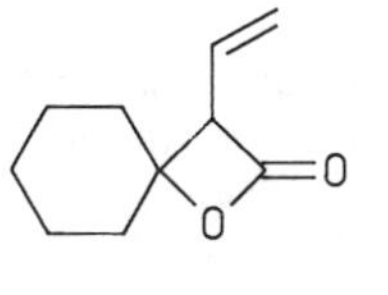

LXXIII 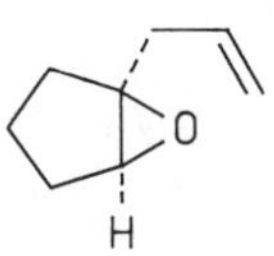 LXXIV

$CH_2CHC_5H_7OC{=}OFe(CO)_3$ (Table **3**, No. **85**). Irradiation of $Fe(CO)_5$ and LXXIV in C_6H_6 [39, 56, 69] as described in Method Id (p. 44) [69] gives after chromatography on Florisil [69] two isomeric ferralactones: the anti isomer LXXVb (11%) and the syn isomer LXXVa (64%) [56, 69]. The physical properties of the two isomers are shown in the following table [69]:

References on pp. 106/8

isomers of No. 85 physical properties

LXXV a

m.p. 96 to 97° (dec.)
^{1}H NMR ($CDCl_3$): 1.52 to 2.38 (m, 4H), 2.42 to 2.78 (m, 3H), 3.49 (dd, 1H; J=2), 4.26 (t, 1H; J=6), 5.34 (dd, 1H; J=8, J=12)
IR: 670, 945, 990, 1010, 1085, 1095, 1375, 1455, 1465, 1655, 1990, 2000, 2075, 2925

LXXV b

m.p. 88 to 89° (dec.)
^{1}H NMR ($CDCl_3$): 1.2 to 1.64 (br,m, 1H), 1.96 to 2.42 (m, 4H), 2.44 to 2.76 (br,m, 1H), 2.99 (dd, 1H; J=3, J=12), 3.5 (dd, 1H; J=9, J=2), 4.22 to 4.44 (br,m, 1H), 4.61 (dd, 1H; J=9, J=12)
IR: 670, 940, 995, 1072, 1115, 1135, 1390, 1455, 1680, 2000, 2075, 2925

The syn isomer LXXVa crystallizes in the triclinic space group $P\bar{1}-C_i^1$, S_2^1 with the unit cell parameters a=13.737(2), b=13.128(2), c=6.733(1) Å; α=85.08(2)°, β=96.15(2)°, γ=107.67(2)°; Z=4 molecules per unit cell. The anti isomer LXXVb crystallizes in the monoclinic space group $P2_1/c-C_{2h}^5$ with the unit cell parameters a=7.617(1), b=12.560(2), c=12.990 Å, β=109.44(2)°; Z=4 molecules per unit cell. The main bond distances are shown in **Fig. 13**.

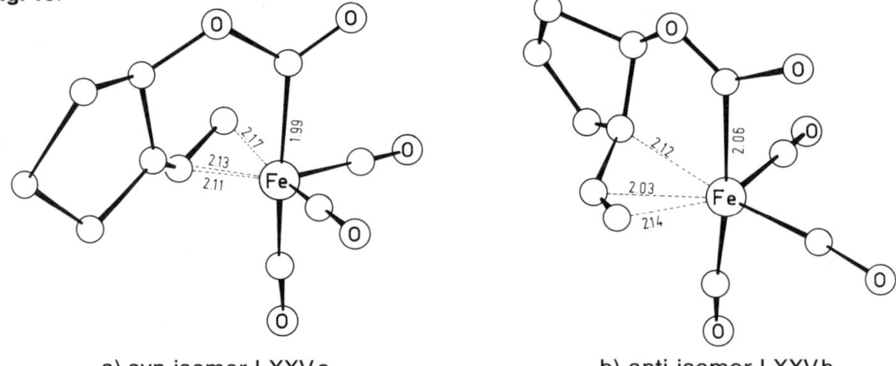

a) syn isomer LXXVa b) anti isomer LXXVb

Fig. 13. Molecular structure of the two isomers of $CH_2CHC_5H_7OC{=}OFe(CO)_3$ (No. 85) [56].

Refluxing the syn isomer in tetrahydrofuran or C_6H_6 for 2.5 h gives LXXVI (38 to 40%) as the only isolable product. However, the anti isomer in tetrahydrofuran affords LXXVI (30%) and the two lactones LXXVII (16%) and LXXVIII (40%). On warming the anti isomer at 70 °C in C_6H_6 for 12 h, LXXVI (7%), LXXVII (9%), LXXVIII (50%), and LXXIX (5%) are isolated [58, 76].

$Fe(CO)_3$ LXXVI LXXVII LXXVIII $Fe(CO)_4$ LXXIX

The syn isomer gives on oxidation with $[NH_4]_2[Ce(NO_3)_6]$ in acetonitrile at −5 °C the organic compounds LXXX (51%) and LXXVII (29%) [39, 69]. The anti isomer gives LXXVII only [69]; a mixture of the syn and anti isomer of No. 85 gives No. 65 on reaction with $C_6H_5CH_2NH_2/ZnCl_2$ in ether at room temperature (48% yield) [59, 62]. Similar reaction in tetrahydrofuran/ether (1:4) for 2.5 h gives a 36% yield. Use of $ClAl(C_2H_5)_2$ (0.75 h) gives a 68% yield [88].

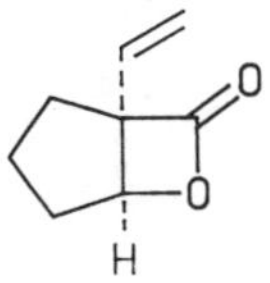

LXXX

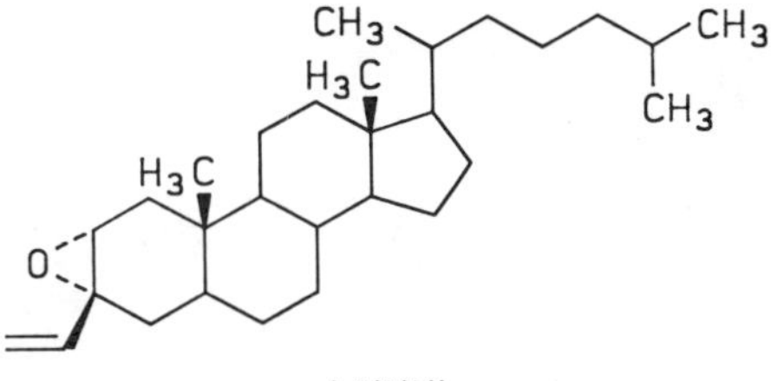

LXXXI

$CH_2CHC_{27}H_{45}OC{=}OFe(CO)_3$ (Table **3**, No. **86**). Irradiation of $Fe(CO)_5$ with LXXXI in C_6H_6 as in Method Id (p. 44) gives after column chromatography the anti isomer LXXXIIb (29%) and the syn isomer LXXXIIa (27%). Their physical properties are shown in the following table [69]:

isomers of No. 86	physical properties
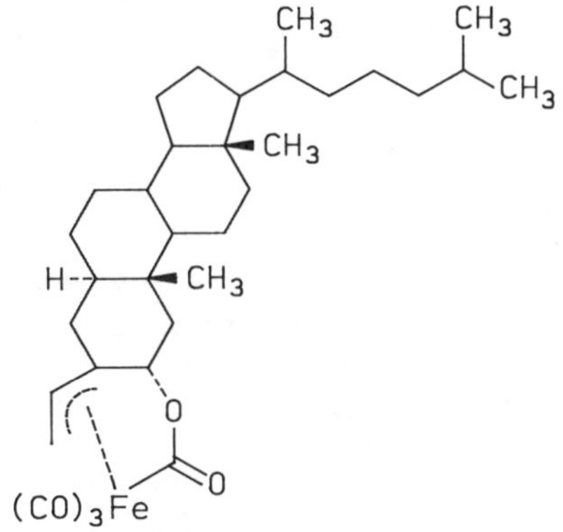 LXXXII a	m.p. 90° (dec., from ether) ^{1}H NMR ($CDCl_3$): 0.9 to 2.3 (m, 38H), 0.76 (s, 3H), 0.94 (s, 3H), 3.0 (dd, 1H; J=12, J=2), 3.68 (dd, 1H; J=9, J=2), 4.4 (dd, 1H; J=12, J=5), 4.76 (dd, 1H; J=12, J=9) IR ($CDCl_3$): 1660, 2000, 2080
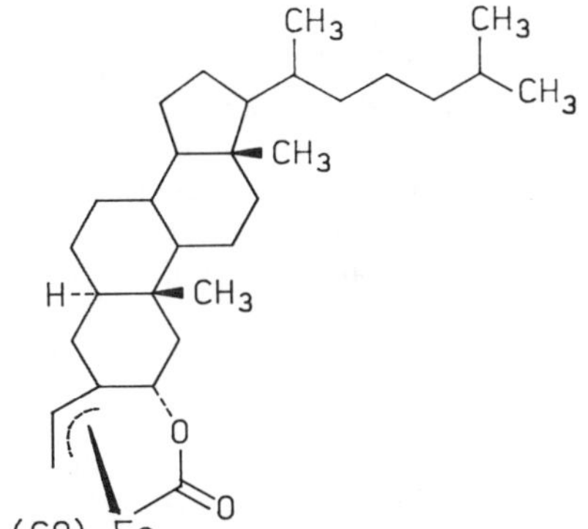 LXXXII b	m.p. 95° (dec.) ^{1}H NMR ($CDCl_3$): 0.9 to 2.2 (m, 38H), 0.66 (s, 3H), 0.83 (s, 3H), 3.05 (dd, 1H; J=13, J=2), 3.57 (dd, 1H; J=10, J=2), 4.39 (dd, 1H; J=3.5, J=12), 4.57 (dd, 1H; J=13, J=10) IR ($CHCl_3$): 1665, 2000, 2080

References on pp. 106/8

Refluxing the syn isomer in tetrahydrofuran for 3 h gives the alcohol LXXXIII (35%) and LXXXIV (24%). Analogous treatment of the anti isomer gives LXXXIII (36%) and LXXXIV (37%) [76]. Oxidation of the syn isomer LXXXIIa with $[NH_4]_2[Ce(NO_3)_6]$ in CH_3CN/C_6H_6 (1:1) at room temperature gives LXXXV (12%) and LXXXIV (27%). Analogous oxidation of the anti isomer LXXXIIb gives LXXXIV (25%).

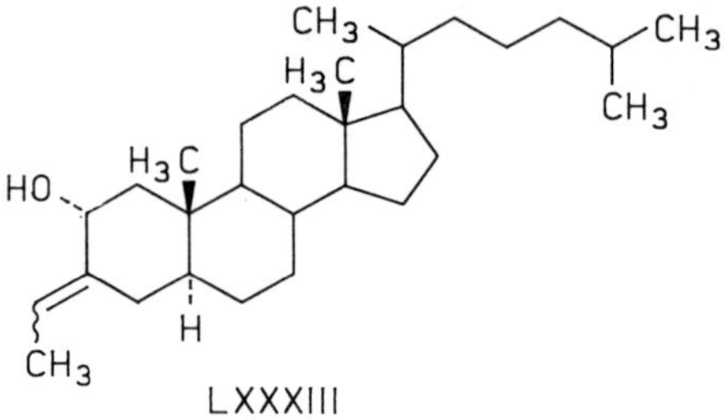

LXXXIII

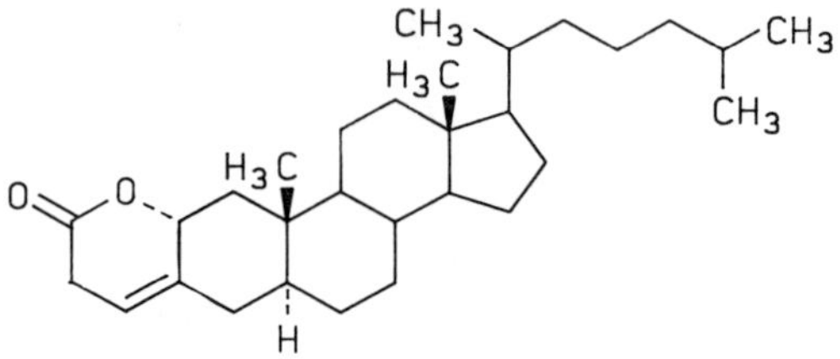

LXXXIV

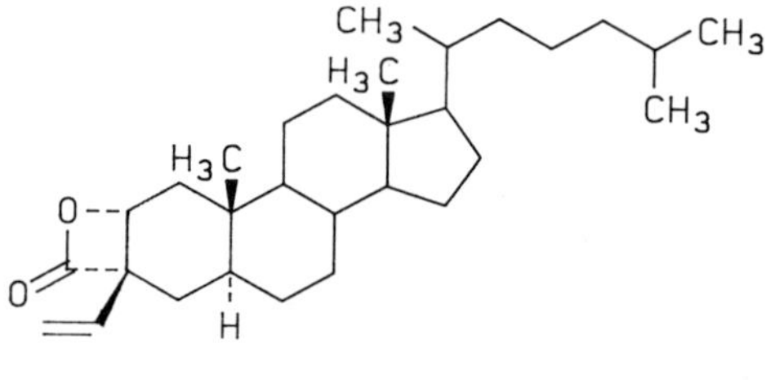

LXXXV

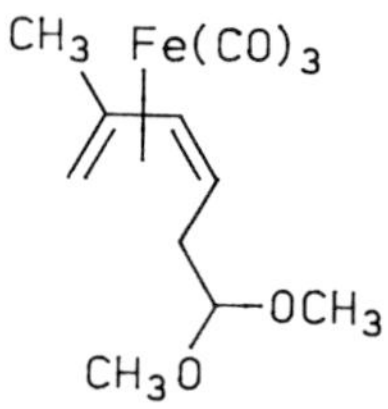

LXXXVI

$CH_3CHCHCHCH(CH_3)OC{=}OFe(CO)_3$ (Table **3**, No. **87**). $Fe(CO)_5$ and LXXXVIIa ($R{=}CH_3$) are irradiated at 20 °C for 6 h. After 2 h further $Fe(CO)_5$ is added, and the irradiation continued for another hour. A 1:2 mixture of the isomers LXXXVIIIa and b is obtained from CH_2Cl_2/pentane (97% yield). Separation is achieved by chromatography on Kieselgel (Merck 100) with ether as eluent. Isomer LXXXVIIIb is first eluted. Irradiation of $Fe(CO)_5$ with LXXXVIIb (contaminated with all in all ca. 15% diene and CH_2Cl_2) in C_6H_6 at 20 °C for 5 h followed by recrystallization from CH_2Cl_2/pentane gives isomer LXXXVIIIc (40%) [55]. According to an earlier report [27] the same isomer is obtained under similar conditions (C_6H_6, 10 °C, 2 h) from $Fe(CO)_5$ and LXXXVII. Thus, reaction of LXXXVIIa or b gives the isomers LXXXVIIIa (93%) or LXXXVIIIb (84%), and a mixture of LXXXVIIc and d gives a mixture of LXXXVIIIa and c in a completely stereospecific manner [27]. This stereospecifity is not confirmed in [55]. Thermal reaction (ether, 36 °C, 2 h) of $Fe_2(CO)_9$ with LXXXVIIa gives LXXXVIIIa and b in a 1:6 ratio (total yield 83%), and with LXXXVIIb gives LXXXVIIIa and c in a 5:1 ratio (total yield 51%) [27]. Reaction of $Fe_2(CO)_9$ with hex-3-en-2,4-dione in petroleum ether at 40 °C followed by filtration (to separate from considerable amounts of $Fe_3(CO)_{12}$), and chromatography on Al_2O_3 with ether gives isomer LXXXVIIIa (by comparison of ^1H NMR spectra) [1]. The physical properties of these isomers are shown in the following table:

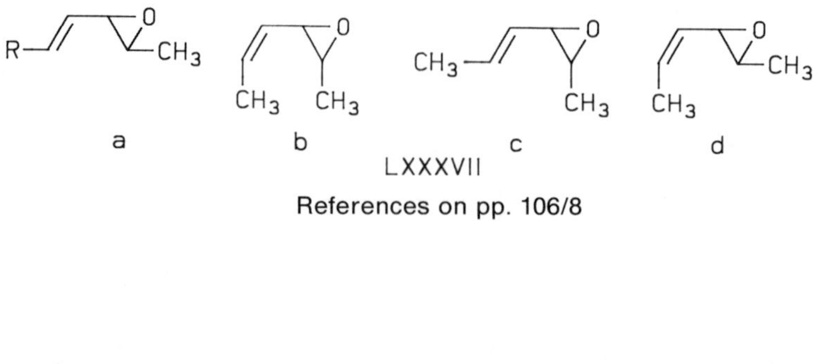

a b c d

LXXXVII

References on pp. 106/8

isomers of No. 87 | physical properties

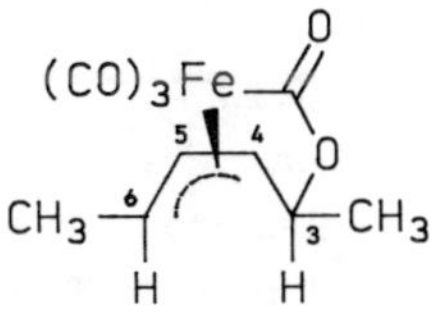

LXXXVIII a

m.p. 85 to 86° [1], m.p. 102 to 103° (dec.) [55], m.p. 108 to 109° [27], colorless needles [55], pale yellow needles (from ether) [1]
^{1}H NMR ($CDCl_3$): 1.35 (CH_3-3; J(H-3, CH_3-3) = 6.5), 1.87 (CH_3-6; J(H-6, CH_3-6) = 6.0), 4.21 (H-6; J(H-5,6) = 12.0), 4.30 (H-3; J(H-3,4) = 1.5), 4.44 (H-4; J(H-4,5) = 7.5), 4.82 (H-5) [55]
^{1}H NMR ($CDCl_3$): 1.32 (CH_3-3; J(H-3, CH_3-3) = 6.5), 1.84 (CH_3-6; J(H-6, CH_3-6) = 6.0), 4.03 (H-6; J(H-5,6) = 12.0), 4.25 (H-3; J(H-3,4) = 1.2), 4.38 (H-4; J(H-4,5) = 7.5), 4.77 (H-5) [27]
^{1}H NMR ($CDCl_3$): 1.27 (CH_3), 1.88 (CH_3), 4.13 (m), 4.28 (m), 4.8 (m) [1]
^{13}C NMR ($CDCl_3$): 20.88, 23.76 (CH_3); 71.49 (C-3), 76.51 (C-4), 79.03 (C-6), 93.52 (C-5); 205.40, 207.74, 210.70 (CO) [55]
IR (KBr): 765, 828, 878, 950, 1000, 1042, 1080, 1092, 1120, 1190, 1318, 1348, 1365, 1390, 1420, 1450, 1460, 1520, 1660, 2890, 2930, 2955, 2980, 3025, 3290 [1]
IR (KBr): 540, 570, 610, 650, 665, 950, 990, 1045, 1070, 1115, 1210, 1235, 1380, 1450, 1650, 2000 [55]
IR (hexane): 2015, 2025, 2080 [1]
IR (hexane): 2006, 2020, 2079 (CO) [55]
IR: 1670 (C=O); 1995, 2005, 2035 (CO) [27]
mass spectrum: $[M]^+$ and fragments given [55]

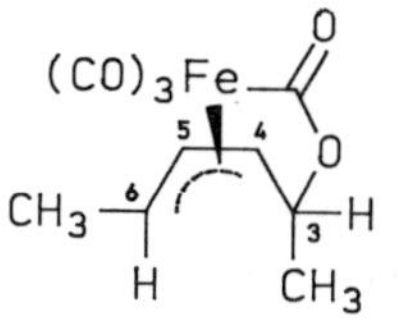

LXXXVIII b

m.p. 115 to 116° (dec.) [55], m.p. 130 to 131° [27], pale yellow crystals
^{1}H NMR ($CDCl_3$): 1.33 (CH_3-3; J(H-3, CH_3-3) = 6.5), 1.85 (CH_3-6; J(H-6, CH_3-6) = 6.0), 4.16 (H-6), 4.39 (H-3), 4.62 (H-4,5) [55]
^{1}H NMR ($CDCl_3$): 1.34 (CH_3-3; J(H-3, CH_3-3) = 6.5), 1.88 (CH_3-6; J(H-6, CH_3-6) = 6.0), 4.14 (H-6), 4.39 (H-3; J(H-3,4) = 4.0), 4.58 (H-4), 4.63 (H-5; J(H-5,6) = 12.0) [27]
^{13}C NMR ($CDCl_3$): 21.19, 21.81 (CH_3); 73.61 (C-3), 77.30 (C-4), 79.62 (C-6), 92.42 (C-5); 203.94, 206.60, 209.14 (CO) [55]
IR (KBr): 580, 610, 655, 940, 995, 1035, 1085, 1350, 1375, 1450, 1655, 2000 [55]
IR (hexane): 2006, 2022, 2079 (CO) [55]
IR: 1672 (C=O); 1990, 2004, 2034 (CO) [27]
mass spectrum: $[M]^+$ and fragments given [55]

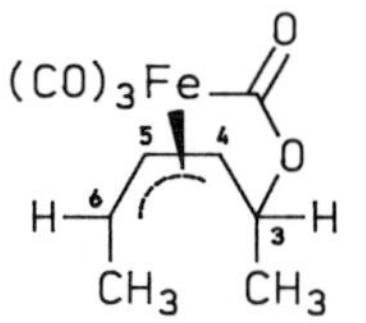

LXXXVIII c

m.p. 91 to 92° (dec.) [55], 103 to 104° [27], colorless plates [55]
^{1}H NMR ($CDCl_3$): 1.43 (CH_3-3; J(H-3, CH_3-3) = 6.5), 1.80 (CH_3-6; J(H-6, CH_3-6) = 7.0), 4.36 (H-5), 4.60 (H-3; J(H-3,4) = 5.0), 4.84 (H-6; J(H-5,6) = 8), 5.18 (H-4; J(H-4,5) = 9) [55]
^{1}H NMR ($CDCl_3$): 1.43 (CH_3-3; J(H-3, CH_3-3) = 6.5), 1.80 (CH_3-6; J(H-6, CH_3-6) = 7.4), 4.31 (H-5), 4.60 (H-3; J(H-3,4) = 5.2), 4.81 (H-6; J(H-5,6) = 9.7), 5.13 (H-4; J(H-4,5) = 9) [27]
^{13}C NMR ($CDCl_3$): 18.46, 22.36 (CH_3); 72.40 (C-3), 77.08 (C-4), 87.54 (C-5,6); 202.70, 206.28, 207.84, 209.14 (CO) [55]
IR (KBr): 590, 610, 660, 800, 870, 940, 980, 1030, 1070, 1085, 1195, 1360, 1420, 1480, 1650, 2000 [55]
IR (hexane): 2006, 2023, 2081 (CO) [55]
IR: 1674 (C=O); 1999, 2008, 2040 (CO) [27]
mass spectrum: $[M]^+$ and fragments given [55]

References on pp. 106/8

The induced shifts for H-6 in LXXXVIII a to c in the presence of $Eu(fod)_3$ (fod = 1,1,1,2,2,3,3-heptafluoro-7,7-dimethyl-4,6-octanedione) are 3.8, 10.0, and 9.9 ppm, respectively. A model in which the shift reagent is coordinated with the lactone ring is assumed [27].

The isomer LXXXVIII a crystallizes in the orthorhombic space group Pbca-D^{15}_{2h}, V^{15}_{h} (No. 61) with the unit cell parameters a = 12.0732(16), b = 15.2272(22), c = 12.2010(15) Å; Z = 8 molecules per unit cell. D_{calc} = 1.576 g/cm. The main bond distances and angles are shown in **Fig. 14** [27, 38]. Note that the complex crystallizes in a space group containing operations of the second kind (i.e., inversion centers and glide planes) and that the crystals therefore contain an ordered racemic array of enantiomers LXXXIX a and b [38].

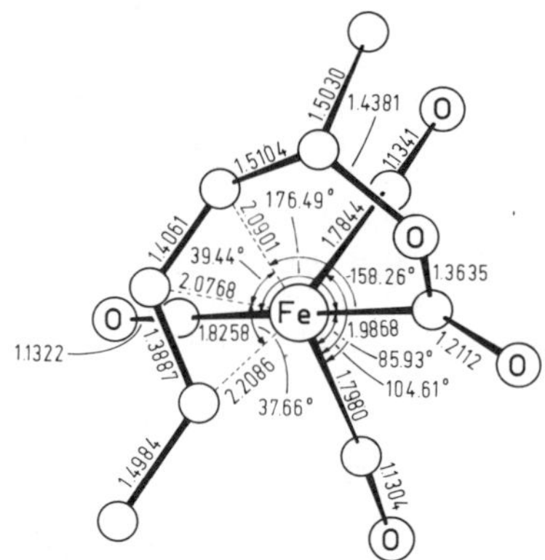

Fig. 14. Molecular structure of isomer LXXXVIII a of No. 87 [38].

The isomer LXXXVIII a is easily soluble in polar solvents; the solutions are air-sensitive; however, crystals are air-stable [1]. Reaction of isomer LXXXVIII b or c with CH_3OH/saturated aqueous $Ba(OH)_2$ (2:1) at 20 °C for 15 h gives syn, syn- or syn, anti-$CH_3CH{=}CHCH{=}CH$-$CH_3Fe(CO)_3$ in 55 or 40% yield, respectively. Reaction of LXXXVIII a with CO (300 bar) in C_6H_6 at 90 °C for 17 h in an autoclave gives LIX (R = R'''' = CH_3, R' = R'' = R''' = H) (p. 86) in 75% yield. Similar reaction in CH_3OH at 75 °C for 15 h gives trans-5-methoxy-hex-3-en-2-ol in 75% yield. The reaction of isomers LXXXVIII b and c with CH_3NH_2 in ether at 20 °C gives isomers XLVIII b and a of No. 60, respectively, in 95% yield each. The mixture is evaporated and the residue is treated with pentane at −20 °C to give the isomers XLVIII a or b. No reaction of LXXXVIII a with CH_3NH_2 is observed [55].

a b

LXXXIX XC

$CH_3O_2CCHCHCHCH(CH_3)OC{=}OFe(CO)_3$ (Table **3**, No. **88**). $Fe(CO)_5$ is irradiated in C_6H_6 in the presence of XC at 20 °C for 6 h. Recrystallization from CH_2Cl_2/pentane gives a 1:2 mixture of XCI a and b (55% yield). Chromatography on Kieselgel (Merck 100) with ether elutes XCI b first. The physical properties are shown in the following table:

isomers of No. 88	physical properties
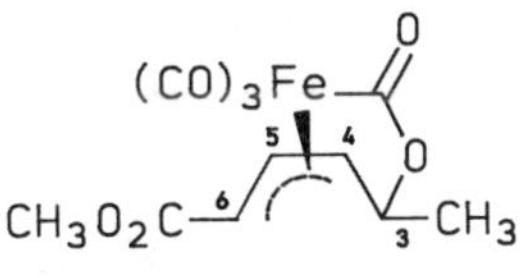 XCI a	m.p. 119 to 120° (dec.), pale yellow needles ^{1}H NMR ($CDCl_3$): 1.42 (CH_3-3; J(H-3, CH_3-3) = 6.5), 3.55 (H-6; J(H-5,6) = 11.0), 3.79 (OCH_3), 4.32 (H-3; J(H-3,4) = 1.0), 4.86 (H-4; J(H-4,5) = 9.0), 5.74 (H-5) ^{13}C NMR ($CDCl_3$): 24.11 (CH_3-3), 52.71 (OCH_3), 58.49 (C-4), 70.84 (C-3), 84.29 (C-6), 95.08 (C-5), 171.25 (C=O), 200.49, 202.96, 204.46, 208.10 (CO) IR (KBr): 560, 575, 605, 615, 655, 932, 950, 1005, 1055, 1085, 1175, 1195, 1355, 1370, 1440, 1500, 1660, 1705, 2000 IR (hexane): 2022, 2043, 2093 (CO) mass spectrum: $[M]^+$ and fragments given
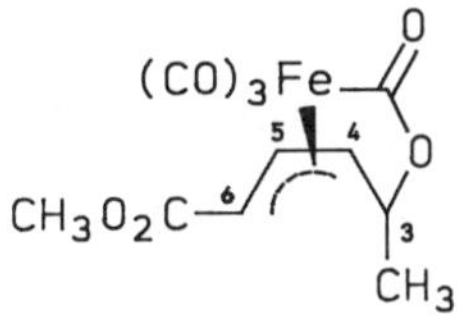 XCI b	m.p. 97 to 98°, yellow crystals ^{1}H NMR ($CDCl_3$): 1.37 (CH_3-3; J(H-3, CH_3-3) = 6.0), 3.65 (H-6; J(H-5,6) = 11.0), 3.81 (OCH_3), 4.53 (H-3; J(H-3,4) = 5.0), 5.17 (H-4; J(H-4,5) = 9.0), 5.61 (H-5) ^{13}C NMR ($CDCl_3$): 21.77 (CH_3-3), 59.01 (C-4), 62.64 (OCH_3), 72.92 (C-3), 85.46 (C-6), 93.72 (C-5), 171.30 (C=O); 200.30, 203.03, 204.78, 207.90 (CO) IR (KBr): 560, 600, 615, 625, 650, 940, 950, 1000, 1020, 1040, 1090, 1120, 1165, 1195, 1325, 1375, 1400, 1435, 1655, 1715, 2000 IR (hexane): 2025, 2043, 2093 (CO) mass spectrum: $[M-CO]^+$ and other fragments

Reaction of isomer XCIb with CO (300 bar) in C_6H_6 at 120 °C in an autoclave for 18 h gives syn,syn-$CH_3O_2CCH{=}CHCH{=}CHCH_3Fe(CO)_3$ (ca. 20% yield) and (E,E)-$CH_3O_2CCH{=}CHCH{=}CHCH_3$ (45%) [55].

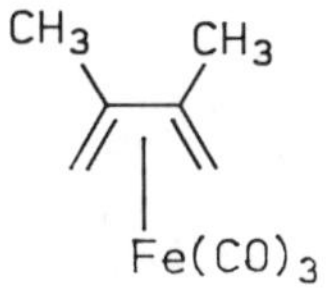

XCII 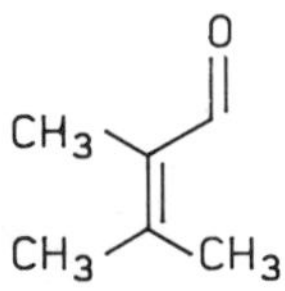 XCIII 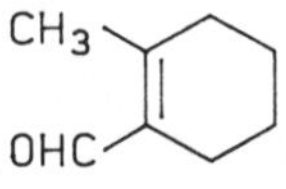 XCIV 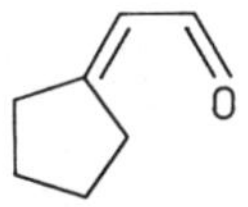 XCV 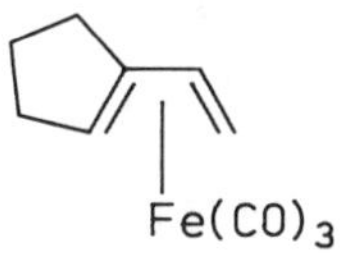 XCVI

$CH_2CRCR'CH_2OC{=}OFe(CO)_3$ (Table **3**, Nos. **89**, **90** with R = R' = CH_3, R,R' = -$(CH_2)_4$-), **$C_5H_7CHCH_2OC{=}OFe(CO)_3$** (Table **3**, No. **91**), and **$C_6H_{10}CHCHCH_2OC{=}OFe(CO)_3$** (Table **3**, No. **92**). Compound No. 89 (R = R' = CH_3) is sparingly soluble in nonpolar solvents but very soluble in polar solvents. The crystals are air-stable but the solutions are air-sensitive [1]. The products and reaction conditions of the thermolysis of all four compounds are shown in the following table:

compound No.	solvent	reaction time (h)/temp.	products (yield in %) remarks	Ref.
89	ether	4/80°	XCII (6.9) + XCIII (59.7)	[76]
90	C_6H_6	4/60°	XCIV (69) after oxidative work-up with $ON(CH_3)_3$	[58, 76]

References on pp. 106/8

compound No.	solvent	reaction time (h)/temp.	products (yield in %) remarks	Ref.
91	C_6H_6	20/60°	XCV (19)	[76]
	tetrahydrofuran	3/reflux	XCVI (54)+XCV (23)	[58, 76]
92	tetrahydrofuran	3/reflux	XCVII (82)	[58, 76]

All four compounds are oxidized by $[NH_4]_2[Ce(NO_3)_6]$ within 2 to 6 h. The reaction products and conditions are shown in the following table:

compound No.	solvent	reaction temp.	products (yield in %)	Ref.
89	C_2H_5OH/H_2O (1:1)	room temp.	LIX (R=R‴=R⁗=H, R′=R″=CH_3) (38)	[39, 69]
	CH_3CN	−5°	XCVIII (42)	[69]
90	CH_3CN	−5°	IC (15)+C (55)	[69]
91	C_2H_5OH	−15°	CI (100)	[69]
92	CH_3OH	−78°	CIIa which is immediately reduced to give CIIb (43.2)	[69]

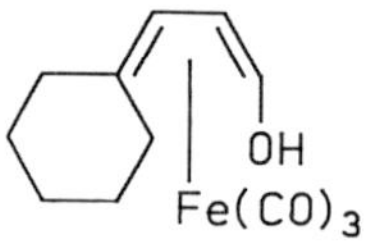

XCVII 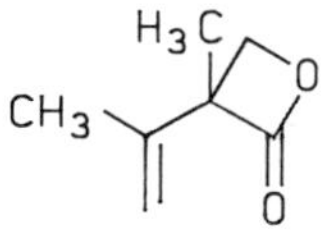 XCVIII 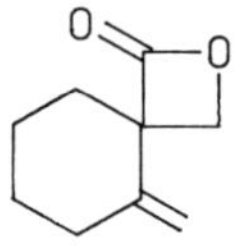 IC 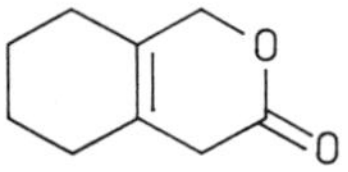 C

Compound No. 89 reacts with $C_6H_5CH_2NH_2/ZnCl_2$ in ether [59, 62] for 1.5 h [88]. Chromatography on Kieselgel (Merck 60) [59, 62] with ether [88] gives No. 64 (75% yield) [59, 62, 88]. Similar reaction of compound No. 91 in tetrahydrofuran/ether (1:4) [59, 62] for 3.0 h followed by chromatography with ether/light petroleum ether (1:1) [88] gives No. 65 (7% [59, 62] or 12% [88]) and No. 66 (39% [59, 62] or 23% [88]) [59, 62, 88]. Reaction of No. 92 with CO (60 atm) in C_6H_6 at 195 °C for 4 h in a bomb affords LIX (R=R′=R″=R‴=H, R⁗=C_5H_{11}), p. 86, in 49.5% yield [76]. Similar reaction of No. 89 with CO (300 bar) in C_6H_6 at 75 °C for 16 h gives a yellow oil (90% yield) consisting of a 1:1 mixture of LIX and LX (R=R‴=R⁗=H, R′=R″=CH_3) [55].

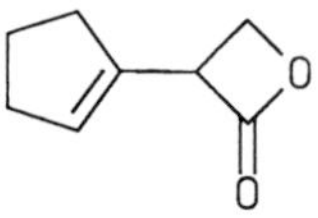

CI CII (a, b) CIII

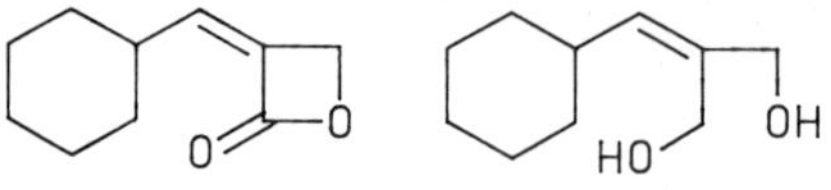

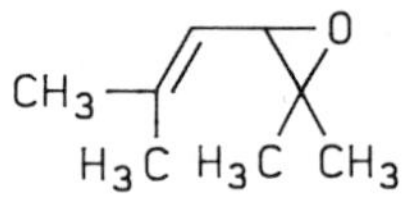

$(CH_3)_2CCHCHC(CH_3)_2OC{=}OFe(CO)_3$ (Table **3**, No. **93**). $Fe(CO)_5$ and CIII are irradiated in C_6H_6 at 20 °C for 6 h. After this time, the reaction mixture is distilled and another portions of CIII and $Fe(CO)_5$ are added. Irradiation is continued for another 7 h. Recrystallization from $CHCl_3$/pentane (1:10) at −20 °C gives a 4:1 mixture of isomers CIVa and b (74%). Slow crystallization of this mixture from C_6H_6/pentane gives pure crystals of CIVb, separated mechanically from CIVa. Heating isomer CIVb in C_6H_6 at 50 °C for 2.5 h gives CIVa; higher temperatures lead to decomposition. To obtain larger amounts of isomer CIVa, a mixture of both isomers is heated in C_6H_6 at 45 °C for several h, followed by concentration and addition of pentane. Cooling to −20 °C gives isomer CIVa in 45% yield. The physical properties of both isomers are shown in the following table [55]:

isomers of No. 93	physical properties
CIVa	m.p. 78 to 80° (dec.), light yellow crystals ^{1}H NMR ($CDCl_3$): 1.49 (CH_3-3′), 1.56 (CH_3-3), 1.95 (CH_3-6′), 2.04 (CH_3-6), 4.21 (H-4; J(H-4,5) = 9.0), 4.86 (H-5) ^{13}C NMR ($CDCl_3$): 24.63, 29.77, 31.78, 34.05 (CH_3); 78.12 (C-3), 86.69 (C-4), 89.10 (C-5), 98.20 (C-6); 203.55, 206.54, 207.97, 210.11 (CO) IR (KBr): 545, 600, 615, 660, 930, 955, 1030, 1065, 1150, 1190, 1260, 1370, 1455, 1645, 2000 IR (hexane): 1997, 2018, 2075 (CO) mass spectrum: $[M]^+$ and fragments given
CIVb	m.p. 82 to 84° (dec.), roundish gold-yellow crystals ^{1}H NMR ($CDCl_3$): 1.55 (CH_3-3), 1.62 (CH_3-6′), 1.66 (CH_3-3′), 2.02 (CH_3-6), 3.36 (H-4; J(H-4,5) = 9.0), 4.52 (H-5) ^{13}C NMR ($CDCl_3$): 26.26, 29.36, 29.79, 30.16 (CH_3); 65.32 (C-4), 81.56 (C-3), 88.19 (C-5), 120.49 (C-6); 204.20, 204.98, 207.25, 210.18 (CO) IR (KBr): 580, 610, 620, 650, 660, 930, 950, 1012, 1065, 1145, 1240, 1315, 1370, 1460, 1535, 1650, 2000 IR (hexane): 1990, 2007, 2069 (CO) mass spectrum: $[M-CO]^+$ and other fragments

Isomer CIVa crystallizes in the monoclinic space group $P2_1/n-C^5_{2h}$ with the unit cell parameters a = 8.806(1), b = 12.547(1), c = 12.110(1) Å, β = 96.32(1)°; Z = 4 molecules per unit cell; D_{calc} = 1.469 g/cm³. Isomer CIVb crystallizes in the same space group with the unit cell parameters a = 10.489(2), b = 12.967(2), c = 10.660(1) Å, β = 69.12(1)°; Z = 4 molecules per unit cell; D_{calc} = 1.442 g/cm³. The main bond distances and angles are shown in **Fig. 15**.

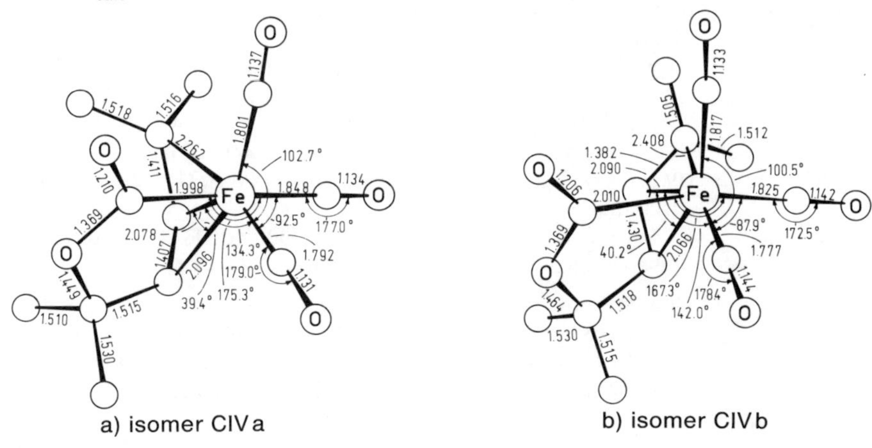

Fig. 15. Molecular structure of the two isomers of No. 93 [55].

References on pp. 106/8

Reaction of isomer CIVa with CO (300 bar) in C_6H_6 at 75 °C for 5 h in an autoclave gives 2,5-dimethylhexa-2,4-diene (89% yield). Similar reaction of this isomer in CH_3OH with CO (220 bar) at 70 °C for 16 h affords (E)-$CH_3O(CH_3)_2CCH{=}CHC(CH_3)_2OH$ in 90% yield [55].

$CH_2CRCRCH_2C(CF_3){=}C(CF_3)Fe(CO)_3$ (Table **3**, Nos. **97**, **99** with R = H or CH_3). Refluxing compound No. 97 in hexane [25] or heptane [40] for 24 h gives CV (R = H) [40]. Refluxing No. 99 in hexane for 4 d gives CV (R = CH_3) [25, 40]. Reaction of No. 99 with CO (50 atm) in hexane at 80 °C for 4 d in an autoclave gives CVIa. Formation of isomer CVIb is excluded by comparison of ^{1}H NMR spectra with that of No. 99 [40].

CV CVI (a, b) CVII

$CH_2CHCH(CH_2)_2CH_2Fe(CO)_3$ (Table **3**, No. **100**) is obtained by irradiation of $^4LFe(CO)_3$ (^4L = butadiene) and $CH_2{=}CH_2$. Oxidation of No. 100 with Ce^{IV}, Fe^{III}, Cu^{II}, or O_2 gives CVII in moderate yield [22].

$CH_2CHCH(CH_2)_2CHCO_2CH_3Fe(CO)_3$ (Table **3**, No. **101**). Irradiation of a mixture of $Fe(CO)_5$, $CH_2{=}CHCO_2CH_3$, and butadiene (molar ratio 1:5:5) in ether for 13 h, followed by evaporation gives a mixture of compounds. Sublimation gives $^4LFe(CO)_3$ (^4L = butadiene) in 52.8% yield, and chromatography of the residue on Kieselgel (Merck 60) with ether/C_6H_6 (1:9) gives a mixture of isomers No. 101. Recrystallization from hexane at −78 °C gives pure CVIIIa. The mother liquor contains CVIIIa and b in the ratio ≈40:60 [37, 42]. For a discussion of the reaction path see [37, 42, 60]. Similar irradiation of $^2LFe(CO)_4$ ($^2L = CH_2{=}CHCO_2CH_3$) or $^4LFe(CO)_3$ (^4L = butadiene) in the presence of butadiene or $CH_2{=}CHCO_2CH_3$, respectively, gives also No. 101. Warming of $^2L_2Fe(CO)_3$ up to 20 °C in the presence of butadiene gives also No. 101 [18]. The physical properties of the two isomers of No. 101 are shown in the following table [37]:

isomers of No. 101	physical properties
CVIIIa	m.p. 61 to 64°, yellow crystals (from hexane) monomeric in C_6H_6 (by cryoscopy) ^{1}H NMR (benzene-d_6): 1.07 (H-3′), 1.19 (H-3), 1.80 (H-1), 1.94 (H-2), 2.02 (H-6′; J(H-5,6′) = 13.5), 2.10 (H-2′), 3.07 (H-6; J(H-6,6′) ≲ 2.0), 3.60 (CH_3), 4.07 (H-5; J(H-5,6) = 8.0), 4.47 (H-4; J(H-4,5) = 8.0) IR (n-hexane): 1705 (C=O); 1996, 2014, 2069 (CO) UV (ether): $\lambda_{max}(\varepsilon)$ = 30000 sh (1370), 40000 sh (11650), 48500 (20100) cm^{-1} UV (n-hexane): $\lambda_{max}(\varepsilon)$ = 30000 sh (1300), 40000 sh (12700) cm^{-1} mass spectrum: $[M - nCO]^+$ (n = 0 to 3)
CVIIIb	all data from mixtures with CVIIIa: ^{1}H NMR (benzene-d_6): 1.38 (H-6′), 3.02 (H-6), 3.58 (CH_3), 3.92 (H-5), 4.1 to 4.4 (H-4) IR (n-hexane): 1701 (C=O); 1995, 2006, 2065 (CO)

Isomer CVIIIa crystallizes in the orthorhombic space group Pbca-D_{2h}^{15}, V_h^{15} with the unit cell parameters a = 7.4525(7), b = 15.786(1), c = 21.020(2) Å; Z = 8 molecules per unit cell. D_{calc} = 1.505 g/cm³. The main bond distances and angles are shown in **Fig. 16**. The X-ray structure confirms that the allyl group is connected with the chain by the anti position at C-4 in CVIIIa. Isomer CVIIIb differs from isomer CVIIIa by exchange of H-1 with the CO_2CH_3 group as shown by the ^{1}H NMR spectra [37].

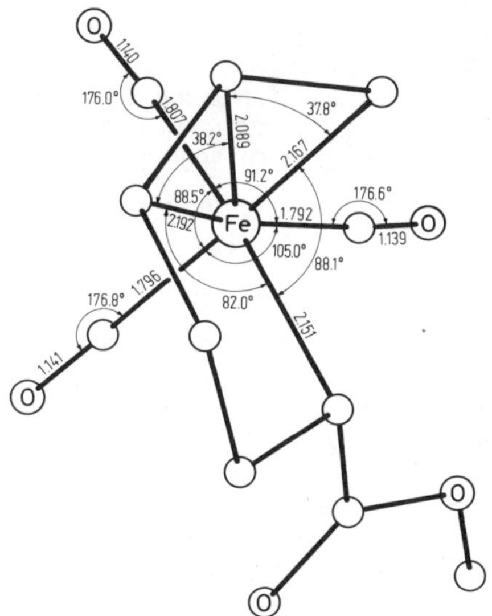

Fig. 16. Molecular structure of $CH_2CHCH(CH_2)_2CHCO_2CH_3Fe(CO)_3$ (No. 101, isomer CVIIIa) [37].

$CH_2CHCHCH_2CH(CO_2CH_3)CH_2Fe(CO)_3$ (Table **3**, No. **102**) is prepared from $^4LFe(CO)_2{}^2L$ (^4L = butadiene, $^2L = CH_2{=}CHCO_2CH_3$) and CO [33].

$CH_2CHCH(CH_2)_2COFe(CO)_3$ (Table **3**, No. **103**) is stable at 25 °C for several hours in hexane solution saturated with CO [12]. Warming crystals of No. 103 at 25 °C for 1 h gives No. 52 [12], cf. [41]. Flash distillation of No. 103 at 15 °C/0.0001 Torr gives No. 52 in almost quantitative yield. Bubbling N_2 through a solution of No. 103 at 25 °C gives No. 52 within seconds. Treating a solution of No. 103 with air or CO (20 atm) gives cyclohex-2-en-1-one in 80% yield [12], cf. [68].

$CH_2CHC(CH_3)(CH_2)_2CH(CO_2CH_3)Fe(CO)_3$ (Table **3**, No. **104**). A mixture of $Fe(CO)_5$, $CH_2{=}CHCO_2CH_3$, and isoprene (molar ratio ca. 1:3.6:3.6) is irradiated in ether for 18 h as described for No. 101. Sublimation gives $^4LFe(CO)_3$ (^4L = isoprene) in 45.9% yield. Chromatography of the residue gives a mixture of isomers of Nos. 104 and 105, namely CIXb/CXb in 11.1% yield, and CIXa/CXb (ratio ca. 60:40) in 39.2% yield. Recrystallization of the latter mixture from hexane at −30 °C gives pure isomer CIXa. The physical properties of the two isomers of No. 104 are shown in the following table [37]:

 References on pp. 106/8

isomers of No. 104	physical properties
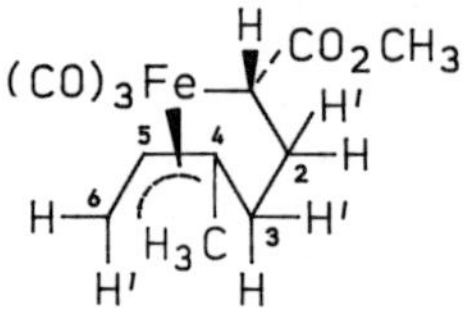 CIX a	m.p. 67° monomeric in C_6H_6 (by cryoscopy) ^{1}H NMR (benzene-d_6): 1.02 (H-3′), 1.40 (H-3), 1.47 (CH_3-4), 1.70 (H-1 or 2), 1.75 (H-1 or 2), 1.78 (H-6′; J(H-5,6′) = 12.0), 2.32 (H-2′), 2.81 (H-6; J(H-6,6′) $\lessapprox$ 1.8), 3.60 (OCH_3), 3.98 (H-5; J(H-5,6) = 8.0) IR (n-hexane): 1707 (C=O); 1992, 2010, 2066 (CO) UV (ether): $\lambda_{max}(\varepsilon)$ = 30000 sh (1490), 39500 sh (9060) cm^{-1} UV (n-hexane): $\lambda_{max}(\varepsilon)$ = 30000 sh (1320), 39500 sh (10400) cm^{-1}
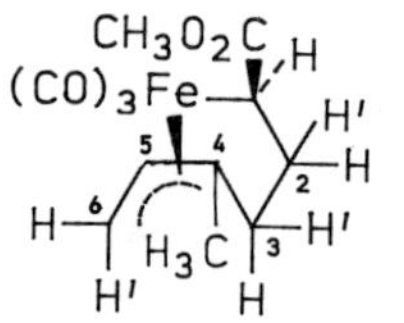 CIX b	mass spectrum: $[M-nCO]^+$ (n = 0 to 3) all data from mixtures with isomer CXb of No. 105: ^{1}H NMR (benzene-d_6): 1.53 (CH_3-4), 3.58 (OCH_3), 3.92 (H-5) IR (n-hexane): 1698 (C=O); 1988, 2001, 2061 (CO)

$CH_2C(CH_3)CH(CH_2)_2CH(CO_2CH_3)Fe(CO)_3$ (Table **3**, No. **105**). For the preparation see "Further information" for compound No. 104. The physical properties of the two isomers CXa and b are given in the following table [37]:

isomers of No. 105	physical properties
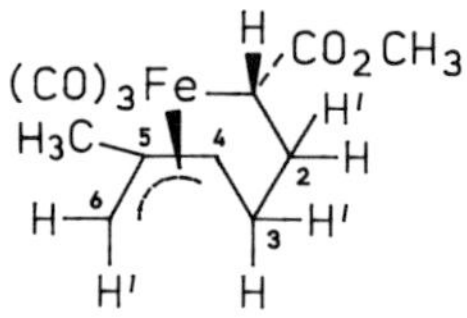 CXa	all data from mixtures with isomer CIXa of No. 104: ^{1}H NMR (benzene-d_6): 1.33 (CH_3-5), ca. 1.85 (H-6′), 3.08 (H-6), 3.65 (OCH_3), ca. 4.40 (H-4) IR (n-hexane): 1707 (C=O); 1992, 2010, 2066 (CO)
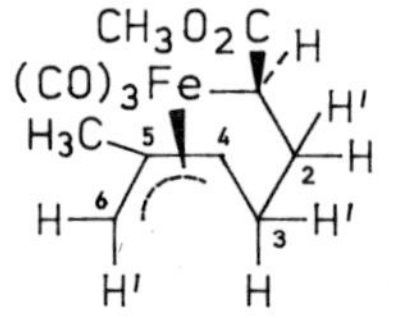 CXb	all data from mixtures with isomer CIXb of No. 104: ^{1}H NMR (benzene-d_6): 1.37 (CH_3-5), 3.59 (OCH_3), 4.0 to 4.3 (H-4) IR (n-hexane): 1698 (C=O); 1988, 2001, 2061 (CO)

$CH_2C(CH_3)C(CH_3)(CH_2)_2CH(CO_2CH_3)Fe(CO)_3$ (Table **3**, No. **106**). Irradiation of a mixture of $Fe(CO)_5$, CH_2=$CHCO_2CH_3$, and 2,3-dimethylbutadiene in ether for 10 h as described for No. 101 gives $^4LFe(CO)_3$ (^4L = 2,3-dimethylbutadiene) after sublimation (64% yield), and a mixture of isomers CXIa and b (32.8%) and CXII (2.6%) by chromatographic work-up. Recrystallization of the mixture of isomers from hexane at −78 °C gives pure isomer CXIa. The mother liquor contains isomer CXIa and b in the ratio ≈55:45 [42], cf. [57, 66]. Stirring a mixture of $^2L_2Fe(CO)_3$ (^2L = CH_2=$CHCO_2CH_3$) and 2,3-dimethylbutadiene in hexane in the dark at 20 °C for 20 h, followed by evaporation and chromatography of the residue gives $^4LFe(CO)_3$ (^4L = 2,3-dimethylbutadiene) in 5% yield, $^2LFe(CO)_4$ (^2L = CH_2=$CHCO_2CH_3$) in 17% yield and a mixture of isomers CXIa and b (19.3%) [37, 47]. An isomeric mixture of No. 106 is also obtained by the reaction of $^4LFe(CO)_2{}^2L$ (^4L = 2,3-dimethylbutadiene, ^2L =

CH_2=$CHCO_2CH_3$) with CO at ≧ −30 °C. The ratio of the isomers is temperature dependent [65]. Complex CXIII is converted into a ca. 5 to 10:1 mixture of CXIa and b on standing in hexane under CO. At ambient temperature the reaction is complete after 4 to 5 h; at 0 °C, 1.5 d are required [9]. The physical properties of both isomers CXIa and b are shown in the following table [37]:

isomers of No. 106	physical properties
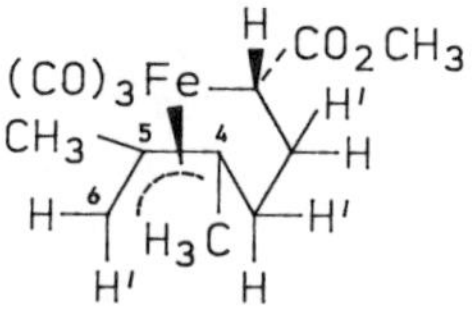 CXI a	m.p. 71 to 72° monomeric in C_6H_6 (by cryoscopy) ^{1}H NMR (benzene-d_6): 0.63 (H-3'; J(H-3,3') = 12.5), 1.44 (H-3; J(H-2,3) = 2.2, J(H-2,3') = 3.6), 1.47 (CH_3), 1.52 (CH_3), 1.79 (H-1; J(H-1,2') = 12.0, J(H-1,2) = 4.3), 1.89 (H-6'; J(H-6,6') = 1.8), 1.93 (H-2; J(H-2,2') = 13.0), 2.75 (H-2'; J(H-2',3) = 5.2, J(H-2',3') = 14.0), 3.03 (H-6), 3.68 (OCH_3) IR (n-hexane): 1709 (C=O); 1985, 2007, 2062 (CO) UV (ether): $\lambda_{max}(\varepsilon)$ = 29500 (1700), 38500 sh (11500) cm^{-1} UV (n-hexane): $\lambda_{max}(\varepsilon)$ = 30000 (1600), 38500 sh (10950) cm^{-1} mass spectrum: $[M - nCO]^+$ (n = 0 to 3)
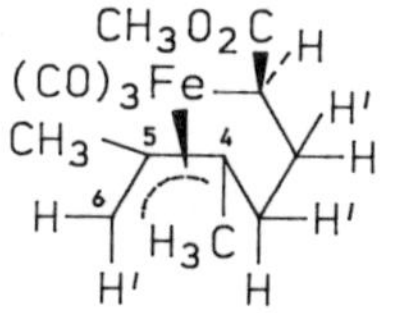 CXI b	all data from mixtures with CXIa: ^{1}H NMR (benzene-d_6): 1.60 (CH_3-4,5), 1.70 (H-6'; J(H-6,6') ≲ 1.9), 2.70 (H-6), 3.50 (OCH_3) IR (n-hexane): 1694 (C=O); 1985, 2000, 2058 (CO)

For the IR and UV spectra of isomer CXIa in Ar, Kr, Xe, N_2, ^{12}CO, and ^{13}CO matrices see [67]. Photolysis of the two isomers in n-hexane results in the interconversion of the two isomers and in the cleavage of the initially formed C-C bond leading to $^4LFe(CO)_3$ (^4L = 2,3-dimethylbutadiene) and CH_2=$CHCO_2CH_3$. The quantum yields are in the range of 0.03 (for No. 106 → $^4LFe(CO)_3$), 0.07 (for CXIa → b), and 0.20 (for CXIb → a). $^4LFe(CO)_2{}^2L$ (^4L = 2,3-dimethylbutadiene, ^2L = CH_2=$CHCO_2CH_3$) is the key intermediate in these processes [9, 57, 66, 70]. Irradiation of CXIa in ether at −30 °C for 14.5 h and removing the CO by an Ar stream gives $^4LFe(CO)_2{}^2L$ [9]. Photochemical reactions of the isomers of No. 106 in inert (Ar, Kr, Xe) and reactive (CO, N_2) matrices at 10 to 12 K show that detachment of CO is the only primary photoreaction. Subsequent reactions depend on the orientation of the CO_2CH_3 group in the two isomers: Either the CO_2CH_3 group is photoreversibly coordinated or the ejected CO is recaptured with no isomerization (CXIa ⇆ b) seen. Effective, stereospecific incorporation of a ligand from the matrix environment is observed for the CXIb isomer in experiments with ^{13}CO and N_2 matrices. From these results it becomes evident that isomerization and C-C bond cleavage observed in solution are due to secondary thermal reactions of the primary photoproducts [66, 67]. Reaction of No. 106 with CO (50 bar) at 70 °C gives CXIV [65].

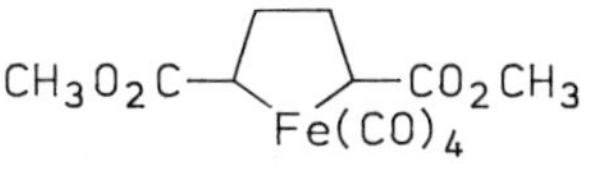

CXII 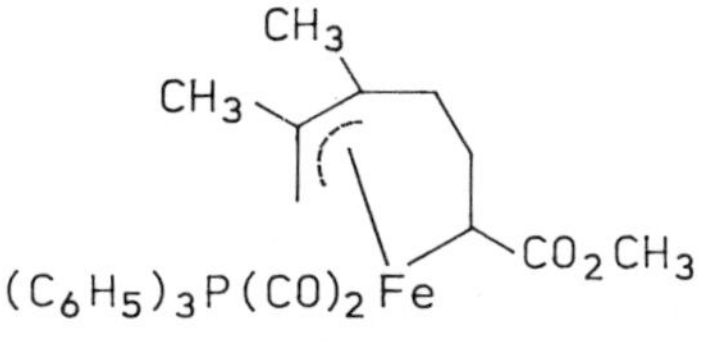 CXIII 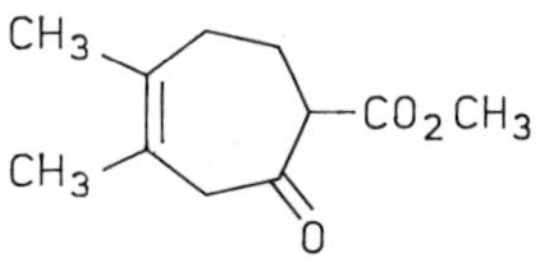 CXIV

References on pp. 106/8

$CH_2C(CH_3)CHCH_2CF_2CFXFe(CO)_3$ (Table **3**, Nos. **111**, **112** with X = F or CF_3). Compound No. 112 (X = CF_3) crystallizes in the orthorhombic space group $P2_12_12_1$-D_2^4, V^4 with the unit cell parameters a = 23.020(12), b = 8.497(5), c = 6.546(4) Å; Z = 4 molecules per unit cell. D_{calc} = 1857 kg/m. The main bond distances and angles are given in **Fig. 17** [19]. This clearly shows that structure CXV as first postulated [5] is wrong.

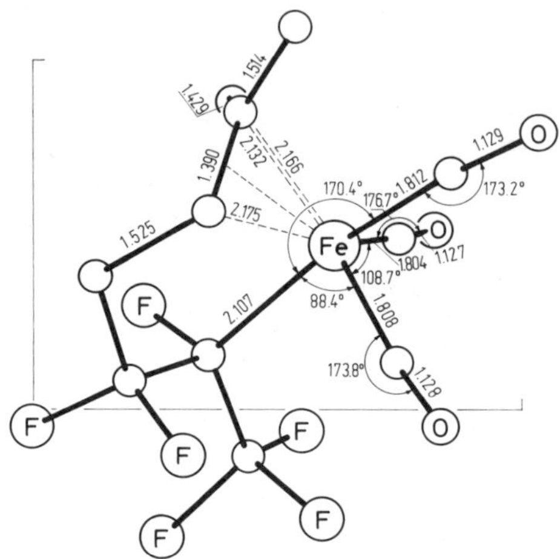

Fig. 17. Molecular structure of $CH_2C(CH_3)CHCH_2CF_2CF(CF_3)Fe(CO)_3$ (No. 112) [19].

Compound No. 111 (X = F) reacts with $P(OCH_3)_3$ in boiling hexane to give $CH_2C(CH_3)CHCH_2CF_2CF_2Fe(CO)_2P(OCH_3)_3$ (see Formula CXVI) [24].

CXV CXVI CXVII

$(CH_2)_2CCH_2OC{=}OFe(CO)_3$ (Table **3**, No. **119**) reacts with NH_3 in ether in the presence of Al_2O_3 at room temperature to give CXVII (R = H). Similar reaction with $C_6H_5NH_2$ gives No. 118 in 65% yield [31].

CXVIII CXIX a b

$C_6H_5CHCHC(R')N(R'')C{=}OFe(CO)_3$ (Table **3**, Nos. **123** to **126** with R'' = H, R' = C_6H_5; R'' = CH_3, R' = C_3H_7-i, C_4H_9-n, C_4H_9-t, and Nos. **130** to **135** with R'' = C_3H_7-i, R' = C_2H_5; R'' =

C_4H_9-n, C_4H_9-i, C_4H_9-t, R'=CH_3; R''=C_6H_{11}-cyclo, R'=C_2H_5, C_4H_9-n). Reversible isomerization to the tautomers XXXIX (R=C_6H_5, p. 83) occurs for all complexes in nonpolar solvents. The following table shows the tautomeric equilibrium constant K_T for these complexes in hexane, as determined from the integrated intensities of the corresponding IR absorption bands of XXXIX and XL (R=C_6H_5, p. 83):

No.	$C_6H_5CHCHC(R')N(R'')C{=}OFe(CO)_3$ R'	R''	K_T
123	C_6H_5	H	—
124	C_3H_7-i	CH_3	<0.05
125	C_4H_9-n	CH_3	0.12±0.02
126	C_4H_9-t	CH_3	<0.05
130	C_2H_5	C_3H_7-i	0.35±0.03
131	CH_3	C_4H_9-n	0.79±0.02
132	CH_3	C_4H_9-i	2.33±0.11
133	CH_3	C_4H_9-t	—
134	C_2H_5	C_6H_{11}-cyclo	0.36±0.03
135	C_4H_9-n	C_6H_{11}-cyclo	0.63±0.02

The IR spectra of hexane solutions of Nos. 123, 125, 132, and 135 also show bands of another isomer XLI (R=C_6H_5, p. 83) and of its decarbonylation product CXVIII. The isomerization depends on the polarity of the solvent and the bulkiness of R' and/or R'' [90].

$CH_2C(CH_3)CHCH(CH_2C(OCH_3)_2H)N(CH(CH_3)C_6H_5)C{=}OFe(CO)_3$ (Table **3**, No. **137**) exists in two diastereoisomeric forms CXIXa and b. Oxidation of CXIXa with $[NH_4]_2[Ce(NO_3)_6]$ in CH_3OH at −30 °C gives the cis-β-lactam CXX (87% yield). Analogous reaction of CXIXb gives the antipode of CXX [89].

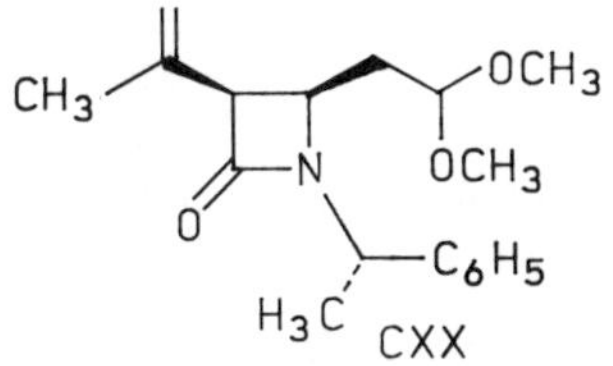

CXX 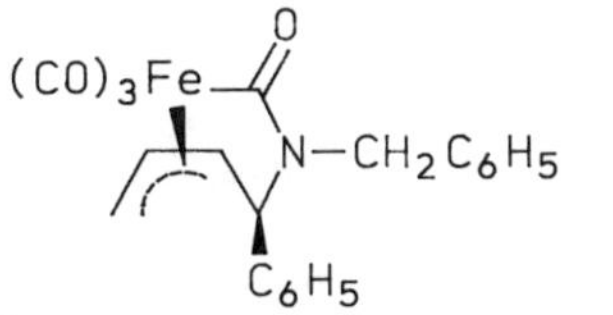 CXXI 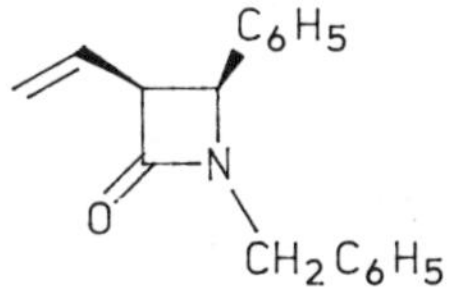 CXXII

$CH_2CHCHCH(C_6H_5)N(CH_2C_6H_5)C{=}OFe(CO)_3$ (Table **3**, No. **138**) is isolated as isomer CXXI. Oxidation with $[NH_4]_2[Ce(NO_3)_6]$ in C_2H_5OH at −30 °C to room temperature gives CXXII (78% yield); a small amount of 1-benzyl-3,6-dihydro-6-phenylpyridone (<1%) is also isolated from this reaction but not fully characterized [88].

$C_6H_5CHCHCHCH_2OC{=}OFe(CO)_3$ (Table **3**, No. **139**). A solution of excess $Fe(CO)_5$ and CXXIII in C_6H_6 is irradiated for 12 min. The green reaction mixture is filtered, frozen, then

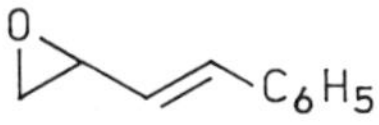

CXXIII 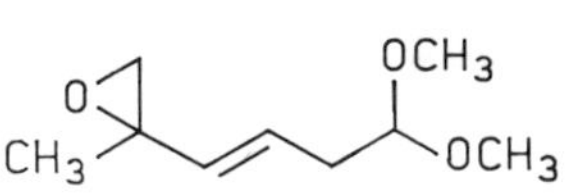 CXXIV 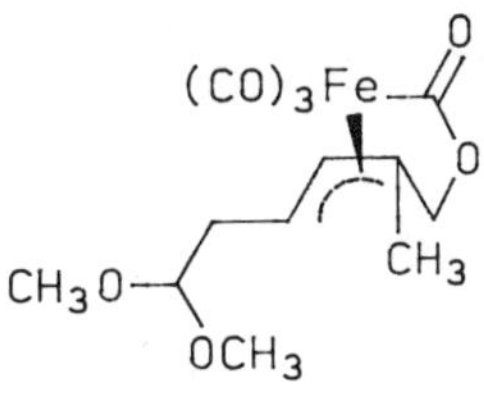 CXXV

References on pp. 106/8

freeze-dried on a rotary evaporator whereby the volatiles are trapped on a cold (−78 °C) condensor. The residue is stirred with ether and filtered through Celite. Removal of the solvent gives No. 139 in 68% yield. Treatment with $ClAl(C_2H_5)_2$ and $C_6H_5CH_2NH_2$ in hexane for 0.5 h followed by chromatography on silica gel (Merck 9385) with ether/light petroleum ether (3:1) gives No. 138 in 80% yield [88].

$H(CH_3O)_2CCH_2CHCHC(CH_3)CH_2OC{=}OFe(CO)_3$ (Table **3**, No. **140**). Irradiation of excess $Fe(CO)_5$ with CXXIV in C_6H_6 gives isomer CXXV of No. 140 in 90% yield. Alternatively, a more convenient procedure involves reaction of CXXIV with excess $Fe_2(CO)_9$ in tetrahydrofuran at room temperature to give CXXV in 84% yield. The syn, anti configuration of CXXV is assigned on the basis of the observed coupling constant of 12 Hz. Insertion of (−)-(1S)-α-methylbenzylamine into CXXV mediated by $ZnCl_2 \cdot HN(CH_2CH_2)_2NH$ proceeds slowly (7 h) to give the two isomers CXIXa (29%) and b (30%) of No. 137. This reaction is also accompanied by a small amount (16%) of LXXXVI (p. 94). The isomers CXIXa and b are readily separated by chromatography [89].

References:

[1] H.D. Murdoch (Helv. Chim. Acta **47** [1964] 936/41). — [2] R.F. Heck, C.R. Boss (J. Am. Chem. Soc. **86** [1964] 2580/3). — [3] R.F. Heck (unpublished results from [4]). — [4] J.P. Candlin, K.P. Taylor, D.T. Thompson (Reactions of Transition Metal Complexes, Elsevier, New York 1968, pp. 1/483, 135). — [5] A. Bond, M. Green, B. Lewis, S.F.W. Lowrie (Chem. Commun. **1971** 1230/1).

[6] A.E. Hill, H.M.R. Hoffmann (J. Chem. Soc. Chem. Commun. **1972** 574/5). — [7] D.H. Gibson, R.L. Vonnahme (J. Am. Chem. Soc. **94** [1972] 5090/2). — [8] R. Aumann (J. Organometal. Chem. **47** [1973] C29/C32). — [9] T. Akiyama, F.-W. Grevels, J.G.A. Reuvers, P. Ritterskamp (Organometallics **2** [1983] 157/60). — [10] A.S. Batsanov, Yu.T. Struchkov (J. Organometal. Chem. **248** [1983] 101/6).

[11] Y. Becker, A. Eisenstadt, Y. Shvo (Tetrahedron **30** [1974] 839/47). — [12] R. Aumann (J. Am. Chem. Soc. **96** [1974] 2631/2). — [13] R. Aumann, K. Fröhlich, H. Ring (Angew. Chem. **86** [1974] 309/10). — [14] T.H. Whiteside, R.W. Arhart (unpublished results from [15]). — [15] M. Brookhart, D.L. Harris (Inorg. Chem. **13** [1974] 1540/1).

[16] S. Sarel, A. Felzenstein, R. Victor, J. Yovell (J. Chem. Soc. Chem. Commun. **1974** 1025/6). — [17] T.H. Whitesides, R.W. Slaven (J. Organometal. Chem. **67** [1974] 99/108). — [18] F.-W. Grevels, U. Feldhoff, K. Schneider (7th Intern. Conf. Organometal. Chem., Venice 1975, Abstr. Papers No. 192). — [19] M. Green, B. Lewis, J.J. Daly, F. Sanz (J. Chem. Soc. Dalton Trans. **1975** 1118/27). — [20] M. Green, B. Lewis (unpublished results from [24]).

[21] T.H. Whitesides, R.W. Arhart (Inorg. Chem. **14** [1975] 209/12). — [22] T.H. Whitesides, J. Shelly (Abstr. Papers 169th Natl. Meeting Am. Chem. Soc., Philadelphia 1975, ORGN-86 from [34]). — [23] M. Bottrill, R. Goddard, M. Green, R.P. Hughes, M.K. Lloyd, B. Lewis, P. Woodward (J. Chem. Soc. Chem. Commun. **1975** 253/5). — [24] A. Bond, B. Lewis, M. Green (J. Chem. Soc. Dalton Trans. **1975** 1109/18). — [25] R. Davis, M. Green, R.P. Hughes (J. Chem. Soc. Chem. Commun. **1975** 405/6).

[26] D.H. Gibson, D.K. Erwin (J. Organometal. Chem. **86** [1975] C31/C33). — [27] K.-N. Chen, R.M. Moriarty, B.G. DeBoer, M.R. Churchill, H.J.C. Herman (J. Am. Chem. Soc. **97** [1975] 5602/3). — [28] P.S. Braterman (unpublished results from [29]). — [29] P.S. Braterman (J. Organometal. Chem. **123** [1976] 75/204, 127). — [30] M. Brookhart, T.H. Whitesides, J.M. Crockett (Inorg. Chem. **15** [1976] 1550/4).

[31] Y. Becker, A. Eisenstadt, Y. Shvo (Tetrahedron **32** [1976] 2123/6). — [32] T.H. Whitesides, J.P. Neilan (J. Am. Chem. Soc. **98** [1976] 63/73). — [33] D. Schultz, F.-W. Grevels, E.A. Koerner von Gustorf (unpublished results from [34]). — [34] R.C. Kerber, E.A. Koerner von Gustorf (J. Organometal. Chem. **110** [1976] 345/57). — [35] G.A. Olah, G. Liang (J. Org. Chem. **41** [1976] 2659/61).

[36] G.A. Olah, G. Liang, S.H. Yu (J. Org. Chem. **41** [1976] 2227/8). — [37] F.-W. Grevels, U. Feldhoff, J. Leitich, C. Krüger (J. Organometal. Chem. **118** [1976] 79/92). — [38] M.R. Churchill, K.-N. Chen (Inorg. Chem. **15** [1976] 788/92). — [39] G.D. Annis, S.V. Ley (J. Chem. Soc. Chem. Commun. **1977** 581/2). — [40] M. Bottrill, R. Davies, R. Goddard, M. Green, R.P. Hughes, B. Lewis, P. Woodward (J. Chem. Soc. Dalton Trans. **1977** 1252/61).

[41] K. Hayakawa, H. Schmid (Helv. Chim. Acta **60** [1977] 1942/60). — [42] I. Fischler, W. Gerhartz, F.-W. Grevels (8th Intern. Conf. Organometal. Chem., Kyoto 1977, Abstr. p. 239). — [43] Y. Becker, A. Eisenstadt, Y. Shvo (Tetrahedron **34** [1978] 799/806). — [44] J.M. Williams, R.K. Brown, A.J. Schultz, G.D. Stucky, S.D. Ittel (J. Am. Chem. Soc. **100** [1978] 7407/9). — [45] A.N. Nesmeyanov, M.I. Rybinskaya, L.V. Rybin, N.T. Gubenko, N.G. Bokii, A.S. Batsanov, Yu.T. Struchkov (J. Organometal. Chem. **149** [1978] 177/83).

[46] T.-A. Mitsudo, T. Sasaki, Y. Watanabe, Y. Takegami, S. Nishigaki, K. Nakatsu (J. Chem. Soc. Chem. Commun. **1978** 252/3). — [47] G. Dettlaf, U. Behrens, E. Weiss (Chem. Ber. **111** [1978] 3019/28). — [48] M. Yokota, T. Sato (Kokagaku Toronkai Koen Yoshishu [Abstr. Lect. Symp. Photochem.], Tokyo 1979, pp. 192/3; C.A. **93** [1980] No. 25561). — [49] M. Franck-Neumann, C. Dietrich-Buchecker, A. Khémiss (Communication aux Journées de Chimie Organique de la Société Chimique de France, Palaiseau 1979 from [83]). — [50] T.-A. Mitsudo, T. Sasaki, Y. Watanabe, Y. Takegami, K. Nakatsu, K. Kinoshita, Y. Miyagawa (J. Chem. Soc. Chem. Commun. **1979** 579/80).

[51] T. Mitsudo, T. Sasaki, Y. Watanabe, K. Nakatsu, S. Sishigaki, K. Kinoshita, Y. Miyagawa (Abstr. Papers Am. Chem. Soc./Chem. Soc. Japan Chem. Congr., Honolulu 1979, INOR 228). — [52] A.N. Nesmeyanov, M.I. Rybinskaya, L.V. Rybin, A.A. Pogrebnyak, N.A. Stelzer (9th Intern. Conf. Organometal. Chem., Dijon 1979, Abstr. Papers B65). — [53] A.N. Nesmeyanov, L.V. Rybin, N.A. Stelzer, M.I. Rybinskaya (J. Organometal. Chem. **182** [1979] 393/8). — [54] A.N. Nesmeyanov, L.V. Rybin, N.A. Stelzer, Yu. Struchkov, A.S. Batsanov, M.I. Rybinskaya (J. Organometal. Chem. **182** [1979] 399/408). — [55] R. Aumann, H. Ring, C. Krüger, R. Goddard (Chem. Ber. **112** [1979] 3644/71).

[56] G.D. Annis, S.V. Ley, R. Sivaramakrishnan, A.M. Atkinson, D. Rogers, D.J. Williams (J. Organometal. Chem. **182** [1979] C11/C14). — [57] T. Akiyama, F.-W. Grevels, I. Salama (9th Intern. Conf. Organometal. Chem., Dijon 1979, Abstr. Papers D2). — [58] G.D. Annis, S.V. Ley, C.R. Self, R. Sivaramakrishnan (J. Chem. Soc. Chem. Commun. **1980** 299). — [59] G.D. Annis, E.M. Hebblethwaite, S.V. Ley (J. Chem. Soc. Chem. Commun. **1980** 297/8). — [60] A. Stockis, R. Hoffmann (J. Am. Chem. Soc. **102** [1980] 2952/62).

[61] M. Franck-Neumann, C. Dietrich-Buchecker, A.K. Khémiss (J. Organometal. Chem. **220** [1981] 187/99). — [62] S.V. Ley, E.M. Hebblethwaite, Beecham Group, Ltd. (Eur. Appl. 38661 [1981] 1/28; C.A. **96** [1982] No. 181064). — [63] W.A. Herrmann, J. Gimeno, J. Weichmann, M.L. Ziegler, B. Balbach (J. Organometal. Chem. **213** [1981] C26/C30). — [64] M. Franck-Neumann, C. Dietrich-Buchecker, A. Khémiss (Tetrahedron Letters **22** [1981] 2307/10). — [65] F.-W. Grevels, K. Schneider (Angew. Chem. **93** [1981] 417/8).

[66] F.-W. Grevels, W.E. Klotzbücher (10th Intern. Conf. Organometal. Chem. Proc., Toronto 1981, Abstr., p. 43). — [67] F.-W. Grevels, W.E. Klotzbücher (Inorg. Chem. **20** [1981] 3002/6). — [68] R.C. Burnier, G.D. Byrd, B.S. Freiser (J. Am. Chem. Soc. **103** [1981] 4360/7). —

[69] G.D. Annis, S.V. Ley, C.R. Self, R. Sivaramakrishnan (J. Chem. Soc. Perkin Trans. I **1981** 270/7). — [70] T. Akiyama, F.-W. Grevels (unpublished results from [67]).

[71] J. Weichmann (Dipl.-Arbeit Univ. Regensburg 1981, pp. 1/111). — [72] A.N. Nesmeyanov, L.V. Rybin, N.A. Shtel'tser, M.I. Rybinskaya (Izv. Akad. Nauk SSSR Ser. Khim. **1981** 1607/11; Bull. Acad. Sci. USSR Div. Chem. Sci. **30** [1981] 1299/302). — [73] M.I. Rybinskaya (10th Intern. Conf. Organometal. Chem. Proc., Toronto 1981, Abstr. p. 14). — [74] M.I. Rybinskaya, L.V. Rybin, A.A. Pogrebnyak, N.A. Shtel'tser (Koord. Khim. **6** [1980] 1475/84; Soviet J. Coord. Chem. **6** [1980] 725/32). — [75] B.L. Booth, S. Casey, R.N. Haszeldine (J. Organometal. Chem. **226** [1982] 289/99).

[76] G.D. Annis, S.V. Ley, C.R. Self, R. Sivaramakrishnan, D.J. Williams (J. Chem. Soc. Perkin Trans. I **1982** 1355/61). — [77] B.R. Bonazza, C.P. Lillya, G. Scholes (Organometallics **1** [1982] 137/42). — [78] M.I. Rybinskaya (Pure Appl. Chem. **54** [1982] 145/59). — [79] W.A. Herrmann, J. Weichmann, B. Balbach, M.L. Ziegler (J. Organometal. Chem. **231** [1982] C69/C72). — [80] W.A. Donaldson, R.P. Hughes (J. Am. Chem. Soc. **104** [1982] 4846/59).

[81] R.P. Hughes, J.M.J. Lambert (unpublished results from [80]). — [82] J. Klimes, E. Weiss (Chem. Ber. **115** [1982] 2606/14). — [83] M. Franck-Neumann, C. Dietrich-Buchecker, A.K. Khémiss (J. Organometal. Chem. **224** [1982] 133/46). — [84] W.A. Herrmann (J. Organometal. Chem. **250** [1983] 319/43). — [85] T. Akiyama, F.-W. Grevels, J.G.A. Reuvers, P. Ritterskamp (Organometallics **2** [1983] 157/60).

[86] A.S. Batsanov, Yu.T. Struchkov (J. Organometal. Chem. **248** [1983] 101/6). — [87] M.I. Rybinskaya, L.V. Rybin, N.A. Stelzer, I.A. Garbusova, B.V. Lokshin (J. Organometal. Chem. **265** [1984] 291/4). — [88] G.D. Annis, E.M. Hebblethwaite, S.T. Hodgson, D.M. Hollinshead, S.V. Ley (J. Chem. Soc. Perkin Trans. I **1983** 2851/6). — [89] S.T. Hodgson, D.M. Hollinshead, S.V. Ley (J. Chem. Soc. Chem. Commun. **1984** 494/6). — [90] L.V. Rybin, N.A. Stelzer, I.A. Garbusova, B.V. Lokshin, M.I. Rybinskaya (J. Organometal. Chem. **265** [1984] 295/303).

1.4.1.4.3.2 ^{4}L is a Cyclic Ligand

The compounds described in Table 4, pp. 119/55, can be prepared by the following methods.

Method I: $Fe(CO)_5$ reacts thermally or photochemically with an organic substrate. The products can be purified by recrystallization or by chromatography followed by recrystallization.

a. $Fe(CO)_5$ and (+)-d-pinene (see Formula I) react as neat liquids in a pressure bottle (22 atm CO) at 160±0.5 °C for 68 h. Compounds No. 26 and 27 are described as intermediates [68].

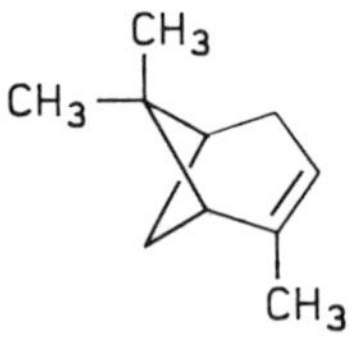

I 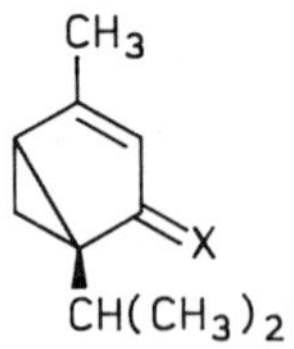 II 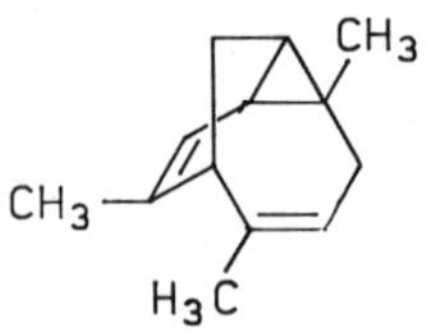 III 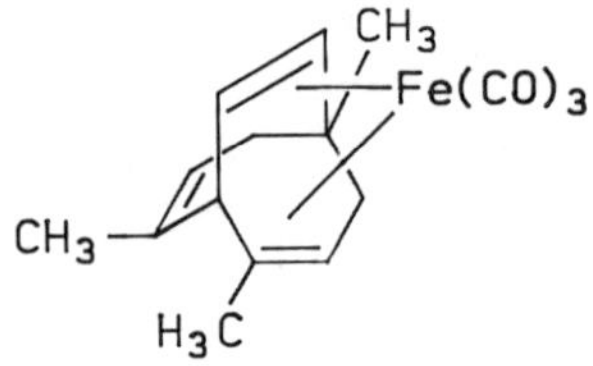 IV

b. $Fe(CO)_5$ and (−)-umbellulone (see Formula II with X=O) are irradiated in petroleum ether (b.p. 40 to 60 °C) for 6 h, and the resulting solution is chromatographed on Florisil. The first eluted fraction contains a very labile organometallic species which cannot be analyzed. The second eluted fraction contains an inseparable mixture of Nos. 29 and 30. On repeated thin layer chromatography this mixture decomposes to form (±)-umbellulone (85%). Similar irradiation with (+)-α-thujene (see Formula II with $X=H_2$) for 1 h followed by filtration through Celite and evaporation of the filtrate gives No. 28 [101].

CH_3 CH_3 CH_3 $Fe(CO)_4$ $(CO)_3Fe$—$Fe(CO)_3$ O

V VI VII VIII IX

c. $Fe(CO)_5$ and III in pentane are irradiated with a 150 W high-pressure mercury lamp through pyrex for 1 h. After filtration the solvent is removed and the residue is chromatographed on silica gel with hexane to give No. 106 (22% yield). The yields are dependent on the irradiation time. Irradiation for 3 h followed by work-up as above, but distillation of the residue to recover III, gives a mixture of No. 106 (major) and IV (minor). The residue is recrystallized from C_2H_5OH at −20 °C to give No. 106 (21.2% yield). Distillation of the mother liquor at 118 to 120 °C/0.008 Torr gives first No. 106 (4% yield) and then IV. Similar irradiation with V for 3 h followed by filtration, and removal of the solvent gives a yellow precipitate of No. 131. The combined filtrates are distilled to recover V (20%), and further distillation gives No. 129 [90]. Similar irradiation with homosemibullvalene (see Formula VI) in ether for 3 h at 20 °C followed by chromatography on silica gel (Woelm neutral) gives VII, VIII, upon elution with hexane, and No. 76 upon elution with ether in the ratio 7:1:10 (total yield 88%) [49].

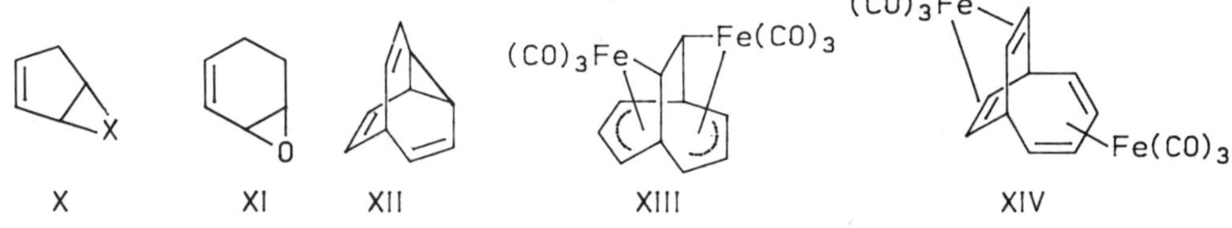

X XI XII XIII XIV

d. Low temperature photolysis of $Fe(CO)_5$ (1% C_6H_6 solution) with IX (ratio 5:1) gives No. 64 [63]. Irradiation of $Fe(CO)_5$ and X (X=O) in C_6H_6 at 20 °C for 2 h gives No. 10 [48]. Similar irradiation with XI and removal of the solvent, taking care not to heat the mixture above 10 °C, gives crude No. 14. The crude product is stirred with ether and filtered through Celite. Trituration with petroleum ether affords the pure complex No. 14 [48, 116]. Similar irradiation with bullvalene (see Formula XII) in humid C_6H_6 for 5 h followed by chromatography on Kieselgel (Woelm, neutral) with C_6H_6 gives a fraction (10%) containing XIII to XVI which are disproportionation products since they are not detected by irradiation at −60 °C. Elution with CH_2Cl_2 gives No. 107 and with ether No. 110. The molar ratio of the products is independent of the time of irradiation. Nos. 109 and 112 are formulated as intermediates in this reaction [66].

References on pp. 172/5

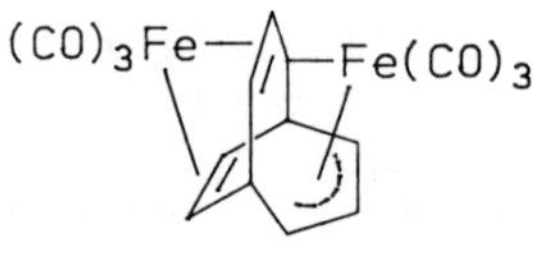

XV 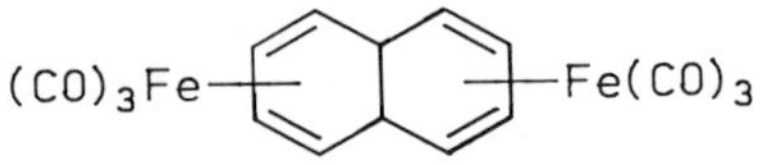 XVI

e. $Fe(CO)_5$ and XVII are irradiated in ether at −60 °C for 8 h. The reaction mixture is filtered at −60 °C and concentrated at −10 °C/15 Torr (finally 0.1 Torr to remove the rest of $Fe(CO)_5$). Addition of hexane to the residue (16 h, −80 °C) affords No. 67, and concentration of the mother liquor gives XVIII [83], cf. [77]. Similar irradiation with XIX at −20 °C for 2.5 h, followed by filtration and evaporation, gives a residual reddish oil. This is diluted with CH_2Cl_2 and chromatographed on Kieselgel (Merck) with CH_2Cl_2. The first fraction is not examined further. The second fraction gives a yellow oil containing XIX (18%) and No. 49, separated by elution with C_6H_6 on Kieselgel. The third fraction gives No. 46, and the fourth fraction affords No. 48. The last fraction gives No. 47 [94]. Similar irradiation with semibullvalene (see Formula XX) in the ratio 3:1 at −50 °C for 2.5 h gives XXI and No. 69 in the ratio 2:1 (total yield 95%). The product ratio is dependent upon the reaction time. In hexane, XXI is the preferred product. Irradiation at 25 °C gives No. 69 and XXII. Longer reaction times increase the yield of XXII [52].

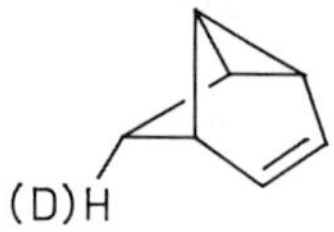

XVII 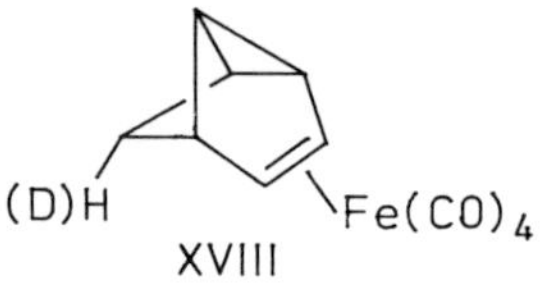 XVIII 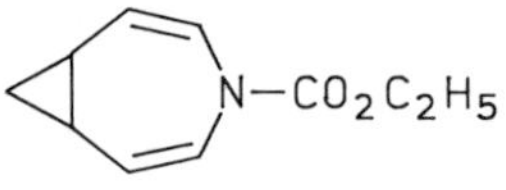 XIX

f. $Fe(CO)_5$ and XXIII are irradiated in tetrahydrofuran for 10 h. Filtration through basic alumina and removal of the solvent affords No. 74 [109].

XX 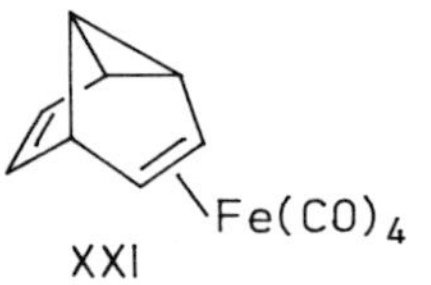 XXI 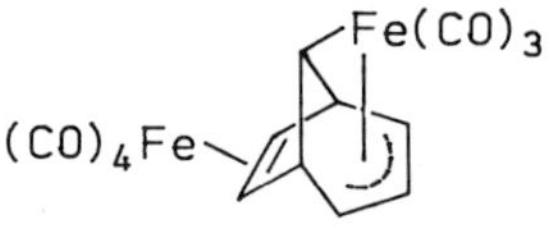 XXII 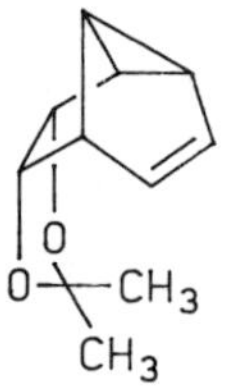 XXIII

Method II: $Fe_2(CO)_9$ reacts with an organic substrate. The products are usually purified by chromatography.

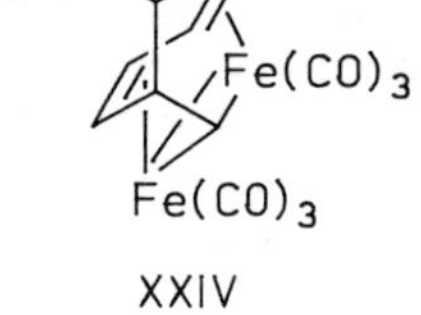

XXIV 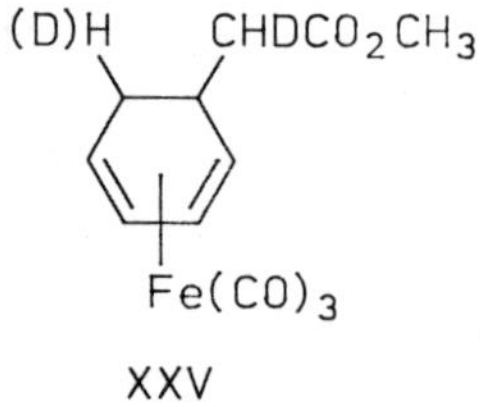 XXV 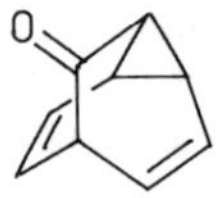 XXVI 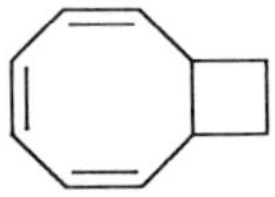 XXVII

References on pp. 172/5

a. $Fe_2(CO)_9$ and XVII are stirred in CH_3OH/conc. HCl for 3 d. The reaction mixture is filtered, and the solvent is removed (20 °C/15 Torr). The residue is chromatographed on Kieselgel. Elution with petroleum ether gives XVIII, $Fe_3(CO)_{12}$, and XXIV. Elution with C_6H_6 gives XXV; CH_2Cl_2 elutes No. 15, and finally ether affords No. 81 [83], cf. [76].

(CO)3Fe—Fe(CO)3 (CO)3Fe (CO)3Fe Fe(CO)3

XXVIII XXIX XXX

b. $Fe_2(CO)_9$ reacts with barbaralone (see Formula XXVI) [23, 30, 35, 54, 62, 101] in n-hexane at 55 °C [30], or in ether at 20 °C [62] or at 25 °C [54, 62, 101] for 30 h [54] to give No. 91 [23, 30, 35, 54, 62, 101] and other products [30, 62]. The compound No. 91 can be separated by chromatography on Kieselgel [30, 54] with n-hexane [30], ether [30], or $CHCl_3$ [54], or on basic [23] alumina [23, 62] (Woelm activity 2 [62]) by elution first with hexane, then with hexane/ether, followed by rechromatography with hexane. Compound No. 91 elutes with ether/acetone (2:1) [62]. Similar reaction of $Fe_2(CO)_9$ with XXVII gives XXVIII to XXXI, No. 91, XXXII, and XXXIII. The compounds trans-$C_8H_8(Fe(CO)_3)_2$ and $C_8H_8Fe_2(CO)_5$ are also isolated, which presumably arise from traces of C_8H_8 present as an impurity in XXVII [47]. Refluxing $Fe_2(CO)_9$ and octafluorocyclooctatetraene in hexane gives No. 61 [117]. Refluxing with homosemibullvalene (see Formula VI) in hexane for 10 h gives VIII, No. 75, and $^4LFe(CO)_3$ (^4L=bicyclo[4.2.1]nonadiene) [49], cf. [30]. Similar reaction with semibullvalene (see Formula XX) in hexane at 70 °C for 14 h followed by filtration and chromatography on alumina (activity 2) with Skelly B gives No. 69 and XXXIV [43]. Similar reaction with XX in hexane at room temperature for 18 h followed by work-up as before gives a very unstable oil (not characterized) which decomposes rapidly at room temperature or on exposure to air (IR spectrum (film): 1972, 1996, 2083 cm^{-1}), and a very air-sensitive yellow oil which is approximately a 1:1 mixture of XXXV, XXXVI, and No. 69 [43].

((CO)2Fe)2 (CO)3Fe Fe(CO)3 Fe(CO)3 Fe(CO)3 (CO)3Fe

XXXI XXXII XXXIII XXXIV

c. $Fe_2(CO)_9$ and semibullvalene (see Formula XX) are refluxed in C_6H_6 for 1 h to give No. 69, purified by repeated bulb-to-bulb distillation [16]. Similar reaction with benzobarbaralone (see Formula XXXVII) in C_6H_6 at 55 °C for 1.5 h followed by filtration, evaporation of the filtrate, and chromatography on basic alumina with CH_2Cl_2/hexane gives the starting material XXXVII. Further elution with CH_2Cl_2 affords No. 102 [59], cf. [42]. Similar reaction

References on pp. 172/5

with XXXVIII in C_6H_6 at 40 °C followed by chromatography on basic alumina gives No. 108 and $^4LFe(CO)_3$ (^4L = carbomethoxycyclooctatetraene) [26]. Refluxing $Fe_2(CO)_9$ and XXXIX (n = 1, 2) gives Nos. 70 and 71. Refluxing with XL for 3 h gives No. 72 [22].

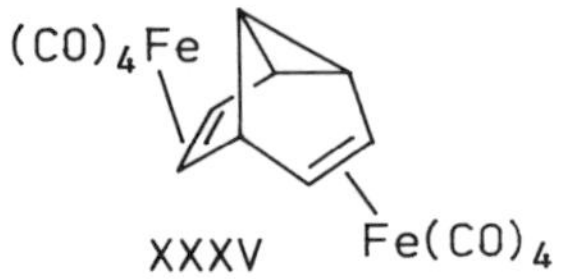

XXXV 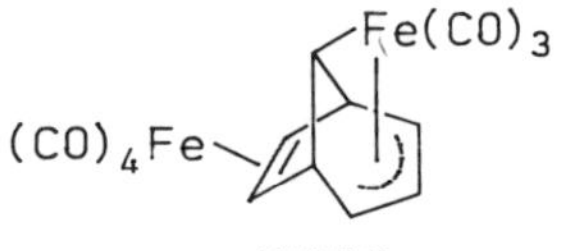 XXXVI 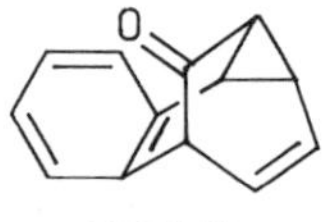 XXXVII

d. $Fe_2(CO)_9$ reacts with X (X = CH_2) in ether at 30 °C for 4 h to give XLI and No. 17 in the ratio 1:10 (total yield 85%) and are separated by chromatography on Kieselgel with C_6H_6 as eluent [38]. Similar reaction with XVII at 20 °C for 20 h followed by filtration, evaporation, and chromatography on Kieselgel with petroleum ether gives XVIII, $Fe_3(CO)_{12}$, and XXIV. Further elution with ether gives No. 81 [83]. Similar reaction with XLII at 25 °C for 24 h, work-up as before but elution with C_6H_6, gives $Fe_3(CO)_{12}$. Elution with ether affords No. 86 [84]. Similar reaction with bicyclo[4.1.0]hept-2-ene gives No. 33 [38]. Reaction with XVII in ether in the presence of $NaOCH_3/CH_3OH$ as in Method II a gives products in the following sequence: XVIII, XXIV, XXV, No. 16, and finally No. 81 [83].

XXXVIII 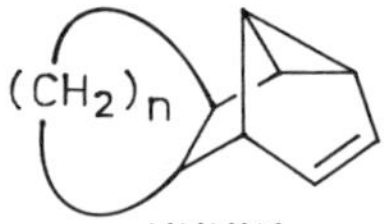 XXXIX 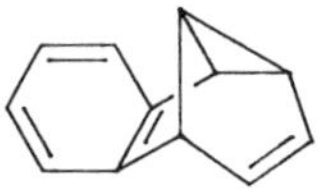 XL

e. $Fe_2(CO)_9$ and XLIII react in C_6H_6/petroleum ether (1:1) at 33 °C for 3 h. The reaction mixture is filtered, and the solvent is evaporated. The residue is chromatographed on basic alumina at 0 °C. The column is washed with petroleum ether, and CH_2Cl_2 elutes No. 21. Further elution with CH_2Cl_2 gives XLIV [96].

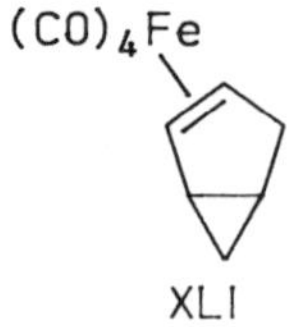

XLI 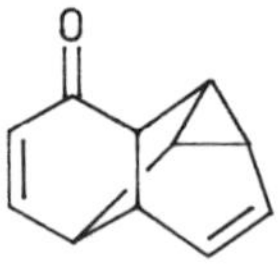 XLII 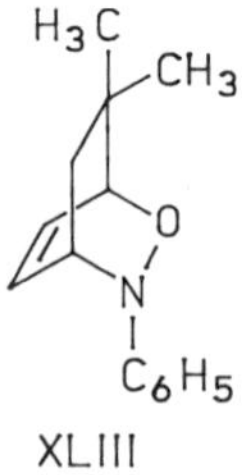 XLIII 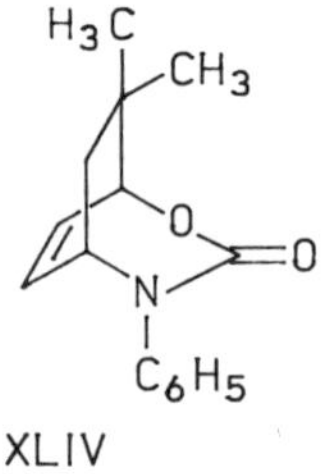 XLIV

Method III: $Fe_3(CO)_{12}$ and II (X = H_2) are refluxed in C_6H_6 or tetrahydrofuran for 3 h. Work-up as in Method I b gives No. 28. The minor products cannot be isolated [101].

Method IV: $^2LFe(CO)_4$ compounds are thermally or photochemically decomposed, or $^2LFe(CO)_4$ or $^2L_2Fe(CO)_3$ compounds react with alkenes.

References on pp. 172/5

a. $^2LFe(CO)_4$ (see Formula XLV) is dissolved in acetone, and the solvent is evaporated under a water-jet vacuum. This procedure is repeated for two or three times to give No. 11 [108]. Thermal decomposition of XLI ($t_{1/2}$ at 30 °C ca. 2 d) gives X (X = CH_2) as the main product and No. 17 [38]. Photolysis of XVIII in ether at −60 °C for 4 h followed by work-up as in Method Ie gives No. 67 [77, 83].

b. $^2LFe(CO)_4$ ($^2L = CH_2{=}CHCO_2CH_3$) and cycloocta-1,3,5-triene are irradiated in the presence of $CH_2{=}CHCO_2CH_3$ to give No. 63. Similar reaction with cyclooctatetraene gives No. 65. Heating $^2L_2Fe(CO)_3$ ($^2L = CH_2{=}CHCO_2CH_3$) and cycloocta-1,3,5-triene in the presence of $CH_2{=}CHCO_2CH_3$ gives No. 63. Similar reaction with cyclooctatetraene gives No. 65 [124].

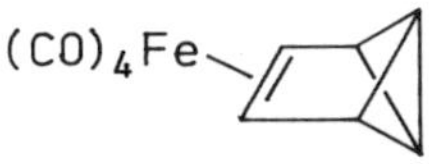

XLV 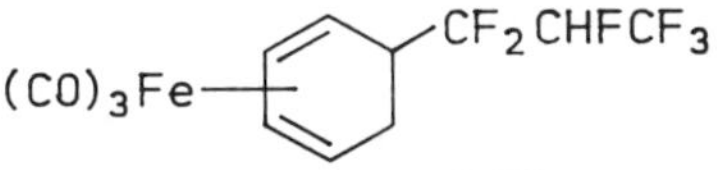 XLVI 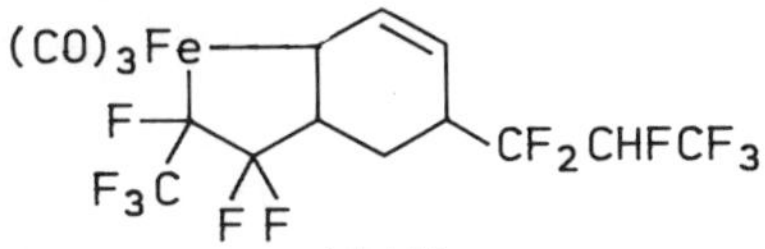 XLVII

Method V: $^4LFe(CO)_3$ compounds react with an organic substrate or with $AlCl_3$. ^4L is an isocyclic or heterocyclic ligand.

a. $R_4C_4Fe(CO)_3$ (R = H or CH_3) and excess $CF_2{=}CF_2$ are irradiated in hexane in a Pyrex Carius tube for 24 h. The solvent is removed and the residue is chromatographed on alumina. Elution with hexane gives $R_4C_4Fe(CO)_3$ (R = H or CH_3), and elution with hexane/C_6H_6 (3:2) gives No. 3 or No. 7, respectively [14, 25]. Analogous reactions with $F_2C{=}CFCF{=}CF_2$, or $CF_3CF{=}CF_2$ gives Nos. 2, 9, or 4, 8 [14, 25]. Similar irradiation with $CF_2{=}CFH$ gives No. 6 (eluent hexane/CH_2Cl_2 (1:1)), but prolonged irradiation gives a mixture of isomeric Nos. 5 and 6 [39, 86].

b. $^4LFe(CO)_3$ (^4L = cyclohexa-1,3-diene) and $CF_2{=}CFH$ [71], $CF_2{=}CF_2$ [64], or $CF_3CF{=}CF_2$ [71] are irradiated in hexane for 80 h [71], 60 h [64], and 3 d [71]. Removal of the solvent gives the compounds No. 18 to 20, respectively. The residue containing No. 20 is chromatographed on alumina with hexane to give XLVI. Elution with CH_2Cl_2/hexane (1:2) gives No. 20, and further elution with CH_2Cl_2/hexane gives XLVII [71]. $^4LFe(CO)_3$ (^4L = cyclohexadiene) reacts with AlX_3 (X = Cl or Br) in the ratio 1:2 in CH_2Cl_2 at room temperature for 2 d to give No. 38 (ca. 9% yield). The same reaction under 1 atm of CO gives No. 38 in ca. 55% yield [102, 104]. $^4LFe(CO)_3$ (^4L = 2-ethylcyclohexa-1,3-diene) and $CF_3CF{=}CF_2$ are irradiated in hexane in a Carius tube for 5 d. Evaporation followed by chromatography on alumina with CH_2Cl_2/hexane (1:2) gives No. 22 [71].

c. XLVIII reacts with excess $OC(CF_3)_2$ in C_6H_6 in a Carius tube at 80 °C for 24 h followed by evaporation to give No. 113 [88]. Similar reaction with

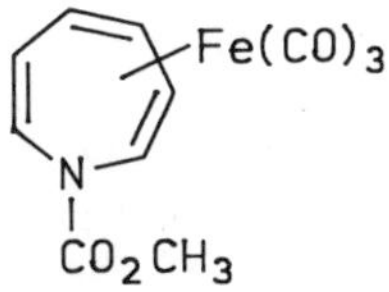

XLVIII 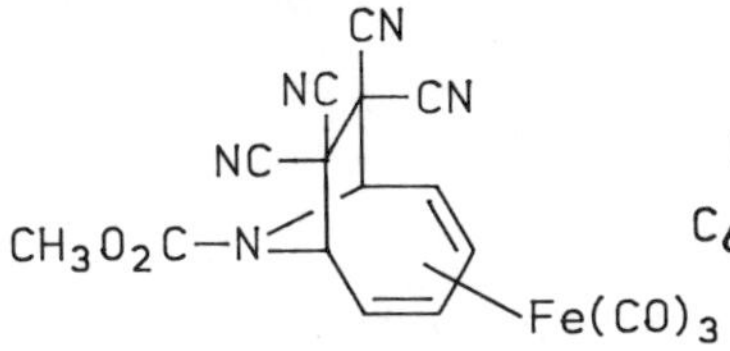 XLIX 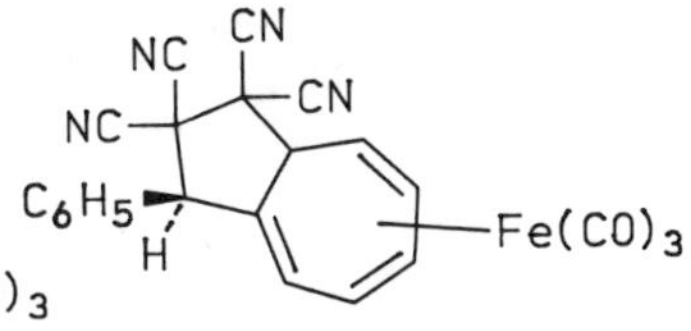 L

References on pp. 172/5

$(NC)_2C{=}C(CN)_2$ in CH_2Cl_2 at room temperature for 0.5 h gives a ca. 1:6 mixture of No. 114 and XLIX [19, 88].

d. $^4LFe(CO)_3$ (^4L = cycloheptatriene) reacts with $OC(CF_3)_2$ in hexane in a Carius tube at room temperature for 2 d to give No. 115 [44], cf. [19, 20]. Reaction with $(NC)_2C{=}C(CN)_2$ [18 to 20, 24, 123] in O_2NCH_3 [123], in C_6H_6 for 2 min [24] or in CH_2Cl_2 at room temperature for 0.5 h [19, 44] gives No. 117. Similar reaction with $NCC(CF_3){=}C(CF_3)CN$ in CH_2Cl_2 at room temperature for 4 h gives three isomers of No. 118 [44], cf. [19, 20]. Analogous reaction with $(NC)_2C{=}C(CF_3)_2$ in hexane at room temperature for 0.5 h gives No. 119 [44], cf. [19, 20]; reaction for 10 h gives a higher yield of No. 119 [12]. Reaction with $O{=}C{=}C(C_6H_5)_2$ in boiling C_6H_6 gives No. 120 [105]. A mixture of $^4LFe(CO)_3$ (^4L = cycloheptatriene), $Fe_2(CO)_9$, and $(CH_3)_2CBrCOC(CH_3)_2Br$ in C_6H_6 is stirred at 50 °C for 8 h [118] to 10 h [119]. Filtration, evaporation of the filtrate, and chromatography of the residue on silica gel with C_6H_6 gives No. 130 [118, 119]. Similar addition reactions are also observed with $^4LFe(CO)_3$ (^4L = cycloheptatrienone). Thus, reaction with $OC(CF_3)_2$ in hexane/C_6H_6 at 80 °C for 36 h gives No. 116 [88]. Reaction with $(NC)_2C{=}C(CN)_2$ in CH_2Cl_2 at room temperature for 4 h gives No. 105 [81]; however, upon extended reaction time (12 h) No. 126 slowly precipitates [88]. For a similar reaction in CH_3NO_2 see [123]. Reaction with N-methyltriazolinedione [97, 106] or N-phenyltriazolinedione in C_6H_6 [92] gives Nos. 87 [97, 106] and 88 [92], respectively. Analogous reactions are also observed with other $^4LFe(CO)_3$ derivatives. Thus, $^4LFe(CO)_3$ (^4L = 7-phenylmethylenecycloheptatriene) and $(NC)_2C{=}C(CN)_2$ are dissolved in CH_2Cl_2 or CH_3NO_2 and frozen immediately with liquid N_2. Rapid removal of the solvent gives No. 127 (minor) and L (major) [125]. Similar reaction of $^4LFe(CO)_3$ (^4L = 2-acetylcyclohepta-1,3,5-triene) with $(NC)_2C{=}C(CN)_2$ gives No. 121 [91, 123], and $^4LFe(CO)_3$ (^4L = 1-acetylcyclohepta-1,3,5-triene) gives No. 123 [91, 123]. Treatment of $^4LFe(CO)_3$ (^4L = 1-formylcyclohepta-1,3,5-triene) with $(NC)_2C{=}C(CN)_2$ in CH_2Cl_2 slowly gives No. 125 [92]. In acetone-d_6 with a threefold excess of $(NC)_2C{=}C(CN)_2$ the ^1H NMR spectrum of $^4LFe(CO)_3$ is immediately replaced by that of No. 122. After 0.5 h new absorptions appear which are assigned to No. 125. Pure No. 122 is obtained in CH_2Cl_2 after 0.5 h. The solution is frozen with liquid N_2, and the solvent is rapidly removed [123]. Reaction of LI with $(NC)_2C{=}C(CN)_2$ gives No. 124 [31, 123]. Refluxing LII in C_6H_6 gives No. 120 [105], cf. [122]. Reaction of $^4LFe(CO)_3$ (^4L = 3,7,7-trimethylcyclohepta-1,3,5-triene) with $(NC)_2C{=}C(CN)_2$ at room temperature for 10 min followed by removal of the solvent gives a 4:1 mixture of No. 128 and LIII (total yield 35%) [128]. Reaction of $C_6H_5CH{=}CHC_7H_7Fe(CO)_3$ with $(NC)_2C{=}C(CN)_2$ for 5 h at room temperature, followed by filtration and precipitation with hexane, gives No. 151 [130].

e. $^4LFe(CO)_3$ (^4L = cyclooctatetraene) reacts with AlX_3 (X = Cl, Br) in C_6H_6 at 10 °C for 3.5 h. The reaction mixture is hydrolyzed with H_2O/conc. HCl and extracted with CH_2Cl_2. The organic layer is dried ($MgSO_4$) and chromatographed on silica gel with 10% $CH_3CO_2C_2H_5/C_6H_6$ to give No. 91 [61, 69, 110]. The conversion is dependent upon CO. The yield increases from 35 to 65% by applying 1 atm of CO [104]. Similar reaction of $^4LFe(CO)_3$ (^4L = methylcyclooctatetraene) with $AlCl_3$ gives No. 100 [61, 69]. Refluxing $^4LFe(CO)_3$ (^4L = cyclooctatetraene) with $Ru_3(CO)_{12}$ in xylene for 12 h gives compound No. 152 [13]. $^4LFe(CO)_3$ (^4L = cyclooctatetraene) reacts with $(NC)_2C{=}C(CN)_2$ [1 to 3, 7, 10, 24, 34, 114] in C_6H_6 at room temperature [1, 3, 24, 34] for 1 h [3] or 15 h [24] to give No. 134, cf. [123]. Similar reaction with $(NC)_2C{=}C(CF_3)_2$ in hexane at room temperature for 10 h gives No. 135 [12]. Compounds $RC_8H_7Fe(CO)_3$ (C_8H_7 = cyclooctatetraenyl) reacts with $(NC)_2C{=}C(CN)_2$ in CH_2Cl_2 [44, 67, 92] or in C_6H_6 [67] at room temperature to give 1:1 addition compounds. Thus, reaction of $BrC_8H_7Fe(CO)_3$ with $(NC)_2C{=}C(CN)_2$ gives No. 141 within 0.5 h [44]. Similar reaction of $CH_3OC_8H_7Fe(CO)_3$ for 1 h followed by evaporation of the reaction mixture gives a residue which is slurried in ether and filtered to leave No. 140. LIV can be isolated from the filtrate [67]. Similar reaction of $CH_3C_8H_7Fe(CO)_3$ for 1 h followed by filtration gives a residue identified as No. 136 [44, 67]. Evaporation of the mother liquor followed by addition of CH_2Cl_2 and filtration gives additional No. 136, and the soluble fraction contains No. 138 [67]. Similar reaction of $CH_3O_2CC_8H_7Fe(CO)_3$ with $(NC)_2C{=}C(CN)_2$ in CH_2Cl_2 for 1 h followed by evaporation gives a yellow solid as a mixture of two isomers in the ratio 1:3.7. Crystallization from hexane/acetone gives No. 142. Addition of more hexane causes precipitation of a 1:1 mixture of Nos. 142 and 144 separated by thin layer chromatography on silica gel with ether/pentane. Analogous reaction in C_6H_6 gives Nos. 142 and 144 in the ratio 3:1 [67]. Similar reaction of $C_6H_5C_8H_7Fe(CO)_3$ in CH_2Cl_2 for 2 h gives No. 143 [44]. Analogous reaction of $(C_6H_5)_3CC_8H_7Fe(CO)_3$ at room temperature with $(NC)_2C{=}C(CN)_2$ in CH_2Cl_2 for 0.5 h or with $(NC)_2C{=}C(CF_3)_2$ in hexane for 24 h gives Nos. 149 and 150 [44]. Similar reaction of $^4LFe(CO)_3$ (^4L = cyclooctatrienone) with $(NC)_2C{=}C(CN)_2$ in CH_2Cl_2 gives No. 145 [92]. $C_6H_5C_8H_7Fe(CO)_3$ reacts with $(NC)_2C{=}C(CN)_2$ in C_6H_6 for 50 min. The solid isolated by filtration is identified as No. 143. The mother liquor is evaporated, and the residue is briefly refluxed and filtered to yield further No. 143. Evaporation of the mother liquor and warming the residue in C_2H_5OH near reflux followed by standing at room temperature separate a solid (process repeated a further

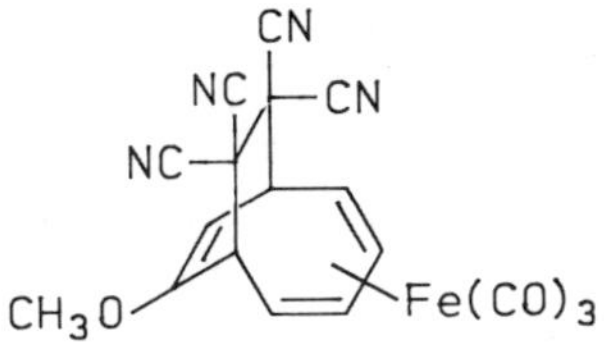

LIV 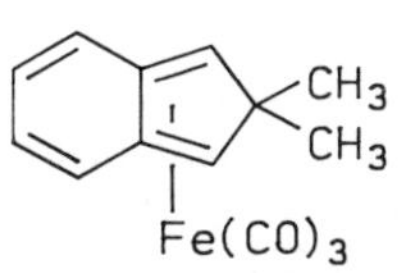 LV 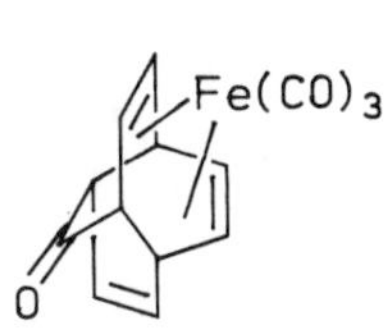 LVI

References on pp. 172/5

two times). The total C_2H_5OH insoluble fraction gives No. 139. The evaporation of the alcoholic mother liquor followed by trituration of the residue with hexane and washing with a few mL C_2H_5OH gives No. 137 [67].

f. $^4LFe(CO)_3$ (see Formula LV) reacts with $AlCl_3$ in CH_2Cl_2 at 0 °C to give No. 31 [55]. Irradiation of $^4LFe(CO)_3$ (^4L = bicyclo[4.2.0]octa-2,4-diene) in hexane in the presence of HCF=CF_2 for 70 h [71], CF_2=CF_2 for 55 h [64], or CF_3CF=CF_2 for 6 d [71] gives the compounds No. 23 to 25 [64, 71] purified by chromatography on alumina with hexane as eluent [71]. Heating LVI in C_6H_6 in an ampule at 100 °C for 3 d (the solution must remain clear, and no solid may be deposited) followed by chromatography on Kieselgel with ether gives a yellow oil, which is treated with petroleum ether to give No. 103 [75, 84].

Method VI: $[^5LFe(CO)_3]X$ compounds (X = BF_4, PF_6, ^5L = cycloheptadienyl or cycloocta-2,4-dienyl or their derivatives) react with various nucleophiles in H_2O or other polar solvents.

a. Reduction of $[^5LFe(CO)_3]BF_4$ (^5L = cycloheptadienyl) with $NaBH_4$ in ether/H_2O at 0 °C for 1 h gives No. 32 and $^4LFe(CO)_3$ (^4L = cyclohepta-1,3-diene) [37, 38, 56, 58, 73] in the ratio 2:1 [37, 56, 58, 73] in a total yield of ca. 90% [37, 58], or 84% [58]. Similar reaction in CH_2Cl_2/C_2H_5OH gives a mixture of $^4LFe(CO)_3$ (55%) and No. 32 (45%) [58]. The organic phase is separated, washed with H_2O, dried ($MgSO_4$) [58, 73], and chromatographed on silica gel with C_6H_6 [58], pentane [38], or petroleum ether (b.p. 30 to 40 °C) [58], cf. [99]. Similar reduction with Na[BH_3CN] gives No. 34 [99]. $[^5LFe(CO)_3]BF_4$ (^5L = cycloheptadienyl) reacts with KCN in acetone/H_2O (1:1) at room temperature for 15 min. H_2O is added to the reaction mixture followed by extraction with ether. The organic layer is dried ($MgSO_4$) and chromatographed on alumina with 5% $CH_3CO_2C_2H_5$/C_6H_6 to give a yellow oil. Distillation gives a mixture of LVII (R = CN) and No. 34 in the ratio 1:3.4 (total yield 53%). Similar reaction with C_5H_5Tl in CH_2Cl_2 in the dark for 12 h followed by work-up as before gives a mixture of LVII (R = C_5H_5) and No. 36 in the ratio 1.5:1 (total yield 55%) [58]. Treatment with $LiCH_3$ in CH_2Cl_2 at −78 °C gives No. 35 in low yield. Treatment of $[^5LFe(CO)_3]BF_4$ (^5L = 7-methylcyclohepta-2,4-dienyl) with $LiCu(CH_3)_2$ in CH_2Cl_2 at −78 °C gives a 3:1 mixture of $^4LFe(CO)_3$ (^4L = 5,7-dimethylcyclohepta-1,3-diene) and No. 37. However, reaction with $LiCH_3$ gives only No. 37. Both reactions are accompanied by dimer formation and proceed in low overall yield (20%) [126].

b. Reduction of LVIII (X = O, R = H) with $NaBH_4$/H_2O, covered with an ether layer, at 0 °C for 10 min followed by extraction with ether, drying, evaporation, and chromatography on silica gel gives LIX (R = H) and No. 39. Similar reaction with NaCN in H_2O, covered with an ether layer, at 0 °C for 30 min, followed by work-up as before, gives LIX (R = CN) and No. 40. Analogous reaction of LVIII ($X = CH_2$, $R = H$) with NaCN gives No. 41, LX, and minor amounts of a dimeric complex [80].

$[LFe(CO)_3]^+ [BF_4]^-$ (CO)$_3$Fe X H

LXI LXII

c. Reduction of LXI with $NaBH_4$ in ice-water [8, 40] gives LXII with X=H, OH and No. 42 (total yield 90%) [40], separable by chromatography [74]. Similar reduction in ice water/ether within 1 h gives different products. The organic phase is diluted with ether, washed with H_2O, dried ($MgSO_4$), and evaporated. Chromatography of the residue on Kieselgel with petroleum ether gives LXII (X=H) and No. 43 in the ratio 5:1 (total yield 90%) [73].

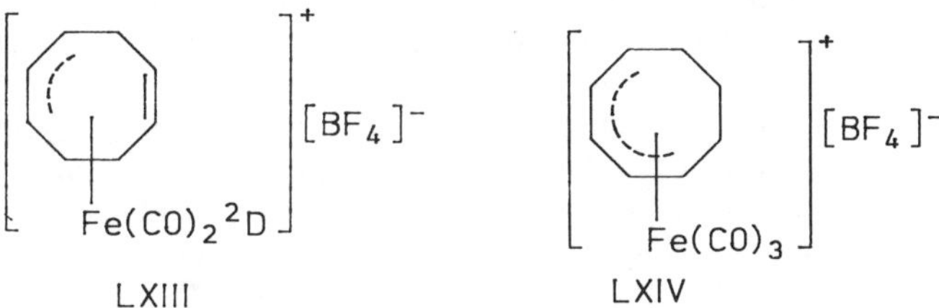

LXIII LXIV

d. Complex LXIII is reduced with $NaBH_4$ in H_2O at 0 °C within 5 min. Extraction with ether followed by chromatography on alumina gives a 1:9 mixture of $^4LFe(CO)_3$ (^4L=cycloocta-1,5-diene) and No. 50 [17, 60]. Treatment of LXIII (^2D=CO) with NaI in acetone followed by filtration through alumina, removal of the solvent, and thin layer chromatography on silica gel with $CHCl_3$ shows the presence of two compounds, probably No. 54 and $^5LFe(CO)_2I$ (^5L=cycloocta-2,5-dienyl). But chromatography on a silica column gives only $^5LFe(CO)_2I$ [127]. Treatment of the salt LXIII (^2D=CO) in acetone with solid NaN_3 followed by filtration after 40 min, evaporation of the filtrate, and chromatography of the residue on alumina with pentane gives No. 51 as the sole product [70, 127]. Treatment of LXIV with N_3^- gives No. 153 [70]. Stirring a mixture of aqueous solutions of LXIII (^2D=CO) and NaCN [60, 127] or KCN [127] for 15 min followed by extraction with ether, drying of the extracts ($MgSO_4$), and removal of the solvent gives No. 55 [60, 127] and $^4LFe(CO)_3$ (^4L=3-cyano-cycloocta-1,5-diene) [60] in the ratio 7:3 [127]. Similar treatment with $NaOCH_3$ in CH_2Cl_2 for 1 h, followed by filtration, evaporation, extraction of the residue with pentane, filtration through alumina followed by evaporation, and sublimation gives No. 53 [127], cf. [70]. The above salt also reacts with $P(C_6H_5)_3$ in CH_2Cl_2. On addition of ether, a bulky yellow precipitate forms which is redissolved in CH_2Cl_2. Chromatography on alumina with $CHCl_3$ gives LXIII ($^2D = P(C_6H_5)_3$), and elution with acetone gives No. 52 [127]. Treatment of the salt LXIII (^2D=CO) with CH_3MgI or CH_2=$CHCH_2MgBr$ in ether gives Nos. 56 or 57, respectively. The products are separated from the appropriate $^4LFe(CO)_3$ compounds (see Formula LXV with R=CH_3 or CH_2CH=CH_2) by chromatography. Reaction with $NaC(COCH_3)_2H$ in H_2O gives No. 58 and

References on pp. 172/5

LXV ($R=C(COCH_3)_2H$) in the ratio 1:1, and reaction with $NaC(CO_2C_2H_5)_2R'$ ($R'=H$, C_6H_5) in tetrahydrofuran gives Nos. 59 or 60 and the appropriate LXV ($R=C(CO_2C_2H_5)_2R'$) in the ratio 3:2 [60].

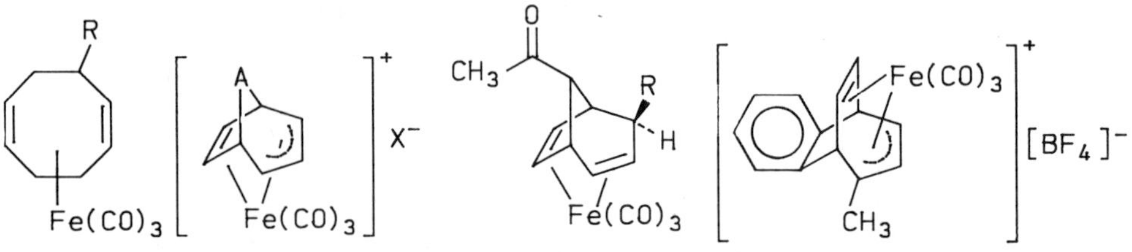

LXV LXVI LXVII LXVIII

e. LXVI ($A=CH_2$, $X=BF_4$) reacts with $NaBH_4$ or $NaBD_4$ to give No. 78 [93]. Similar reaction with a saturated solution of KCN in H_2O or ether gives No. 79 after extraction with ether and drying the organic layer ($MgSO_4$) [41, 93]. Similarly, LXVI ($A=CHCOCH_3$, $X=PF_6$) is added in small portions to a 10% solution of $NaBH_4$ in ether at 0 °C. After 1 h H_2O is added and the layers are separated. The organic layer is dried ($MgSO_4$), and the solvent is removed. Chromatography (2 ×) on silica gel with 10% ether/C_6H_6 gives two major bands (in addition to several minor ones), and the faster running band gives No. 80. The slower running band gives the unstable isomer compound, probably LXVII (R=H). Similar reaction with $LiBD_4$ in D_2O for 30 min gives the deuterated derivative of No. 80. Reaction of LXVI ($A=CHCOCH_3$, $X=PF_6$) with a saturated aqueous solution of NaCN followed by work-up as above but elution with 20% $CH_3CO_2C_2H_5/C_6H_6$ gives No. 82 and LXVII (R=CN) in the ratio 14:1. A suspension of LXVI ($A=CHCHOHCH_3$, $CHCHOHC_6H_5$, $CHC(CH_3)_2OH$, $X=PF_6$) reacts with $N(C_2H_5)_3$ in CH_2Cl_2 at 0 °C for 2 min, and the mixture is hydrolyzed with 1% aqueous HCl and H_2O. The organic layer is separated, dried (K_2CO_3), and the solvent is removed to give Nos. 83, 85, or 84, respectively [121]. Similar reaction of LXVI (A=-CH=CH-, -o-C_6H_4-, $X=PF_6$) with KCN saturated in H_2O or ether for 30 min, followed by extraction with ether, and drying the organic layer ($MgSO_4$) gives Nos. 90 or 101, respectively [41]. Addition of small portions of solid LXVI (A=-$C(CH_3)$=CH-, $X=PF_6$) to a saturated aqueous solution of NaCN in ether followed by separation of the two layers after 30 min, drying the organic layer ($MgSO_4$), and concentration gives an inseparable (neutral alumina column) mixture of Nos. 96 and 97. Similar reaction of XLVIII (A=-$C(C_6H_5)$=CH-) yields an inseparable 1:1 mixture of Nos. 98 and 99. Similar treatment of LXVIII gives No. 104 [113]. Reaction of CXXII (p. 167) with OH^- gives No. 110 [66]. Treatment of LXIX with $NaBH_4$ in tetrahydrofuran yields No. 147 [115].

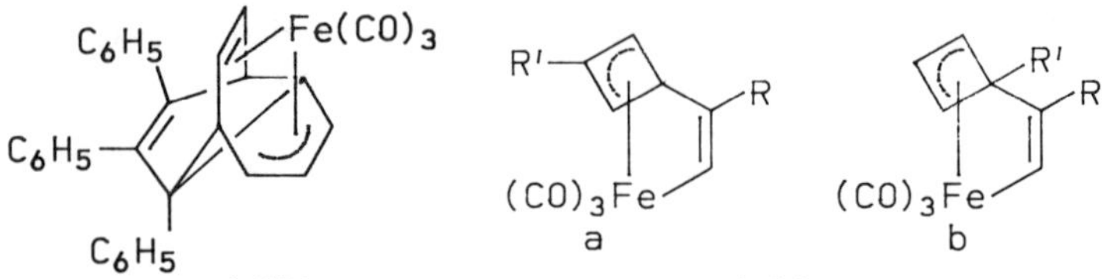

LXIX LXX

Irradiation of $RC_4H_3Fe(CO)_3$ complexes and propyne gives substituted toluenes. Reaction of $RC_4H_3Fe(CO)_3$ ($R=CH_3$, i-C_3H_7) with substituted acetylenes gives complex mixtures of isomeric benzene compounds. Complexes of the Type LXX a or b are discussed as intermediates in these reactions [120]. N-benzoyloxypiperidine exhibits an unexpected type of reactivity toward $Fe_2(CO)_9$ (cf. Method II e). Benzaldehyde and N-formylpiperidine are formed, and LXXI is formulated as a possible intermediate [112]. $^4LFe(CO)_3$ reacts with CF_2=CF_2 or CF_3CF=CF_2 to form π-allylic species, in which the fluoroolefin formally links the Fe and the organic ligand (no details, cf. Method V d) [45]. The reaction path for the interconversion of $^4LFe(CO)_3$ (^4L=cycloocta-1,3-diene) and CF_3CF=CF_2 is discussed in [111]. Irradiation of $Fe(CO)_5$ with LXXII in ether at 20 °C for 3 h gives LXXIII and a compound of the composition $C_{11}H_6FeO_5$ (>90 °C slow decomposition, orange plates, ^1H NMR ($CDCl_3$): δ=2.75 (dd, 2H), 3.36 (ddd, 2H), 4.18 (d, 2H) ppm, IR (KBr): 1695 (>C=O) cm^{-1}, IR (hexane): 1998, 2008, 2060 cm^{-1}) and unknown structure [50].

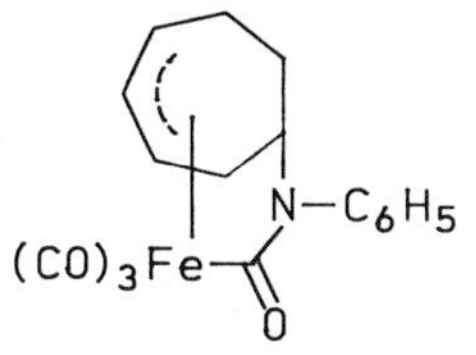

(CO)$_3$Fe, N−C_6H_5, O — LXXI

O — LXXII

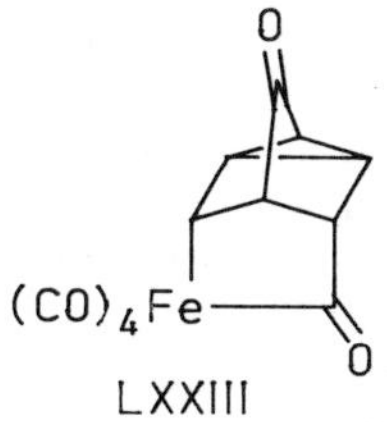

O, (CO)$_4$Fe, O — LXXIII

For the arrangement of the compounds in Table 4, see also 1.4.1.4.3 (p. 43).

Table 4
Compounds of the Type $^4LFe(CO)_3$ where ^4L is a Cyclic σ,π-Allyl System.
Further information on numbers preceded by an asterisk is given at the end of the table, pp. 155/72.
For abbreviations and dimensions see p. VIII.

No.	compound	method of preparation (yield in %), properties and remarks	Ref.
compounds with a four-membered ring system:			
*1	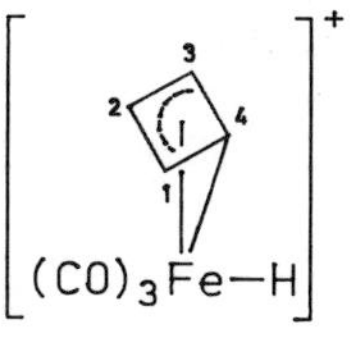 [(CO)$_3$Fe−H, ring positions 1, 2, 3, 4]$^+$	^1H NMR ($HOSO_2F$/liquid SO_2 at −85°): −11.16 (d, Fe-H; J(Fe-H, 4)=29.0), 3.45 (br,d, H-4), 4.86 (br,s, H-1,3), 6.44 (br,s, H-2) ^{13}C NMR ($HOSO_2F$/liquid SO_2 at −90°): 9.0 (dd, C-4; J(^{13}C-4, Fe-H)=81.2, J(^{13}C,H)=191.4), 65.8 (d, C-1,3; J(^{13}C,H)=202.9), 117.3 (d, C-2; J(^{13}C,H)=191.2), 200.0 (s, 2CO), 202.7 (s, 1CO)	[78]

References on pp. 172/5

Table 4 [continued]

No.	compound	method of preparation (yield in %), properties and remarks	Ref.
2	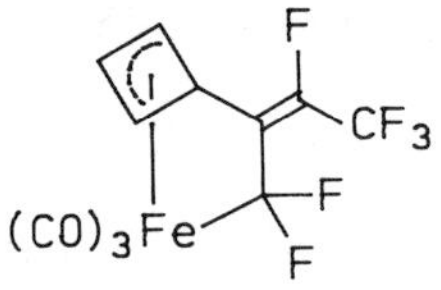 $(CO)_3Fe$ complex (structure: F, CF_3, F, F)	Va (47) m.p. 101°, pale yellows needles (from hexane at −78°) ^{1}H NMR ($CDCl_3$): 3.66 (s, 1H), 4.64 (s, 2H), 6.38 (s, 1H) ^{19}F NMR ($CDCl_3$): 68.1 (CF_3; J(CF_3, CF) = 11.0), 94.8 (CF; J(CF, CF_2) = 5.0), 95.8 (CF_2; J(CF_2, CF_3) = 15.0) IR (hexane): 1727 ν(C=C); 2011, 2047, 2089 (CO) mass spectrum: $[M-nCO]^+$ (n = 0 to 3), $[M-2CO-F]^+$	[25]
3	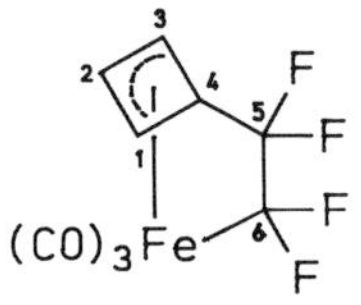 $(CO)_3Fe$ complex (structure: positions 1–6, F, F, F, F)	Va (82) m.p. 81°, pale yellow crystals (from hexane at −78°) ^{1}H NMR ($CDCl_3$): 3.59 (t, H-4; J(H, F) = 5.0), 4.44 (s, H-1, 3), 6.34 (s, H-2) ^{19}F NMR ($CDCl_3$): 66.1 (d, F-5; J(F, H) = 5.0), 118.0 (br, s, F-6) IR (hexane): 2016, 2037, 2091 (CO) mass spectrum: $[M-nCO]^+$ (n = 0 to 3)	[25]
4	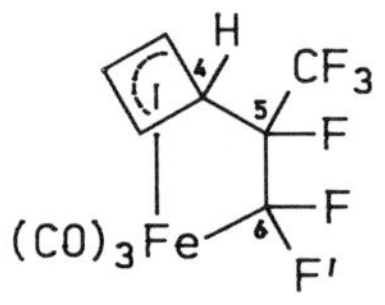 $(CO)_3Fe$ complex (structure: H, CF_3, F, F, F′; positions 4, 5, 6)	Va (27) m.p. 104°, pale yellow crystals ^{1}H NMR ($CDCl_3$): 3.70 (s, 1H), 4.15 (d, 1H; J(H, H) = 10.0 (transannular)), 5.08 (d, 1H), 6.38 (d, 1H; J(H, F) = 3.0) ^{19}F NMR ($CDCl_3$): 69.3 (CF_3; J(CF_3, F-5) = 11.0, J(CF_3, F-6′) = 3.0, J(CF_3, F-6) = 16.0), 111.8 (F-6′; J(F-6, 6′) = 244), 120.2 (F-6; J(F-5, 6) = 9.0), 143.5 (F-5; J(F-5, H-4) = 3.0) IR (hexane): 2015, 2040, 2093 (CO) mass spectrum: $[M-nCO]^+$, $[M-nCO-F]^+$ (n = 0 to 3), $[M-F]^+$	[25]
*5	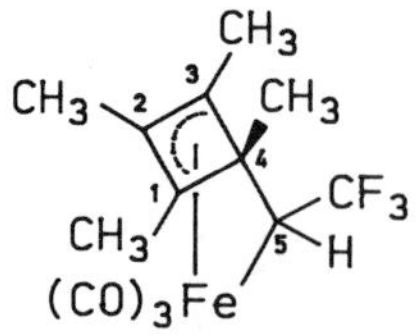 $(CO)_3Fe$ complex (structure: CH_3 groups at 1, 2, 3, 4; CF_3, H at 5)	Va (20) m.p. 100°, pale yellow crystals (from hexane at −78°) ^{1}H NMR ($CDCl_3$): −0.2 (q, H-5; J(H, F) = 12.0), 0.84 (s, CH_3-4), 1.86 (s, CH_3-1, 3), 1.88 (s, CH_3-2) ^{19}F NMR ($CDCl_3$): 55.7 (d, CF_3; J(F, H) = 12.0) IR (hexane): 1977, 1995, 2056 mass spectrum: $[M-nCO]^+$ (n = 0 to 3)	[39, 86]

References on pp. 172/5

Table 4 [continued]

No.	compound	method of preparation (yield in %), properties and remarks	Ref.
6	CH_3 CH_3 CH_3 F 5 H F CH_3 $(CO)_3Fe$ 6 F'	Va (60) m.p. 120°, yellow crystals ^{1}H NMR ($CDCl_3$): 1.12 (s, CH_3), 1.67 (s, CH_3), 1.76 (s, CH_3), 2.16 (s, CH_3), 4.13 (m, H-5; J(H-5, F-6) = 6.0, J(H-5, F-6′) = 15.0, J(H-5, F-5) = 54.0) ^{19}F NMR ($CDCl_3$): 51.4 (m, F-6′; J(H-5, F-6′) = 15.0, J(F-6, 6′) = 220, J(F-5, 6′) = 11.0), 69.0 (m, F-6; J(F-5, 6) = 2.0, J(F-6, H-5) = 6.0), 191.0 (m, F-5; J(F-5, H-5) = 54.0) IR (hexane): 1993, 2015, 2071 (CO) mass spectrum: $[M-nCO]^+$ (n = 0 to 3), $[M-nCO-F]^+$ (n = 1 to 3), $[M-F]^+$	[39, 86]
7	CH_3 CH_3 CH_3 F 5 F F CH_3 $(CO)_3Fe$ 6 F	Va (93) m.p. 148 to 150°, pale yellow crystals (from hexane at −78°) ^{1}H NMR ($CDCl_3$): 1.10 (s, CH_3), 1.76 (s, 2 CH_3), 2.20 (s, CH_3) ^{19}F NMR ($CDCl_3$): 70.8 (t, F-5; J(F, F) = 3.0), 120.8 (t, F-6; J(F, F) = 3.0) IR (hexane): 1999, 2011, 2076 (CO) mass spectrum: $[M-nCO]^+$ (n = 0 to 3), $[M-3CO-F]^+$	[14, 25]
8	CH_3 CH_3 CH_3 CF_3 5 F F CH_3 $(CO)_3Fe$ 6 F'	Va (72) m.p. 123°, pale yellow crystals ^{1}H NMR ($CDCl_3$): 1.12 (s, CH_3), 1.62 (s, CH_3), 1.79 (s, CH_3), 2.20 (s, CH_3) ^{19}F NMR ($CDCl_3$): 69.3 (CF_3; J(CF_3, F-5) = 11.0, J(CF_3, F-6′) = 3.0, J(CF_3, F-6) = 17.0), 107.5 (F-6′; J(F-6, 6′) = 240, J(F-5, 6′) = 11.0), 123.0 (F-6; J(F-6, CF_3) = 11.0), 155.0 (F-5; J(F-5, 6) = 8.0) IR (hexane): 1996, 2023, 2076 (CO) mass spectrum: $[M-nCO]^+$, $[M-nCO-F]^+$ (n = 0 to 3)	[14, 25]
9	CH_3 CH_3 CH_3 F CF_3 CH_3 F $(CO)_3Fe$ F	Va (37) m.p. 136°, pale yellow crystals (from hexane at −78°) ^{1}H NMR ($CDCl_3$): 1.10 (s, CH_3), 1.68 (s, 2 CH_3), 2.20 (s, CH_3) ^{19}F NMR ($CDCl_3$): 68.2 (CF_3; J(CF_3, CF_2) = 16), 94.8 (CF; J(CF_3, CF) = 11.0), 96.8 (CF_2; J(CF_2, CF) = 8.0) IR (hexane): 1727 (C=C); 2001, 2021, 2074 (CO) mass spectrum: $[M-nCO]^+$ (n = 0 to 3), $[M-nCO-F]^+$ (n = 1 to 3)	[25]

References on pp. 172/5

Table 4 [continued]

No.	compound	method of preparation (yield in %), properties and remarks	Ref.
compounds with a five-membered ring system:			
*10	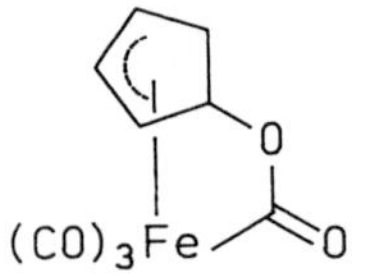	Id (48) m.p. 107° (dec.), pale yellow IR (KBr): 1650 (C=O) IR (hexane): 1993, 2003, 2073 (CO) mass spectrum: $[M]^+$	[48]
*11	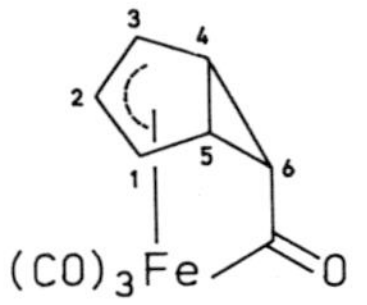	IVa (80% for the crude product) dec. >70°, yellow, air-sensitive crystals (from acetone/pentane (20:1) at −78° with high losses) ^{1}H NMR ($CDCl_3$ at −10°): 1.90 (H-4,5; J(H-5,6) = J(H-4,6) = 5.5), 2.53 (H-6; J(H-2,6) = 2.5), 4.95 (H-1,3; J(H-1,5) = J(H-3,4) = ca. 1.8), 5.90 (H-2; J(H-2,3) = J(H-1,2) = 3) ^{13}C NMR ($CDCl_3$ at −20°): 28.8 (C-4,5), 77.3 (C-1,3), 77.9 (C-6 in acetone-d_6), 97.7 (C-2); 207.2 (1 CO), 210.0 (2 CO), 252.7 (C=O) IR (KBr): 1652 (C=O) IR (hexane): 1992, 2008, 2065 (CO) mass spectrum: $[M-nCO]^+$ (n = 0 to 4)	[107, 108]
*12	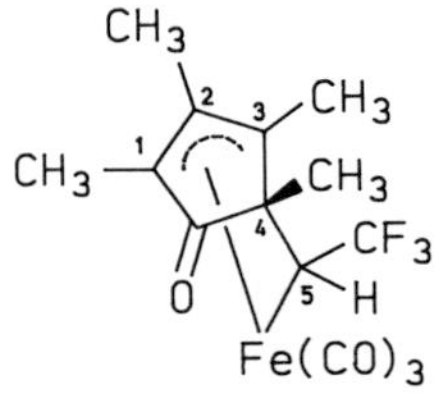	see "Further information" for No. 5, p. 155 m.p. 105 to 106°, pale yellow needles (from hexane at −78°) ^{1}H NMR ($CDCl_3$): −0.14 (q, H-5; J(H,F) = 13.0), 0.92 (q, CH_3-4), 1.85 (s, CH_3-3), 1.99 (s, CH_3-1), 2.18 (s, CH_3-2) ^{19}F NMR ($CDCl_3$): 53.9 (dq, CF_3; J(CF_3,H-5) = 13.0), J(CF_3,CH_3) = 1.5) IR (hexane): 1731, 1988, 2001, 2059 mass spectrum: $[M-nCO]^+$ (n = 0 to 3)	[39, 86]
compounds with a six-membered ring system:			
*13	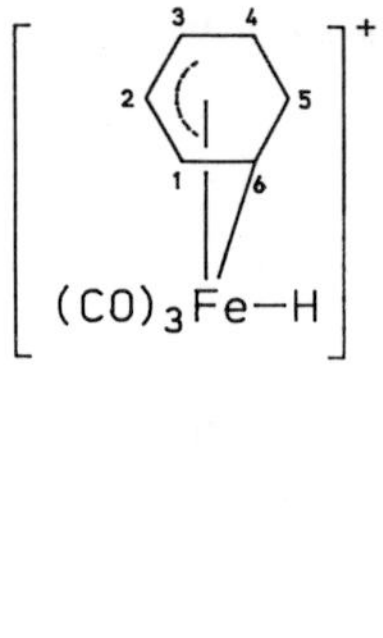	^{13}C NMR ($HOSO_2F$/liquid SO_2 at −80°): 13.8 (C-5); 19.0, 26.4 (C-4,6); 81.2, 83.6 (C-1,3); 96.4 (C-2); 196.2, 200.3, 203.0 (CO) ^{13}C NMR ($HOSO_2F$/liquid SO_2 at −20°): 14.4 (C-5), 23.5 (C-4,6), 82.9 (C-1,3), 97.0 (C-2) ^{13}C NMR (CF_3CO_2H/HBF_4 at 25°): 16.0 (t; J = 133), 24.6 (dd, C-4,6; J = 109, J = 142), 84.5 (d, C-1,3; J = 178), 99.0 (d, C-2; J = 179), 201.0 (2 CO), 201.5 (1 CO) IR (CF_3CO_2H/HBF_4): 2084, 2128 (CO)	[82]

References on pp. 172/5

Table 4 [continued]

No.	compound	method of preparation (yield in %), properties and remarks	Ref.
*14	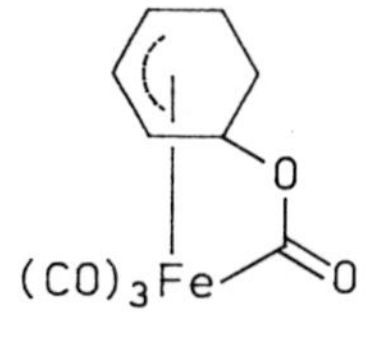 (CO)$_3$Fe, O, O	I d (32 to 60) m.p. 74° (dec.), m.p. 80° (dec.), pale yellow crystals ^{1}H NMR ($CDCl_3$): 1.08 to 1.46 (m, 1H), 1.5 to 1.78 (m, 1H), 1.92 to 2.64 (m, 2H), 4.44 to 4.57 (m, 1H), 4.7 (t, 1H; J=6), 5.44 (t, 1H; J=6), 5.68 to 5.84 (m, 1H) IR: 665, 970, 995, 1010, 1060, 1325, 1340, 1375, 1460, 1650, 1990, 2100, 2900 IR (KBr): 1630 (C=O) IR (hexane): 1993, 2002, 2072 (CO) mass spectrum: $[M]^+$	[48, 85, 116]
*15	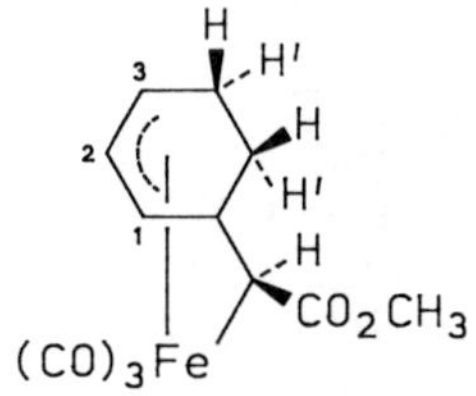 H, H′, H, H′, H, CO_2CH_3, 3, 2, 1, (CO)$_3$Fe	IIa (7.5) m.p. 101°, pale yellow prisms (from hexane at 0°) ^{1}H NMR (C_6H_6): −0.1 (H-7; J(H-6,7)=2); 0.7, 1.0 (H-5,5′; J(H-5,5′)=5, J(H-4,5′)=9), 1.4 (H-4; J(H-4,4′)=−17, J(H-4,5)=8), 1.7 (H-4′; J(H-4′,5)=6), 3.05 (H-6; J(H-1,6)=5), 3.76 (H-3, CH_3; J(H-3,4′)=6, J(H-1,3)=−2), 4.05 (H-2; J(H-2,3)=6), 4.38 (H-1; J(H-1,2)=6) ^{13}C NMR (benzene-d_6): −9.0 (C-7); 21.5, 24.4 (C-4,5); 31.6 (C-6), 50.9 (CH_3); 62.1, 64.8, 92.4 (C-1,2,3); 181.1 (CO_2); 206.4, 211.1, 213.6 (CO) IR (KBr): 1685 (C=O) IR (hexane): 1987, 2000, 2060 (CO) mass spectrum: $[M-nCO]^+$ (n=0 to 3), $[C_6H_6Fe]^+$	[83]
*16	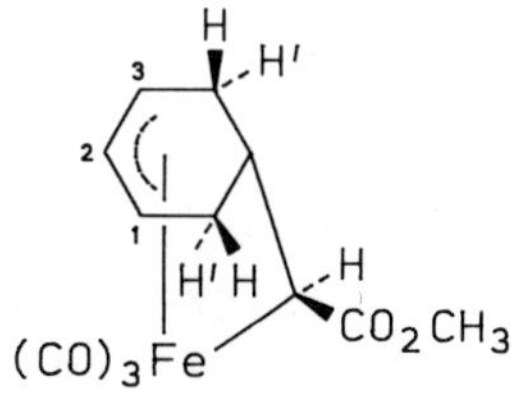 H, H′, H′ H, H, CO_2CH_3, 3, 2, 1, (CO)$_3$Fe	II d (4.5), see also No. 67, p. 163 m.p. 55°, pale yellow crystals (from hexane at −60°) ^{1}H NMR (benzene-d_6): 1.20 (H-4′; J(H-3,4′)=4), 1.72 (H-6; J(H-6,6′)=−14), 1.76 (H-4; J(H-4,4′)=−14), 2.03 (H-5; J(H-4′,5)=4), 2.32 (H-7; J(H-5,7)=1), 2.50 (H-6′; J(H-5,6′)=4), 3.8 (CH_3), 4.7 (H-1,3; J(H-2,3)=7, J(H-1,6′)=4), 4.86 (H-2; J(H-1,2)=7) ^{13}C NMR (benzene-d_6): 30.1 (C-6), 33.3 (C-4), 39.9 (C-5), 46.9 (C-7), 50.3 (CH_3); 78.5, 78.9 (C-1,3); 98.2 (C-2), 179.8 (CO_2); 207.7, 213.6, 214.1 (CO) IR (KBr): 1685 (C=O) IR (hexane): 1987, 2002, 2059 (CO) mass spectrum: $[M-nCO]^+$ (n=0 to 3)	[83]

References on pp. 172/5

Table 4 [continued]

No.	compound	method of preparation (yield in %), properties and remarks	Ref.
*17	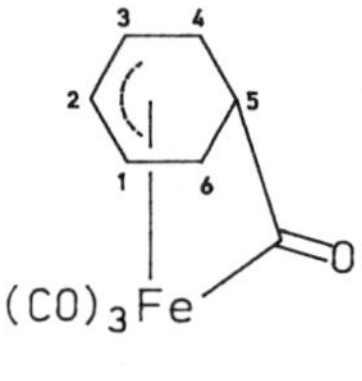	II d (7.7), IV a (traces) m.p. 90 to 101°, subl. p. 40°/0.01 Torr, long (2 to 3 cm) pale yellow crystals (from C_6H_6/pentane (1:20)) ^{1}H NMR (C_6H_6): 1.03 (H-4, 4′, H-6, 6′), 2.11 (H-5), 4.02 (H-1, 3), 4.43 (H-2) IR (KBr): 1670 (C=O) IR (hexane): 2005, 2063 (CO)	[38]
18	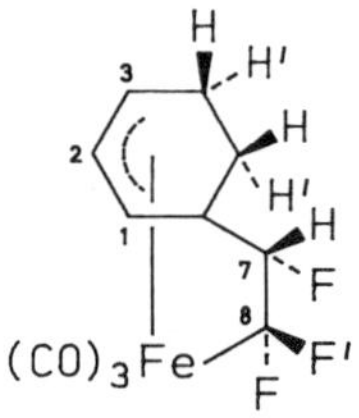	V b (33) m.p. 94 to 96°, pale yellow crystals (from CH_2Cl_2/hexane) ^{1}H NMR ($CFCl_3$ (?)): 1.29 (m, H-4′, 5′; J (H-4′, F-8′) = 7.5, J (H-3, 4′) = 3.0), 2.23 (m, H-4, 5), 2.60 (dddd, H-6; J (H-1, 6) = J (H-5, 6) = J (H-5′, 6) = 7.5, J (H-6, F-7) = 16.0), 4.52 (dddd, H-3; J (H-2, 3) = 6.5, J (H-3, 4) = J (H-3, 4′) = 3.0, J (H-1, 3) = 2.0), 4.51 (dd, H-2; JH-1, 2) = 6.5), 5.60 (ddd, H-1; J (H-1, 3) = 2.0) ^{19}F NMR ($CHCl_3$): 44.9 (dddd, F-8; J (F-8, 8′) = 252, J (F-7, 8) = 17.0, J (F-8, H-4′) = 7.5, J (F-8, H-7) = 3.0), 87.3 (ddd, F-8′; J (F-7, 8′) = 15.0, J (F-8′, H-7) = 12.0), 177.8 (dddd, F-7; J (F-7, 8′) = 15.0, J (F-7, H-7) = 52.0, J (F-7, H-6) = 16.0) IR (hexane): 2007, 2024, 2080 (CO). mass spectrum: $[M - nCO]^+$ (n = 0 to 3), $[M - nCO - HF]^+$ (n = 0 to 2)	[71]
19	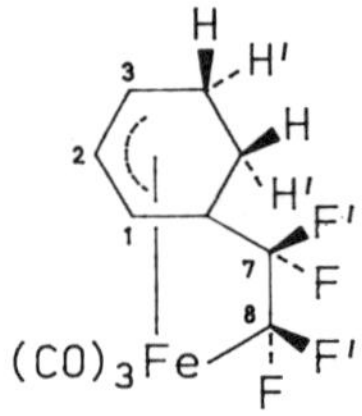	V b (26) m.p. 111 to 114°, yellow needles (from CH_2Cl_2/n-hexane at 0°) ^{1}H NMR ($CDCl_3$): 1.40 (m, H-4′, 5′; J (H-5′, 6) = 7.0), 2.16 (m, H-4, 5; J (H-5, 6) = 7.0), 2.74 (ddddd, H-6; J (H-5, 6) = J (H-4′, 6) = J (H-6, F-8′) = 7.0), 4.42 (dddd, H-3; J (H-2, 3) = 6.5, J (H-3, 4 or 4′) = 5.0, J (H-3, 4 or 4′) = 2.0), 4.97 (ddd, H-1; J (H-1, 6) = 7.0, J (H-1, 2) = 6.5, J (H-1, 3) = 2.0), 5.48 (ddd, H-2; J (H-2, F-7) = 3.5) ^{19}F NMR ($CDCl_3$): 63.0 (F-8′; J (F-7, 8′) = 14.0, J (F-8′, H-6) = 7.0), 97.2 (F-8; J (F-8, 8′) = 247), 110.5 (F-7′; J (F-7′, 8′) = 7.0), 114.8 (F-7; J (F-7, 7′) = 226, J (F-7, 8) = J (F-7, H-6) = 9.0, J (F-7, H-2) = 3.5) IR (n-hexane): 2009, 2029, 2084 (CO) mass spectrum: $[M - nCO]^+$ (n = 0 to 3)	[64]

References on pp. 172/5

Table 4 [continued]

No.	compound	method of preparation (yield in %), properties and remarks	Ref.
*20	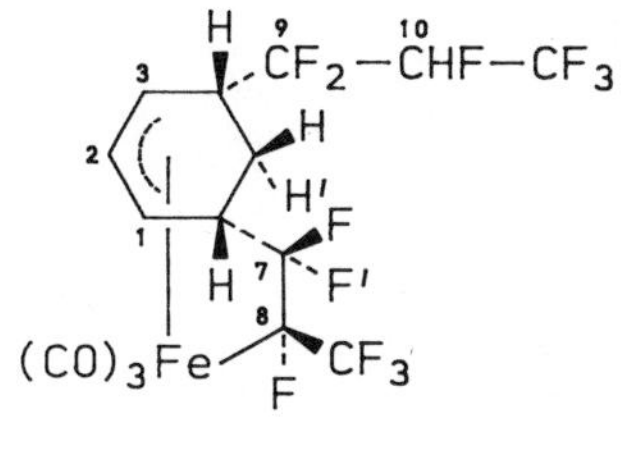	Vb (3) m.p. 129 to 131°, yellow needles (from CH_2Cl_2/hexane) ^{1}H NMR (acetone-d_6): 2.38 to 2.97 (compl. m, H-4, 5, 5′), 3.50 (m, H-6; J(H-1, 6) = 5.0), 4.90 m, H-3; J(H-2, 3) = 5.0, J(H-1, 3) = 1.5), 5.25 (ddd, H-1; J(H-1, 2) = 5.0), 5.48 (dm, H-10; J(H-10, F-10) = 43.0, J(H-10, F-9 or 9′) = 16.0, J(H-10, CF_3-10) = 11.0), 6.04 (ddd, H-2; J(H-2, F-7′) = 5.0) ^{19}F NMR (acetone): 65.4 (ddd, CF_3-8; J(CF_3-8, F-8) = J(CF_3-8, F-7) = 9.0, J(CF_3-8, F-7′) = 17.0), 73.5 (ddd, CF_3-10; J(CF_3-10, F-9′) = J(CF_3-10, F-10) = J(CF_3-10, H-10) = 11.0, J(CF_3-10, F-9) = 6.0), 90.3 (dm, F-7; J(F-7, H-2) = 5.0), 108.9 (dm, F-7′; J(F-7, 7′) = 248), 112.8 (m, F-9, 9′; J(F-9 or 9′, H-10) = 16.0), 188.5 (m, F-8), 211.2 (dm, F-10; J(F-10, H-10) = 43.0) IR (hexane): 2020, 2037, 2090 (CO) mass spectrum: $[M-nCO]^+$ (n = 0 to 3), $[M-nCO-F]^+$ (n = 1, 2), $[M-CO-nF]^+$ (n = 1, 2), $[M-C_3F_6-CO]^+$, $(M-C_3F_6CO-F]^+$	[71, 89]
*21	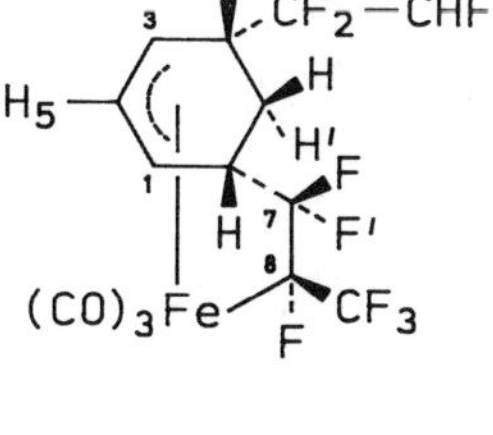	II e (30) oil, stable when kept at 0° under N_2	[96]
22		Vb (9) m.p. 116 to 119°, pale yellow crystals ^{1}H NMR (acetone-d_6): 1.37 (t, CH_3; J(H, H) = 7.5), 2.48 (q, CH_2; J(H, H) = 7.5), 2.24 to 2.80 (compl. m, H-4, 5, 5′), 3.56 (m, H-6; J(H-1, 6) = 8.0), 4.78 (m, H-3; J(H-1, 3) = 2.0), 5.13 (dd, H-1), 5.50 (dm, H-10; J(H-10, F-10) = 43.0, J(H-10, F-9 or 9′) = 18.0, J(H-10, CF_3-10) = 12.0) ^{19}F NMR (acetone): 65.7 (ddd, CF_3-8; J(CF_3-8, F-7′) = 16,5, J(CF_3-8, F-8) = J(CF_3-8, F-7) = 9.5), 73.7 (dddd, CF_3-10; J(CF_3-10, F-9′) = J(CF_3-10, F-10) = J(CF_3-10, H-10) = 12.0, J(CF_3-10, F-9) = 6.0), 94.3 (dm, F-7; J(F-7, 7′) = 255), 109.3 (dm, F-7′), 113.7 (compl. m, F-9, 9′; J(F-9 or 9′, H-10) = 18.0), 186.0 (m, F-8), 209.5 (dm, F-10; J(F-10, H-10) = 43.0)	[71]

References on pp. 172/5

Table 4 [continued]

No.	compound	method of preparation (yield in %), properties and remarks	Ref.
		IR (hexane): 2015, 2032, 2086 (CO) mass spectrum: $[M-nCO]^+$ (n = 1 to 3), $[M-nCO-F]^+$ (n = 2, 3), $[M-C_3F_6-CO-F]^+$ (F = 0, 1) (structure drawn in analogy to No. 20)	
*23	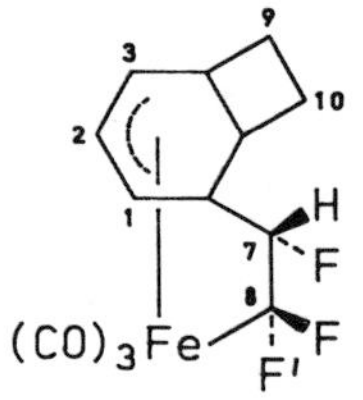	Vf (66) m.p. 110 to 113°, yellow needles ^{1}H NMR ($CDCl_3$): 1.46 to 2.40 (compl. m, H-4, 5, 9, 10; J(H-4, F-8′) = 7.5), 2.83 (m, H-6; J(H-6, F-7) = 18.0), 4.27 (d, H-3), 4.45 (ddd, H-7; J(H-7, F-7) = 54.0, J(H-7, F-8) = 12.5, J(H-7, F-8′) = 3.5), 5.46 (dd, H-1; J(H-1, 6) = 7.0), 5.61 (dd, H-2; J(H-1, 2) = J(H-2, 3) = 7.0) ^{19}F NMR (acetone): 44.6 (dddd, F-8′; J(F-8, 8′) = 250, J(F-8′, H-4) = 7.5, J(F-8′, H-7) = 3.5), 80.1 (ddd, F-8; J(F-7, 8) = J(F-8, H-7) = 12.5), 176.7 (dddd, F-7; J(F-7, 8′) = J(F-7, H-6) = 18.0, J(F-7, H-7) = 54.0) IR (hexane): 2008, 2026, 2080 (CO) mass spectrum: $[M-nCO]^+$ (n = 0 to 3), $[C_8H_{10}Fe(CO)_n]^+$ (n = 0 to 2)	[71]
24	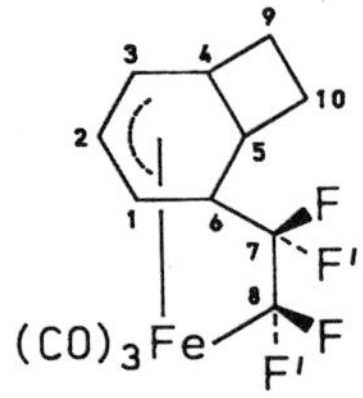	Vf (40) m.p. 134 to 136°, yellow crystals (from CH_2Cl_2/n-hexane at 0°) ^{1}H NMR ($CDCl_3$): 1.45 to 2.62 (compl. m, H-4, 5, 9, 10; J(H-4, F-8′) = 7.5), 2.78 (m, H-6; J(H-6, F-7′) = 8.5, J(H-6, F-7) = 3.0), 4.31 (d, H-3), 4.85 (dd, H-1; J(H-1, 6) = 7.0), 5.65 (dd, H-2; J(H-1, 2) = J(H-2, 3) = 7.0, J(H-2, F-7) = 3.5) ^{19}F NMR (acetone: 64.7 (dddd, F-8′; J(F-8, 8′) = 248, J(F-7′, 8′) = 13.0, J(F-7, 8′) = J(F-8′, H-4) = 7.5), 91.3 (dm, F-8; J(F-7′, 8) = 8.5), 111.4 (ddddd, F-7; J(F-7, 7′) = 222, J(F-7, H-6) = 3.0, J(F-7, H-2) = 3.5), 114.5 (dddd, F-7′; J(F-7′, H-6) = 8.5) IR (hexane): 2013, 2030, 2085 (CO) mass spectrum: $[M-nCO]^+$ (n = 0 to 3)	[64]
25	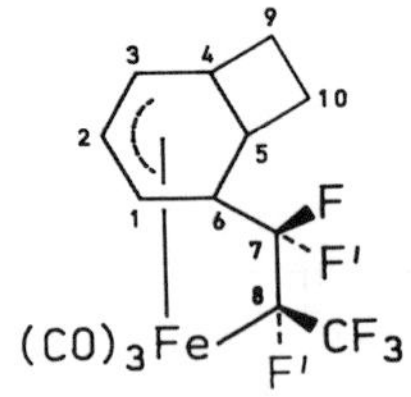	Vf (58) m.p. 139 to 142°, yellow platelets ^{1}H NMR ($CDCl_3$): 1.44 to 2.76 (compl. m, H-4, 5, 9, 10), 2.91 (m, H-6; J(H-6, F-7′) = 15.0), 4.41 (d, H-3), 4.82 (dd, H-1; J(H-1, 6) = 7.0), 5.75 (ddd, H-2; J(H-1, 2) = J(H-2, 3) = 7.0, J(H-2, F-7) = 4.0)	[71]

Table 4 [continued]

No.	compound	method of preparation (yield in %), properties and remarks	Ref.
		^{19}F NMR (acetone): 65.5 (ddd, CF_3-8; J(CF_3-8, F-7′) = 15.0, J(CF_3-8, F-7) = J(CF_3-8, F-8) = 9.0), 98.5 (dm, F-7; J(F-7, H-2) = 4.0), 114.6 (dm, F-7′; J(F-7,7′) = 232, J(F-7′,8) = J(F-7′, H-6) = 15.0), 186.3 (m, F-8) IR (hexane): 2012, 2030, 2083 (CO) mass spectrum: $[M-nCO]^+$ (n = 0 to 3), $[M-2CO-F]^+$	
26	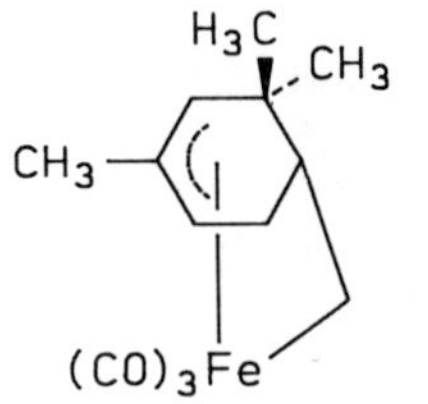	I a	[68]
27	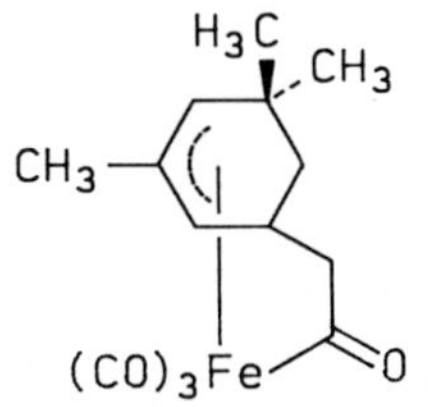	I a	[68]
*28	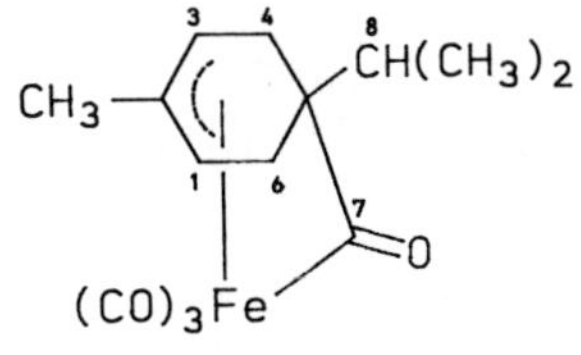	I b (62), III a (55) m.p. 106 to 107°, light yellow crystals (from tetrahydrofuran) ^{1}H NMR ($CDCl_3$): δ = 45 (d, 2CH_3-8; J = 7.5), 78 to 108 (m, H-4,6,8), 120 (s, CH_3-2), 276 (t, H-1,3) ^{13}C NMR: 19.56 (2CH_3-8), 27.33 (CH_3-2), 30.3 (C-8), 30.83 (C-4,6), 74.08 (C-5), 75.0 (C-1,3), 211.9 (C=O, CO) IR: 1665 (C=O); 2005, 2020, 2080 (CO) mass spectrum: $[M-nCO]^+$ (n = 0 to 3), $[M-Fe(CO)_3]^+$	[101]
*29	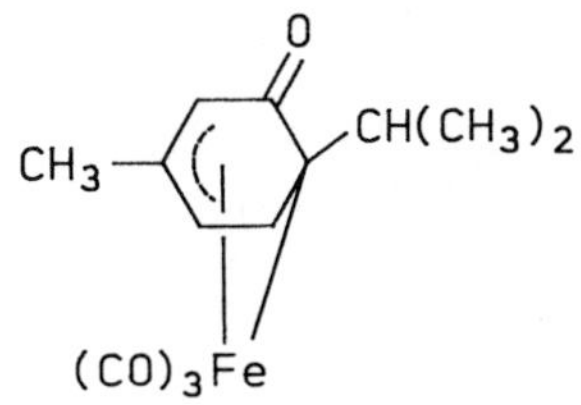	I b	[101]

References on pp. 172/5

Table 4 [continued]

No.	compound	method of preparation (yield in %), properties and remarks	Ref.
*30	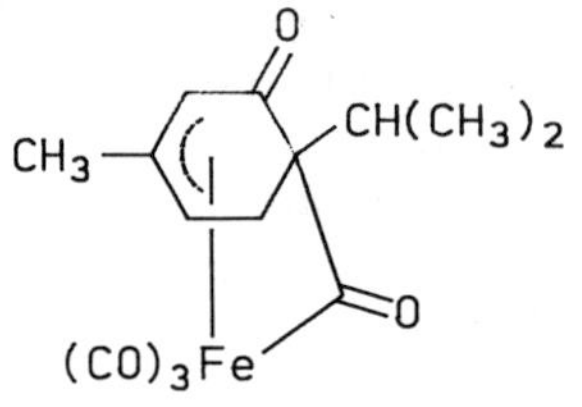	I b	[101]
*31	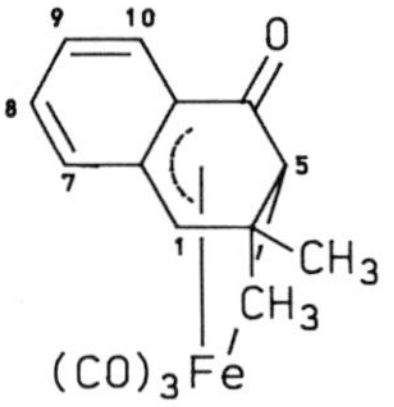	V f (46) m.p. 100 to 102° ^{1}H NMR ($CDCl_3$, shift on addition of Eu (fod) (cf. No. 15, p. 156) is shown in brackets): 0.6 (0.32) (s, CH_3-6), 1.4 (0.37) (s, CH_3-6), 2.75 (0.9) (s, H-5), 4.25 (0.3) (s, H-1), 7.3 (0.2) (m, H-7), 7.6 (0.24) (m, H-8, 9), 8.6 (1.2) (m, H-10) IR (hexane): 1645, 1992, 2012, 2066	[55]

compounds with a seven-membered ring system:

No.	compound	method of preparation (yield in %), properties and remarks	Ref.
*32	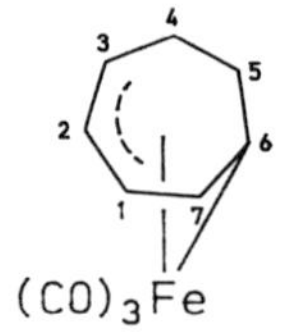	VI a (45 to 75), see also No. 33, p. 158 m.p. 61 to 63°, subl. 30°/0.01 Torr, yellow crystals (from CH_3OH at −60°) ^{1}H NMR (C_6H_6): 1.04 (m, H-6), 1.7 to 2.1 (m, H-4, 4′, 5, 5′, 7 exo), 2.55 (m, H-7 endo), 3.75 (m, H-1, 2), 4.55 (m, H-3) ^{13}C NMR (benzene-d_6): 14.0 (C-6); 25.2, 35.2, 44.8 (C-4, 5, 7); 60.9, 78.6, 97.0 (C-1, 2, 3); 204.0, 215.6, 216.2 (CO) IR (hexane): 1988, 1996, 2048 (CO) mass spectrum: $[M-nCO]^+$ (n = 0 to 3), $[M-nCO-2H]^+$ (n = 2, 3), $[M-3CO-C_2H_2]^+$, $[M-Fe(CO)_3]^+$	[37, 38, 56, 58, 73]
*33		II d, see also No. 32, p. 158 sinters or decomposes at ca. 60°, long pale yellow crystals, yellow laminas ^{1}H NMR (C_6H_6 at 6°): 1.51 (dt, H-4 endo); 1.25, 1.6 to 2.1 (m, H-5, 5′, 6, 4 exo, 7, 7′), 3.85 (m, H-1), 4.07 (t, H-2), 4.59 (m, H-3) ^{13}C NMR (benzene-d_6): 27.0, 29.3, 36.2 (C-4, 5, 7); 54.4 (C-6); 73.2, 84.3, 91.5 (C-1, 2, 3); 205.0, 211.21, 212.1 (CO); 242.2 (C-8) IR (KBr): 1635 (C=O) IR (hexane): 1997, 2061 (CO)	[38, 73]

Table 4 [continued]

No.	compound	method of preparation (yield in %), properties and remarks	Ref.
34	$(CO)_3Fe$ … CN	VI a	[58, 99]
35	$(CO)_3Fe$ … CH_3	VI a (low yield)	[126]
36	$(CO)_3Fe$ … C_2H_5	VI a	[58]
37	CH_3, $(CO)_3Fe$ … CH_3	VI a (<20%) ^{1}H NMR ($CDCl_3$): 0.85 (d, CH_3; J=6.3), 0.86 (d, CH_3; J=7.2), 1.39 (d, 1H; J=8.8), 1.68 (m, 1H), 2.35 (m, 2H), 2.79 (m, H-7), 4.3 (m, H-1,3), 4.59 (t, H-2; J=8) IR ($CHCl_3$): 1975, 2045	[126]
*38	$(CO)_3Fe$ … O	V b (ca. 9 to 55) subl. 40°/0.001 Torr, pale yellow crystalline solid ^{1}H NMR (benzene-d_6): 1.6 (m, H-4,5), 2.82 (br, s, H-6), 3.59 (br,t, H-3; J(H-2,3)=8), 4.45 (t, H-2), 5.15 (d, H-1; J(H-1,2)=8) ^{13}C NMR ($CDCl_3$): 26.3, 34.5 (C-4,5); 56.6 (C-6); 78.4, 89.0, 100.2 (C-1,2,3); 182.4 (C-7); 201.5, 206.5, 212.1 (CO) IR (C_8H_{12}): 1650 (C=O); 1999, 2010, 2067 (CO)	[102, 104]
39	O, $(CO)_3Fe$	VI b m.p. 116° ^{1}H NMR ($CDCl_3$): 1.36 (m, H-6), 2.68 (H-4 exo; J(H-4 exo, 4 endo)=15.7), 2.68 (H-7), 3.23 (H-4 endo), 4.2 (H-1), 4.47 (H-2), 4.68 (H-3; J(H-3,4) =3.0) mass spectrum: $[M-nCO]^+$ (n=1 to 3), $[M-Fe(CO)_3]^+$	[80]

References on pp. 172/5

Table 4 [continued]

No.	compound	method of preparation (yield in %), properties and remarks	Ref.
*40	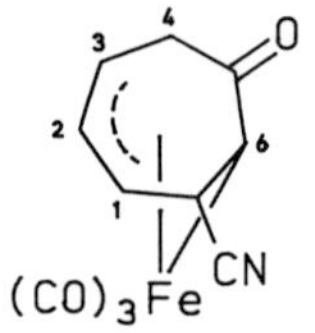 (CO)$_3$Fe, CN, O (structure)	VIb m.p. 166 to 170° (dec., from $CH_3CO_2C_2H_5$) ^{1}H NMR ($CDCl_3$): 1.72 (H-6; J(H-4,6) = 2.0, J(H-6,7) = 9.0), 2.91 (H-4 exo; J(H-3,4 exo) = 8.0, J(H-2,4 exo) = 2.0), 3.42 (H-4 endo; J(H-4 endo, 4 exo) = 16.5, J(H-3,4 endo) = 2.5), 4.3 (H-1), 4.69 (H-2; J(H-1,2) = 6.5), 4.73 (H-7; J(H-1,7) = 7.0), 4.85 (H-3; J(H-2,3) = 8.5) IR ($CHCl_3$): 1670 (C=O); 2000, 2080 (CO); 2220 (CN) mass spectrum: $[M-nCO]^+$, $[M-nCO-HCN]^+$ (n = 0 to 3)	[80]
41	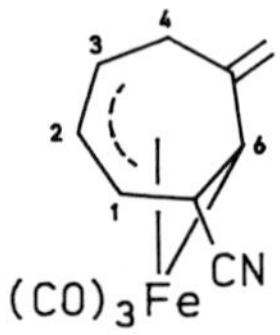 (CO)$_3$Fe, CN (structure)	VIb m.p. 100 to 104° (dec.), yellow crystals (from hexane) ^{1}H NMR ($CDCl_3$): 1.73 (H-6), 2.95 (H-4 exo, 4 endo), 3.76 (H-7), 4.22 (H-1); 4.33, 4.55 (CH_2); 4.55 (H-2), 5.04 (H-3) (all J values as in No. 40) IR ($CHCl_3$): 1630 (C=C), 2220 (CN) IR (hexane): 1994, 1999, 2058 (CO) mass spectrum: $[M-nCO]^+$ (n = 1 to 3), $[M-3CO-HCN]^+$, $[M-Fe(CO)_3-CN]^+$, $[M-3CO-HCN-HC-C]^+$	[80]
*42	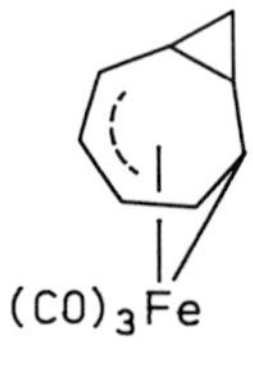 (CO)$_3$Fe (structure)	VIc (18 to 20), see No. 43, p. 159	[8, 40, 74]
*43	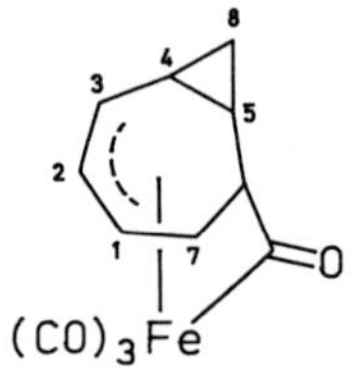 (CO)$_3$Fe, O (structure)	see No. 42, p. 159, VIc (18) m.p. 86 to 87° (dec., from pentane at −60°) ^{1}H NMR (C_6H_6): −0.40 (ddd, H-8 syn), 0.16 (ddd, H-8 anti), 0.95 (m, H-4,5), 1.04 (m, H-7 exo), 1.50 (ddd, H-7 endo), 2.50 (t, H-6), 3.85 (t, H-2), 4.45 (m, H-1,3) ^{13}C NMR (benzene-d_6): 11.9 (C-4 or 5), 15.6 (C-8), 21.2 (C-4 or 5), 27.3 (C-7), 58.2 (C-6); 77.4, 84.8 (C-1,2,3); 205.8, 211.7 (d) (CO); 248.5 (C=O) IR (KBr): 1640 (C=O) IR (hexane): 2004, 2062 (CO) mass spectrum: $[M-nCO]^+$ (n = 0, 2 to 4), $[M-3CO-Fe]^+$, $[C_5H_5Fe]^+$, $[C_7H_7]^+$	[40, 73]

References on pp. 172/5

Table 4 [continued]

No.	compound	method of preparation (yield in %), properties and remarks	Ref.
44	(CO)$_3$Fe, OH	–	[40]
45	(CO)$_3$Fe, O, OH	–	[40]
compounds with an eight-membered ring system:			
*46	$CO_2C_2H_5$, N, (CO)$_3$Fe	I e (19) yellow oil ^{1}H NMR (toluene-d_8 at 213 K): 1.0 (H-7; J(H-7,8 endo)=9, J(H-7,8 exo)=3), 1.44 (H-8 exo; J(H-8 exo, 8 endo)= −15), 1.93 (H-8 endo; J(H-1,8 endo)=9, J(H-1,8 exo)=3), 2.92 (H-1; J(H-1,2)=7.5), 3.38 (H-2; J(H-2,3)=7.5), 5.02 (H-6; J(H-6,7)=8), 6.25 (H-3), 7.19 (H-5; J(H-5,6)=8) ^{13}C NMR ($CDCl_3$ at −20°): −11.1 (d, C-7), 14.6 (9CH_3), 31.1 (t, C-8), 62.9 (t, OCH_2); 64.2 (d), 78.1 (d), 83.0 (d) (C-1,2,3); 121.6 (d), 126.9 (d) (C-5,6); 155.3 (s, C=O); 203.7 (s), 211.0 (s), 214.6 (s) (CO) IR (film): 1635 (C=C); 1705, 1730 (C=O) IR (hexane): 1982, 1994, 2059 (CO) mass spectrum: $[M-nCO]^+$ (n=0 to 3) and other fragments given	[94]
*47	N, $CO_2C_2H_5$, (CO)$_3$Fe	I e (27) yellow oil ^{1}H NMR (toluene-d_8 at 335 K): 1.96 (H-4 exo; J(H-4 exo, 4 endo)= −17), 2.44 (H-4 endo; J(H-4 endo, 5)=7), 2.90 (H-7; J(H-6,7)=6), 3.62 (H-2; J(H-2,3)=9), 4.13 (H-3; J(H-3,4 exo)=3, J(H-3,4 endo)=9), 5.60 (H-1; J(H-1,2)=4), 6.0 (H-5,6) ^{13}C NMR ($CDCl_3$ at −20°): 14.5 (q, CH_3), 17.4 (d, C-7), 24.4 (t, C-4), 61.3 (t, OCH_2); 66.1 (d), 80.4 (d), 83.4 (d) (C-1,2,3); 132.6 (d), 134.5 (d) (C-5,6); 154.7 (s, C=O); 205.5 (s), 213.7 (s), 214.5 (s) (CO)	[94]

References on pp. 172/5

Table 4 [continued]

No.	compound	method of preparation (yield in %), properties and remarks	Ref.
		IR (film): 1700 (C=O) IR (hexane): 1986, 1996, 2060 (CO) mass spectrum: $[M-nCO]^+$ (n=1 to 3) and other fragments given	
*48	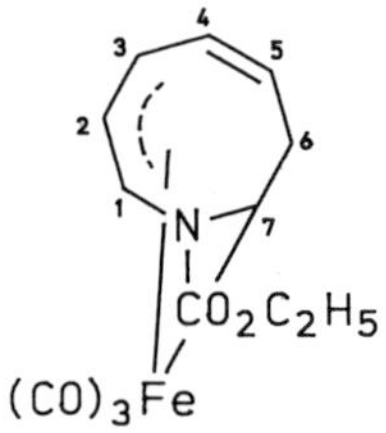 $CO_2C_2H_5$, $(CO)_3Fe$	I e (22) pale yellow oil ^{1}H NMR (CS_2 at 300 K): 1.6 (H-6 endo; J(H-6 endo, 7)=small), 2.2 (H-6 exo; J(H-6 exo, 7)=small), 2.72 (H-7), 3.34 (H-2; J(H-2,3)=9), 4.18 (H-3; J(H-3,4)=8), 5.54 (H-5; J(H-5,6)=3), 5.95 (H-4; J(H-4,5)=11), 6.05 (H-1; J(H-1,2)=5) ^{13}C NMR ($CDCl_3$ at −20°): 14.6 (q, CH_3), 17.7 (d, C-7), 29.2 (t, C-6), 61.5 (t, OCH_2); 56.2 (d), 75.6 (d), 84.9 (d) (C-1,2,3); 121.9 (d), 129.7 (d) (C-4,5); 203.9 (s), 212.2 (s), 212.9 (s) (CO) IR (film): 1700 (C=O) IR (hexane): 1983, 1987, 2061 (CO) mass spectrum: $[M-nCO]^+$ (n=0 to 3) and other fragments given	[94]
*49	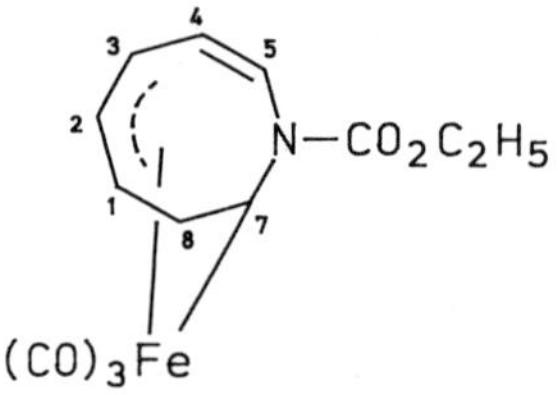 N–$CO_2C_2H_5$, $(CO)_3Fe$	I e (12) yellow oil ^{1}H NMR (CS_2 at 30°): 2.03 (H-8 exo; J(H-8 exo, H-8 endo)=−15), 2.77 (H-8 endo), 3.91 (H-1; J(H-1,8 endo)=7.5, J(H-1,8 exo)=3.5), 4.14 (H-2; J(H-1,2)=7.5), 4.24 (H-7; J(H-7,8 endo)=11, J(H-7,8 exo)=3.5), 4.79 (H-3; J(H-2,3)=9), 5.40 (H-4; J(H-3,4)=7.5), 6.83 (H-5; J(H-4,5)=9.5) ^{13}C NMR ($CDCl_3$ at −20°): 8.6 (d, C-7), 14.4 (q, CH_3), 31.7 (t, C-8), 62.4 (t, OCH_2); 59.9 (d), 60.1 (d), 88.6 (d) (C-1,2,3); 102.6 (d), 123.9 (d) (C-4,5); 152.3 (s, C=O); 204.5 (s), 209.9 (s), 215.6 (s) (CO) IR (film): 1640 (C=C); 1705, 1730 (C=O) IR (hexane): 1987, 1992, 2052 (CO) mass spectrum: $[M-nCO]^+$ (n=0 to 3) and other fragments given	[94]
*50	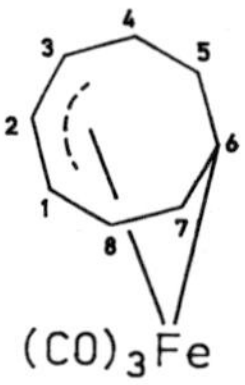 $(CO)_3Fe$	VI d (90) m.p. 105° ^{1}H NMR ($CDCl_3$): 1.2 to 2.7 (m, H-4 to 8), 4.35 (t, H-2; J(H-1,2)=J(H-2,3)=8.0), 5.20 (m, H-1,3) ^{13}C NMR ($CDCl_3$): 27.4 (C-5,7 or C-4,8), 51.2 (C-5,7 or C-4,8), 51.8 (C-6), 88.6 (C-1,3), 93.8 (C-2), 216.0 (CO)	[17, 57 60, 91]

Table 4 [continued]

No.	compound	method of preparation (yield in %), properties and remarks	Ref.
		^{13}C NMR (liquid SO_2): 27.3 (C-5,7 or C-4,8), 51.3 (C-5,7 or C-4,8), 51.9 (C-6), 89.1 (C-1,3), 94.1 (C-2), 215.6 (CO) ^{57}Fe-γ (−196°): $\delta=0.10$ (rel. stainless steel), $\Delta=0.62$ (±0.05) IR (cyclohexane): 1981, 2046 (CO)	
51	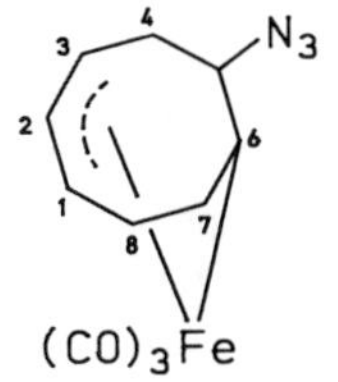 $(CO)_3Fe$ complex of 5-N_3-cyclooctadienyl (positions 1–8)	VId yellowish solid ^1H NMR: 1.1 to 3 (H-4,6,7,8), 4.2 (H-5), 4.9 (H-2), 5.49 (H-1,3) IR (pentane): 1990, 2045 (CO); 2085 (N_3) mass spectrum: $[M-nCO]^+$ (n=0 to 3)	[70, 127]
52	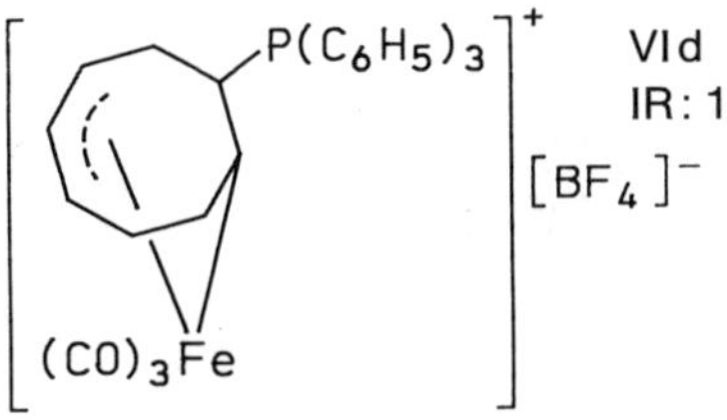 $[(CO)_3Fe$ complex with $P(C_6H_5)_3]^+[BF_4]^-$	VId IR: 1970, 2050 (CO)	[127]
53	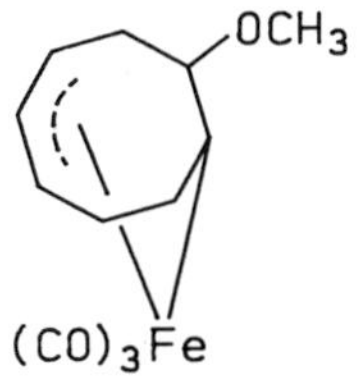 $(CO)_3Fe$ complex with OCH_3	VId subl. p. 0°/0.005 Torr (−30° cold finger), yellow crystals IR (CH_2Cl_2): 1080 (OCH_3); 1980, 2030, 2050 (CO) mass spectrum: $[M-nCO]^+$ (n=0 to 3)	[70, 127]
*54	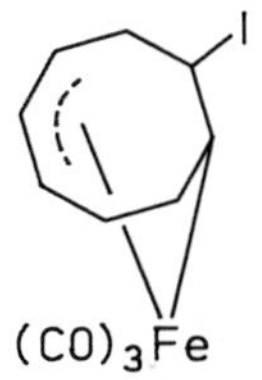 $(CO)_3Fe$ complex with I	VId IR: 1980, 2020 (CO)	[127]
*55	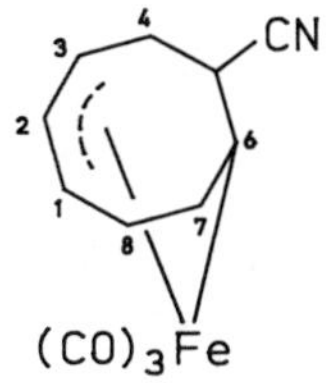 $(CO)_3Fe$ complex with CN (positions 1–8)	VId (7 to 20), see also No. 54, p. 162 m.p. 120°, white crystals (from CH_2Cl_2), white solid ^1H NMR ($CDCl_3$): 1.6 to 2.7 (m, H-4,6,7,8), 3.40 (m, H-5), 4.45 (t, H-2; J(H-2,3)=J(H-1,2)=8.2), 4.99 (m, H-1), 5.37 (m, H-3) IR (cyclohexane): 1991, 2055 (CO) IR (CH_2Cl_2): 1980, 2050 (CO); 2225 (CN) mass spectrum: $[M-nCO]^+$ (n=0 to 3)	[60, 127]

References on pp. 172/5

Table 4 [continued]

No.	compound	method of preparation (yield in %), properties and remarks	Ref.
56	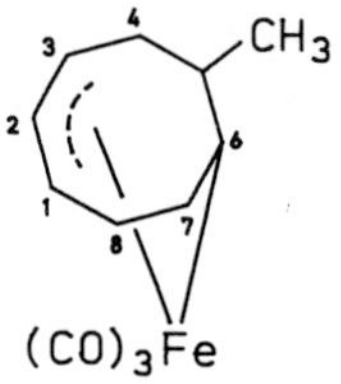 5-methyl compound, $(CO)_3Fe$	VI d (30) yellow oil ^1H NMR (C_6H_6): 0.65 (d, CH_3-5; J(CH_3-5, H-5) = 6.5), 0.9 to 2.3 (m, H-4, 6, 7, 8), 2.79 (m, H-5), 3.92 (t, H-2; J(H-1,2) = J(H-2,3) = 8.1), 4.70 (m, H-1,3) IR (cyclohexane): 1972, 1980, 2046 (CO)	[60]
57	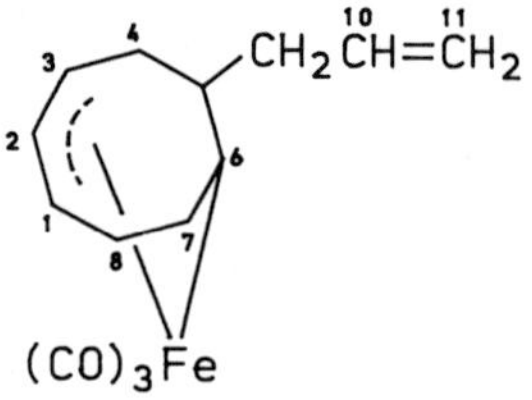 $CH_2\overset{10}{C}H{=}\overset{11}{C}H_2$, $(CO)_3Fe$	VI d (23) viscous orange oil ^1H NMR (C_6H_6): 1.05 to 2.78 (m, H-4 to 8, CH_2-5), 3.94 (t, H-2; J(H-1,2) = J(H-2,3) = 8.1), 4.65 (m, H-1,3), 4.9 (m, H-11), 5.55 (m, H-10) IR (cyclohexane): 1977, 2043 (CO)	[60]
58	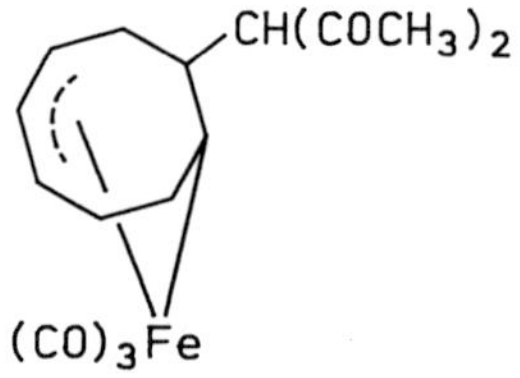 $CH(COCH_3)_2$, $(CO)_3Fe$	VI d IR (cyclohexane): 1981, 2048 (CO)	[60]
59	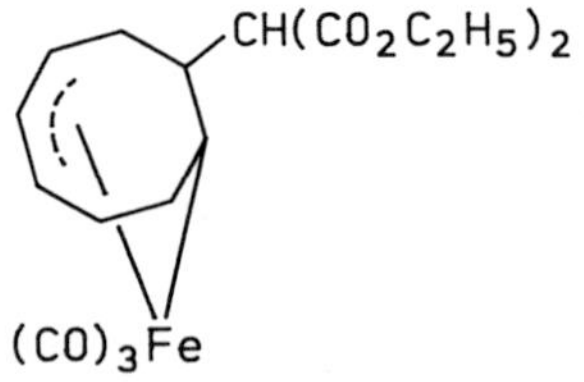 $CH(CO_2C_2H_5)_2$, $(CO)_3Fe$	VI d IR (cyclohexane): 1982, 2051 (CO)	[60]
60	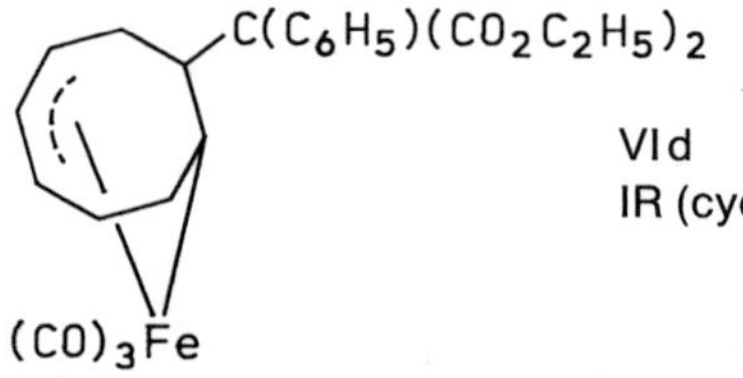 $C(C_6H_5)(CO_2C_2H_5)_2$, $(CO)_3Fe$	VI d IR (cyclohexane): 1980, 2051 (CO)	[60]
*61	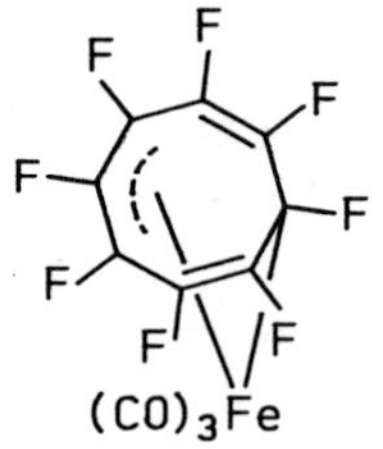 octafluoro compound, $(CO)_3Fe$	II b (38) m.p. 126 to 127.5°, pale yellow sublimable, air-stable crystals ^{19}F NMR (1:1 $CDCl_3/C_6H_6$): 113.0 (2F), 127.9 (3F), 161.9 (2F), 170.4 (1F) IR (C_6H_6): 1745 (C=C); 2060, 2105 (CO)	[117]

References on pp. 172/5

Table 4 [continued]

No.	compound	method of preparation (yield in %), properties and remarks	Ref.
*62	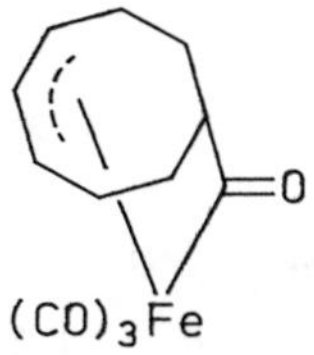 $(CO)_3Fe$	—	[17]
63	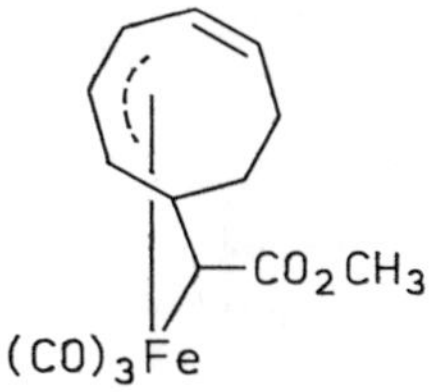 $(CO)_3Fe$, CO_2CH_3	IVb	[124]
*64	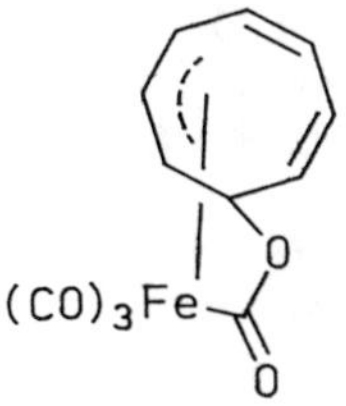 $(CO)_3Fe$	Id (ca. 5) m.p. 106° (dec.), colorless crystals (from ether) IR (KBr): 1638 (C=O) IR (hexane): 2012, 2027, 2081 (CO) mass spectrum: $[M]^+$	[63]
65	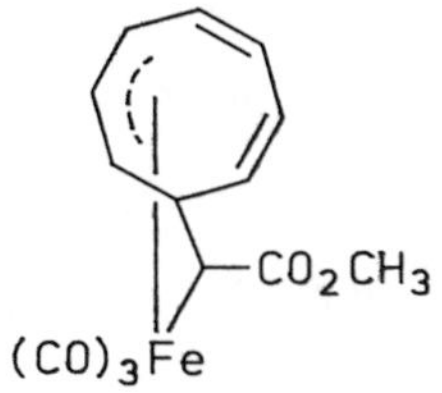 $(CO)_3Fe$, CO_2CH_3	IVb	[124]

compounds of the type $^4LFe(CO)_3$ where ^4L = bicyclo[3.1.1]hept-3-en-2,6-diyl or its derivatives:

No.	compound	method of preparation (yield in %), properties and remarks	Ref.
66	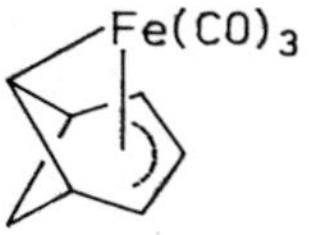 $Fe(CO)_3$	see No. 67, p. 163 IR (hexane): 1974, 2048 (CO)	[77, 83]
*67	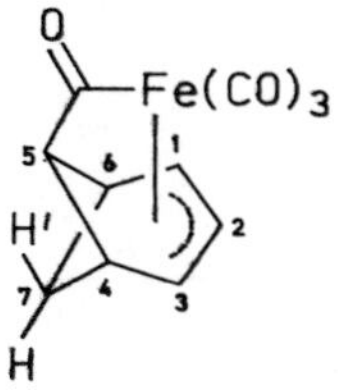 $Fe(CO)_3$	Ie (39), IVa (21) m.p. 60° (dec. even under Ar), pale yellow, air-sensitive needles ^{1}H NMR (CS_2 at −20°): 1.95 (H-7; J(H-7,7′) = −7), 2.39 (H-7′; J(H-6,7′) = 4), 2.67 (H-4,6), 3.15 (H-5; J(H-5,6) = 5), 5.35 (H-1,3; J(H-1,6) = 6), 5.64 (H-2; J(H-1,2) = 6) ^{13}C NMR ($CDCl_3$ at −10°): 35.7 (C-4,6), 43.6 (C-7), 84.4 (C-1,3), 85.0 (C-5), 90.4 (C-2); 205.9, 210.6 (CO); 248.2 (C=O)	[77, 83]

References on pp. 172/5

Table 4 [continued]

No.	compound	method of preparation (yield in %), properties and remarks	Ref.
		IR (KBr): 1650 (C=O) IR (hexane): 1995, 2003, 2062 (CO) mass spectrum: $[M-nCO]^+$ (n=0 to 4)	
compounds of the type $^4LFe(CO)_3$ where ^4L = bicyclo[3.2.1]oct-3-en-2,8-diyl or its derivatives:			
68	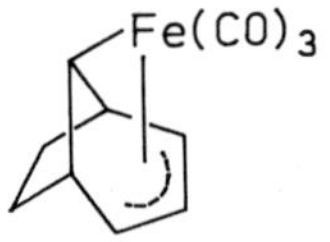 $Fe(CO)_3$	see No. 69, p. 163	[16]
*69	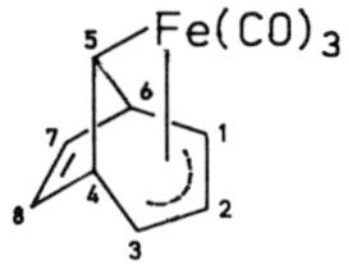 $Fe(CO)_3$	Ie, IIb (ca. 2 or 10), IIc (70), see No. 73, p. 163 b.p. 30 to 35°/0.04 Torr, b.p. 60°/0.25 Torr, yellow oil ^1H NMR ($CDCl_3$): 0.60 (H-5; J(H-5,6)=6.3), 3.10 (H-4,6; J(H-6,7)=1.5, J(H-6,8)=1.1), 4.30 (H-2; J(H-2,6)=1.2), 4.70 (H-1,3; J(H-1,2)=6.1, J(H-1,5)=1.4, J(H-1,6)=5.5), 6.30 (H-7,8) ^1H NMR (CS_2): 0.39 (t, H-5; J=6), 3.05 (t, H-4,6; J=5.5), 4.18 (t, H-2; J=6), 4.59 (t, H-1,3; J=5.5), 6.16 (t, H-7,8; J=1.4) IR: 1570 (C=C); 1975, 2035, 2070 (CO) IR (hexane): 1967, 2000, 2070 (CO) IR (hexane): 1990, 1995, 2052 (CO) mass spectrum: $[M-nCO]^+$ (n=0 to 3), $[C_8H_8]^+$	[16, 43, 52]
70	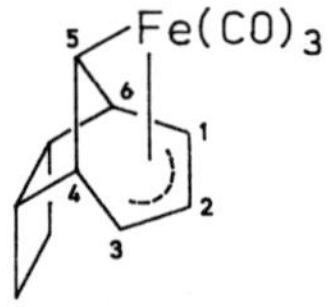 $Fe(CO)_3$	IIc (80) solidifies below 20°, oil ^1H NMR ($CDCl_3$): 0.50 (H-5), 2.00 (m, other H), 3.23 (H-4,6), 4.33 (H-1,3), 4.78 (H-2) IR: 1983, 2054 (CO) mass spectrum: $[M-nCO]^+$ (n=0 to 3)	[22]
71	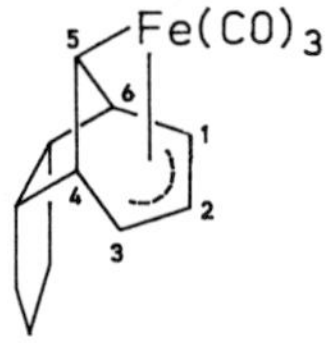 $Fe(CO)_3$	IIc (75) m.p. 51 to 52.5° (from pentane) ^1H NMR ($CDCl_3$): 0.85 (H-5), 2.76 (H-4,6), 4.25 (H-1,3), 4.60 (H-2) IR: 1982, 2050 (CO) mass spectrum: $[M-nCO]^+$ (n=0 to 3)	[22]
72	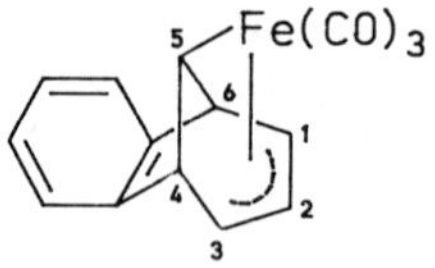 $Fe(CO)_3$	IIc (75) m.p. 89.5 to 90.5° ^1H NMR (CS_2): 0.75 (H-5), 3.59 (H-4,6), 4.12 (H-2), 4.62 (H-1,3), 7.00 (s, aromat. H) IR: 1955 to 2040 (CO) mass spectrum: $[M-nCO]^+$ (n=0 to 3), $[C_{12}H_{10}]^+$	[22]

References on pp. 172/5

Table 4 [continued]

No.	compound	method of preparation (yield in %), properties and remarks	Ref.
*73	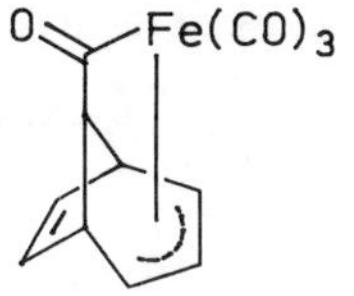 O=C–Fe(CO)$_3$	see No. 69, p. 163 m.p. 55° (dec.)	[27, 65]
74	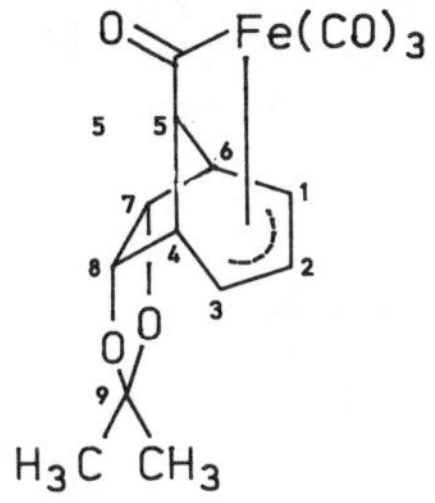 O=C–Fe(CO)$_3$; 5, 6, 7, 8, 4, 1, 2, 3, O, O, 9, H_3C CH_3	If (40) m.p. 108 to 109°, yellow crystals (from CH_2Cl_2/C_6H_6 at −70°) ^{1}H NMR ($CDCl_3$): 1.27 (s, CH_3), 1.31 (s, CH_3), 2.47 (m, H-4,6), 2.97 (m, H-5), 4.69 (m, H-1,3), 4.80 (s, H-7,8), 5.30 (m, H-2) ^{13}C NMR ($CDCl_3$): 23.97, 26.23 (2CH_3); 45.51 (C-4,6), 76.03 (C-1,3), 78.97 (C-7,8), 85.37 (C-2), 87.28 (C-5), 110.66 (C-9); 205.08, 206.84, 209.94 (CO); 252.43 (C=O) IR (cyclohexane): 1470, 1480 (C=C); 1670 (C=O); 1984, 1986, 1999, 2004, 2050, 2063 (CO) mass spectrum: $[M-nCO]^+$ (n = 0 to 4), $[M-Fe(CO)_4]^+$, $[M-3CO-OC(CH_3)_2]^+$, $[C_6H_6Fe(CO)_n]^+$ (n = 0 to 2), $[C_5H_8O_2]^+$, $[C_7H_7]^+$, $[C_6H_6]^+$, $[CO]^+$, $[Fe]^+$	[109]

compounds of the type $^4LFe(CO)_3$ where ^4L = bicyclo[3.3.1]non-3-en-2,9-diyl or its derivatives:

No.	compound	method of preparation (yield in %), properties and remarks	Ref.
*75	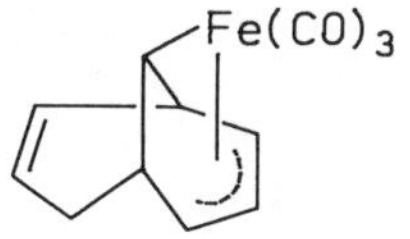 Fe(CO)$_3$	IIb, see No. 76, p. 163 yellow oil IR (hexane) 1989, 1998, 2048 (CO)	[27, 49]
*76	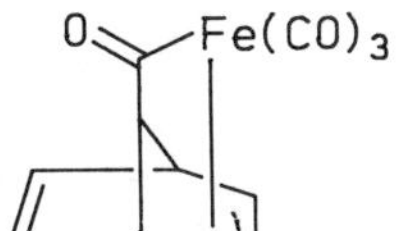 O=C–Fe(CO)$_3$	Ic (48), see also No. 75, p. 163 m.p. 68 to 70°, pale yellow crystals (from hexane) IR (KBr): 1670 (C=O) IR (hexane): 2002, 2062 (CO) mass spectrum: $[M]^+$	[27, 49]

compounds of the type $^4LFe(CO)_3$ where ^4L = bicyclo[4.3.1]dec-8-en-7,10-diyl or its derivatives:

No.	compound	method of preparation (yield in %), properties and remarks	Ref.
*77	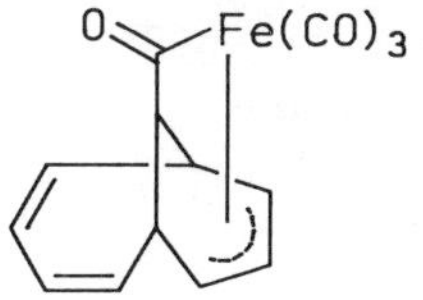 O=C–Fe(CO)$_3$	–	[84]

References on pp. 172/5

Table 4 [continued]

No.	compound	method of preparation (yield in %), properties and remarks	Ref.
	compounds of the type $^4LFe(CO)_3$ where ^4L = bicyclo[3.2.1]oct-3-en-2,6-diyl or its derivatives:		
78	(D)H, H, $Fe(CO)_3$	VIe	[93]
79	NC, H, H′, $Fe(CO)_3$	VIe (>70) m.p. 65° (from n-hexane) ^1H NMR ($CDCl_3$): 0.7 to 1.22 (H-6, 8), 1.62 (H-8′), 2.80 (H-4, 7), 2.97 (H-5), 4.2 (H-1), 4.78 (H-2), 5.19 (H-3) IR (hexane): 1993, 1955, 2063 (CO); 2210 (CN) mass spectrum: $[M-nCO]^+$ (n = 0 to 3)	[41, 93]
80	endoH, exoH, CH_3, O, $Fe(CO)_3$	VIe (25) ^1H NMR ($CDCl_3$): 0.88 (m, H-6), 1.56 (m, H-5 exo), 1.88 (s, H-8), 1.95 to 2.20 (m, H-5 endo), 1.98 (s, CH_3), 3.27 (t, H-7; J = 7), 3.82 (m, H-4), 4.08 (m, H-1), 4.72 (t, H-2; J = 6), 5.4 (td, H-3; J = 2.5, J = 6) (identical with the deuterated derivative except H-5 exo) IR (cyclohexane): 1716 (C=O); 1979, 1984, 2051 (CO) mass spectrum: $[M-nCO]^+$ (n = 1 to 3)	[121]
*81	O, H, H′(D), $Fe(CO)_3$	IIa (2.6), IId (16), also see No. 67, p. 163 m.p. 87.5°, pale yellow crystals (from hexane) ^1H NMR (C_6H_6): 0.51 (H-8′; J(H-8, 8′) = −11), 1.0 (H-8; J(H-7, 8) = 5, J(H-4, 8) = 3), 1.45 (H-6; J(H-6, 7) = 7), 2.0 (H-7; J(H-1, 7) = 6), 2.35 (H-4), 3.10 (H-1; J(H-1, 2) = 5), 4.25 (H-2; J(H-2, 3) = 6), 4.38 (H-3; J(H-3, 4) = 7) ^{13}C NMR (benzene-d_6): 11.1 (C-6); 34.4, 38.2 (C-4, 7); 35.0 (C-8); 51.1, 61.4, 89.7 (C-1, 2, 3); 191.0 (C=O) IR (KBr): 1670 (C=O) IR (hexane): 2010, 2067 (CO) mass spectrum: $[M-nCO]^+$ (n = 0 to 4), $[M-Fe(CO)_3]^+$, $[C_6H_6Fe]^+$, $[C_7H_7]^+$	[76, 77, 83]
82	H, NC, CH_3, O, $Fe(CO)_3$	VIe (13) almost colorless crystals (from hexane/CH_2Cl_2) ^1H NMR ($CDCl_3$): 1.14 (dd, H-6; J = 2, J = 7), 2.0 (s, H-8), 2.16 (s, CH_3), 2.80 (br, s, H-5 endo), 3.32 (br, dd, H-4; J = 2, J = 6), 3.56 (t, H-7; J = 7), 4.14 (m, H-1), 4.81 (t, H-2), 5.38 (td, J = 2.5, J = 6)	[121]

References on pp. 172/5

Table 4 [continued]

No.	compound	method of preparation (yield in %), properties and remarks	Ref.
		IR (cyclohexane): 1715 (C=O); 1931, 1996, 2061 (CO) mass spectrum: $[M-nCO]^+$ (n=0 to 3)	
*83	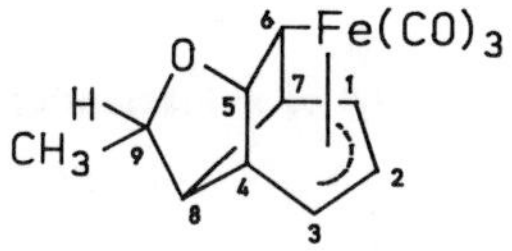	VIe (100) diastereoisomer A: ^1H NMR (acetone-d_6): 0.39 (dd, H-6; J=7.3), 0.95 (d, CH_3; J=6), 1.55 (d, H-8; J=2), 2.59 (H-4), 2.94 (t, H-7; J=7), 3.76 (qd, H-9; J=2, J=6), 4.00 (m, H-1,5), 4.60 (m, H-2,3) diastereoisomer B: ^1H NMR (acetone-d_6): 0.39 (dd, H-6; J=7.3), 0.83 (d, CH_3; J=6), 1.47 (s, H-8), 2.59 (m, H-4,7), 3.34 (q, H-9; J=6), 4.00 (m, H-1,5), 4.60 (m, H-2,3) both isomers: IR (cyclohexane): 1980, 1982, 2048 (CO) mass spectrum: $[M-nCO]^+$ (n=0 to 3)	[121]
*84	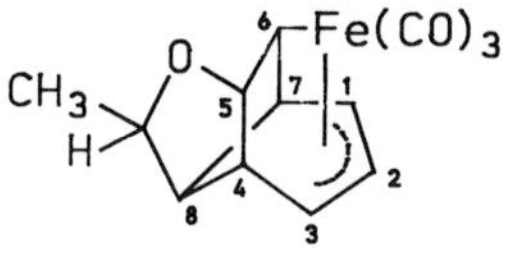	VIe (100) ^1H NMR (acetone-d_6): 0.42 (m, H-6), 0.94 (s, CH_3), 1.01 (s, CH_3), 1.48 (m, H-8), 2.82 (m, H-4), 2.98 (t, H-7; J=7), 3.92 (d, H-5; J=3), 4.04 (m, H-1), 4.60 (m, H-2,3) IR (cyclohexane): 1979, 1981, 2050 (CO) mass spectrum: $[M-nCO]^+$ (n=0 to 3)	[121]
*85	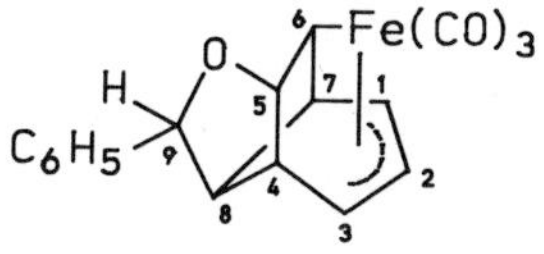	VIe (100) diastereoisomers A and B: ^1H NMR: 0.57 (d, H-6; J=7), 1.40 (m, H-8, A+B), 2.18 to 2.46 (m, H-4,7, A+B); 3.30, 3.52 (m, H-1, A+B), 3.92 (m, H-2,3), 4.30 (s, H-9, B), 4.34 (dd, H-5; J=1, J=4), 4.77 (d, H-9, A; J=3), 7.20 (m, C_6H_5) IR (cyclohexane): 1983, 2051 (CO) mass spectrum: $[M-nCO]^+$ (n=0 to 3)	[121]
*86	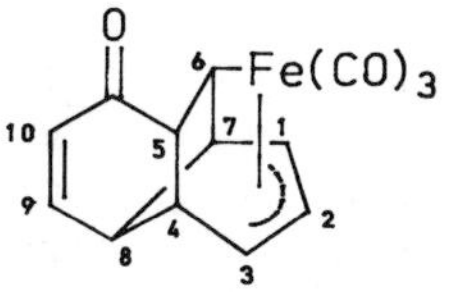	IId (24) m.p. 109°, yellow crystals (from C_6H_6/petroleum ether) ^1H NMR (C_6H_6): 0.33 (H-6; J(H-6,7)=7), 1.26 (H-8; J(H-8,9)=7), 2.10 (H-4; J(H-3,4)=6.5), 2.20 (H-7; J(H-1,7)=7), 3.27 (H-5), 3.63 (H-1; J(H-1,3)=2.5), 4.10 (H-2; J(H-1,2)=5), 4.46 (H-3; J(H-2,3)=6), 5.84 (H-10; J(H-5,10)=1.5), 6.06 (H-9; J(H-9,10)=9.5)	[84]

References on pp. 172/5

Table 4 [continued]

No.	compound	method of preparation (yield in %), properties and remarks	Ref.
		IR (KBr): 1660 (C=O) IR (hexane): 1984 1990, 2057 (CO) mass spectrum: $[M-nCO]^+$ (n=0 to 4), $[M-4CO-C_2H_2]^+$, $[M-4CO-C_4H_4]^+$	
compounds of the type $^4LFe(CO)_3$ where ^4L = bicyclo[3.2.2]non-3-en-2,6-diyl or its derivatives:			
*87	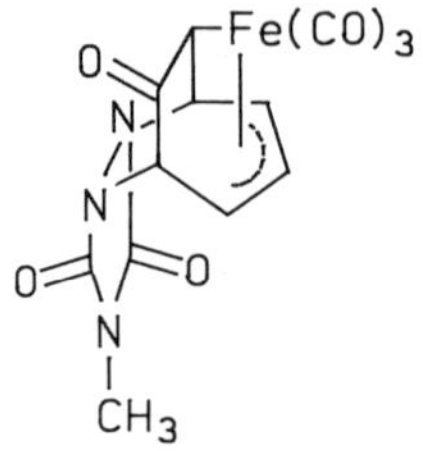	Vd prismatic yellow crystals (from CH_3OH)	[97, 106]
*88	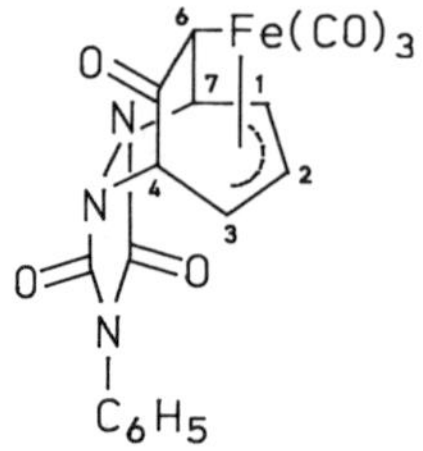	Vd (90) m.p. 153° (dec., from CH_2Cl_2/pentane) ^1H NMR: 2.54 (dd, H-6; J=2, J=9), 4.3 (ddd, H-7; J=2, J=6, J=7), 4.88 (dd, H-1; J=7, J=9), 5.0 to 5.2 (m, H-2,3,4)	[92]
89	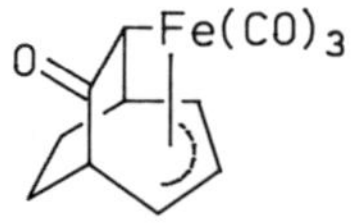	printing error in [104], is identical with No. 91	
90	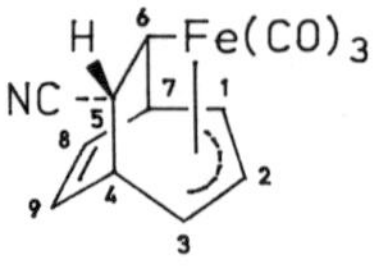	VIe (>70) m.p. 99 to 105°, yellow crystals (from n-hexane) ^1H NMR ($CDCl_3$): 1.25 (H-6), 3.07 (H-5), 3.24 (H-4,7), 4.31 (H-2), 4.43 (H-1), 5.18 (H-3), 6.23 (H-8,9) IR (hexane): 1996, 2062 (CO); 2208 (CN) mass spectrum: $[M-nCO]^+$ (n=0 to 3)	[41]
*91	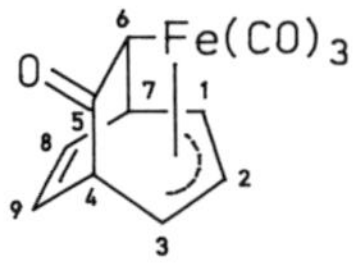	IIb (56 or 70), Ve (35 to 65), also see Nos. 92 to 95, p. 166 m.p. 132 to 133°, m.p. 133° (from hexane), m.p. 135 to 136° (from ether or $CHCl_3$/ether), colorless crystals, very pale yellow crystals (from $CHCl_3$/ether), good light yellow crystals (from pentane at −5°), yellow complex (from hexane)	[23, 27, 30, 35, 47, 54, 59, 61, 62, 69, 101, 104, 110,

Table 4 [continued]

No.	compound	method of preparation (yield in %), properties and remarks	Ref.
		^{1}H NMR ($CDCl_3$): 1.96 (dd, H-6; J(H-6,7) = 8.5), 2.88 (q, H-7; J(H-1,7) = 7.0), 3.11 (t, H-4; J(H-4,9) = 5.5, J(H-4,6) = 2.5), 4.15 (t, H-1; J(H-1,3) = 1.5, J(H-1,2) = 6.0), 4.45 (t, H-2; J(H-2,3) = 8.0), 5.16 (t, H-3; J(H-3,4) = 7.5), 6.07 (m, H-8; J(H-7,8) = 2.5), 6.25 (m, H-9; J(H-8,9) = 8.5) ^{13}C NMR (toluene-d_8, −7°): 191.8 (C=O); 205.8, 208.8, 214.4 (CO) IR (hexane): 1670 (C=O); 1922, 2000, 2058 (CO) IR (CS_2): 1683 (C=O); 1993, 2015, 2065 (CO) mass spectrum: $[M-nCO]^+$ (n = 0 to 4), $[M-4CO-C_2H_2]^+$	129]
*92	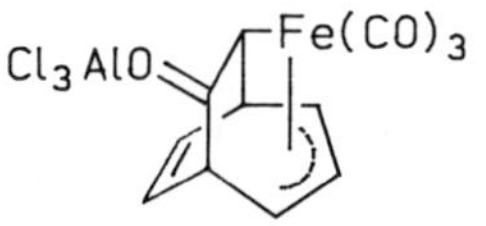 Cl_3AlO … $Fe(CO)_3$	see No. 91, p. 165 IR (CH_2Cl_2): 1520 (C=O); 2035, 2095 (CO)	[103, 110]
*93	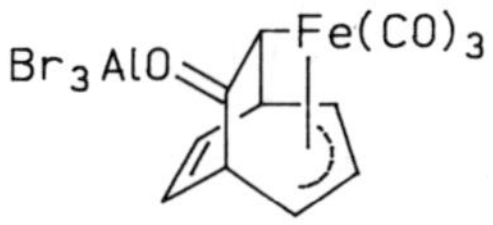 Br_3AlO … $Fe(CO)_3$	see No. 91, p. 165 IR (CH_2Cl_2): 1520 (C=O); 2039, 2094 (CO)	[103, 110]
*94	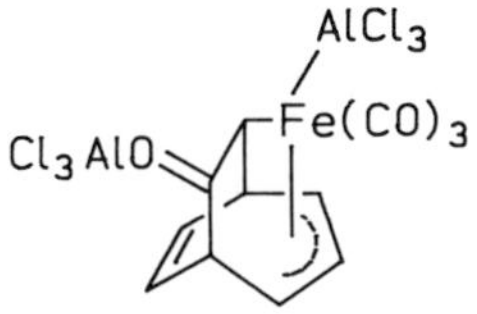 $AlCl_3$ / $Fe(CO)_3$, Cl_3AlO	see No. 91, p. 165 IR (CH_2Cl_2): 1610 (C=O); 2088, 2127 (CO)	[103, 110]
*95	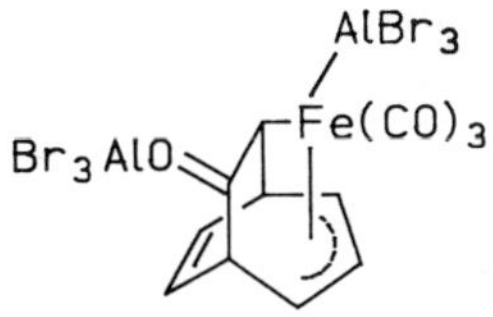 $AlBr_3$ / $Fe(CO)_3$, Br_3AlO	see No. 91, p. 165 IR (CH_2Cl_2): 1610 (C=O); 2095, 2130 (CO)	[103, 110]
*96	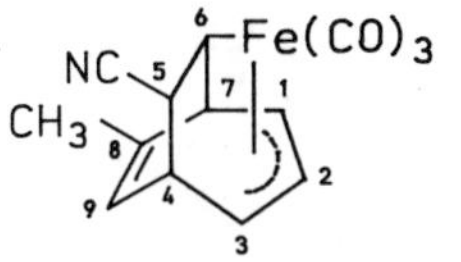 NC, CH_3, $Fe(CO)_3$ (positions 1–9)	all data for mixtures of Nos. 96 and 97: VIe (87) m.p. 53°, yellow crystals (from petroleum ether) ^{1}H NMR ($CDCl_3$): 1.20 (dm, H-6,6′); 1.73 (d), 1.87 (d) (CH_3); 3.08 (m, 3H), 4.3 (m, 2H); 5.14 (ddd), 5.21 (ddd) (H-3,3′); 5.83 (dm, H-9,8′)	[113]
*97	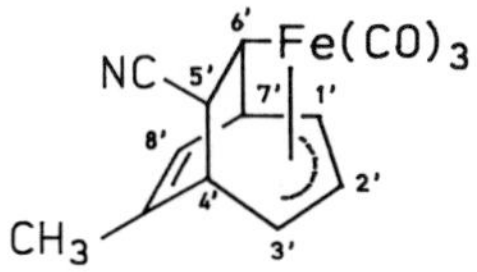 NC, CH_3, $Fe(CO)_3$ (positions 1′–8′)	IR (hexane): 1995, 2050 (CO); 2224 (CN) mass spectrum: $[M-nCO]^+$ (n = 0 to 3)	[113]

References on pp. 172/5

Table 4 [continued]

No.	compound	method of preparation (yield in %), properties and remarks	Ref.
*98	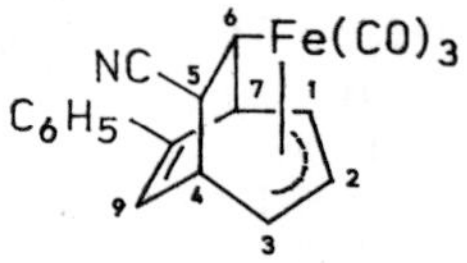	all data for mixtures of Nos. 98 and 99: VI e (83) m.p. 127° ^{1}H NMR ($CDCl_3$): 1.35 (dm, H-6,6′), 3.24 (m, 2H), 3.75 (m, 1H), 4.24 (m, 2H), 5.22 (ddd, 1H), 5.35 (dd+dd, $^1/_2$H each, H-9,8′)	[113]
*99	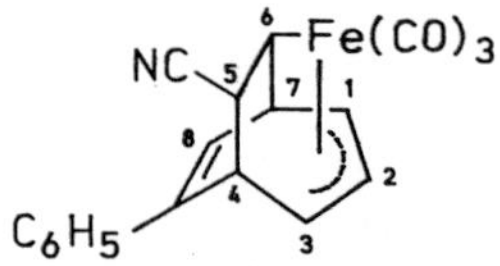	IR (hexane): 1995, 2050 (CO); 2220 (CN) mass spectrum: $[M-nCO]^+$ (n=0 to 3)	
100	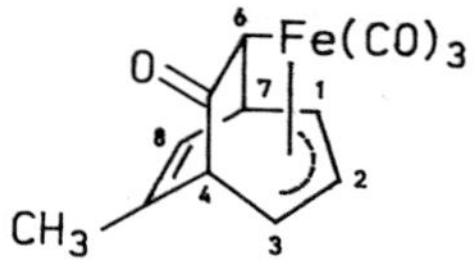	V e (ca. 85) yellow oil ^{1}H NMR ($CDCl_3$): 0.62 (s, CH_3), 1.92 (br,d, H-6; J=8), 2.86 (q, H-7; J=7), 2.94 (d, H-4; J=8), 4.18 (t, H-1; J=7), 4.45 (t, H-2; J=7.5), 5.13 (t, H-3; J=8), 5.74 (d, H-8; J=7) (in the presence of Eu(dpm)$_3$, see No. 91, p. 165) the greatest downfield shifts occur for H-4,6) IR (hexane): 1693 (C=O); 1992, 2009, 2065 (CO) mass spectrum: $[M-nCO]^+$ (n=0 to 4) and further fragments	[61, 69]
*101	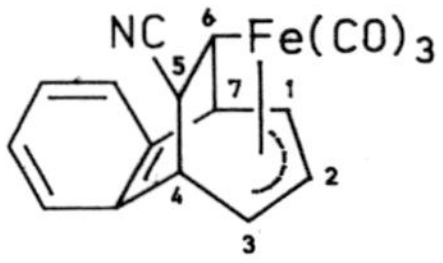	VI e (>70) m.p. 134 to 135° (from n-hexane) ^{1}H NMR ($CDCl_3$): 1.52 (H-6), 3.31 (H-5; J(H-5,6)=1.4), 3.77 (H-4,7; J(H-4,5)=5.5), 4.25 (H-2), 4.42 (H-1), 5.28 (H-3) IR (hexane): 1994, 2060 (CO); 2210 (CN) mass spectrum: $[M-nCO]^+$ (n=0 to 3)	[41]
102	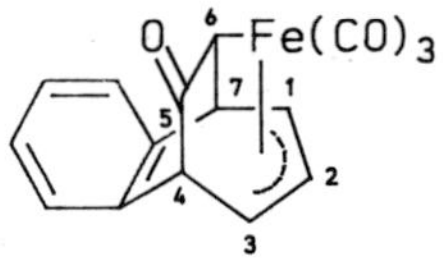	II c (40) m.p. 125 to 127° ^{1}H NMR ($CDCl_3$): 2.12 (dd, H-6; J(H-6,7)=8.5), 3.45 (dt, H-7; J(H-1,7)=7.0), 3.67 (dd, H-4; J(H-3,4)=8.0), 4.17 (t, H-1), 4.38 (t, H-2), 5.23 (t, H-3) IR ($CHCl_3$): 1665 (C=O) IR (hexane): 1985, 2060, 2065 (CO); 1695 (C=O) mass spectrum: $[M-nCO]^+$ (n=0 to 4)	[42, 59]
*103	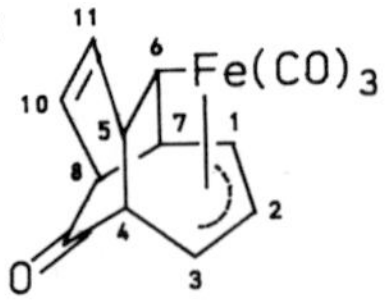	V f (95) m.p. 115 to 117°, almost colorless mushroom-shaped crystals (from C_6H_6/petroleum ether)	[75, 84]

References on pp. 172/5

Table 4 [continued]

No.	compound	method of preparation (yield in %), properties and remarks	Ref.
		^{1}H NMR (C_6H_6): 0.05 (H-6; J(H-4,6) = 2.5), 1.9 (H-7; J(H-1,7) = 6, J(H-6,7) = 10), 2.32 (H-4; J(H-4,5) = 6), 2.75 (H-8; J(H-8,10) = 7, J(H-7,8) = 4), 2.85 (H-5; J(H-5,6) = 2.5, J(H-5,10) = 1.5), 3.9 (H-1,2; J(H-1,2) = 6, J(H-1,3) = 2.5, J(H-2,3) = 6.5), 4.32 (H-3; J(H-3,4) = 8), 5.48 (H-10; J(H-10,11) = 8), 5.78 (H-11; J(H-5,11) = 6) ^{13}C NMR (benzene-d_6): 10.0 (C-6); 36.8, 44.9, 56.1, 56.4, 66.7, 75.9, 94.0, 124.2, 138.8 (C-1 to 5, 7, 8, 10, 11); 203.0, 203.2, 213.9, 214.6 (CO, C-9) IR (KBr): 600; 1715 (C=O); 2960, 2980, 3060 IR (hexane): 1986, 2054 (CO) mass spectrum: $[M-nCO]^+$ (n = 0 to 4), $[M-4CO-C_2H_2]^+$	
*104	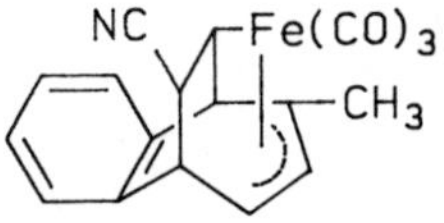 NC, $Fe(CO)_3$, CH_3	VIe (64) m.p. 104 to 106°, yellow needles (from petroleum ether) IR (hexane): 1980, 2050 (CO); 2235 (CN) mass spectrum: $[M-nCO]^+$ (n = 0 to 3)	[113]
*105	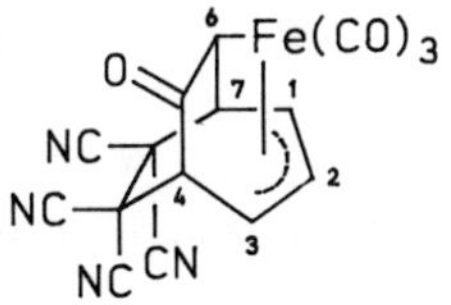 O, NC, NC, NC, CN, $Fe(CO)_3$	Vd (90), also see No. 126 m.p. 144 to 146°, m.p. 148 to 150° (dec.), yellow powder (from acetone/pentane or acetone/hexane) ^{1}H NMR (acetone-d_6): 2.30 (H-6; J(H-6,7) = 9.0), 3.99 (H-4; J(H-4,6) = 2.2), 4.24 (H-7; J(H-1,7) = 7.6), 4.58 (H-1; J(H-1,2) = 7.3), 5.31 (H-3; J(H-3,4) = 8.5, J(H-1,3) = 2.0), 5,67 (H-2; J(H-2,3) = 8.0) ^{13}C NMR (acetone-d_6): 17.11 (C-6), 47.69 (C-4); 45.93, 52.60, 57.09 (C-1,3,7), 96.29 (C-2) IR (CH_2Cl_2): 1683 (C=O); 2030, 2090 (CO) mass spectrum: $[M-(CN)_2C{=}C(CN)_2-nCO]^+$ (n = 0 to 3), $[C_7H_6O]^+$	[81, 88]

compounds of the type $^4LFe(CO)_3$ where ^4L = bicyclo[3.3.2]dec-3-en-2,10-diyl or its derivatives:

No.	compound	method of preparation (yield in %), properties and remarks	Ref.
*106	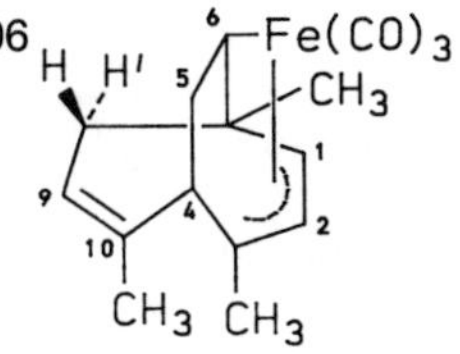 H, H′, $Fe(CO)_3$, CH_3, CH_3, CH_3	Ic (ca. 22) m.p. 88 to 89°, pale yellow prisms (from C_2H_5OH) ^{1}H NMR (benzene-d_6): 0.86 (m, H-6), 1.08 (s, CH_3-7), 1.64 (small m, CH_3-10), 1.85 (s, CH_3), 1.48 to 2.40 (m, superimposed by s and small m, 4H), 2.47 (dm, H-4; J(H-4,5) = 9.0), 3.40 (dd, H-1; J(H-1,6) = 1.0), 3.98 (d, H-2; J(H-1,2) = 7.0), 4.76 (dm, H-9; J(H-8′,9) = 8.0)	[90]

References on pp. 172/5

Table 4 [continued]

No.	compound	method of preparation (yield in %), properties and remarks	Ref.
		^{1}H NMR ($CDCl_3$): 1.04 (m, H-6), 1.30 (s, CH_3-7), 1.89 (br,s, CH_3-10), 2.21 (s, CH_3-3), 1.66 to 2.58 (m, superimposed by s and br,s, 4H), 2.83 (dm, H-4; J(H-4,5) = 8.5), 3.89 (d, H-1), 4.42 (d, H-2; J(H-1,2) = 7.5), 4.92 (dm, H-9; J(H-8′,9) = 7.0) IR (KBr): 778, 1028, 1378, 1385, 1455; 1970, 1980, 2040 (CO); 2860, 2900, 2940, 2970 UV (hexane): $\lambda_{max}(\varepsilon)$ = 212 (20600), 263 (sh, 7850) nm mass spectrum: $[M-nCO]^+$ (n = 0 to 3) and further fragments given	
*107	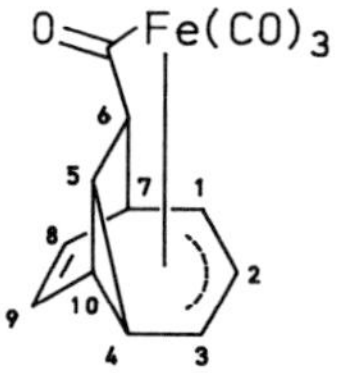 O=, $Fe(CO)_3$	I d (59) m.p. 78 to 80° (dec.), yellow plates (from C_6H_6/pentane) ^{1}H NMR (C_6H_6): ca. 1.1 (H-10; J(H-5,10) is strong but not measurable), ca. 1.2 (H-5; J(H-4,5) = 8), 1.60 (H-4; J(H-4,10) = 8), 2.00 (H-7; J(H-6,7) = 8), 2.38 (H-6; J(H-5,6) = 5), 3.78 (H-2; J(H-2,3) = 8), 4.19 (H-3; J(H-3,4) = 8), 4.78 (H-1; J(H-1,2) = 8, J(H-1,7) = 8, J(H-1,3) = 2), 5.46 (H-9; J(H-9,10) = 6), 5.97 (H-8; J(H-8,9) = 9.5, J(H-7,8) = 6) IR (KBr): 1640 (C=O) IR (hexane): 2001, 2058 (CO)	[66]
compounds of the type $^4LFe(CO)_3$ where ^4L = bicyclo[3.3.2]dec-3-en-2,6-diyl or its derivatives:			
108	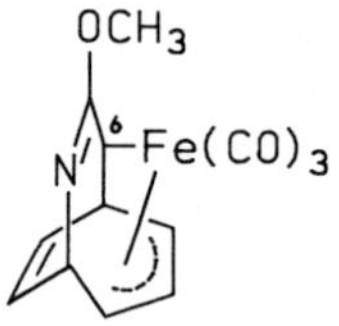 OCH_3, N, $Fe(CO)_3$	II c (40) yellow-brown oil ^{1}H NMR: 1.32 (d, H-6) IR (hexane): 1942, 2000, 2016 (CO) IR ($CHCl_3$): 1625 (C=N) mass spectrum: $[M]^+$	
*109	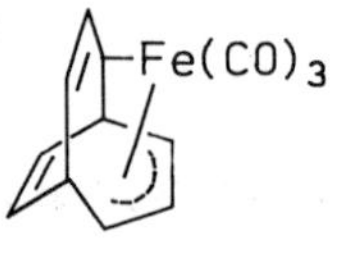 $Fe(CO)_3$	I d not isolated	[66, 95]
*110	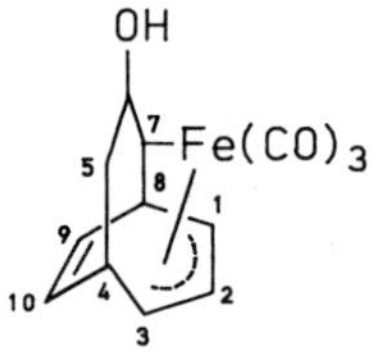 OH, $Fe(CO)_3$	I d (29), VI e m.p. 94 to 95°, pale yellow, matted crystals (from C_6H_6/pentane at −28°) ^{1}H NMR (CS_2): ca. 2.1 (H-5,7, OH; J(H-4,5) = 8), 2.81 (H-8; J(H-7,8) strong but not measurable), 3.04 (H-4; J(H-3,4) = 9), 4.0 (H-6; J(H-5,6) = 8), 4.57 (H-2; J(H-1,2) = 7.5),	[66]

References on pp. 172/5

Table 4 [continued]

No.	compound	method of preparation (yield in %), properties and remarks	Ref.
		4.88 (H-3; J(H-2,3) = 7.5), 5.37 (H-1; J(H-1,8) = 9, J(H-1,3) = 2), 5.65 (H-10; J(H-4,10) = 8), 6.11 (H-9; J(H-9,10) = 9, J(H-8,9) = 8) IR (KBr): 1015 (C–O), 1650 (C=C), 3340 (OH) IR (hexane): 1993, 2053 (CO) mass spectrum: $[M-nCO]^+$ (n = 0, 1, 3), $[M-nCO-OH]^+$ (n = 1 to 3), $[M-3CO-OH-C_2H_2]^+$	
111	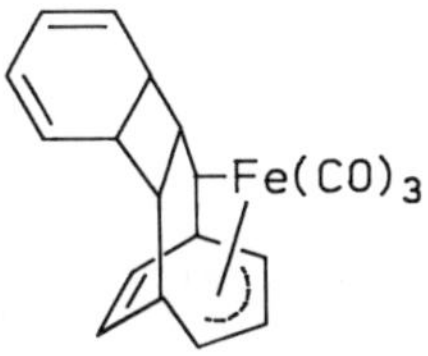	not isolated, mentioned together with No. 109	[95]
112	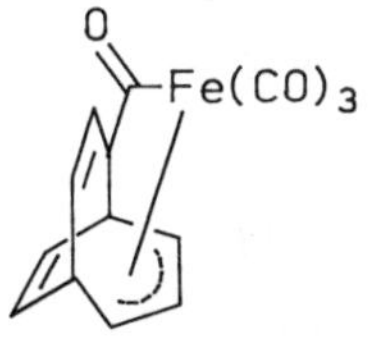	Id not isolated	[66]

compounds of the type $^4LFe(CO)_3$ where ^4L = bicyclo[4.2.1]non-3-en-2,9-diyl or its derivatives (continued at No. 151):

No.	compound	method of preparation (yield in %), properties and remarks	Ref.
113	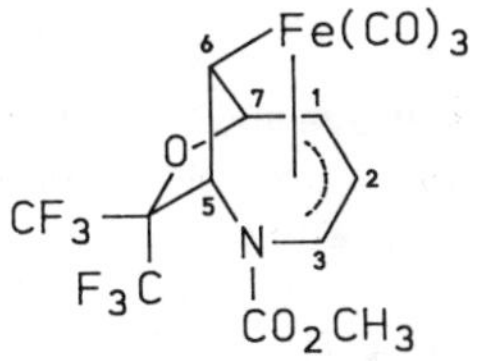	Vc (38) m.p. 111°, pale yellow crystals (from CH_2Cl_2/hexane) ^1H NMR ($CDCl_3$): 1.8 (H-6; J(H-6,7) = 8.0), 3.78 (H-2; J(H-2,7) = 2.0, J(H-1,2) = 6.5), 3.80 (OCH_3), 4.7 (H-1; J(H-1,6) = 2.0), 4.96 (H-5,7; J(H-1,7) = 6.0, J(H-5,6) = 8.0), 7.12 (H-3; J(H-1,3) = 1.0, J(H-2,3) = 8.0) ^{19}F NMR ($CDCl_3$): 71.65 (q, CF_3; $^4J(F,F)$ = 10), 76.80 (q, CF_3) IR (CS_2): 1737 (C=O); 2069, 2007 (CO) mass spectrum: $[M-nCO]^+$ (n = 0 to 3)	[88]
114	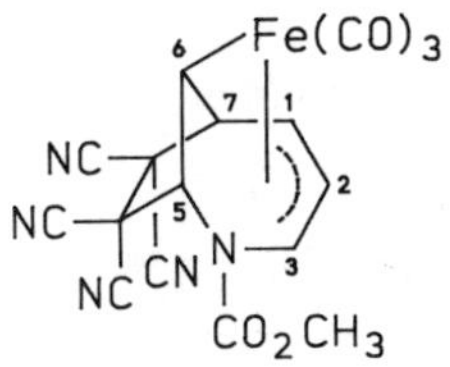	all data from mixtures with XLIX (p. 113): Vc (14) pale yellow crystals (from CH_2Cl_2/hexane at 0°) ^1H NMR (acetone-d_6): 2.02 (H-6; J(H-6,7) = 9.5), 3.71 (OCH_3), 4.51 (H-7; J(H-1,7) = 6.5), 4.53 (H-2; J(H-2,3) = 7.5), 5.10 (H-1; J(H-1,2) = 6.5), 5.26 (H-5; J(H-5,6) = 6.0), 7.12 (H-3) IR (CH_2Cl_2): 1727 (C=O); 2015, 2019, 2078 (CO) mass spectrum: $[M-nCO]^+$ (n = 1 to 3)	[19, 88]

References on pp. 172/5

Table 4 [continued]

No.	compound	method of preparation (yield in %), properties and remarks	Ref.
*115	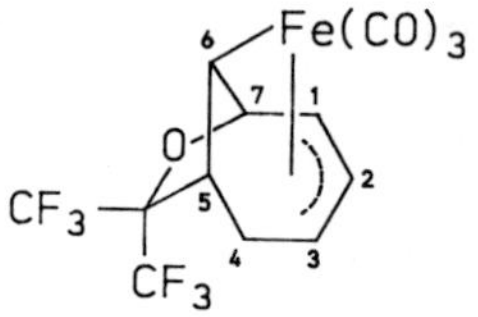	Vd (69) m.p. 75 to 76°, pale yellow crystals (from hexane) ^{1}H NMR ($CDCl_3$): 1.66 (H-6; J(H-6,7) = 8.0), 2.68 (H-4,4′; J(H-4,5) = 4.0, J(H-4′,5) = 8.0), 2.92 (H-5; J(H-5,6) = 8.0), 4.42 (H-2,3; J(H-1,2) = 4.0), 4.66 (H-1; J(H-1,7) = 7.0), 4.84 (H-7) ^{19}F NMR ($CDCl_3$): 70.9 (CF_3; J(F,F) = 10.7), 77.1 (CF_3; J(H,F) = 1.7) IR (Nujol): 2001, 2066 (CO) mass spectrum: $[M-nCO]^+$ (n = 0 to 3)	[19, 20, 44]
116	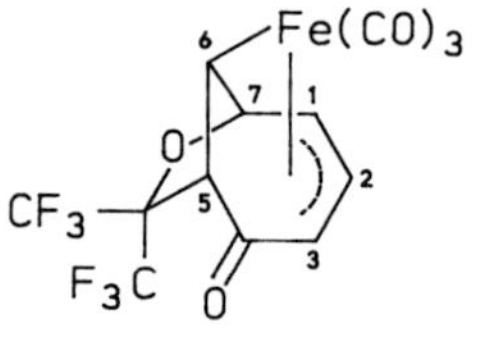	Vd (35.8) m.p. 119° (dec.), white crystals (from hexane at −78°) ^{1}H NMR ($CDCl_3$): 1.7 (H-6; J(H-5,6) = 8.0), 3.04 (H-5), 4.42 (H-3; J(H-1,3) = 4.0), 4.84 (H-2; J(H-2,3) = 7.0), 4.9 (H-1; J(H-1,6) = 1.0), 4.98 (H-7; J(H-6,7) = 7.0) ^{19}F NMR ($CDCl_3$): 69.0 (q, CF_3; ^{4}J(F,F) = 10.0), 76.0 (dq, CF_3) IR (hexane): 1715 (C=O); 2017, 2026, 2081 (CO) mass spectrum: $[M-nCO]^+$ (n = 0 to 3)	[88]
*117	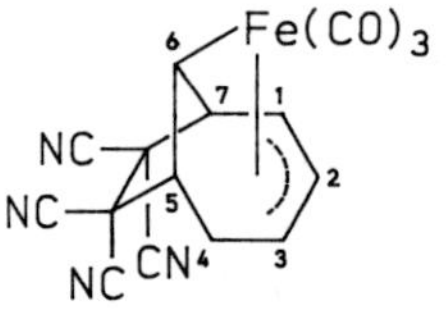	Vd (84 to 92) m.p. 160°, m.p. 160° (dec.), darkens above 160°, does not melt below 350°, pale yellow crystals (from CH_2Cl_2/hexane), yellow solid, yellow flattened needles ^{1}H NMR (acetone-d_6): 1.68 (H-6; J(H-6,7) = 10.0), 2.38, 3.15 (H-4,4′; J(H-4,4′) = 16.0, J(H-4,5) = 2.0, J(H-4,5) = 12.0), 3.88 (H-5; J(H-5,6) = 6.0), 4.43 (H-7; J(H-1,7) = 8.32), 4.76 (H-3), 5.10 (H-1,2; J(H-2,4) = 6.0) ^{1}H NMR (liquid SO_2): 1.69 (H-6); 2.42, 3.18 (H-4,4′), 3.9 (H-5), 4.5 (H-7), 4.80 (H-3), 5.16 (H-1,2) ^{57}Fe-γ (−196°): $\delta = 0.11 \pm 0.05$ (rel. to stainless steel), $\Delta = 0.83$ IR (KBr): 681, 779, 834, 845, 886, 911, 929, 995, 1033, 1086, 1111(?), 1145, 1209, 1239, 1285, 1323, 1342, 1381, 1453, 1504; 2004, 2028, 2083 (CO); 2257 (CN), 2849, 2899, 2976 IR ($CHCl_3$): 2011, 2073 (CO) mass spectrum: $[M-nCO]^+$ (n = 0 to 3)	[15, 18 to 20, 24, 44, 91, 123]

References on pp. 172/5

Table 4 [continued]

No. compound	method of preparation (yield in %), properties and remarks	Ref.
*118	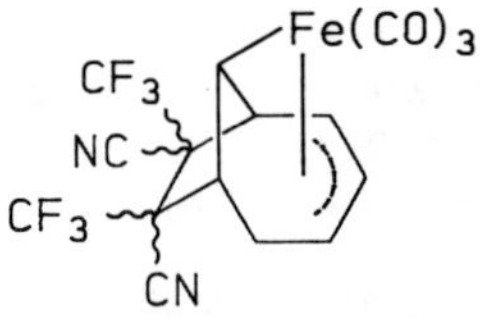 Vd	[19, 20, 44]
*119	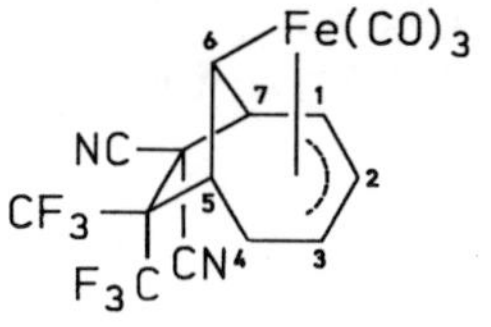 Vd (79 or 95) m.p. 120° (dec.), pale yellow crystals (from CH_2Cl_2/hexane) ^{1}H NMR ($CDCl_3$): 1.65 (H-6; J(H-6,7) = 9.0), 2.89 (H-4,4′; J(H-4,5) = 6.0, J(H-4,5) = 3.0), 3.26 (H-5; J(H-5,6) = 6.0), 4.02 (H-7), 4.56 (H-3), 4.60 (H-1; J(H-1,7) = 8.0), 4.72 (H-2; J(H-1,2) = 4.0) ^{19}F NMR ($CDCl_3$): 60.1 (CF_3; J(F,F) = 12.8), 63.8 (CF_3) IR (cyclohexane): 2005, 2067 (CO) mass spectrum: $[M-nCO]^+$ (n = 0 to 3)	[19, 44]
*120	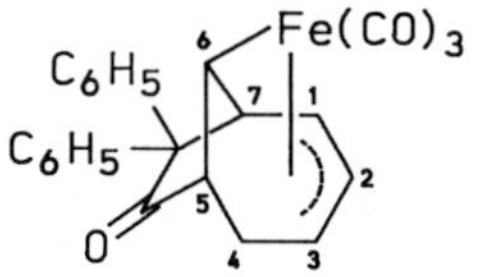 Vd m.p. 155 to 156° (from hexane) ^{1}H NMR (benzene-d_6): 1.66 (dd, H-6; J = 8, J = 9), 2.02 (br,t, H-5; J = 9, J = 10), 2.39 (ddd, H-4; J = 3.5, J = 10, J = 16), 2.69 (br,d, H-4′, J = 16), 3.35 (m, H-1,2), 3.71 (m, H-3,7), 7.3 (m, C_6H_5) ^{13}C NMR: 22.11 (C-6), 40.49 (C-4), 49.56 (C-5), 55.63 (C-7); 66.41, 72.27 (C-1,3); 69.03 (C-8), 98.89 (C-2); 126.4, 127.3, 127.9, 129.9, 139.0, 139.6 (C_6H_5); 203.1, 212.8, 214.1 (CO); 218.3 (C=O)	[105, 122]
121	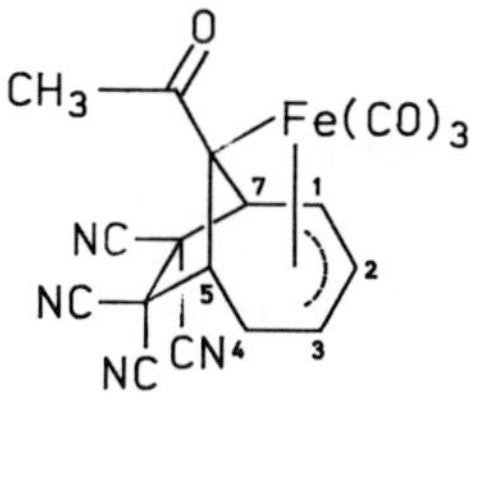 Vd white crystals (from toluene/n-hexane) ^{1}H NMR (acetone-d_6): 2.34 (s, CH_3), 2.75 (d, H-4′; J = 16), 3.48 (m, H-4), 4.39 (d, H-5; J = 12); 4.92 (m, H-7), 5.21 (m, H-3), 5.36 (m, H-1,2) ^{1}H NMR (liquid SO_2): 2.34 (CH_3); 2.75, 3.48 (H-4,4′), 4.39 (H-5), 4.96 (H-7), 5.21 (H-3), 5.36 (H-1,2) IR (Nujol): 1740 (C=O); 2020, 2090 (CO), 2260 (CN)	[91, 123]
122	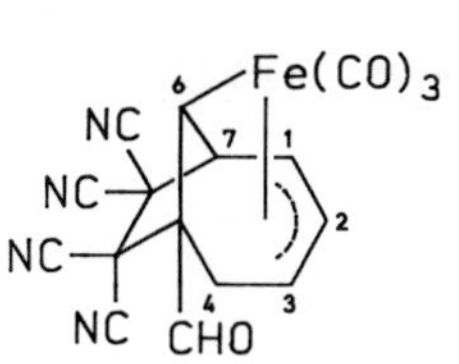 Vd (80) ^{1}H NMR (acetone-d_6): 1.82 (d, H-6; J = 9); 2.6 (d; J = 16), 3.5 (dd; J = 6, J = 16) (H-4, 4′); 4.64 (m, H-7), 5.0 (m, H-3), 5.24 (m, H-1,2), 9.74 (s, CHO) IR (CH_2Cl_2): 2014, 2076 (CO)	[123]

References on pp. 172/5

Table 4 [continued]

No.	compound	method of preparation (yield in %), properties and remarks	Ref.
123	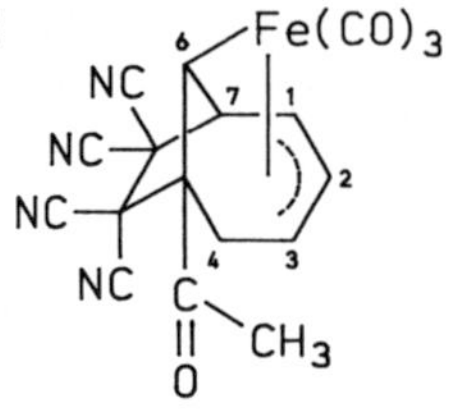	Vd white needles (from toluene/n-hexane) ^1H NMR (acetone-d_6): 2.1 (H-6), 2.54 (s, CH_3), 3.15 (m, H-4,4'), 4.57 (t, H-7; J=8), 4.83 (t, H-3; J=8), 5.15 (H-1,2) ^1H NMR (liquid SO_2): 2.1 (H-6), 2.51 (CH_3), 2.87 (H-4,4'), 4.57 (H-7), 4.83 (H-3), 5.15 (H-1,2) IR (Nujol): 1760 (C=O); 2040, 2093 (CO); 2260 (CN)	[91, 123]
124	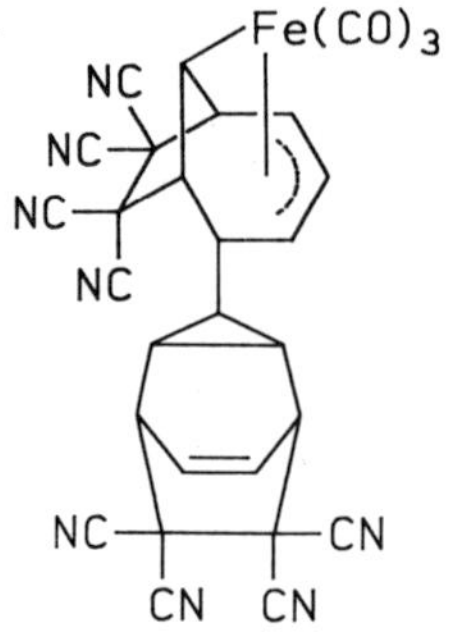	Vd IR ($CHCl_3$): 2013, 2080 (CO)	[31, 123]
*125	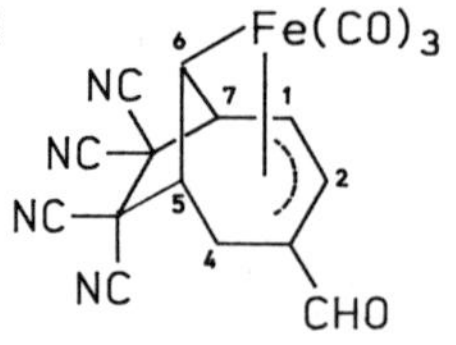	Vd (44) m.p. 165° (dec., from acetone/pentane) ^1H NMR (acetone-d_6): 3.71 (dd, H-4 endo; J=13, J=16.5), 4.20 (dd, H-5; J=13, J=6, J=2), 4.70 (dd, H-7; J=9.5, J=7), 5.52 (t, H-1; J=7), 5.78 (d, H-2; J=7), 9.36 (s, CHO) ^1H NMR ($OS(CD_3)_2$): 1.80 (d, H-4 exo; J=16.5) IR (CH_2Cl_2): 2021, 2082 (CO)	[92, 123]
*126	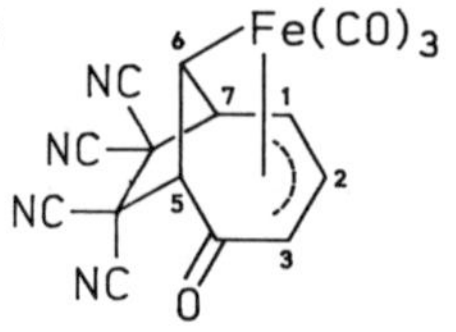	Vd (42) m.p. 110° (dec.), pale yellow crystals (from CH_2Cl_2/hexane at 0°) ^1H NMR (acetone-d_6 at −20°): 2.18 (H-6), 4.00 (H-5; J(H-5,6)=5.9), 4.76 (H-3; J(H-2,3)=9.1), 4.85 (H-7), 5.43 (H-1; J(H-1,7)=10.7), 5.75 (H-2; J(H-1,2)=7.5) ^{13}C NMR (acetone-d_6 at −60°): 5.16 (C-6), 61.95 (C-5); 57.15, 65.10, 73.17 (C-1,3,7); 99.74 (C-2) IR (CH_2Cl_2): 1692 (C=O); 2034, 2090 (CO)	[88, 123]
*127	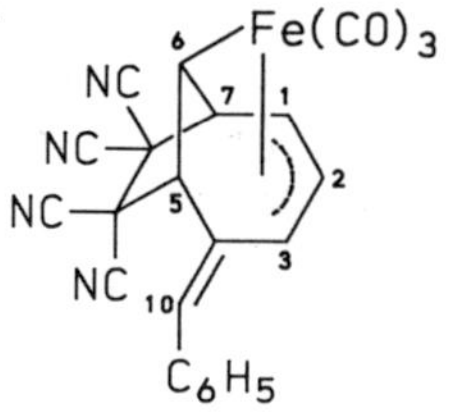	Vd (15) isolated by fractional crystallization from CH_2Cl_2/hexane ^{57}Fe-γ: δ=0.26, Δ=0.76 IR (CH_2Cl_2): 2009.6, 2071.9 (CO)	[125]

References on pp. 172/5

Table 4 [continued]

No.	compound	method of preparation (yield in %), properties and remarks	Ref.
128	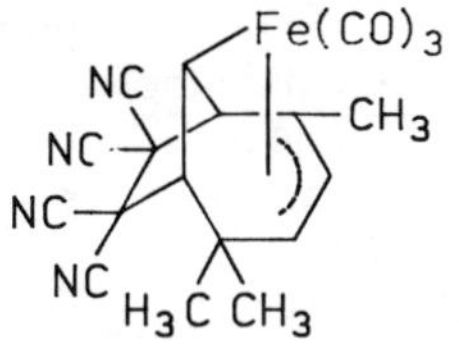	all data for mixtures with LIII (p. 114) Vd m.p. 135° (dec.), bright yellow crystals (from CH_2Cl_2/hexane) ^{1}H NMR ($CDCl_3$): 1.53 (s, CH_3), 1.66 (s, CH_3), 2.15 (s, CH_3), 1.84 (dd, 1H; J=9.3, J=6.9), 3.00 (d, 1H; J=6.9), 4.13 (d, 1H; J=9.6), 4.21 (d, 1H; J=9.6), 4.47 (d, 1H; J=9.3) mass spectrum: $[M-nCO]^+$ (n=0 to 3)	[128]

compounds of the type $^4LFe(CO)_3$ where ^4L=bicyclo[4.3.1]dec-3-en-2,10-diyl or its derivatives:

No.	compound	method of preparation (yield in %), properties and remarks	Ref.
129	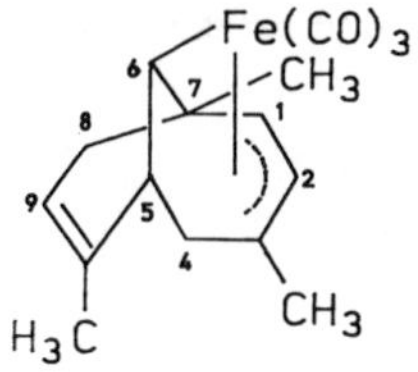	Ic (25), see also No. 131, p. 169 b.p. 95 to 100°/0.04 Torr, pale yellow oil ^{1}H NMR (benzene-d_6): 1.20 (s, CH_3-7), 1.08 to 1.24 (m, superimposed by s, H-6), 1.64 (s, CH_3-10), 1.79 (br,s, CH_3-3), 1.40 to 2.86 (m, superimposed by br,s and s, 5H), 3.74 (d, H-1; J(H-1,2)=7.0), 4.06 (dd, H-2; J(H-2,4)=2.0), 5.15 (dm, H-9; J(H-8,9)=8.0) IR (film): 768, 1025, 1040, 1375, 1453; 1965, 2040 (CO); 2855, 2930, 2960 UV (hexane): $\lambda_{max}(\varepsilon)$=210 (19300), 229 (sh, 17300), 273 (2930) nm mass spectrum: $[M-nCO]^+$ (n=0 to 3) and further fragments given	[90]
*130	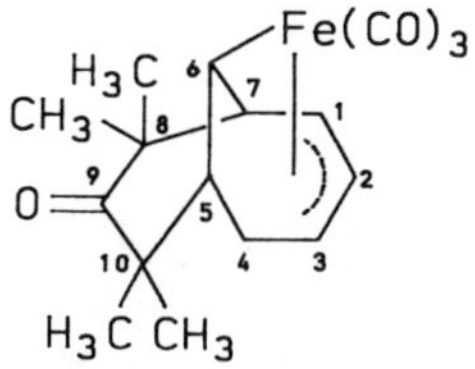	Vd (20) m.p. 88 to 89°, pale yellow crystals (from n-hexane or C_6H_6/n-hexane) ^{1}H NMR ($CDCl_3$): 0.92, 1.00, 1.14, 1.16 (s, 4CH_3), 1.28 (dd, H-6; J(H-5,6)=7, J(H-6,7)=14), 1.96 to 2.16 (m, H-4), 2.28 (ddd, H-5; J(H-4,5)=1, J(H-4′,5)=6), 2.36 to 2.72 (m, H-7), 3.05 (ddd, H-4′; J(H-4,4′)=1, J(H-3,4′)=9.5), 4.24 to 4.56 (m, H-1,2,3) ^{13}C NMR ($CDCl_3$): 19.7 (d, C-6); 22.5, 25.1, 25.6, 31.7 (q, 4CH_3); 33.2 (t, C-4); 47.5, 49.7 (s, C-8,10); 52.3, 59.5 (d, C-5,7); 66.0, 71.5 (d, C-1,3); 97.0 (d, C-2); 202.7, 213.3, 214.3 (s, 3CO); 220.2 (s, C-9) IR (Nujol): 1695 (C=O); 1980, 2080 (CO) mass spectrum: $[M-nCO]^+$ (n=0 to 3)	[118, 119]

References on pp. 172/5

Table 4 [continued]

No.	compound	method of preparation (yield in %), properties and remarks	Ref.
*131	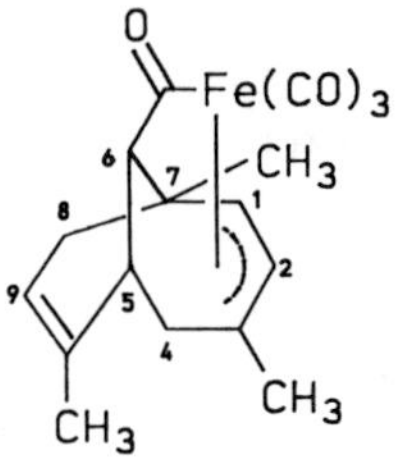	Ic (34.6) m.p. 75 to 77° (dec.), yellow prisms (from pentane) ^{1}H NMR (benzene-d_6): 1.06 (s, CH_3-7), 1.44 (br,s, CH_3-10), 1.50 (s, CH_3-3), 1.82 to 2.12 (m, 5H), 2.37 (dm, H-4; J(H-4,4′) = 18.0), 3.80 (d, H-1; J(H-1,2) = 9.0), 4.08 (dd, H-2; J(H-2,4) = 2.0), 5.04 to 5.18 (m, H-9) IR (KBr): 660, 1050, 1375, 1445, 1455; 1643 (C=O); 1980, 2000, 2060 (CO) UV (hexane): $\lambda_{max}(\varepsilon)$ = 208 (24600), 251 (sh, 9230) nm mass spectrum: $[M - nCO]^+$ (n = 1 to 4) and further fragments given	[90]
compounds of the type $^4LFe(CO)_3$ where ^4L = bicyclo[5.2.1]dec-3-en-2,10-diyl or its derivatives:			
132	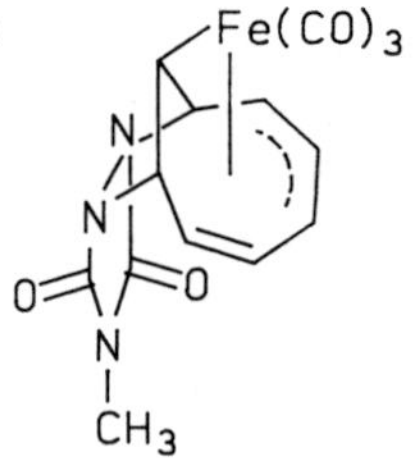	not isolated formulated as an intermediate in the reaction of $^4LFe(CO)_3$ (^4L = cyclooctatetraene) with N-methyltriazolinedione	[87]
133	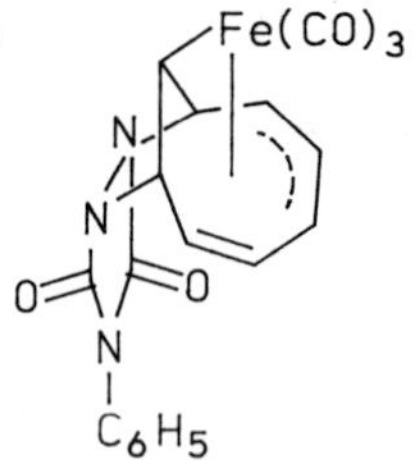	not isolated formulated as an intermediate in the reaction of $^4LFe(CO)_3$ (^4L = cyclooctatetraene) with N-phenyltriazolinedione	[87]
*134	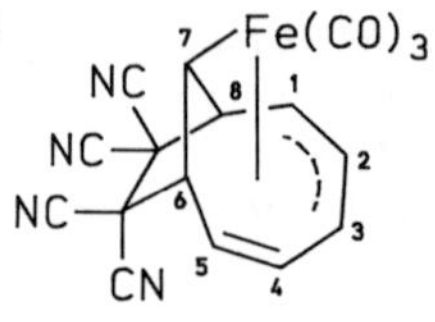	Ve (96) dec. 150°, dec. 176°, darkens above 150 or 160°, does not melt under 350°, pale yellow diamagnetic solid or yellow crystals ^{1}H NMR (acetone-d_6): 2.0 (H-7; J(H-7,8) = 12.0), 3.88 (H-6; J(H-6,7) = 7.0), 4.6 (H-3,8), 5.1 (H-1,2), 5.84 (H-5; J(H-5,6) = 5.0), 6.63 (H-4; J(H-4,6) = 2.0, J(H-4,5) = 12.0) ^{1}H NMR ($OS(CD_3)_2$): 1.56 (dd, H-7; J(H-7,8) = 11.5), 3.94 (ddd, H-6; J(H-6,7) = 6.5), 4.4 to 4.68 (m, H-3,8; J(H-3,4) = 9), 4.9 to 5.2 (m, H-1,2), 5.76 (dd, H-5; J(H-5,6) = 4.5), 6.6 (ddd, H-4; J(H-4,6) = 2, J(H-4,5) = 11)	[1 to 5, 7, 24, 34, 44, 67, 114]

References on pp. 172/5

Table 4 [continued]

No.	compound	method of preparation (yield in %), properties and remarks	Ref.
		^{57}Fe-γ (78 K): $\delta = 0.864$, $\Delta = 0.21$ IR (KBr): 678, 755, 779, 843, 859, 903, 931, 962, 988, 1026, 1065, 1082, 1116, 1142, 1182, 1218, 1238, 1264, 1284, 1305, 1318, 1340, 1401, 1437, 1497, 1658; 1966, 2016, 2066 (CO); 2237 (CN), 2899 IR (Nujol): 693, 750, 774, 827, 838, 855, 898, 928, 960, 985 1024, 1062, (1080), 1113, 1139, 1180, 1222, 1240, 1270, 1286, (1306), 1320, 1344, 1405, 1670, 2012, 2072, 2250, 2950, 2978, 2995, 3030	
*135	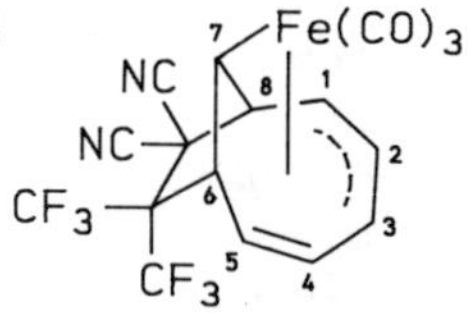	Ve (95) m.p. 130°, pale yellow crystalline solid (from CH_2Cl_2/hexane) ^{1}H NMR ($CDCl_3$): 1.63 (H-7; J(H-7,8) = 11.5), 3.34 (H-6; J(H-6,7) = 8.0), 4.06 (H-8; J(H-1,8) = 7.5), 4.42 (H-3; J(H-2,3) = 10.0), 4.60 (H-1; J(H-1,2) = 8.0), 4.79 (H-2; J(H-2,4) = 9.0), 5.68 (H-5; J(H-5,6) = 6.0), 6.45 (H-4; J(H-4,5) = 11.0) ^{19}F NMR ($CDCl_3$): 58.6 (CF_3; J(F,F) = 13.0), 63.3 (CF_3; J(H,F) = 4.5) IR (cyclohexane): 1999, 2005, 2062 (CO) mass spectrum: $[M - nCO]^+$ (n = 0 to 3)	[12, 24]
*136	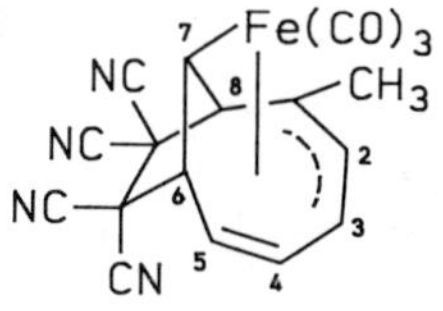	Ve (60) m.p. 162° (dec.), m.p. >270° (dec. starts at 195°), yellow crystals (from CH_2Cl_2/hexane) ^{1}H NMR (acetone-d_6): 1.80 (CH_3), 1.9 (H-7), 4.0 (H-6; J(H-6,7) = 6.0), 4.42 (H-3), 4.68 (H-8; J(H-7,8) = 12.0), 5.1 (H-2), 5.88 (H-5; J(H-5,6) = 5.0), 6.66 (H-4; J(H-4,5) = 9.0) IR (CH_2Cl_2): 2008, 2070 (CO) mass spectrum: $[M - nCO]^+$ (n = 1 to 3), $[M - 3CO - C_6N_4]^+$	[44, 67]
*137	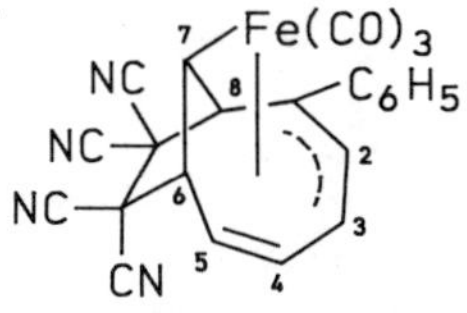	Ve (16) m.p. >260° (dec., from C_6H_6/hexane) ^{1}H NMR (acetone-d_6): 2.14 (dd, H-7; J(H-7,8) = 12), 3.96 to 4.16 (m, H-6; J(H-6,7) = 7), 4.72 (dd, H-3), 5.30 (d, H-8), 5.64 (d, H-2; J(H-2,3) = 10), 6.02 (dd, H-5; J(H-5,6) = 5), 6.77 (ddd, H-4; J(H-4,6) = 2, J(H-4,5) = 11), 7.32 to 7.6 (m, 3H in C_6H_5), 7.74 to 7.94 (m, 2H in C_6H_5)	[67]

References on pp. 172/5

Table 4 [continued]

No.	compound	method of preparation (yield in %), properties and remarks	Ref.
*138	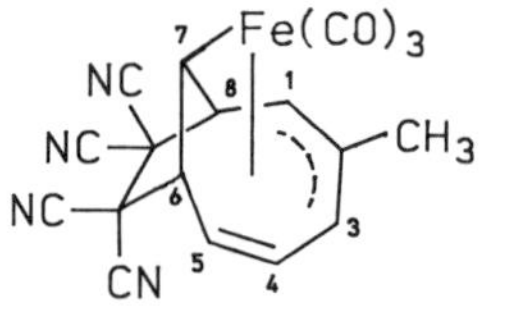	Ve m.p. 120° (dec.), yellow solid (from CH_2Cl_2/pentane) ^{1}H NMR ($OS(CD_3)_2$): 1.7 (dd, H-7; J(H-7,8) = 12), 2.13 (s, CH_3), 3.78 to 3.99 (m, H-6; J(H-6,7) = 6), 4.38 to 4.74 (m, H-3,8), 4.92 to 5.21 (m, H-1), 5.86 (dd, H-5; J(H-5,6) = 5.5), 6.66 (ddd, H-4; J(H-4,6) = 2, J(H-4,5) = 12)	[67]
*139	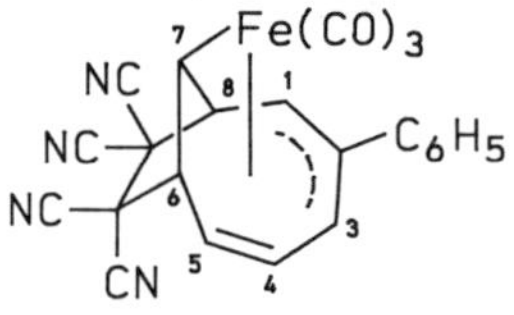	Ve (23) m.p. >260° (dec., from hexane/C_6H_6 (3:1)) ^{1}H NMR (acetone-d_6): 1.90 (dd, H-7; J(H-7,8) = 12), 3.92 to 4.12 (m, H-6; J(H-6,7) = 7), 4.80 to 5.08 (dd, H-8), 5.18 to 5.42 (m, H-3; J(H-3,4) = 8.5), 5.58 (d, H-1; J(H-1,8) = 8), 6.04 (dd, H-5; J(H-5,6) = 4.5), 6.98 (ddd, H-4; J(H-4,6) = 2, J(H-4,5) = 10), 7.30 to 7.70 (m, C_6H_5)	[67]
*140	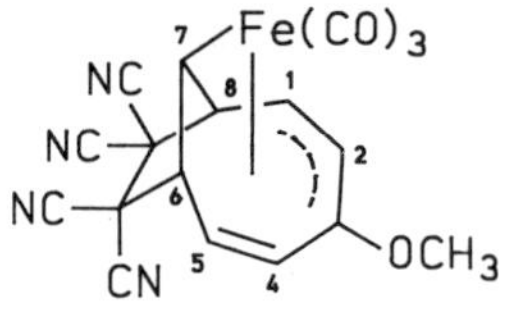	Ve (31.5) m.p. 110° (dec.), pale yellow solid (from CH_2Cl_2) ^{1}H NMR (acetone-d_6): 1.69 (s, H-7; J(H-7,8) = 11), 3.66 (s, CH_3), 3.97 (m, H-6; J(H-6,7) = 7), 4.59 (dd, H-8; J(H-1,8) = 7), 4.96 (dd, H-1; J(H-1,2) = 8.5), 5.36 (d, H-2), 6.01 (dd, H-5; J(H-5,6) = 4.5, J(H-4,5) = 11), 6.73 (dd, H-4; J(H-4,6) = 2)	[67]
141	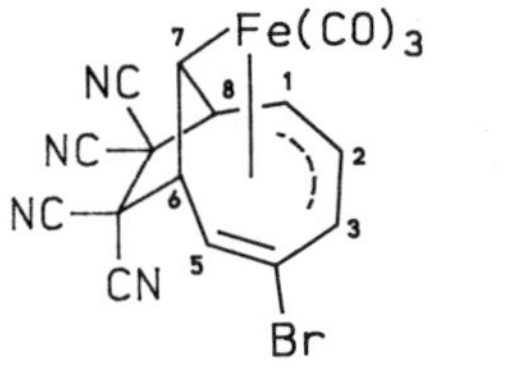	Ve (81) yellow crystals (from CH_2Cl_2/hexane) ^{1}H NMR (acetone-d_6): 1.8 (H-7; J(H-7,8) = 10.0), 4.0 (H-6; J(H-6,7) = 8.0), 4.56 (H-8), 4.77 (H-3), 5.14 (H-2; J(H-2,3) = 10.0), 5.20 (H-1; J(H-1,8) = 8.0), 6.18 (H-5; J(H-5,6) = 4.0) IR (CH_2Cl_2): 2028, 2089 (CO) mass spectrum: fragmentation pattern of C_8H_7Br only	[44]
*142	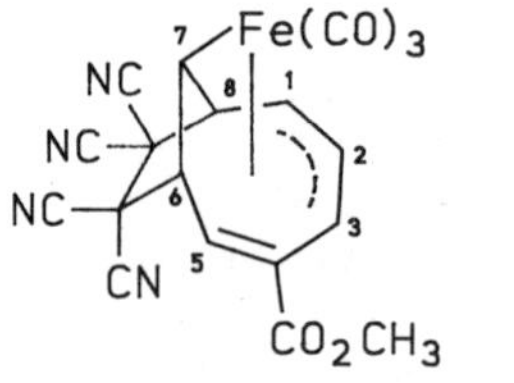	Ve (64) m.p. >260° (with onset of dec. at 160°), pale yellow crystals (from hexane/acetone) ^{1}H NMR ($CDCl_3$): 1.6 (dd, H-7; J(H-7,8) = 12), 3.54 (s, CH_3), 3.98 (dd, H-6; J(H-6,7) = 7), 4.32 (ddd, H-8; J(H-2,8) = 2), 4.73 to 5.2 (m, H-1,2,3; J(H-1,8) = 5.5), 6.78 (d, H-5; J(H-5,6) = 5)	[67]

References on pp. 172/5

Table 4 [continued]

No.	compound	method of preparation (yield in %), properties and remarks	Ref.
*143	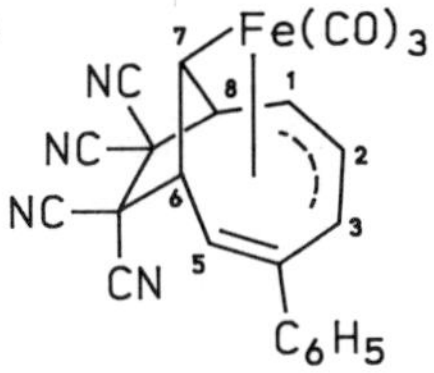 $Fe(CO)_3$ complex; NC, NC, NC, CN; C_6H_5	Ve (39.2), Ve (76) m.p. 120° (dec.), m.p. >250° (dec.), yellow crystals (from CH_2Cl_2/hexane), yellow solid (from aqueous acetone) ^1H NMR (acetone-d_6): 1.76 (H-7; J(H-7,8) = 10.0), 4.10 (H-6; J(H-6,7) = 7.0), 4.52 (H-8; J(H-1,8) = 6.0), 4.70 (H-3; J(H-2,3) = 11.0), 5.26 (H-1,2), 5.90 (H-5; J(H-5,6) = 5.0) ^1H NMR (acetone-d_6): 1.88 (dd, H-7; J(H-7,8) = 11), 4.18 (dd, H-6; J(H-6,7) = 6.5), 4.52 to 4.92 (m, H-3,8), 5.22 to 5.5 (m, H-1,2), 6.0 (d, H-5; J(H-5,6) = 4.8), 7.28 to 7.68 (m, C_6H_5) IR (cyclohexane): 2009, 2016, 2070 mass spectrum: $[M-C_6N_4-nCO]^+$ (n = 0 to 3)	[44, 67]
144	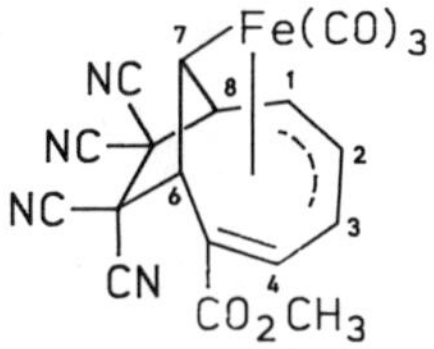 $Fe(CO)_3$ complex; NC, NC, NC, CN; CO_2CH_3	Ve (23) m.p. 120° (dec.) ^1H NMR ($CDCl_3$): 1.97 (dd, H-7; J(H-7,8) = 11), 3.82 (s, CH_3), 4.28 (d, H-6; J(H-6,7) = 7.2), 4.58 (dd, H-8; J(H-1,8) = 7), 4.69 (dd, H-2; J(H-1,2) = J(H-2,3) = 8.5), 5.01 to 5.42 (m, H-1,3; J(H-3,4) = 9), 7.96 (d, H-4)	[67]
*145	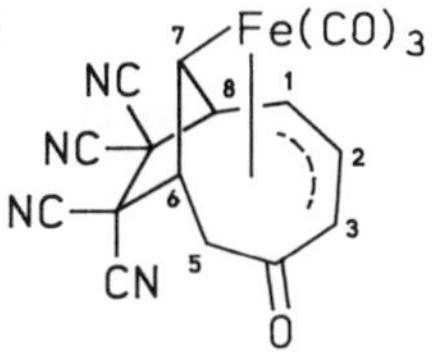 $Fe(CO)_3$ complex; NC, NC, NC, CN; O	Ve (68) 142° (dec., from CH_2Cl_2/pentane) ^1H NMR (acetone-d_6): 3.64 (m, H-5), 4.2 (m, H-6), 4.31 (d, H-3; J = 10), 4.88 (dd, H-8; J = 7, J = 9.5), 5.4 (t, H-1; J = 9.5), 5.8 (br,t, H-2; J = 9.5, J = 10)	[92]
*146	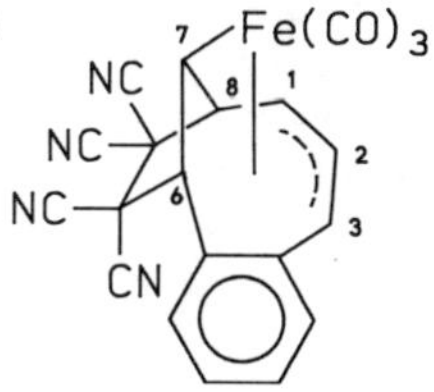 $Fe(CO)_3$ complex; NC, NC, NC, CN; benzo	–	[67]

compounds of the type $^4LFe(CO)_3$ where ^4L = tricyclo[4.3.2.0^{2,7}]undeca-4-en-3,11-diyl or its derivatives:

No.	compound	method of preparation (yield in %), properties and remarks	Ref.
*147	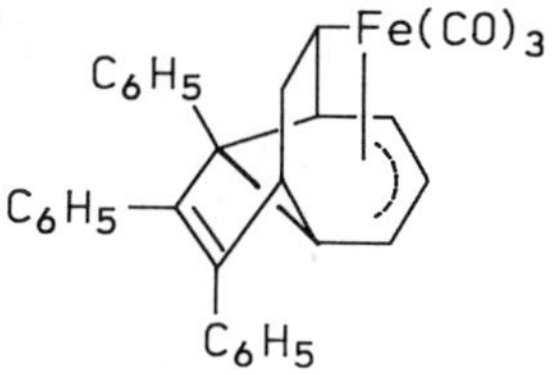 $Fe(CO)_3$ complex; C_6H_5, C_6H_5, C_6H_5	VIe (25) yellow crystals (from n-hexane at −78°) IR (hexane): 1985, 1989, 2052 (CO)	[115]

References on pp. 172/5

Table 4 [continued]

No.	compound	method of preparation (yield in %), properties and remarks	Ref.
other compounds:			
*148	$[(CO)_3Fe{-}H$ complex$]^+$ (structural formula)	—	[78]
149	$(C_6H_5)_3C(NC)_4C_{10}H_7Fe(CO)_3$	Ve (75) m.p. 170° (dec.), pale yellow crystals of low solubility IR (CH_2Cl_2): 2016, 2076 (CO) mass spectrum: $[M-C_6N_4-nCO]^+$ (n=1 to 3) (structure unknown)	[44]
150	$(C_6H_5)_3C(CF_3)_2(NC)_2C_{10}H_7Fe(CO)_3$	Ve (25) m.p. 210° (dec.), pale yellow crystals (from CH_2Cl_2/hexane) of low solubility IR (cyclohexane): 2003, 2013, 2068 (CO) mass spectrum: $[M-nCO]^+$ (n=0 to 3) (structure unknown)	[44]
supplement: **compound of the type $^4LFe(CO)_3$ where ^4L = bicyclo[4.2.1]non-3-en-2,9-diyl or its derivatives:**			
151	(structural formula: Fe(CO)₃, NC, NC, NC, NC, C₆H₅; positions 1–7, 10, 11)	Vd (33) m.p. 158 to 160° (dec.), white crystals (from CH_2Cl_2/n-hexane) ^1H NMR (acetone-d_6): 1.97 (m, H-6), 4.04 (dd, H-5), 4.29 (m, H-4), 4.64 (dd, H-7; J(H-6,7)=9), 4.89 (dd, H-3; J(H-2,3)=9, J(H-3,4)=4), 5.08 (dd, H-1; J(H-1,7)=7), 5.25 (dd, H-2; J(H-1,2)=7), 6.47 (dd, H-10; J(H-4,10)=8), 6.87 (d, H-11; J(H-10,11)=16), 7.37 (m, C_6H_5) IR (CH_2Cl_2): 2009, 2070 (CO) mass spectrum: $[M]^+$	[130]
compounds with an eight-membered ring system:			
*152	$(CO)_2Ru{-}Fe(CO)_2$ with bridging CO and C_8 ring (structural formula)	Ve IR: 1815, 1968, 2004, 2035 mass spectrum: $[M-nCO]^+$ (n=0 to 5)	[13]

Table 4 [continued]

No.	compound	method of preparation (yield in %), properties and remarks	Ref.
153	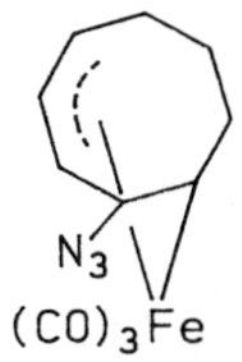 N_3 ... $(CO)_3Fe$	VId	[70]

* Further information:

$[C_4H_4Fe(CO)_3H]^+$ (Table **4**, No. **1**) is obtained from $C_4H_4Fe(CO)_3$ and excess $HOSO_2F$ in liquid SO_2 at −80 to −78 °C. The ^1H NMR spectrum of $C_4H_4Fe(CO)_3$ in $DOSO_2F$/liquid SO_2 at −85 °C shows chemical shifts at $\delta=3.42$ (br,s, H-4), 4.86 (br,s, H-1,3), and 6.44 (br,s, H-2) ppm consistent with $[C_4H_4Fe(CO)_3D]^+$. Its ^{13}C NMR spectrum in $DOSO_2F$/liquid SO_2 at −90 °C is the same as that for the undeuterated ion. The NMR spectra are not temperature dependent, and the ion slowly decomposes above −50 °C [78].

$(CH_3)_4C_4CH(CF_3)Fe(CO)_3$ (Table **4**, No. **5**). The structure of this compound given in the table is confirmed by preliminary X-ray crystallographic studies [36]. An analysis of the electronic origin of geometrical deformations in this compound is made [79]. Heating in hexane at 100 °C in a Carius tube for 6 h gives $[(CF_3)(CH_3)_4C_5Fe(CO)_2]_2$ (28% yield) [86] and No. 12 [39, 86]. Reaction with CO (100 atm) in hexane in an autoclave at 80 °C [39] or 100 °C [86] for 2 h gives compound No. 12 in 80% [39] to 90% yield [86] after chromatography on alumina with hexane/CH_2Cl_2 (1:1) as eluent [86].

$C_5H_6OC{=}OFe(CO)_3$ (Table **4**, No. **10**). The ^1H NMR spectrum is given in a figure in [48]. An X-ray analysis shows the crystals to be disordered possibly due to a loss of CO_2 at room temperature [72]. The compound is soluble in C_6H_6 but scarcely soluble in hexane [48].

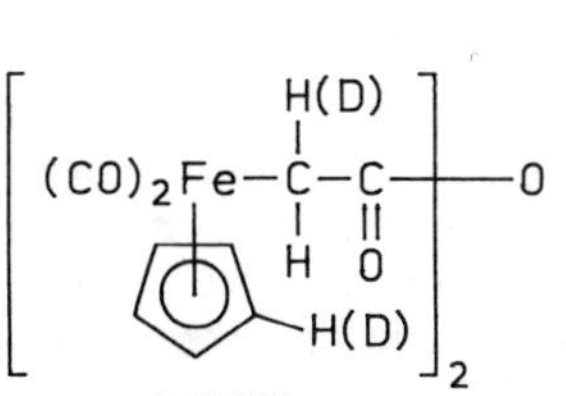

LXXIV 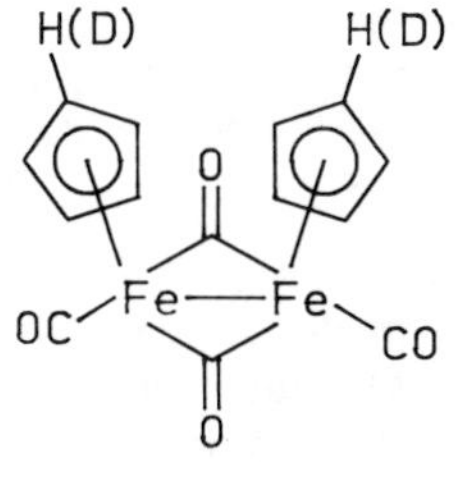 LXXV 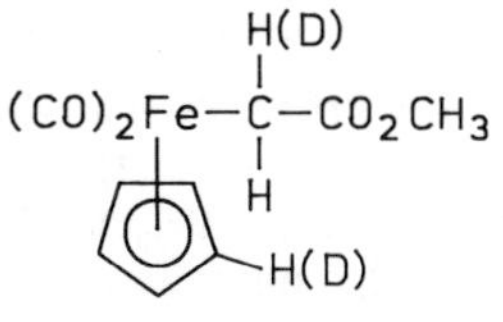 LXXVI

$C_6H_6C{=}OFe(CO)_3$ (Table **4**, No. **11**) is soluble in C_6H_6, $CHCl_3$ and acetone, slightly soluble in ether, scarcely soluble in hexane, and thermally unstable. Warming in acetone-d_6/10% $Si(CH_3)_4$ or in toluene-d_8/10% C_6H_6 at 30 °C (half-life time ca. 40 min) gives C_6H_6, LXXIV and LXXV in the ratio 0.2:1:1; the portion of LXXIV is dependent from the H_2O content of the solvent. Similar treatment in the above solvents in the presence of CH_3OH at 40 °C for 15 min gives C_6H_6 and LXXVI (ratio ca. 1:4) [108]. Protonation with $HOSO_2F$ in $CHFCl_2$/SO_2ClF at −60 °C gives LXXVII [107, 108]. Analogous results are obtained with No. 11 deuter-

References on pp. 172/5

ated at C-5,6. Reaction with $P(C_6H_5)_3$ in acetone gives LXXVIII a ($^2D = P(C_6H_5)_3$). This product may include that LXXVIII b is an intermediate. Reaction with $Fe_2(CO)_9$ gives LXXIX, LXXX, LXXV, and LXXXI in the ratio 4:5:6:4 (total yield ca. 40%) [108].

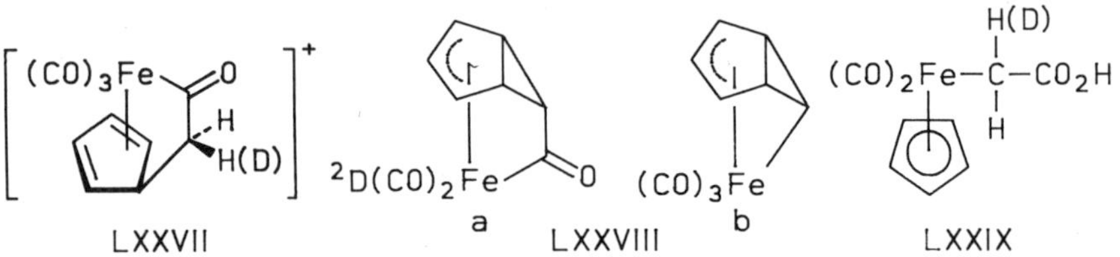

$O{=}(CH_3)_4C_5CH(CF_3)Fe(CO)_3$ (Table **4**, No. **12**) is stable to CO during its formation and does not give $[(CF_3)(CH_3)_4C_5Fe(CO)_2]_2$ on heating [39].

$[C_6H_8Fe(CO)_3H]^+$ (Table **4**, No. **13**) is obtained from $C_6H_8Fe(CO)_3$ (C_6H_8 = cyclohexa-1,3-diene) and $HOSO_2F$/liquid SO_2 or CF_3CO_2H/HBF_4 (ratio ca. 1:15). The temperature dependent ^1H NMR (no data given) and ^{13}C NMR spectra are discussed in terms of a reductive elimination process to give a symmetrical π-allyl species $[\eta^3\text{-}C_6H_9Fe(CO)_3]^+$ as an intermediate in the process by which C-4 and C-6 equilibrate rapidly on the NMR time scale [82]. Formulation as an Fe^{IV} σ-π allyl hydride is questioned in [100].

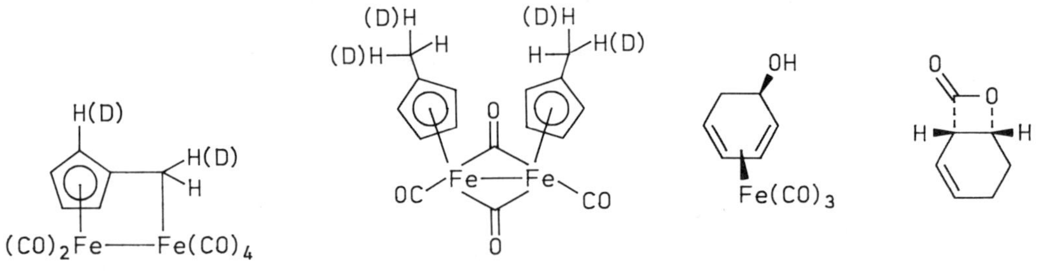

$C_6H_8OC{=}OFe(CO)_3$ (Table **4**, No. **14**) is readily soluble in C_6H_6, but scarcely soluble in hexane. Heating in C_6H_6 gives LXXXII [48]. Oxidation with $[NH_4]_2[Ce(NO_3)_6]$ in C_2H_5OH at −5 °C gives LXXXIII in 86% yield [85, 116].

$C_6H_8CH(CO_2CH_3)Fe(CO)_3$ (Table **4**, No. **15**). The deuterated derivative LXXXIV is obtained by Method II a (p. 111) starting with deuterated XVII (p. 110) in CH_3OD. The deuteration at C-7 stems from CH_3OD. The ^1H NMR spectrum in $CDCl_3$ in the presence of $Eu(fod)_3$ (fod = 1,1,1,2,2,3,3-heptafluoro-7,7-dimethyl-octane-4,6-dionate) shows a strong chemical shift of the H-1,6,7 signals consistent with the given position of the CO_2CH_3 group. The compound crystallizes in the space group Cc-C_s^4, C_{1h}^4 (or C2/c-C_{2h}^6) with the unit cell parameters a = 25.073(6), b = 8.127(2), c = 12.778(2) Å, β = 105.95(2)°; Z = 8 molecules per unit cell. D_{calc} = 1.544 g/cm³. The main bond distances and angles are shown in **Fig. 18** [83].

$C_6H_8CH(CO_2CH_3)Fe(CO)_3$ (Table **4**, No. **16**). The deuterated derivative LXXXV is obtained by Method II d (p. 112) starting with deuterated XVII (p. 110). The ^1H NMR spectrum in the presence of $Eu(fod)_3$ (for "fod" see No. 15) shows a strong chemical shift of the H-5,6′,7′ signals consistent with the given position of the CO_2CH_3 groups. Carbonylation in C_6H_6 in an autoclave with CO (100 atm) at 110 °C for 12 h affords LXXXVI (67% yield) [83].

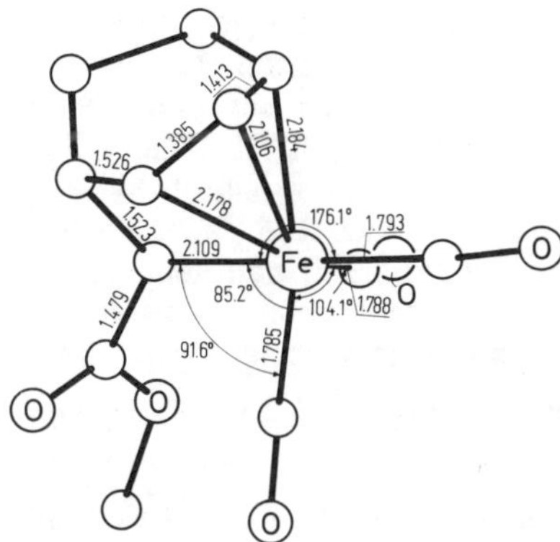

Fig. 18. Molecular structure of $C_6H_8CH(CO_2CH_3)Fe(CO)_3$ (No. 15) [83].

LXXXIV LXXXV LXXXVI LXXXVII

$C_6H_8C{=}OFe(CO)_3$ (Table **4**, No. **17**). Thermal decomposition of this compound in n-decane at 130 °C ($t_{1/2}$ ~ 4 h) gives $^4LFe(CO)_3$ (^4L = cyclohexadiene) presumably by intermediate LXXXVII [38].

$CF_3C(H)FCF_2C_6H_7CF_2CF(CF_3)Fe(CO)_3$ (Table **4**, No. **20**). An X-ray structure investigation of complex XLVII, p. 113, which is formed upon CO addition from No. 20, establishes the structure given in Table 4. It shows that the $CF_2CHF(CF_3)$ side chain has an endo configuration relative to the Fe atom [89].

$(CH_3)_2C_6H_6N(C_6H_5)C{=}OFe(CO)_3$ (Table **4**, No. **21**) shows in the IR spectrum one band at 1575 (C=O) cm^{-1} and three bands in the 2000 (CO) cm^{-1} region. Refluxing in C_6H_6 for 5 min gives LXXXVIII and LXXXIX (5% yield). Reaction with $^2D = P(C_6H_5)_3$ or $P(OCH_3)_3$ substitutes one of the CO groups in the $Fe(CO)_3$ moiety to give $(CH_3)_2C_6H_6N(C_6H_5)C{=}OFe(CO)_2{}^2D$ [96].

LXXXVIII LXXXIX XC

$(CH_2)_2C_6H_6C(H)FCF_2Fe(CO)_3$ (Table **4**, No. **23**) reacts with $P(OCH_3)_3$ in boiling hexane to give the CO substitution product $(CH_2)_2C_6H_6C(H)FCF_2Fe(CO)_2P(OCH_3)_3$ [71].

References on pp. 172/5

$CH_3C_6H_6(C(CH_3)_2H)C{=}OFe(CO)_3$, **$CH_3C_6H_4(O)(C(CH_3)_2H)Fe(CO)_3$**, and **$CH_3C_6H_4(O)$-$(C(CH_3)_2H)C{=}OFe(CO)_3$** (Table **4**, Nos. **28** to **30**). The mixture of compounds No. 29 and 30 shows in the IR spectrum (hexane) bands between 1660, 1690 (C=O), and at 1983, 1995, 2017, 2037, 2060 (CO) cm^{-1}. Both compounds decompose on repeated chromatography to give ($\pm$)-umbellulone (85% yield, cf. Formula II (X=O), p. 108). Thermal decomposition of No. 28 in boiling xylene is very slow. The decomposition products cannot be isolated. On the other hand, solutions of No. 28 in petroleum ether (b.p. 40 to 60 °C) or in ether tend to decompose on standing at room temperature (open to the air) to yield iron oxides mixed with an unidentifiable polymeric material [101].

$(CH)_4C_6H_2(O)(CH_3)_2Fe(CO)_3$ (Table **4**, No. **31**) reacts with CO (80 bar, 30 °C) to yield a new complex XC. This complex is unstable and readily looses CO to reform No. 31 [55].

$C_7H_{10}Fe(CO)_3$ (Table **4**, No. **32**). Reduction of $[^5LFe(CO)_3]BF_4$ (^5L=cycloheptadienyl) with $NaBD_4$ as in Method VIa (p. 116) gives the monodeuterated derivative, with D-7exo (cf. Table 4, p. 128) [73]. Heating No. 32 gives $^4LFe(CO)_3$ (^4L=cyclohepta-1,3-diene) [38]. Reaction of the 2:1 mixture of No. 32 and $^4LFe(CO)_3$ obtained by Method VIa with CO (80 atm) at 84 °C for 4 h gives $^4LFe(CO)_3$ and XCI (96%). Similar reaction of the monodeuterated products $^4LFe(CO)_3$ (^4L=cyclohepta-1,3-diene deuterated at C-5) and No. 32 deuterated at C-7 gives deuterated XCI (90% yield); D is equally distributed at C-2 and C-4 [73]. But reaction of No. 32 in pentane (20 to 30%) with CO (80 to 100 atm) at 20 to 25 °C for 6 d gives No. 33 (61%) [38, 73]. Reaction with $P(C_6H_5)_3$ in boiling C_6H_6 gives the CO substitution product of No. 32, $C_7H_{10}Fe(CO)_2P(C_6H_5)_3$. Reaction of a mixture of $^4LFe(CO)_3$ (^4L= cyclohepta-1,3-diene) and No. 32 with $[C(C_6H_5)_3]BF_4$ gives $[^5LFe(CO)_3]BF_4$ (^5L=cycloheptatrienyl) [58].

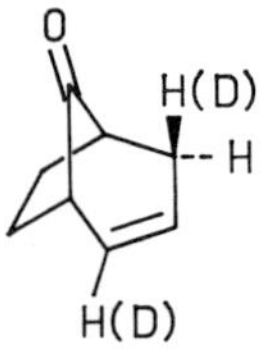

XCI 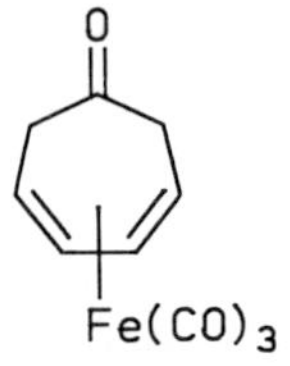 XCII 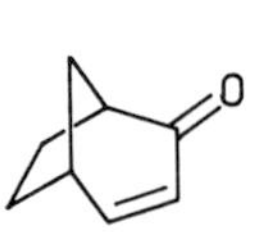 XCIII 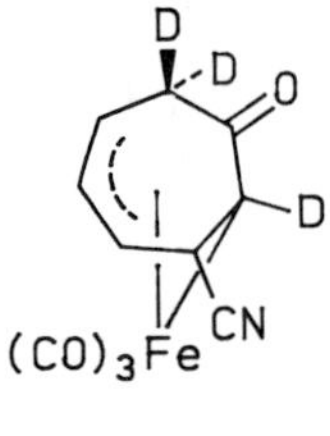 XCIV

$C_7H_{10}C{=}OFe(CO)_3$ (Table **4**, No. **33**) is stable for months, if stored at −15 °C under N_2. It decomposes under vacuum (12 Torr) at 20 °C and in solution at normal pressure at 30 °C to give quantitatively No. 32 [38, 73].

$C_7H_8OFe(CO)_3$ (Table **4**, No. **38**). The structure is confirmed by a preliminary X-ray analysis [98]. Heating the compound in C_6H_6 near reflux temperature gives XCII, and reaction with CO (100 atm) at 110 °C gives XCIII (70% yield) [104].

$NCC_7H_7OFe(CO)_3$ (Table **4**, No. **40**). The deuterated derivative XCIV is prepared from LVIII (R=D, see p. 116) as in Method VIb (pp. 116/7) [80].

XCV 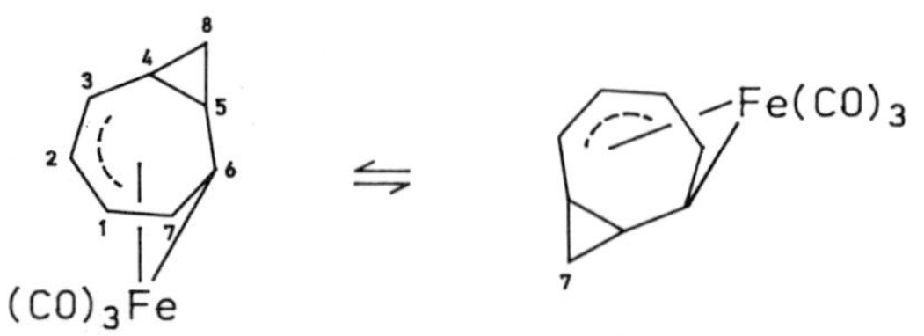 a XCVI b

$CH_2C_7H_8Fe(CO)_3$ (Table **4**, No. **42**). The monodeuterated derivative XCV is prepared from LXI (p. 117) and $NaBD_4$, analogous to Method VIc (p. 117) [40]. Similar reaction of perdeuterated LXI with $NaBH_4$ gives XCV-d_9 with H in the exo position at C-7 instead of D [74]. The ^{1}H NMR spectra are given in a figure in [40]. The above mentioned deuterated derivatives are used to demonstrate the degenerate valence isomerization shown in XCVI. There is a rapid equilibration of XCVIa:b = 1:1 at room temperature. Experiments to examine the H-2 → H-3, H-1 → H-4, H-7 → H-8 and H-6 → H-5 exchange in No. 42 (XCVIa) failed because isomerization from No. 42 to $^4LFe(CO)_3$ (^{4}L = cycloocta-1,3,5-triene) occurs at the temperatures (40 °C) needed [40, 74]. At 120 °C the isomerization of No. 42 gives $^4LFe(CO)_3$ (^{4}L = bicyclo[4.2.0]octa-2,4-diene). See [74] for the intermediates and the reaction mechanism. Oxidation of No. 42 in air at elevated temperatures gives cycloocta-1,3,6-triene. Reaction with CO gives No. 43 [40].

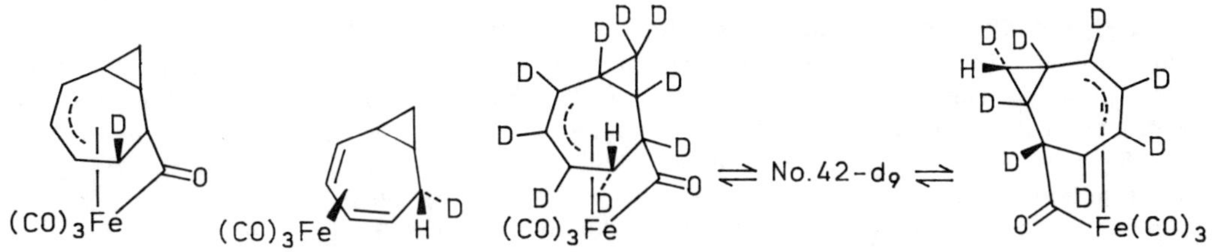

XCVII XCVIII IC

$CH_2C_7H_8C{=}OFe(CO)_3$ (Table **4**, No. **43**). The monodeuterated derivative XCVII is obtained together with XCVIII (ratio 1:4) by reduction of LXI (p. 117) with $NaBD_4$ in ice-water [51]. Similar reaction of perdeuterated LXI (PF_6 salt) with $NaBH_4$ gives XCVII-d_9 with H in the exo position at C-7 instead of D [74]. The ^{1}H NMR spectra are given in a figure in [40]. Flash distillation (40 °C/1 Torr) gives quantitatively No. 42 [51]. The equilibrium No. 43 ⇌ No. 42 + CO in C_6H_6 at 40 °C is reached after ca. 25 min [40]. See [74] for the mechanism of the equilibration shown in IC.

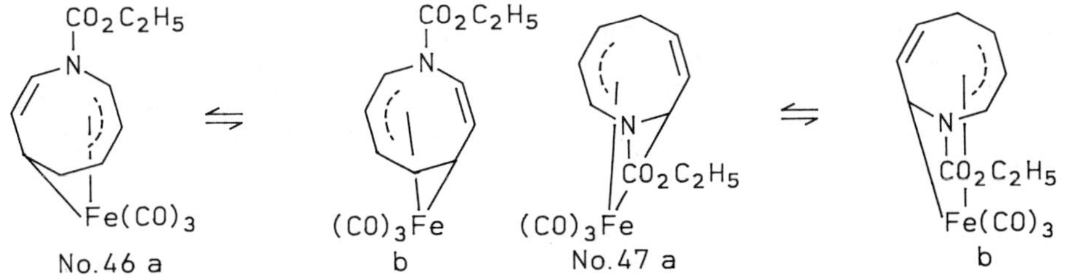

No. 46 a ⇌ b No. 47 a ⇌ b

C CI

$C_2H_5O_2CNC_7H_8Fe(CO)_3$ (Table **4**, Nos. **46** to **49**). The ^{1}H NMR spectra of Nos. 46 and 47 are temperature dependent and show the rapid degenerate valence isomerization shown in C and CI. The rotation of the $CO_2C_2H_5$ group is slow at 213 K, so that both rotamers are observed by the splitting of the H-3 and H-5 signals in No. 46. This is confirmed by the temperature dependent ^{13}C NMR spectrum of No. 46, which also shows that one of the equatorial CO groups remains unchanged in the transformation C. The equilibrium of the rotamers of No. 47 is estimated to be 2:1; the coalescence point is at 335 K. Heating of No. 46 in benzene-d_6 at 65 °C for 50 h gives XIX (p. 110) in 30% yield. But thermal reaction of No. 47 (60 °C, 20 h) or No. 48 (60 °C, 15 h) gives No. 49 in >90% yield. Photolyzation of No. 46 in ether for 30 min gives nearly exclusively No. 47. Similar photolysis of No. 46

for 3 h gives Nos. 47, 48, 49 in the ratio 2:5:1. Carbonylation of Nos. 46 or 47 with CO (160 atm) in benzene-d_6 at 80 °C for 20 h in an autoclave affords exclusively CII. Similar carbonylation of No. 49 for 15 h gives CII (40%) and CIII (57%) [94].

CII CIII CIV CV

$\mathbf{C_8H_{12}Fe(CO)_3}$ (Table **4**, No. **50**) is moderately stable when stored at −10 °C under Ar [57]. Refluxing in cyclohexane for 8 h affords $^4LFe(CO)_3$ (^4L = 5-methylcyclohepta-1,3-diene) [60]. Reaction with $AlCl_3$ in C_6H_6 at 10 °C for 3.5 h gives CIV (ca. 20%) and CV with R=D, if No. 50 is deuterated at C-5 [69]. A mixture of $^4LFe(CO)_3$ (^4L = cycloocta-1,5-diene) and No. 50 gives $[^5LFe(CO)_3]^+$ (^5L = cycloocta-2,5-dienyl) on reaction with $[C(C_6H_5)_3]BF_4$ [17]. No reaction is observed with liquid SO_2 (see ^{13}C NMR spectra) [91], but CO [17, 60, 69] (85 atm) in cyclohexane at 25 °C for 50 h [60] or in C_6H_6 (100 atm CO) at 25 °C for 20 h in a rocking autoclave [69] affords CIV [60, 69] in 80% yield [69]. CVI may be an intermediate in this reaction [17]. Treatment of No. 50 with $^2D = P(C_6H_5)_3$ [17, 32, 57, 60] or $P(OC_6H_5)_3$ [32, 57] or $P(C_6H_5)_2CH_2CH_2P(C_6H_5)_2$ [60] in n-heptane at 40 to 70 °C [32, 57] or refluxing in cyclohexane for 20 min [60] produces the monosubstituted complexes $C_8H_{12}Fe(CO)_2{}^2D$ [17, 32, 57, 60]. Kinetic studies are in agreement with a CO-dissociative mechanism for this reaction [17, 32, 57]:

$$C_8H_{12}Fe(CO)_3 \text{ (No. 50)} \underset{k_{-1}}{\overset{k_1}{\rightleftharpoons}} C_8H_{12}Fe(CO)_2 + CO$$

$$C_8H_{12}Fe(CO)_2 + {}^2D \xrightarrow{k_2} C_8H_{12}Fe(CO)_2{}^2D$$

Assuming stationary-state kinetic conditions for the intermediate $C_8H_{12}Fe(CO)_2$, the following expression is obtained for the observed pseudo-first-order rate constant: $k_{obs} = k_1 \cdot k_2 \cdot [^2D]\ (k_{-1} \cdot [CO] + k_2 \cdot [^2D])^{-1}$. Selected rate data for the reaction of No. 50 (1 to 2.5×10^3 M) with $^2D = P(C_6H_5)_3$ and $P(OC_6H_5)_3$ in n-heptane under Ar are given in the following table [57]:

t in °C	^2D	$[^2D]$ in M	k_{obs} in s^{-1}	k_{calc} in s^{-1}
39.8	$P(C_6H_5)_3$	0.105	5.45×10^{-5}	5.36×10^{-5}
45.0	$P(C_6H_5)_3$	0.150	1.17×10^{-4}	1.18×10^{-4}
50.0	$P(C_6H_5)_3$	0.0777	2.34×10^{-4}	2.44×10^{-4}
55.1	$P(C_6H_5)_3$	0.0906	4.73×10^{-4}	5.02×10^{-4}
59.9	$P(C_6H_5)_3$	0.0844	9.45×10^{-4}	9.70×10^{-4}
60.0	$P(OC_6H_5)_3$	0.0485	1.01×10^{-3}	9.84×10^{-4}
65.0	$P(OC_6H_5)_3$	0.238	2.03×10^{-3}	1.92×10^{-3}
70.0	$P(OC_6H_5)_3$	0.201	3.84×10^{-3}	3.66×10^{-3}

The calculated values are derived from the activation parameters of $\Delta H^{\ddagger} = 29.21 \pm 0.16$ kcal/mol and $\Delta S^{\ddagger} = 15.2 \pm 0.5\ cal \cdot K^{-1} \cdot mol^{-1}$ [32, 57].

References on pp. 172/5

Observed pseudo-first-order rate constants for reactions with a range of more nucleophilic $^2D = PR_3$ ligands (e.g., $P(OC_2H_5)_3$ are given by the relation $k_{obs} = k'_1 + k'_2[^2D]$; k'_1 corresponds to formation of $C_8H_{12}Fe(CO)_2{}^2D$ and the ligand dependent term to formation of $(CO)_2Fe(^2D)_3$ and the trans-annular ketone CIV. From plots of k_{obs} against $[P(OC_2H_5)_3]$, the values of k'_1 and k'_2 for reaction of No. 50 with $P(OC_2H_5)_3$ are derived in n-heptane and are shown in the following table [57]:

t in °C	$k'_2 \cdot 10^3$ in $L \cdot mol^{-1} \cdot s^{-1}$	k'_1 in s^{-1}	k_1 (calc) in s^{-1}
40.0	2.38 ± 0.09	$(4.86 \pm 0.94) \times 10^{-5}$	5.36×10^{-5}
50.0	5.19 ± 0.16	$(2.57 \pm 0.14) \times 10^{-4}$	2.44×10^{-4}
60.0	9.50 ± 0.32	$(1.01 \pm 0.03) \times 10^{-3}$	0.98×10^{-3}

The values k_1(calc) are calculated from data for the reaction of No. 50 with PR_3 ($R = C_6H_5$, OC_6H_5) for loss of CO from No. 50, and they agree with k'_1 to within one standard deviation. The activation parameters for the ligand dependent process (k'_2), $\Delta H^{\neq} = 13.7 \pm 0.5$ kcal/mol and $\Delta S^{\neq} = -27.0 \pm 1.6$ cal $\cdot K^{-1} \cdot mol^{-1}$, are typical for a bimolecular reaction. This reaction path results in removal of the organic moiety from the iron center as the trans annular ketone CIV and formation of $(CO)_2Fe(^2D)_3$. It is likely that a CO bound to iron becomes inserted into the iron-carbon bond during or after the bimolecular process.

The same observed rate constant ($k'_1 + k'_2[^2D]$) is observed if the reaction is monitored by following the loss of reactant, or the appearance of either product. It could be shown that the ratio $(k_{obs} - k'_1):k'_1$ equals the product concentration ratio $(CO)_2Fe(^2D)_3/C_8H_{12}$-$Fe(CO)_2P(OC_2H_5)_3$ for a series of experiments at 50 °C. This is again consistent with complex $(CO)_2Fe(^2D)_3$ being formed exclusively by the bimolecular reaction and C_8H_{12}-$Fe(CO)_2P(OC_2H_5)_3$ by the CO-dissociative pathway [32, 57].

The kinetics of reactions of No. 50 with a variety of substituted phosphines and phosphites ^2D at 60 °C are studied to obtain information of the bimolecular path, and whether this reaction always leads to formation of CIV. The next table contains second-order rate constants for reactions of No. 50 with ^2D ligands in n-heptane at 60 °C. The products are identified by their IR bands.

ligand ^2D	$(k'_2 \pm 5\%) \cdot 10^3$ in $L \cdot mol^{-1} \cdot s^{-1}$	$C_8H_{12}Fe(CO)_2{}^2D$	$(CO)_2Fe(^2D)_3$
$P(C_6H_5)_3$	—	yes	—
$P(OC_6H_5)_3$	estimated upper limit 0.2×10^{-3}	yes	—
$P(C_6H_5)_2CH_3$	1.5	yes	yes
$P(C_2H_5)_2C_6H_6$	2.4	yes	yes
$P(C_4H_9\text{-}t)_3$	4.7	yes	yes
$P(OCH_2)_3CC_2H_5$	ca. 5	yes	—
$P(OC_3H_7\text{-}i)_3$	5.4	yes	yes
$P(CH_3)_2C_6H_5$	7.0	yes	yes
$P(OC_2H_5)_3$	9.5	yes	yes
$P(OCH_3)_3$	10.2	yes	yes

k'_2 values are estimated from the always linear plots of k_{obs} against $[^2D]$. Inspection of these data suggests that both steric and electronic factors are important in determining

References on pp. 172/5

the value of k'_2. The sterically crowded $P(C_6H_5)_3$, and the less hindered but less basic $P(OC_6H_5)_3$, have no detectable second-order reaction. The increased second-order reactivity of $P(CH_3)_2C_6H_5$ over that of $P(C_2H_5)_2C_6H_5$ may be assigned to steric factors as may be the decrease of k'_2 along the series $P(OR)_3$ ($R = CH_3$, C_2H_5, C_3H_7-i). $P(OCH_2)_3CC_2H_5$ might be expected to be more reactive than $P(OCH_3)_3$ but this is not the case. This reaction with phosphite is also unusual because no CIV or $(CO)_2Fe(P(OCH_2)_3CC_2H_5)_3$ is formed, although there is a second-order reaction. It is not clear why the second-order rate constant for reaction with $P(OCH_2)_3CC_2H_5$ is less than that for reaction with $P(OCH_3)_3$ [57].

$(CO)_3Fe$ CVI $(CO)_4Fe$ $Fe(CO)_3$ CVII $Fe(CO)_3$ CVIII $(CO)_4Fe$ $Fe(CO)_3$ CIX

$IC_8H_{11}Fe(CO)_3$ (Table **4**, No. **54**). The product obtained by Method VId (pp. 117/8) and formulated as No. 54 is dissolved in acetone and treated with excess NaCN. After 90 min the solvent is removed, and CH_2Cl_2 is added. Chromatography on alumina gives No. 55. Similar treatment of the product obtained by Method VId in CH_2Cl_2 with an approximately stoichiometric amount of solid $AgPF_6$ gives $[^5LFe(CO)_3]^+$ (^5L = cycloocta-2,5-enyl) as the PF_6^- salt [127].

$NCC_8H_{11}Fe(CO)_3$ (Table **4**, No. **55**) is soluble in CH_2Cl_2, acetone, and ether [127].

$C_8F_8Fe(CO)_3$ (Table **4**, No. **61**). $^2LFe(CO)_4$ ($^2L = C_8F_8$) is converted to No. 61 on heating in hexane. The complex crystallizes (at −30 °C) in the orthorhombic space group Pnma-D_{2h}^{16}, V_h^{16} with the unit cell parameters a = 10.432(2), b = 13.262(2), c = 8.599(2) Å; Z = 4 molecules per unit cell. The main bond distances and angles are shown in **Fig. 19** [17].

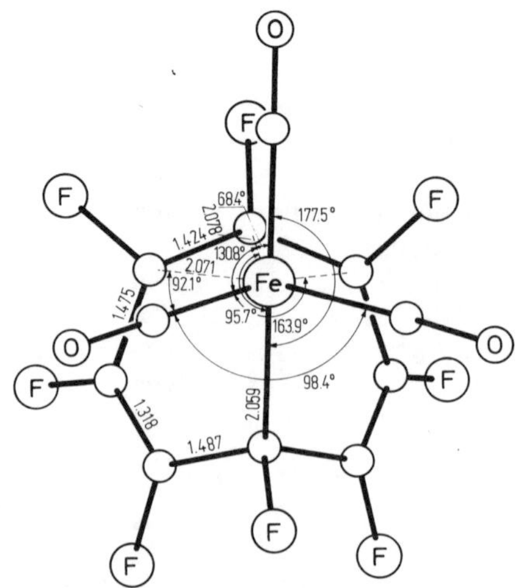

Fig. 19. Molecular structure of $C_8F_8Fe(CO)_3$ (No. 61) [117].

$C_8H_{12}C{=}OFe(CO)_3$ (Table **4**, No. **62**) may be the intermediate product in the reaction of No. 50 with CO affording CIV [17].

$C_8H_8OC{=}OFe(CO)_3$ (Table **4**, No. **64**) is an intermediate in the photoreaction of $Fe(CO)_5$ (1% C_6H_6 solution) with IX (p. 109) at 25 °C yielding $Fe_2(CO)_9$ and CVII within 6 h. The ^1H NMR spectrum of No. 64 is shown in a figure in [63]. Heating in $CHCl_3$ at 40 °C gives CVIII. Irradiation of No. 64 in the presence of $Fe(CO)_5$ or heating with $Fe_2(CO)_9$ at 25 °C gives CVII [63].

$C_7H_8C{=}OFe(CO)_3$ (Table **4**, No. **67**) is soluble in C_6H_6 or CS_2 but scarcely soluble in hexane [83]. Treatment with CO (90 atm) in C_6H_6 at 20 °C for 3 d gives insoluble brown decomposition products, and 67% of No. 67 is isolated; but similar treatment at 80 °C for 12 h gives quantitatively No. 81. This transformation is also observed without CO in an Ar atmosphere in C_6H_6 at 50 °C within ca. 15 min. Fresh solutions of No. 67 in hexane show the typical CO absorption, but within several min the two absorptions of compound No. 66 appear (see Table 4, p. 135). This reaction is reversible. Treatment of No. 66 with CO (90 atm) in hexane at −10 °C for 12 h gives No. 67 in nearly quantitative yield. Experiments to isolate pure No. 66 by evaporation of the solvent of etheral solutions of No. 67 or by bubbling Ar through a $CHCl_3$ solution of No. 67 failed, in that only insoluble brown decomposition products are isolated. Extraction of these products with C_6H_6 affords No. 67. Irradiation of No. 67 with $^2LFe(CO)_4$ (^2L = cyclohexene) gives $Fe_3(CO)_{12}$ and XXIV (p. 110) [83], cf. [77] together with small amounts of No. 81 [77]. Treatment with CH_3OH/C_6H_6 (1:10) at 25 °C for 2 h followed by evaporation gives a mixture of XXV (p. 110) and No. 16. The yields are badly reproducible and differ between 20 to 60% for XXV by a total yield of 60 to 70% [83].

$C_8H_8Fe(CO)_3$ (Table **4**, No. **69**) is also obtained by heating CIX [28]. The simultaneously formed $Fe_3(CO)_{12}$ is responsible for the green color cited in [16] for No. 69 [28]; pure No. 69 is yellow [28]. It is also formed on treating CIX in hexane with CO (ca. 100 atm) at 25 °C or in the reaction of CIX with $P(C_6H_5)_3$ in C_6H_6 at 70 °C in 95% yield together with $(CO)_4FeP(C_6H_5)_3$ [27]. The ^1H NMR spectrum remains unchanged up to 125 °C in $O_2NC_6H_5$ [43]. Catalytic hydrogenation affords compound No. 68 [16]. Treatment with 48% aqueous HBF_4 in $O(COCH_3)_2$ at 0 °C causes vigorous gas evolution, but no product can be isolated [43]. No reaction with HBr in ether at 0 °C [43] or with CO (1 atm) or $P(C_6H_5)_3$ in hexane at 25 °C is observed [27]. But, in pentane at 15 °C with CO (20 atm), compound No. 73 is obtained [65]. Reaction with $Fe_2(CO)_9$ in hexane at room temperature for 17.5 h gives XXXVI (p. 112). Similar treatment of a 1:1 mixture of No. 69 and XXXVI with $Fe_2(CO)_9$ in hexane (62 °C, 1.5 h) affords XXXIV (p. 111) [43].

$C_8H_8C{=}OFe(CO)_3$ (Table **4**, No. **73**) gives No. 69 on slight warming [65].

$C_9H_{10}Fe(CO)_3$ (Table **4**, No. **75**). The ^1H NMR spectrum is given in a figure in [49]. Reaction with CO in hexane at 25 °C [27] or in C_6H_6 (20 atm) [49] gives No. 76. Reaction with $P(C_6H_5)_3$ in hexane at 25 °C gives CX [27].

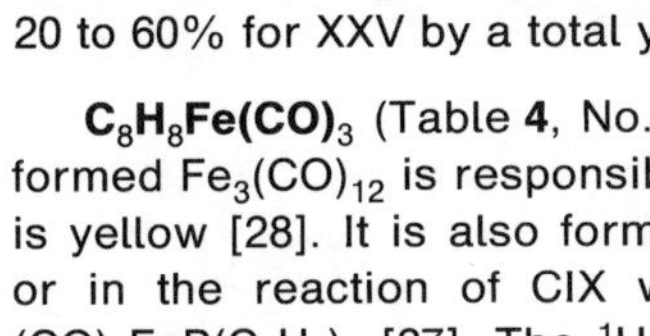

CX 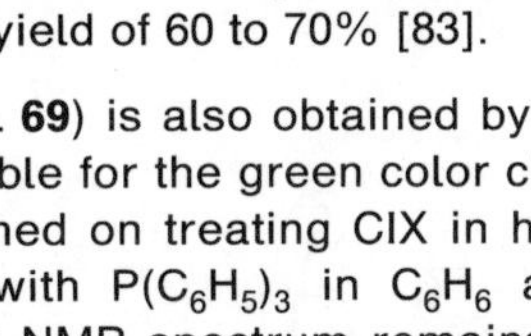 CXI 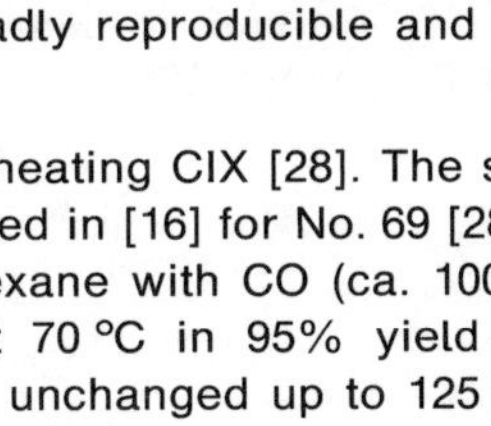 CXII 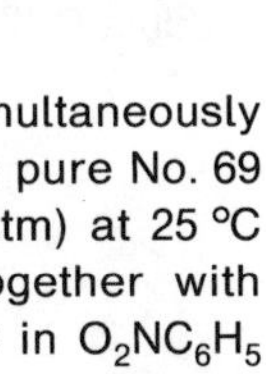 CXIII

$C_9H_{10}C{=}OFe(CO)_3$ (Table **4**, No. **76**). The ^1H NMR spectrum is given in a figure in [49]. Refluxing in hexane gives No. 75 [27, 49].

References on pp. 172/5

$C_{10}H_{10}C$=$OFe(CO)_3$ (Table **4**, No. **77**) is prepared by carbonylation of CXI at >120 °C [84].

$OC_8H_8Fe(CO)_3$ (Table **4**, No. **81**). The ^1H NMR spectrum of the monodeuterated compound is identical with that given in the Table, p. 138, except H-8′. The compound is soluble in C_6H_6 or CS_2 but scarcely soluble in hexane. Reaction with concentrated H_2SO_4 gives cation CXII. Treatment with CO (130 atm) in toluene in an autoclave at 175 °C for 12 h gives CXIII (97% yield) [83].

$RR'COC_8H_8Fe(CO)_3$ (Table **4**, Nos. **83** to **85** with R=H, R′=CH_3; R=H, R′=CH_3; R=H, R′=C_6H_5). The compounds No. 83 and 85 obtained by Method VIe are present as a diastereomeric mixture A and B as revealed by the ^1H NMR spectra. The assignments of structures A and B are purely arbitrary and intended for clarity. The preparation reaction is reversed by the addition of HPF_6 to Nos. 83 to 85 yielding the appropriate LXVI (A=$CHCHOHCH_3$, $CHCHOHC_6H_5$, $CHC(CH_3)_2OH$, X=PF_6, p. 118) [121].

$C_{11}H_{10}OFe(CO)_3$ (Table **4**, No. **86**) reacts with CO (100 atm) in the presence of traces of $(C_2H_5)_2OBF_3$ in C_6H_6 in an autoclave to give XLII (p. 112) in 99% yield after 2 d [84].

$RC_9H_6N_3O_3Fe(CO)_3$ (Table **4**, Nos. **87** and **88** with R=CH_3, C_6H_5) crystallizes in the monoclinic space group $P2_1/n-C^5_{2h}$ with the unit cell parameters a=8.726(2), b=23.169(3), c=6.719(2) Å, β=91.2(1)°; Z=4 molecules per unit cell. D_{calc}=1.75 g/cm³. The main bond distances and angles are shown in **Fig. 20**. In the organic ligand the dihedral angle between two parts of the seven-membered ring is 115.0°. The five-membered ring deviates slightly from planarity [97]. Careful oxidation of No. 88 with $[NH_4]_2[Ce(NO_3)_6]$ gives the rearranged product CXIV in 25% yield [92].

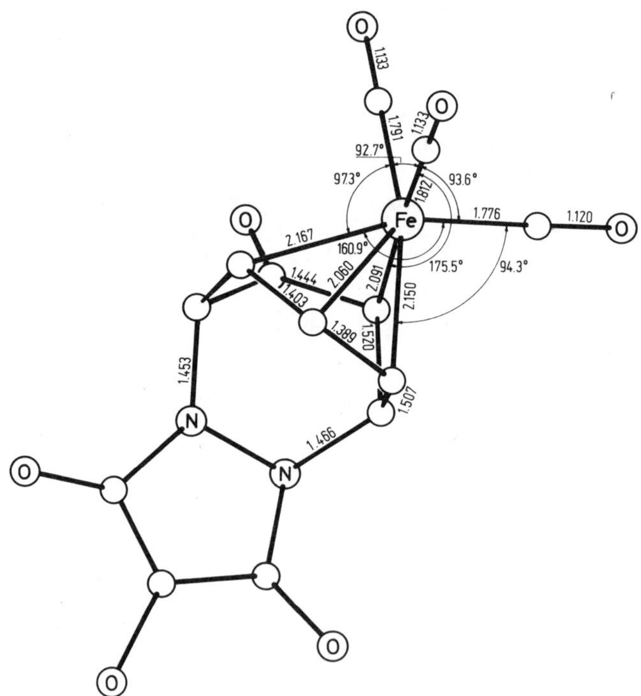

Fig. 20. Molecular structure of $C_{10}H_9N_3O_3Fe(CO)_3$ (No. 87) [97].

References on pp. 172/5

$C_9H_8OFe(CO)_3$ (Table **4**, No. **91**). The ^1H NMR spectrum in the presence of $Eu(dpm)_3$ (dpm = 2,2,6,6-tetramethyl-heptane-3,5-dionate) shows the greatest downfield shifts for the H-4,6 protons [69]. The ^{13}C NMR spectrum in toluene-d_8 in the presence of $Cr(acac)_3$ (acac = acetylacetonate) shows temperature dependence in the range of −7 to +76 °C. Thus, in the slow exchange limit it shows 3 well-separated resonances for the CO groups. At and below ca. −7 °C the rate of scrambling is too slow to influence line-shapes, but above ca. 0 °C all three lines noticeably broaden and disappear into the base line. The coalescence of the CO resonances occurs at 76 °C. Above about 76 °C thermal decomposition is sufficiently rapid to spoil the measurements. The scrambling of the 3CO groups proceeds in a concerted manner with an activation energy of $E_a = 16.5(3)$ kcal/mol, $\Delta G^{\ddagger}_{298} = 16.0(4)$ kcal/mol, $\Delta H^{+} = 15.8(3)$ kcal/mol, $\Delta S^{+} = -0.5(1.0)$ cal · K^{-1} · mol^{-1} [129]. An X-ray analysis [21, 35, 47, 53, 54, 62, 129] shows the compound to crystallize from $CHCl_3$/ether [54] in the orthorhombic [54] space group $Pna2_1$-C^9_{2v} [35, 47, 54] with the unit cell parameters a = 14.800(8), b = 8.775(4), c = 8.412(4) Å; Z = 4 molecules per unit cell [54]. $D_{calc} = 1.66$ g/cm^3 [54]. The main bond distances and angles are shown in **Fig. 21 a** [54]. Recrystallization from pentane at −5 °C gives crystals of the triclinic space group $P\bar{1}$-C^1_i,S^1_2 with the unit cell parameters a = 7.476(2), b = 11.912(4), c = 6.60(2) Å; α = 94.55(2)°, β = 110.17(2)°, γ = 92.38(3)° [47, 62]; Z = 2 molecules per unit cell. $D_{calc} = 1.65$ g/cm^3 [62]. The main bond distances and angles are shown in **Fig. 21 b** [62]. There exists no systematic trend in the crystal structure differences [53, 62]. It appears that the different types of crystals result from the different solvents [62].

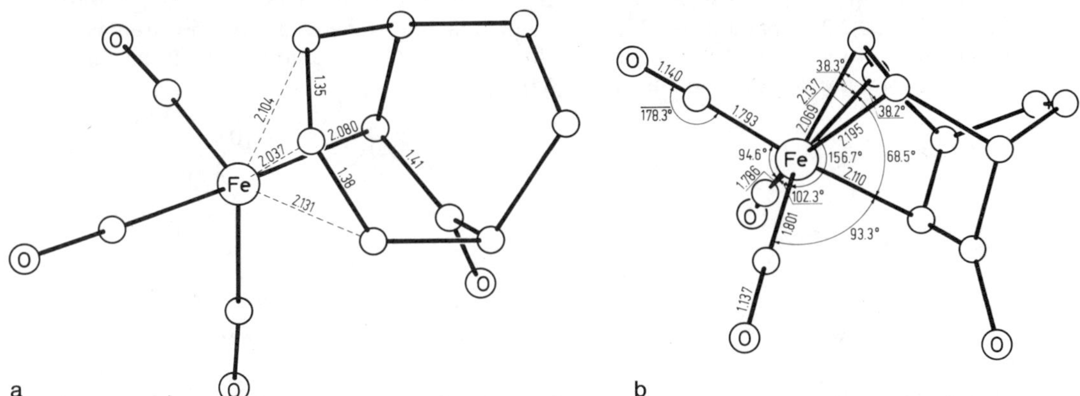

Fig. 21. Molecular structure of $C_9H_8OFe(CO)_3$ (No. 91) with space group $Pna2_1$(a) [54] and $P\bar{1}$(b) [62].

Reaction of compound No. 91 with concentrated H_2SO_4 gives CXV [83]. Treatment with AlX_3 (X = Cl, Br) in CH_2Cl_2 affords the appropriate compounds No. 92 to 95 [103, 110]. The compound No. 91 does not react with CO or $P(C_6H_5)_3$ at 25 °C [27], nor with 100 atm CO at 25 °C within 2 weeks [54]. But treatment with CO (100 atm) in toluene at 120 °C [61, 69, 101, 113] in a rocking autoclave [69] gives barbaralone (see Formula XXVI, p. 110) in 90% [61] or 96% yield [69]. The electron acceptor $(NC)_2C{=}C(CN)_2$ and No. 91 form a 2:1 charge transfer complex. The crystal structure of No. 91 remains almost unchanged upon complexation, and the σ,π-allyl bonding between $Fe(CO)_3$ and the carbocyclic organic moiety is confirmed. But since the four atoms bonded to Fe comprise a moderately good plane, a "homobutadiene" structure cannot be ruled out completely. The short contacts between the oxygen of the ketonic C=O group and the carbon atoms of $(NC)_2C{=}C(CN)_2$ in the crystal indicate an interaction between the carbonyl oxygen and the π-orbitals of the olefin [72].

References on pp. 172/5

$C_9H_8OAlX_3Fe(CO)_3$ and **$C_9H_8OAlX_3Fe(CO)_3AlX_3$** (Table **4**, Nos. **92** to **95** with X=Cl, Br). Hydrolysis of solutions of Nos. 92 to 95 results in quantitative recovery of No. 91 [103, 110].

CXIV CXV CXVI CXVII

$R(NC)C_9H_8Fe(CO)_3$ (Table **4**, Nos. **96** to **99** with $R=CH_3$ and C_6H_5). The mixture of compounds No. 96 and 97 or No. 98 and 99 are oxidized with $ON(CH_3)_3$ in acetone within 24 h to afford 5:1 mixture of CXVI and CXVII ($R=CH_3$) and a mixture of CXVI and CXVII ($R=C_6H_5$). CXVII ($R=CH_3$) and CXVI ($R=C_6H_5$) can be isolated pure [113].

$NCC_{13}H_{11}Fe(CO)_3$ (Table **4**, No. **101**) gives CXVIII with R=H (51% yield) on oxidative degradation with $ON(CH_3)_3$ in acetone [113].

$C_{11}H_{10}OFe(CO)_3$ (Table **4**, No. **103**). The ^{1}H NMR spectrum in the presence of $Eu(fod)_3$ (fod=1,1,1,2,2,3,3-heptafluoro-7,7-dimethyl-octane-4,6-dionate) shows the greatest paramagnetic shift for the H-4,8 protons. Treatment with CO (130 atm) in C_6H_6 in an autoclave at 110 °C for 3 d gives CXIX in 70% yield [84].

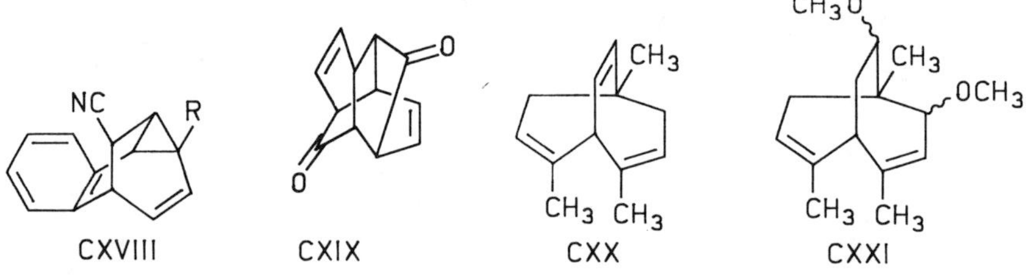

CXVIII CXIX CXX CXXI

$CH_3(NC)C_{13}H_{10}Fe(CO)_3$ (Table **4**, No. **104**). The ^{1}H NMR spectrum in $CDCl_3$ is given in a figure in [113]. Oxidative degradation with $ON(CH_3)_3$ in acetone at room temperature for 24 h gives CXVIII ($R=CH_3$) in 24% yield [113].

$(NC)_4C_9H_6OFe(CO)_3$ (Table **4**, No. **105**) gives CXXIV (X=O) in 71% yield on careful oxidation with $[NH_4]_2[Ce(NO_3)_6]$ [92].

$(CH_3)_3C_{10}H_9Fe(CO)_3$ (Table **4**, No. **106**). Irradiation in pentane followed by filtration, distillation, and chromatography gives III (p. 108) in 23% yield, CXX (27.5%), and three new hydrocarbons (7.5%, 14%, 15.6%) which are not isolated. Oxidation with $[NH_4]_2[Ce(NO_3)_6]$ in CH_3OH at room temperature gives CXXI (58%) within 4 h [90].

$C_{10}H_{10}C{=}OFe(CO)_3$ (Table **4**, No. **107**). The occurrence of two other bands in the IR spectrum at 1980 and 1989 cm^{-1} is not unambiguous since No. 107 rapidly looses CO in solution to give another complex (IR spectrum: 1980, 1989, 1995, 2045 cm^{-1}) of unknown structure [66].

$C_{10}H_{10}Fe(CO)_3$ (Table **4**, No. **109**) gives CXXII on protonation [66].

CXXII CXXIII CXXIV

$HOC_{10}H_{11}Fe(CO)_3$ (Table **4**, No. **110**) affords the PF_6 salt of CXXII on protonation with HPF_6 in ether [66].

$(CF_3)_2C_8H_8OFe(CO)_3$ (Table **4**, No. **115**) reacts with NO in hexane in a Carius tube at room temperature to give CXXIII (X=O) in 42% yield within 4 h. Similar reaction with CO (100 atm) in CH_2Cl_2 in an autoclave at 60 °C for 24 h gives CXXIII (X=O) in 84% yield [44].

$(NC)_4C_9H_8Fe(CO)_3$ (Table **4**, No. **117**) crystallizes in the triclinic space group $P\bar{1}-C_i^1$, S_2^1 with the unit cell parameters a=12.18 (1), b=9.05 (1), c=7.38 (1) Å, α=107°16′(10′), β=82°19′(10′), γ=88°20′(10′); Z=2 molecules per unit cell [15, 19]. D_{meas}=1.55 g/cm³, D_{calc}=1.56 g/cm³ [15]. The main bond distances and angles are shown in **Fig. 22** [15, 19]. The compound is insoluble in organic solvents [24]. Oxidation with $FeCl_3$ in CH_3OH at ca. 100 °C [18, 24] or with $[NH_4]_2[Ce(NO_3)_6]$ [24, 92] in 95% C_2H_5OH for 2 d [24] gives CXXIV ($X=H_2$) [18, 24, 92]. It does not react with liquid SO_2 as shown by the ¹H NMR spectrum (see Table 4, p. 146) [91] but may be readily carbonylated to give organic ketones [20].

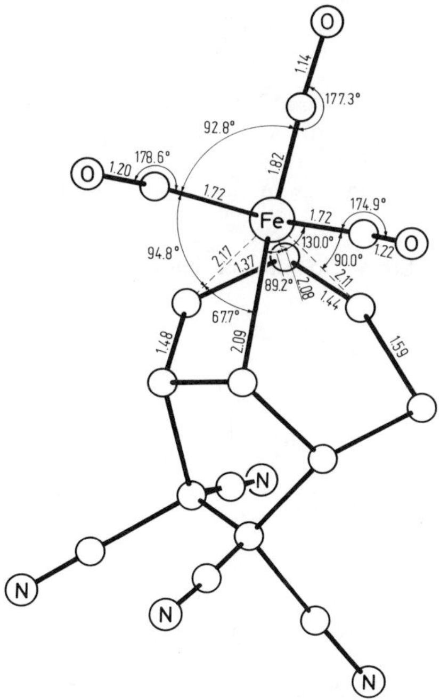

Fig. 22. Molecular structure of $(NC)_4C_9H_8Fe(CO)_3$ (No. 117) [15, 19].

$(CF_3)_2(NC)_2C_9H_8Fe(CO)_3$ (Table **4**, No. **118**). Reaction of 0.99 mmol $^4LFe(CO)_3$ (^4L=cycloheptatriene) with 1.17 mmol (E)-$NCC(CF_3)$=$C(CF_3)CN$ as in Method Vd gives a mixture of CXXVa and b (52% yield) as pale yellow crystals (from CH_2Cl_2/hexane), m.p. 138 to 139 °C.

References on pp. 172/5

Similar reaction with a mixture of (Z)- and (E)-$NCC(CF_3)=C(CF_3)CN$ (ratio 1:3) gives a product showing in the ^{19}F NMR spectrum two sets of a quartet corresponding to CXXVa and b and one set of CXXVc (ratio 3:1). The ^1H NMR spectra (from mixtures) in $CDCl_3$ show chemical shifts at $\delta=1.55$ (H-6), 2.75 (H-4,4′), 3.25 (H-5; J(H-5,6)=6.0 Hz), 3.77 (H-7), 4.54 (H-1,3; J(H-1,7)=8.0 Hz), 4.79 (H-2) ppm for CXXVa, which are identical with that of CXXVb except 3.68 (J(H-6,7)=8.0 Hz) ppm, and with that of CXXVc (J(H-6,7)=8.0 Hz). The ^{19}F NMR spectra in $CDCl_3$ (δ values in ppm, J in Hz) are shown in the following table:

isomer	δ_1	δ_2	J(F,F)	J(H,F_1)	J(H,F_2)
CXXVa	61.7	68.4	4.1	1.5	—
b	62.0	67.1	6.2	—	0.8
c	61.1	66.9	13.0	—	—

The IR spectrum of the mixture of CXXVa and b shows bands (cyclohexane) at 2003, 2065 (CO) cm^{-1}, and the mass spectrum shows the parent ion and successive loss of CO [44].

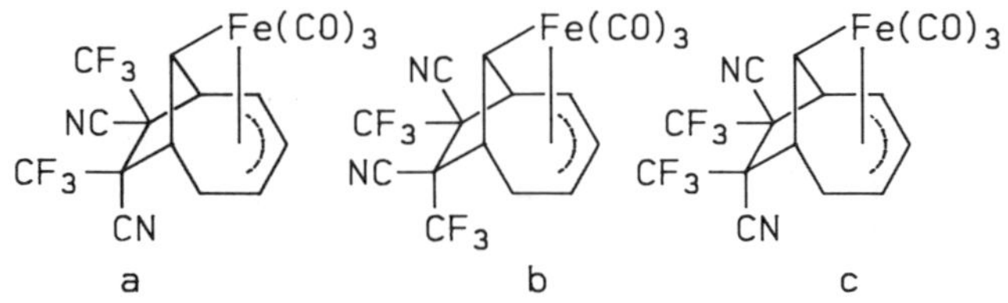

a b c

CXXV

$(CF_3)_2(NC)_2C_9H_8Fe(CO)_3$ (Table **4**, No. **119**) gives CXXIII ($X=C(CN)_2$) in 67% yield on reaction with CO (100 atm) in CH_2Cl_2 in an autoclave at 60 °C for 24 h [44]. The compound obtained from $C_8H_8Fe(CO)_3$ and $(NC)_2C{=}C(CF_3)_2$ at room temperature (cf. Method Vd) was first formulated as CXLI (p. 172) [11], and later found to be No. 119 [19, 44], confirmed by [24].

$(C_6H_5)_2OC_2C_9H_8Fe(CO)_3$ (Table **4**, No. **120**) is oxidized by $[NH_4]_2[Ce(NO_3)_6]$ to give CXXVI [105].

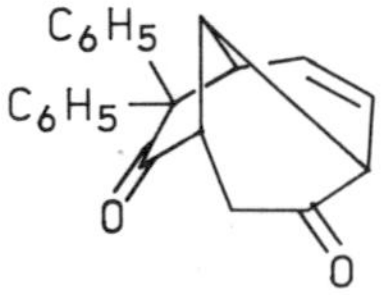

CXXVI 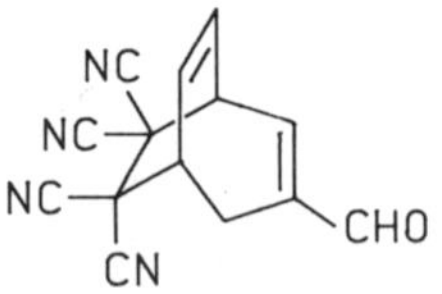 CXXVII 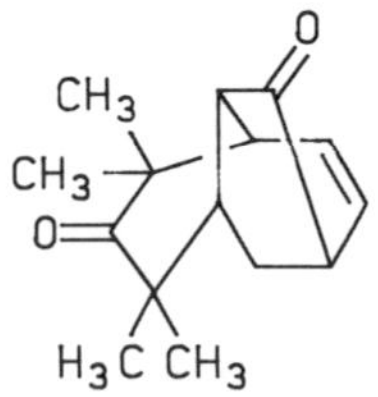 CXXVIII

$OHC(NC)_4C_9H_7Fe(CO)_3$ (Table **4**, No. **125**) is oxidized by $[NH_4]_2[Ce(NO_3)_6]$ to give CXXVII [92].

$(NC)_4C_2C_7H_6OFe(CO)_3$ (Table **4**, No. **126**). Starting the preparation with $^4LFe(CO)_3$ ($^4L=$ cycloheptatrienone-d_1, deuterated at C-1) according to Method Vd and monitoring this reaction by the ^1H NMR spectrum show that H-5 decreases in intensity and H-6 collapses to a broad singlet. After 2 d, H-5 had decreased in intensity with H-6 and H-7 broadening relative to an undeuterated sample. In the ^{13}C NMR spectrum, C-5 becomes very broad and less intense; after 2 d the peak at 57.15 ppm disappeared. Heating in acetone at 30 °C for 2 d followed by addition of hexane precipitates No. 105. This isomerization is rapid in acetone, CH_3NO_2, or CH_3OH but more slowly in CH_2Cl_2 or C_6H_6 [88].

$C_6H_5HC(NC)_4C_9H_6Fe(CO)_3$ (Table **4**, No. **127**) exists as a mixture of 2 isomers with the phenyl group in cis or trans position to the π-allyl group. ^1H NMR (CH_2Cl_2): δ=3.02 (H-10, cis isomer), 3.35 (H-10, trans isomer) ppm. The mixture readily isomerizes to a single isomer L (p. 113) with the phenyl group in endo position to Fe. Reaction of the cis isomer is too fast to be measured by IR. The trans isomer is converted to L with rate constants of $(1.39 \pm 0.05) \times 10^{-4}\ s^{-1}$ in CH_2Cl_2, $(2.49 \pm 0.02) \times 10^{-3}\ s^{-1}$ in CH_2Cl_2/CH_3NO_2 (1:1) and $>5.5 \times 10^{-3}\ s^{-1}$ in CH_3NO_2 at 25 °C [125].

$(CH_3)_4C_{10}H_8OFe(CO)_3$ (Table **4**, No. **130**) crystallizes in the triclinic space group $P\bar{1}-C_i^1, S_2^1$ with the unit cell parameters a=9.840(6), b=12.176(6), c=8.076(4) Å, α=95.43(4)°, β=118.31(3)°, γ=105.91(5)°; Z=2 molecules per unit cell. D_{meas}=1.420 g/cm³ (by flotation with H_2O/KI), D_{calc}=1.445 g/cm³. The main bond distances and angles are shown in **Fig. 23**. Oxidative degradation with o-chloranil in $CHCl_3$ at room temperature for 24 h gives CXXVIII (47% yield) [118, 119].

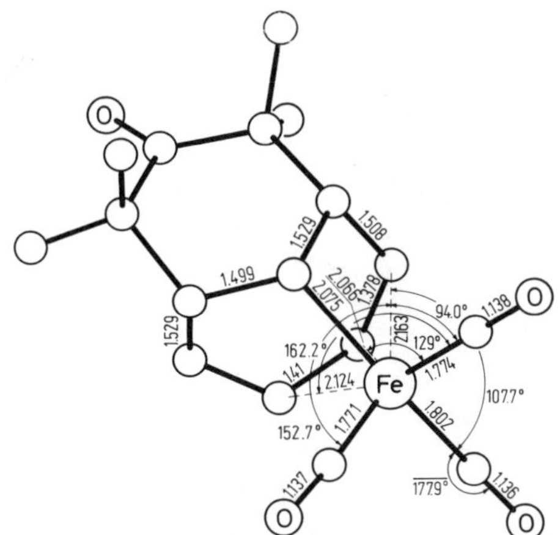

Fig. 23. Molecular structure of $(CH_3)_4C_{10}H_8OFe(CO)_3$ (No. 130) [118, 119].

$(CH_3)_3C_{10}H_9C{=}OFe(CO)_3$ (Table **4**, No. **131**). Refluxing in C_6H_6 for 30 min followed by evaporation and distillation of the residue at 90 to 100 °C/0.008 Torr gives No. 129 in 95% yield. Oxidative degradation with $[NH_4]_2[Ce(NO_3)_6]$ in C_2H_5OH at 0 °C for 30 min gives CXXIX and CXXX (yield 20% each) [90].

CXXIX CXXX CXXXI

References on pp. 172/5

$(NC)_4C_{10}H_8Fe(CO)_3$ (Table **4**, No. **134**) was first considered to be a 1,4- [1 to 3,10], 1,3- [24], or a 1,2-adduct [12]. An X-ray diffraction study confirms the structure as a 1,3-adduct. The compound crystallizes in the triclinic space group $P\bar{1}-C_i^1$, S_2^1 with the unit cell parameters a = 7.231(1), b = 10.052(3), c = 15.046(4) Å, α = 76.60(1)°, β = 124.51(1)°, γ = 117.26(2)°; Z = 2 molecules per unit cell. D_{meas} = 1.54 g/cm³, D_{calc} = 1.54 g/cm³, see **Fig. 24** [34]. The compound is insoluble in common organic solvents [2, 3] and H_2O [3], or scarcely soluble [1], and stable in air [3]. Oxidative degradation with $[NH_4]_2[Ce(NO_3)_6]$ in 95% C_2H_5OH [24, 34] for 2 d [24] gives CXXXI (X = CN) [24, 34, 67].

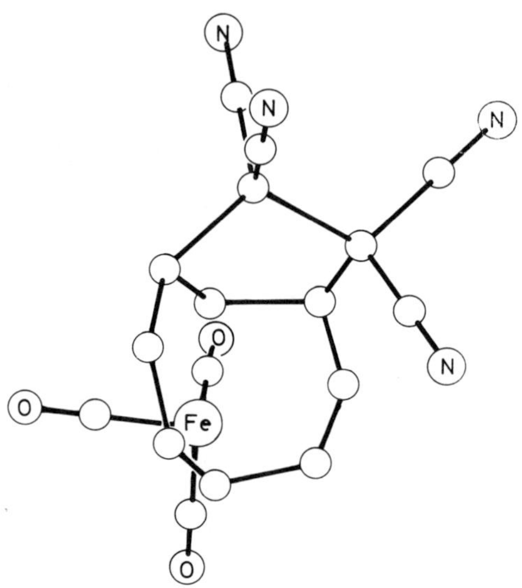

Fig. 24. Molecular structure of $(NC)_4C_{10}H_8Fe(CO)_3$ (No. 134) [34].

$(CF_3)_2(NC)_2C_{10}H_8Fe(CO)_3$ (Table **4**, No. **135**) was at first considered to be a 1,2-adduct [12], but later considered to be a 1,3-adduct in analogy to No. 134 [24]. Reaction with NO in CH_2Cl_2 in a tube at room temperature for 2 h gives CXXXI (X = CF_3) together with $(CO)_2Fe(NO)_2$. CXXXI is also obtained in the reaction of No. 135 with $P(C_6H_5)_2CH_2CH_2P(C_6H_5)_2$ in CH_2Cl_2 at room temperature within 24 h [12].

$R(NC)_4C_{10}H_7Fe(CO)_3$ (Table **4**, Nos. **136** to **140, 142, 143** with R = CH_3, C_6H_5, OCH_3, and CO_2CH_3). An X-ray single crystal structural determination confirms that in No. 136 (R = CH_3) the $Fe(CO)_3$ group is σ,π-allyl bonded as shown in the table [33]. Oxidative degradation of all these compounds with $[NH_4]_2[Ce(NO_3)_6]$ at room temperature gives CXXXII to CXXXIV, respectively. The reaction conditions and yields are shown in the following table [67]:

R, NC, NC, NC, CN — CXXXII

NC, NC, NC, CN, R — CXXXIII

NC, NC, NC, CN, OCH_3 — CXXXIV

References on pp. 172/5

compound No.	solvent	reaction time	product (yield)
136 ($R=CH_3$)	CH_3OH	96 h	CXXXII (70%)
137 ($R=C_6H_5$)	C_2H_5OH	24 h	CXXXII (61.5%)
138 ($R=CH_3$)	CH_3OH	120 h	CXXXIII (47.5%)
139 ($R=C_6H_5$)	C_2H_5OH	24 h	CXXXIII (84%)
140 ($R=OCH_3$)	CH_3OH	36 h	CXXXIV (69%)
142 ($R=CO_2CH_3$)	CH_3OH	48 h	CXXXIII (86%)
143 ($R=C_6H_5$)	C_2H_5OH	24 h	CXXXIII (100%)

$(NC)_4C_{10}H_8OFe(CO)_3$ (Table **4**, No. **145**) is stable in its crystalline form but decomposes rapidly in polar solvents, making measurements difficult. Careful oxidative degradation with $[NH_4]_2[Ce(NO_3)_6]$ gives CXXXV (18% yield) [92].

$(NC)_4C_{14}H_{10}Fe(CO)_3$ (Table **4**, No. **146**). Reaction of CXXXVI with $(NC)_2C{=}C(CN)_2$ in C_6H_6 at room temperature for 16 h, followed by filtration, dissolving the residue in acetone, and decolorization with activated carbon gives a mixture of adducts (87%) as a bright yellow powder. The ^1H NMR spectrum in acetone-d_6 shows chemical shifts at $\delta=1.2$ to 1.7 (m, H-7), 3.75 to 4.2 (m, H-6), 4.55 to 4.9 (m, H-3,8), 5.06 to 5.7 (m, H-1,2), 7.2 to 7.95 (m, 4H aryl) ppm.

CXXXV CXXXVI CXXXVII CXXXVIII CXXXIX

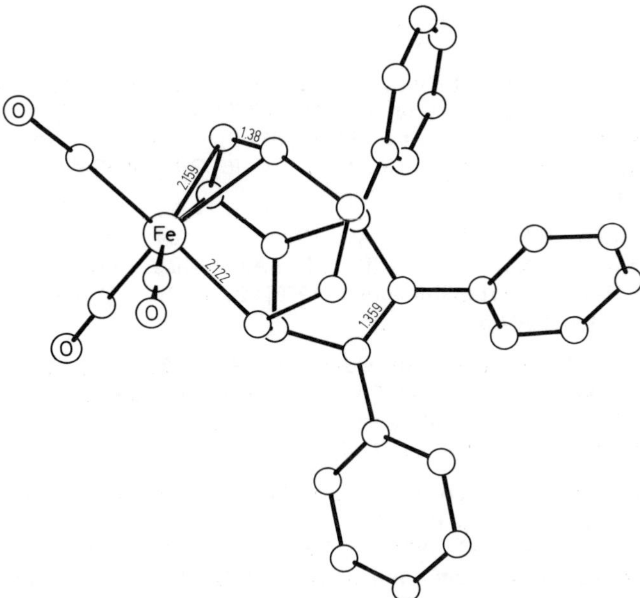

Fig. 25. Molecular structure of $(C_6H_5)_3C_{11}H_9Fe(CO)_3$ (No. 147) [115].

References on pp. 172/5

Attempted recrystallization from acetone leads to decomposition and consequently the mixture is directly oxidized with $[NH_4]_2[Ce(NO_3)_6]$ in 95% C_2H_5OH to give CXXXVII to CXXXIX. From these oxidation products it can be concluded that the mixture contains No. 146 [67].

$(C_6H_5)_3C_{11}H_9Fe(CO)_3$ (Table **4**, No. **147**) crystallizes in the monoclinic space group $P2_1/n-C^5_{2h}$ with the unit cell parameters a = 13.651 (6), b = 12.744 (3), c = 14.932 (6) Å, β = 107.35 (3)°; Z = 4 molecules per unit cell. The main bond distances are shown in **Fig. 25**, p. 171 [115].

$[C_7H_8Fe(CO)_3H]^+$ (Table **4**, No. **148**). Protonation of $^4LFe(CO)_3$ (^4L = norbornadiene) with $HOSO_2F$/liquid SO_2 produces a cation with large coupling constants of (average) J(H-4(1)-Fe-H_x) = 13 Hz and J(^{13}C-4(1)-Fe-H_x) = 37.8 Hz in the ^{1}H NMR spectrum. This leads to the structure of the rapidly equilibrating cation CXL (No. 148) [78].

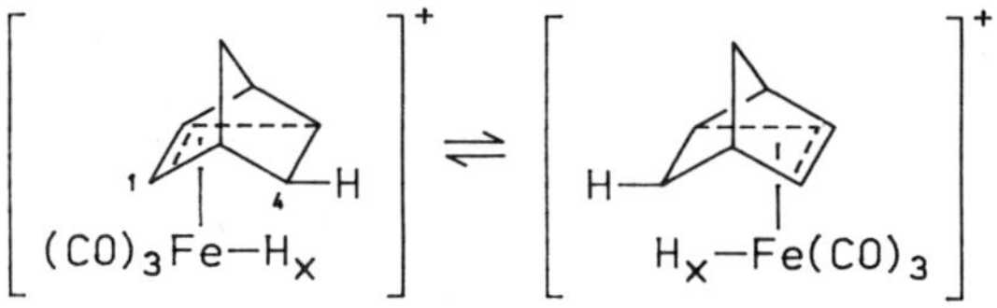

$(CO)_3Fe-H_x$ ⇌ $H_x-Fe(CO)_3$

CXL

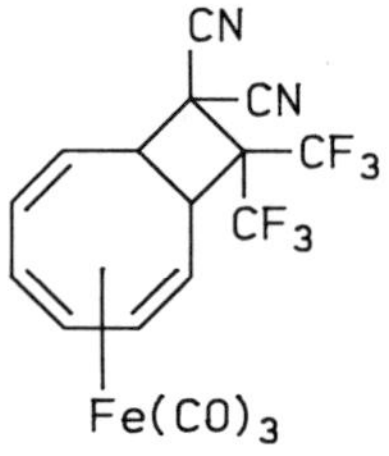

CXLI

$C_8H_8FeRu(CO)_5$ (Table **4**, No. **152**). One sharp peak in the ^{1}H NMR spectrum at ordinary temperatures indicates that this compound is a fluxional system like its di-iron (see 2.8.2.2 in "Organoiron Compounds" C5, 1981, Table 7, No. 1, p. 114) and diruthenium analogues (see 3.21.2 in "Ruthenium" Erg.-Bd., 1970, pp. 484/6) [13].

References:

[1] G.N. Schrauzer, S. Eichler (Angew. Chem. **74** [1962] 585). – [2] A. Davison, W. McFarlane, G. Wilkinson (Chem. Ind. [London] **1962** 820/1). – [3] A. Davison, W. McFarlane, L. Pratt, G. Wilkinson (J. Chem. Soc. **1962** 4821/9). – [4] G.K. Wertheim, R.H. Herber (TID-16124 [1962]; N.S.A. **16** [1962] No. 27037). – [5] G.K. Wertheim, R.H. Herber (unpublished results from [6]).

[6] G.K. Wertheim, R.H. Herber (J. Am. Chem. Soc. **84** [1962] 2274/5). – [7] R.B. King (Organometal. Syn. **1** [1965] 126/7). – [8] C.E. Keller (Diss. Univ. Texas 1966 from [9]). – [9] R. Pettit, L.W. Haynes (Carbonium Ions **5** [1976] 2263/302). – [10] W. McFarlane, G. Wilkinson (Inorg. Syn. **8** [1966] 184/5).

[11] M. Green, D.C. Wood (Chem. Commun. **1967** 1062). – [12] M. Green, D.C. Wood (J. Chem. Soc. A **1969** 1172/5). – [13] E.W. Abel, S. Moorhouse (Inorg. Nucl. Chem. Letters **6** [1970] 621/2). – [14] A. Bond, M. Green (Chem. Commun. **1971** 12/3). – [15] J. Weaver, P. Woodward (J. Chem. Soc. A **1971** 3521/4).

[16] R.M. Moriarty, C.-L. Yeh, K.C. Ramey (J. Am. Chem. Soc. **93** [1971] 6709/10). – [17] F.A. Cotton, A.J. Deeming, P.L. Josty, S.S. Ullah, A.J.P. Domingos, B.F.G. Johnson, J. Lewis (J. Am. Chem. Soc. **93** [1971] 4624/6). – [18] D.J. Ehntholt (Diss. State Univ. New York 1971, pp. 1/118; Diss. Abstr. Intern. B **32** [1972] 4486). – [19] M. Green, S. Tolson, J. Weaver, D.C. Wood, P. Woodward (Chem. Commun. **1971** 222/3). – [20] M. Green, S. Tolson (unpublished results from [19]).

[21] I.C. Paul (unpublished results from [27]). — [22] R.M. Moriarty, C.-L. Yeh, K.-N. Chen, R. Srinivasan (Tetrahedron Letters **1972** 5325/8). — [23] A. Eisenstadt (Tetrahedron Letters **1972** 2005/8). — [24] D.J. Ehntholt, R.C. Kerber (J. Organometal. Chem. **38** [1972] 139/45). — [25] A. Bond, M. Green (J. Chem. Soc. Dalton Trans. **1972** 763/8).

[26] Y. Becker, A. Eisenstadt, Y. Shvo (J. Chem. Soc. Chem. Commun. **1972** 1156). — [27] R. Aumann (Angew. Chem. **84** [1972] 583/4). — [28] R. Aumann (unpublished results from [29]). — [29] G. Deganello, P. Uguagliatti, L. Calligero, L. Sandrini, F. Zingales (Inorg. Chim. Acta **13** [1975] 247/89). — [30] R. Aumann (unpublished results from [27]).

[31] P. McArdle (J. Chem. Soc. Chem. Commun. **1973** 482/3). — [32] B.F.G. Johnson, J. Lewis, M.V. Twigg (J. Organometal. Chem. **52** [1973] C31/C32). — [33] J. Weaver, P. Woodward (unpublished results from [44, 67]). — [34] L.A. Paquette, S.V. Ley, M.J. Broadhurst, D.F. Truesdell, J. Clardy (Tetrahedron Letters **1973** 2943/6). — [35] A.H.-J. Wang, I.C. Paul, R. Aumann (unpublished results from [47]).

[36] J.A. Howard, R.A. Marsh, P. Woodward (unpublished results from [39]). — [37] R. Aumann (unpublished results from [38]). — [38] R. Aumann (J. Organometal. Chem. **47** [1973] C29/C32). — [39] A. Bond, M. Green, S.H. Taylor (J. Chem. Soc. Chem. Commun. **1973** 112/3). — [40] R. Aumann (Angew. Chem. **85** [1973] 628/9).

[41] A. Eisenstadt (J. Organometal. Chem. **60** [1973] 335/42). — [42] A. Eisenstadt (unpublished results from [41]). — [43] D. Ehntholt, A. Rosan, M. Rosenblum (J. Organometal. Chem. **56** [1973] 315/21). — [44] M. Green, S. Heathcock, D.C. Wood (J. Chem. Soc. Dalton Trans. **1973** 1564/9). — [45] M. Green, S. Tolson (unpublished results from [46]).

[46] M. Green, B. Lewis (J. Chem. Soc. Chem. Commun. **1973** 114/5). — [47] F.A. Cotton, J.M. Troup (J. Am. Chem. Soc. **95** [1973] 3798/9). — [48] R. Aumann, K. Fröhlich, H. Ring (Angew. Chem. **86** [1974] 309/10). — [49] R. Aumann (J. Organometal. Chem. **77** [1974] C33/C36). — [50] R. Aumann (J. Organometal. Chem. **76** [1974] C32/C34).

[51] R. Aumann (J. Organometal. Chem. **78** [1974] C31/C34). — [52] R. Aumann (J. Organometal. Chem. **66** [1974] C6/C10). — [53] I.C. Paul (unpublished results from [62]). — [54] A.H.-J. Wang, I.C. Paul, R. Aumann (J. Organometal. Chem. **69** [1974] 301/4). — [55] B.F.G. Johnson, J. Lewis, D.J. Thompson (Tetrahedron Letters **1974** 3789/92).

[56] M. Sternberg (Dipl.-Arbeit Univ. Münster **1974** from [73]). — [57] B.F.G. Johnson, J. Lewis, M.V. Twigg (J. Chem. Soc. Dalton Trans. **1974** 241/6). — [58] R. Edwards, J.A.S. Howell, B.F.G. Johnson, J. Lewis (J. Chem. Soc. Dalton Trans. **1974** 2105/12). — [59] A. Eisenstadt (Tetrahedron **30** [1973] 2353/6). — [60] A.J. Deeming, S.S. Ullah, A.J.P. Domingos, B.F.G. Johnson, J. Lewis (J. Chem. Soc. Dalton Trans. **1974** 2093/104).

[61] V. Heil, B.F.G. Johnson, J. Lewis, D.J. Thompson (J. Chem. Soc. Chem. Commun. **1974** 270/1). — [62] F.A. Cotton, J.M. Troup (J. Organometal. Chem. **76** [1974] 81/8). — [63] R. Aumann, H. Averbeck (J. Organometal. Chem. **85** [1975] C4/C6). — [64] A. Bond, B. Lewis, M. Green (J. Chem. Soc. Dalton Trans. **1975** 1109/18). — [65] R. Aumann (unpublished results from [66]).

[66] R. Aumann (Chem. Ber. **108** [1975] 1974/88). — [67] L.A. Paquette, S.V. Ley, S. Maiorana, D.F. Schneider, M.J. Broadhurst, R.A. Boggs (J. Am. Chem. Soc. **97** [1975] 4658/67). — [68] A. Stockis, E. Weissberger (J. Am. Chem. Soc. **97** [1975] 4288/92). — [69] B.F.G. Johnson, J. Lewis, D.J. Thompson, B. Heil (J. Chem. Soc. Dalton Trans. **1975** 567/71). — [70] G. Schiavon, C. Paradisi, C. Boanini (Inorg. Chim. Acta **14** [1975] L5/L6).

[71] M. Green, B. Lewis, J.J. Daly, F. Sanz (J. Chem. Soc. Dalton Trans. **1975** 1118/27). — [72] C.C.T. Chiang (Diss. Univ. Illinois 1975, pp. 1/243; Diss. Abstr. Intern. B **36** [1975] 2238). —

[73] R. Aumann, J. Knecht (Chem. Ber. **109** [1976] 174/9). – [74] R. Aumann (Chem. Ber. **109** [1976] 168/73). – [75] R. Aumann (Angew. Chem. **88** [1976] 375).

[76] R. Aumann, H. Wörmann (unpublished results from [77]). – [77] R. Aumann, H. Wörmann, C. Krüger (Angew. Chem. **88** [1976] 640/1). – [78] G.A. Olah, G. Liang (J. Org. Chem. **41** [1976] 2659/61). – [79] R. Hoffmann, P. Hofmann (J. Am. Chem. Soc. **98** [1976] 598/604). – [80] A. Eisenstadt (J. Organometal. Chem. **113** [1976] 147/56).

[81] Z. Goldschmidt, Y. Bakal (Tetrahedron Letters **1976** 1229/32). – [82] M. Brookhart, T.H. Whiteside, J.M. Crockett (Inorg. Chem. **15** [1976] 1550/4). – [83] R. Aumann, H. Wörmann, C. Krüger (Chem. Ber. **110** [1977] 1442/61). – [84] R. Aumann (Chem. Ber. **110** [1977] 1432/41). – [85] G.D. Annis, S.V. Ley (J. Chem. Soc. Chem. Commun. **1977** 581/2).

[86] A. Bond, M. Bottrill, M. Green, A.J. Welch (J. Chem. Soc. Dalton Trans. **1977** 2372/81). – [87] H. Olsen (Acta Chem. Scand. B **31** [1977] 635/6). – [88] M. Green, S.M. Heathcock, T.W. Turney, D.M.P. Mingos (J. Chem. Soc. Dalton Trans. **1977** 204/11). – [89] R. Goddard, P. Woodward (J. Chem. Soc. Dalton Trans. **1977** 1181/4). – [90] K. Hayakawa, H. Schmid (Helv. Chim. Acta **60** [1977] 1942/60).

[91] D. Cunningham, P. McArdle, H. Sherlock, B.F.G. Johnson, J. Lewis (J. Chem. Soc. Dalton Trans. **1977** 2340/4). – [92] Z. Goldschmidt, Y. Bakal (Tetrahedron Letters **1977** 955/8). – [93] A.D. Charles, P. Diversi, B.F.G. Johnson, K.D. Karlin, J. Lewis, A.V. Rivera, G.M. Sheldrick (J. Organometal. Chem. **128** [1977] C31/C34). – [94] R. Aumann, J. Knecht (Chem. Ber. **111** [1978] 3429/41). – [95] R. Aumann, H. Averbeck, C. Krüger (J. Organometal. Chem. **160** [1978] 241/53).

[96] Y. Becker, A. Eisenstadt, Y. Shvo (Tetrahedron **34** [1978] 799/806). – [97] G.D. Andretti, G. Bocelli, P. Sgarabotto (J. Organometal. Chem. **150** [1978] 85/92). – [98] P.R. Raithby (unpublished results from [104]). – [99] B.Y. Shu, E.R. Biehl, P.C. Reeves (Syn. Commun. **8** [1978] 523/31). – [100] J.M. Williams, R.K. Brown, H.J. Schultz, G.D. Stucky, S.D. Ittel (J. Am. Chem. Soc. **100** [1978] 7407/9).

[101] S. Sarel, G. Chriki (J. Org. Chem. **43** [1978] 4971/5). – [102] K.D. Karlin, B.F.G. Johnson, J. Lewis (unpublished results from [104]). – [103] K.D. Karlin, B.F.G. Johnson, J. Lewis (J. Organometal. Chem. **160** [1978] C21/C23). – [104] B.F.G. Johnson, K.D. Karlin, J. Lewis (J. Organometal. Chem. **145** [1978] C23/C25). – [105] Z. Goldschmidt, S. Antebi (Tetrahedron Letters **1978** 271/4).

[106] R. Gandolfi (unpublished results from [97]). – [107] R. Aumann, H. Averbeck, H. Wörmann (9th Intern. Conf. Organometal. Chem., Dijon 1979, Abstr. No. D33). – [108] R. Aumann, H. Wörmann (Chem. Ber. **112** [1979] 1233/52). – [109] J. Mattay, H. Leismann, H.-D. Scharf (Chem. Ber. **112** [1979] 577/99). – [110] B.F.G. Johnson, K.D. Karlin, J. Lewis (J. Organometal. Chem. **174** [1979] C29/C31).

[111] A. Stockis, R. Hoffmann (J. Am. Chem. Soc. **102** [1980] 2952/62). – [112] Y. Blum, Y. Becker, Y. Shvo (J. Organometal. Chem. **202** [1980] 65/76). – [113] S. Abramson, A. Eisenstadt (Tetrahedron **36** [1980] 105/15). – [114] K.L. Amos, N.G. Connelly (J. Organometal. Chem. **194** [1980] C57/C59). – [115] K. Broadley, N.G. Connelly, J.A.K. Howard, W. Risse (J. Organometal. Chem. **221** [1981] C29/C32).

[116] G.D. Annis, S.V. Ley, C.R. Self, R. Sivaramakrishnan (J. Chem. Soc. Perkin Trans. I **1981** 270/7). – [117] A.C. Barefoot, E.W. Corcoran, R.P. Hughes, D.M. Lemal, W.D. Saunders, B.B. Laird, R.E. Davis (J. Am. Chem. Soc. **103** [1981] 970/2). – [118] T. Ishizu, K. Harano, M. Yasuda, K. Kanematsu (J. Org. Chem. **46** [1981] 3630/4). – [119] T. Ishizu, K. Harano, M. Yasuda, K. Kanematsu (Tetrahedron Letters **22** [1981] 1601/4). – [120] A.V. Gist, P.C. Reeves (J. Organometal. Chem. **215** [1981] 221/7).

[121] A.D. Charles, P. Diversi, B.F.G. Johnson, J. Lewis (J. Chem. Soc. Dalton Trans. **1981** 1906/17). — [122] Z. Goldschmidt, S. Antebi (J. Organometal. Chem. **206** [1981] C1/C3). — [123] S.K. Chopra, M.J. Hynes, P. McArdle (J. Chem. Soc. Dalton Trans. **1981** 586/9). — [124] R. Benn, F.-W. Grevels, R. Schrader (13th Intern. Conf. Coord. Chem., Budapest 1982, Vol. 1, p. 324). — [125] S.K. Chopra, M.J. Hynes, G. Moran, J. Simmie, P. McArdle (Inorg. Chim. Acta **63** [1982] 177/81).

[126] A.J. Pearson, S.L. Kole, B. Chen (J. Am. Chem. Soc. **105** [1983] 4483/4). — [127] G. Schiavoni, C. Paradisi (J. Organometal. Chem. **243** [1983] 351/8). — [128] Z. Goldschmidt, S. Antebi (J. Organometal. Chem. **259** [1983] 119/25). — [129] F.A. Cotton, B.E. Hanson (Israel J. Chem. **15** [1977] 165/73). — [130] K. Broadley, N.G. Connelly, R.M. Mills, M.W. Whiteley, P. Woodward (J. Chem. Soc. Dalton Trans. **1984** 683/8).

1.4.1.4.4 $^4LFe(CO)_3$ Compounds where ^4L is a σ,σ,π-Bonded Ligand

Only a few complexes of this type are known. The reaction product obtained from $(CH_3)_4C_4Fe(CO)_3$ and $CF_3C{\equiv}CCF_3$ in hexane by irradiation was first considered to be Ia or b [2] but later identified as II [12].

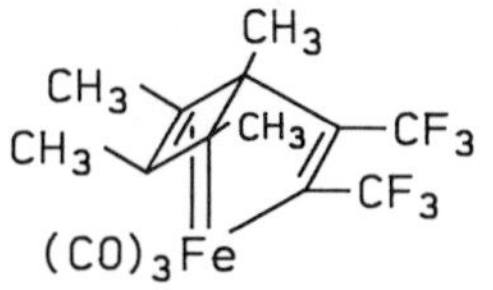

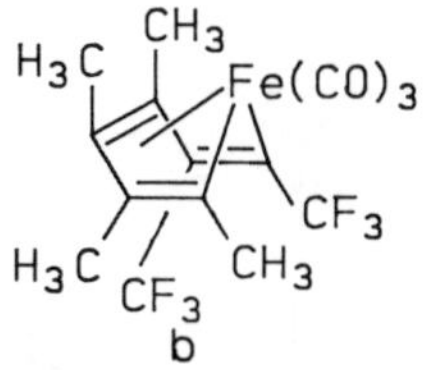

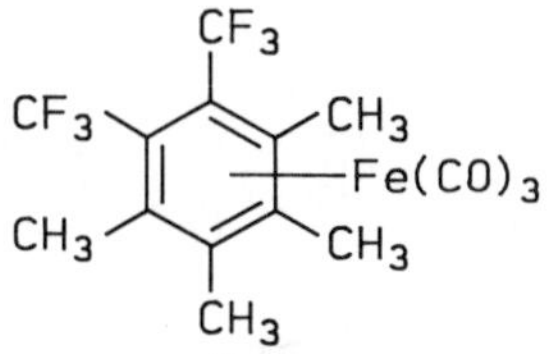

$C_{16}H_{20}O_8Fe(CO)_3$ (Formula III) is prepared by irradiating $C_4H_4Fe(CO)_3$ and dimethyl maleate [1, 7, 13] in pentane [7]. Compound III is the major product and smaller amounts of IV are isolated [7]. Irradiation in ether gives IV as the major product [1, 7]. Compound III is a yellow solid [1] or forms pale yellow plates from Skelly B solution [4, 6]. Subsequent elemental analysis indicates the stoichiometry $C_{19}H_{20}O_{11}Fe$, but the ^1H NMR and IR spectra of this material do not allow an unequivocal assignment of a molecular structure [5]. An X-ray structure determination shows the compound to crystallize in the orthorhombic space group $P2_12_12_1$ (No. 19)-D_2^4,V^4 with the unit cell parameters a = 15.646(2), b = 23.941(2), c = 11.033(1) Å; Z = 8 molecules per unit cell. D_{meas} = 1.551 g/cm³. D_{calc} = 1.550 g/cm³. The crystal structure is composed of discrete independent molecules a and b of $C_{16}H_{20}O_8Fe(CO)_3$ as shown in **Fig. 26**, p. 176, the closest nonbonded approach between nonhydrogen atoms is 2.86(1) Å. Two symmetry-independent molecules are related by a pseudo-b-glide parallel to the crystallographic bc plane, at x ≅ 0.26, approximate space group Pbca. The main bond distances and angles are shown in Fig. 26 [4].

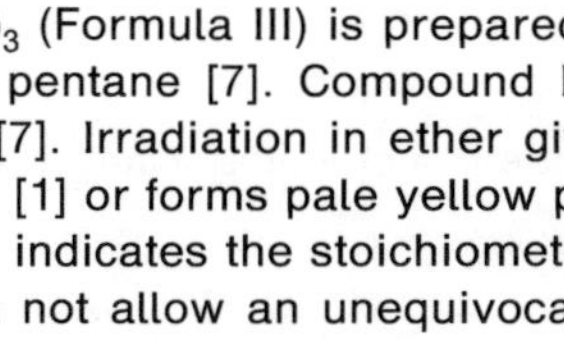

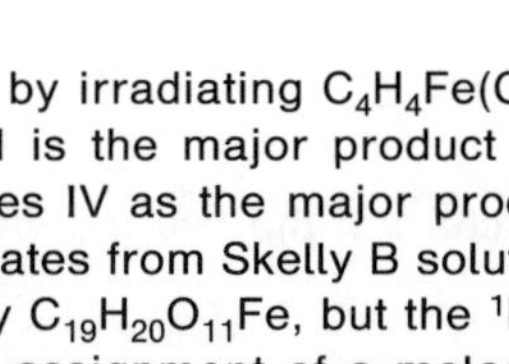

References on p. 178

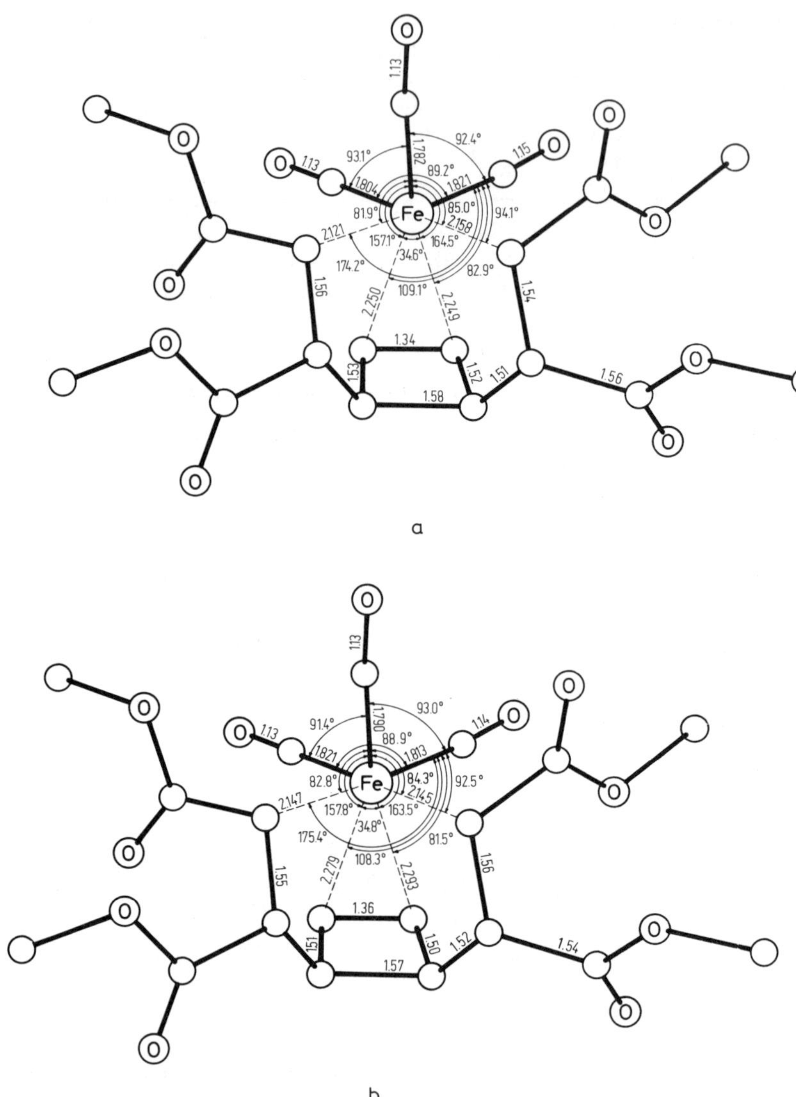

Fig. 26. Molecular structure of the two independent molecules a and b of $C_{16}H_{20}O_8Fe(CO)_3$ [4].

CH3 CH3 H H' 3 O 2 CF3 (CO)3Fe 1 CF3

V

CH3 H H' CH3 CF3 CF3 (CO)3Fe

VI

References on p. 178

$C_{11}H_{10}F_6OFe(CO)_3$ (Formula V) is prepared by the reaction of VI with CO (50 atm) in hexane in an autoclave at 80 °C for 4 d. Cooling the reaction mixture to 0 °C gives V in 30% yield. The yellow crystals melt at 114 to 116 °C. The ^{1}H NMR spectrum ($CDCl_3$) shows chemical shifts at δ=1.10 (d, H-3'; J=14.0 Hz), 1.30 (s, CH_3), 1.68 (d, H-3; J=14.0 Hz), 1.75 (s, CH_3), 2.92 (s, 1H), and 3.38 (s, 1H) ppm, and the ^{19}F NMR spectrum (CH_2Cl_2) shows chemical shifts at δ=54.3 (q, CF_3-1; J(CF_3, CF_3)=9.0 Hz), 58.2 (q, CF_3-2) ppm. The IR spectrum in Nujol shows a band at 1534 (C=C) cm^{-1} and in hexane bands at 1704 (C=O); 1994, 2007, 2068 (CO) cm^{-1} [11].

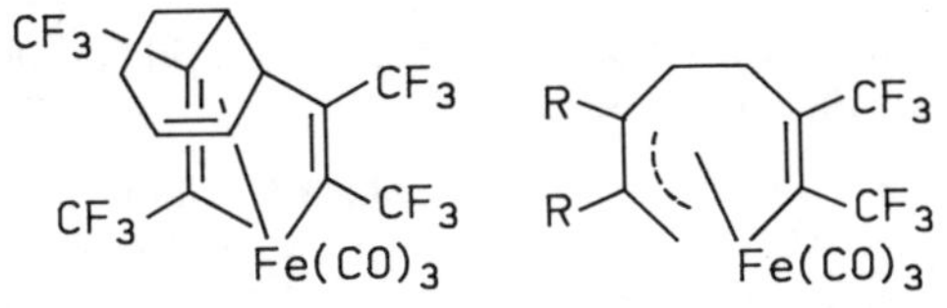

VII VIII

$C_{14}H_8F_{12}Fe(CO)_3$(Formula VII) is prepared from $C_6H_8Fe(CO)_3$ (C_6H_8=cyclohexa-1,3-diene) and $CF_3C{\equiv}CCF_3$ in hexane by irradiation. No evidence for the intermediacy of species analogous to VIII is observed. The compound VII is crystalline [9].

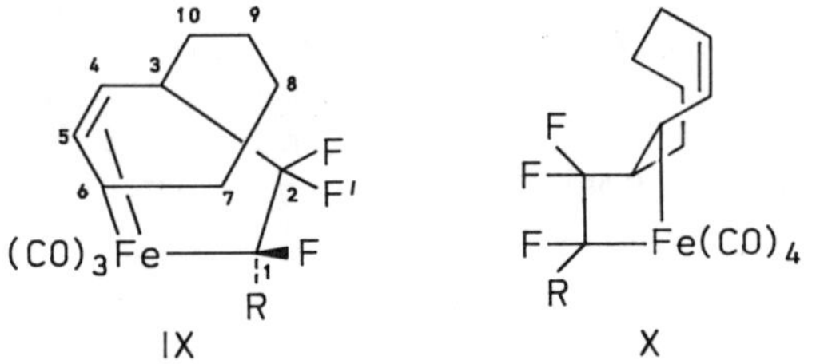

IX X

$C_{10}H_{12}F_4Fe(CO)_3$ (see Formula IX with R=F) is prepared by irradiating $C_8H_{12}Fe(CO)_3$ and excess $CF_2{=}CF_2$ in hexane in a Carius tube for 4 d. Filtration of the resulting solution followed by subsequent cooling to 0 °C for 3 d affords bright yellow needles of IX (30%), m.p. 97 to 100 °C. Further cooling (−30 °C, 3 d) produces X (R=F). The ^{1}H NMR spectrum ($CDCl_3$) shows chemical shifts at δ=0.96 to 1.94 (complex m, H-8 to 10), 2.09 (m, H-7,7'; J(H-6,7)=J(H-6,7')=9.0 Hz), 3.15 (m, H-3), 4.18 (dddd, H-6), 5.00 (m, H-4,5; J(H-5,6)= 7.5 Hz, J(H-4,6)=3.5 Hz) ppm. The ^{19}F NMR spectrum (C_6H_6) shows resonances at δ=56.6 (dm, F-1; J(F-1,1')=258 Hz), 89.8 (dm, F-1), 103.8 (dm, F-2; J(F-2,2')=222 Hz), 107.8 (dm, F-2') ppm. The IR spectrum in hexane shows bands at 2014, 2030, 2084 (CO) cm^{-1}, and the mass spectrum shows the ions $[M-nCO]^+$ (n=0 to 3), $[M-nCO-F]^+$ (n=2,3) [10].

$C_{11}H_{12}F_6Fe(CO)_3$ (see Formula IX with R=CF_3) is prepared by irradiating $C_8H_{12}Fe(CO)_3$ and $CF_3CF{=}CF_2$ (ratio 1:2) in hexane in a Carius tube for 50 h (yield 50%). The yellow crystals melt at 102 to 105 °C. The ^{1}H NMR spectrum ($CDCl_3$) shows chemical shifts at δ=0.86 to 1.90 (complex m, H-7 to 10), 2.07 to 3.15 (m, H-3), 3.44 (m, H-6; J(H-6,7')= J(H-6,7)=7.0 Hz), 4.07 (m, H-4,5; J(H-5,6)=7.0 Hz) ppm. The ^{19}F NMR spectrum ($CHCl_3$) shows δ=65.8 (ddd, CF_3; J(CF_3, F-2')=15.0 Hz, J(CF_3, F-1)=J(CF_3, F-2)=10.0 Hz), 94.3 (dm, F-2; J(F-2,2')=234 Hz), 105.0 (dm, F-2'), 187.8 (m, F-2) ppm. Thermolysis of IX (R=CF_3) in hexane in a Carius tube at 80 °C for 6 h gives X (R=CF_3) in 31% yield. Refluxing IX with $P(OCH_3)_3$ for 4 h gives X (R=CF_3) with 2CO groups substituted by $P(OCH_3)_3$ [3].

References on p. 178

References:

[1] P.C. Reeves (Diss. Univ. Texas 1969 from [4]). — [2] A. Bond, M. Green (Chem. Commun. **1971** 12/3). — [3] M. Green, B. Lewis, J.J. Daly, F. Sanz (J. Chem. Soc. Dalton Trans. **1975** 1118/27). — [4] P.E. Riley, R.E. Davis (Inorg. Chem. **14** [1975] 2507/13). — [5] P.E. Riley, R.E. Davis (Abstr. 25th Meeting Am. Cryst. Assoc., Charlottesville, Va., 1975, No. A8 from [4]).

[6] P.C. Reeves, R. Pettit (unpublished results from [4]). — [7] R.E. Davis, P.C.Reeves, R.Pettit (unpublished results from [8]). — [8] R. Pettit (J. Organometal. Chem. **100** [1975] 205/17). — [9] M. Bottrill, R. Goddard, M. Green, R.P. Hughes, M.K. Lloyd, B. Lewis, P. Woodward (J. Chem. Soc. Chem. Commun. **1975** 253/5). — [10] A. Bond, B. Lewis, M. Green (J. Chem. Soc. Dalton Trans. **1975** 1109/18).

[11] M. Bottrill, R. Davies, R. Goddard, M. Green, R.P. Hughes, B. Lewis, P. Woodward (J. Chem. Soc. Dalton Trans. **1977** 1252/61). — [12] A. Bond, M. Bottrill, M. Green, A.J. Welch (J. Chem. Soc. Dalton Trans. **1977** 2372/81). — [13] P.C. Reeves (Diss. Univ. Texas 1969; Diss. Abstr. Intern. B **30** [1969/70] 5463).

1.4.2 Compounds with One ^{4}L Ligand and One ^{2}L or ^{3}L Ligand or Two ^{2}L Ligands

1.4.2.1 Compounds of the Type $^4LFe^2L$ and $^4LFe(CO)_n(^2D)_{2-n}{}^2L$

Some of the compounds listed in Table 5, pp. 180/3, are only suggested as intermediates, for others no properties are given in the papers. But in some cases compounds first cited as intermediates have been isolated later. Therefore, the intermediates are also shown in this section. The compounds from Table 5 can be prepared by the following methods or are suggested to be formed as intermediates in the following reactions:

Method I: The co-condensation of Fe with cyclohexa-1,3-diene or cyclohexa-1,4-diene on a liquid N_2 cooled surface of pentane over 1.75 h followed by condensation of a $P(OCH_3)_3$ layer and a second layer of pentane gives a mixture of products. The mixture is allowed to warm to room temperature, followed by filtration and the volatiles are removed. The residue is extracted with pentane, filtered, and the filtrate is chromatographed on alumina to remove a green and a brown band. The orange solution thus obtained is concentrated and cooled for 48 h to −40 °C giving orange solids. Sublimation at room temperature onto a liquid N_2 cooled finger gives $C_6H_6FeC_6H_8$. The unsublimed residue is recrystallized from pentane to give No. 3. The filtrate from the low temperature filtration gives a mixture of $C_6H_8Fe(P(OCH_3)_3)_3$ and $C_6H_6Fe(P(OCH_3)_3)_2$, separated by chromatography on silica gel. Similar co-condensation with 2,3-dimethylbutadiene and $P(OC_3H_7\text{-}i)_3$ gives No. 2. Analogous co-condensation with cycloocta-1,3-diene and $P(OC_3H_7\text{-}i)_3$ gives $C_8H_{12}Fe(P(OC_3H_7\text{-}i)_3)_3$ (isolated) and No. 4 detected as a minor component in the reaction mixture [20].

Method II: $^2LFe(CO)_4$ ($^2L = CH_2{=}CHCO_2CH_3$) and 2,3-dimethylbutadiene are irradiated in ether at −40 °C for 2.5 h. The liberated CO is removed by an Ar stream. The solvent is removed and the residue is chromatographed on silica gel at −40 °C. Elution with pentane recovers $^2LFe(CO)_4$, and pentane/ether (4:1) gives No. 10 as a crude oil [23].

Method III: $^4LFe(CO)_n(^2D)_{3-n}$ compounds ($^2D = P(OCH_3)_3$, n = 1 or 3) are irradiated in an inert solvent in the presence of an alkene.

References on p. 185

a. $^4LFe(CO)_3$ (^4L = 1-phenylbutadiene) and 2,3-dimethylbutadiene are irradiated to give No. 8 [6]. Similar irradiation of $^4LFe(CO)_3$ (^4L = 2,3-dimethylbutadiene) in the presence of CH_2=$CHCO_2CH_3$ gives I; No. 10 is formulated as an intermediate in this reaction [19].

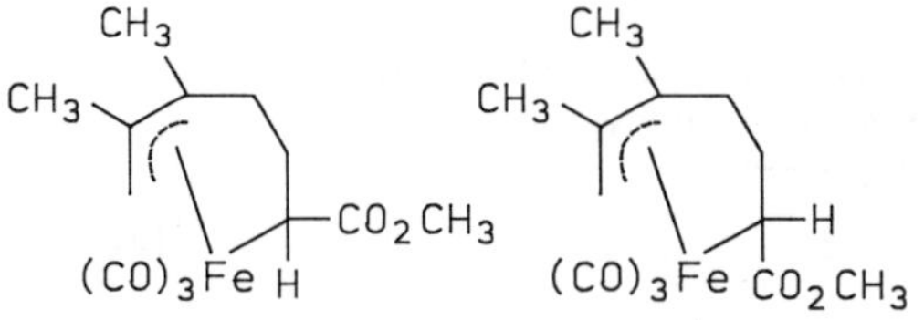

CH_3 CH_3 $-CO_2CH_3$ $(CO)_3Fe$ H
a

I

CH_3 CH_3 -H $(CO)_3Fe$ CO_2CH_3
b

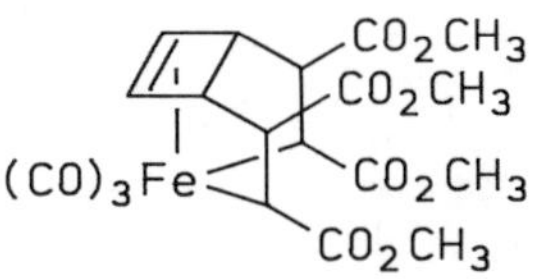

CO_2CH_3 CO_2CH_3 $(CO)_3Fe$ CO_2CH_3 CO_2CH_3

II

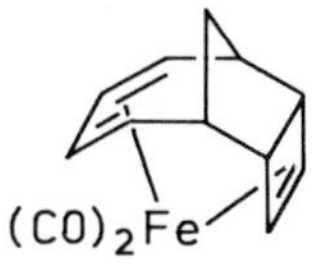

$(CO)_2Fe$

III

b. Irradiation of $C_4H_4Fe(CO)_3$ with vinyl ether at −50 °C in pentane gives an oily brown residue containing $C_4H_4Fe(CO)_3$ and No. 11 [11]. Similar irradiation in the presence of dimethylfumarate gives No. 13 [1, 4]. Analogous irradiation in the presence of dimethyl maleate [1, 3, 12] in ether [12] gives No. 14 [1, 3, 12] (major product [12]) and II [3, 12]. This reaction in pentane gives II as the major product and No. 14 in minor amounts [12]. Irradiation of $C_4H_4Fe(CO)_3$ at −50 °C in $(CH_3)_2C$=$C(CH_3)_2$ produces No. 15 [10, 11] and $C_4H_4Fe[(CH_3)_2C$=$C(CH_3)_2]_2CO$ [11]. Irradiation of $C_4H_4Fe(CO)_3$ in the presence of cycloheptatriene in ether for 4 h gives III; No. 16 is considered to be an intermediate [8].

c. Irradiation of 1,2-$(CH_3)_2C_4H_2Fe(CO)_3$ with dimethyl fumarate in ether at 10 °C for 30 min followed by chromatography with ether/C_6H_6 (2:8) gives No. 17. Similar irradiation with dimethyl maleate in C_6H_6/pentane (1:3) gives No. 18 [14]. Irradiation of $(CH_3)_4C_4Fe(CO)_3$ in the presence of maleic anhydride gives No. 19 [2, 4]. Prolonged irradiation results in decomposition but no Diels-Alder product is detected [2].

d. $(C_6H_5)_4C_4Fe(P(OCH_3)_3)_2CO$ in CH_2Cl_2 reacts with $(NC)_2C$=$C(CN)_2$ for 3 h at room temperature. Filtration, concentration of the filtrate, and addition of hexane precipitates No. 7 [22].

e. Irradiation of Ia in ether at −30 °C for 14.5 h accompanied by removal of CO by an Ar stream followed by evaporation and chromatography of the residue on Kieselgel 60 (Merck, 230 to 240 mesh) gives $^4LFe(CO)_3$ (^4L = 1,3-dimethylbutadiene) by elution with pentane. Elution with pentane/ether (7:3) gives No. 10. Further elution with ether gives Ia and other unidentified products [24].

Method IV: Refluxing $C_8H_8OFe_2(CO)_6$ (see 2.6.1.2.3 in "Organoiron Compounds" C5, 1981, Table 2, No. 15, p. 18) in n-heptane for 15 h followed by evaporation and chromatography on neutral alumina gives $C_8H_8Fe(CO)_3$ and trans-$C_8H_8Fe_2(CO)_6$ by elution with petroleum ether. Further elution with petroleum ether/C_6H_6 (1:1) gives $C_8H_8Fe_2(CO)_5$, $(C_8H_8O)_2Fe_2(CO)_4$, and No. 20 [7].

References on p. 185

Table 5
Compounds of the Type $^4LFe^2L$ and $^4LFe(CO)_n(^2D)_{2-n}{}^2L$.
Further information on numbers preceded by an asterisk is given at the end of the table, pp. 183/4.
For abbreviations and dimensions see p. VIII.

No.	compound	method of preparation (yield in %), properties and remarks	Ref.
*1	$[(C_3H_5)_2Fe(CH_2{=}C{=}O)]^+$ (structure)	not isolated	[21]
*2	$[(i\text{-}C_3H_7O)_3P]_2Fe(CH_2{=}C(CH_3)C(CH_3){=}CH_2)_2$ (structure)	I (traces) ^{31}P NMR (at $-90°$): 177.6 (J(P,P)=25), 192.5	[20]
3	$[(CH_3O)_3P]_2Fe(C_6H_8)_2$ (structure)	I yellow powder (from pentane) ^1H NMR: 1.37, 1.63 (m, H-5,6); 2.50 (m, H-1,4), 2.8 (m, H-3′,6′), 3.24 (t, H-1′,2′), 3.33 (t, CH_3), 4.09 (m, H-2,3), 6.20 (q, H-4′,5′) ^{31}P NMR: 183.2 (2P, single line) mass spectrum: $[M-nC_6H_8]^+$ (n=0,2)	[20]
*4	$[(i\text{-}C_3H_7O)_3P]_2Fe(C_8H_{12})_2$ (structure)	I ^{31}P NMR (at $-90°$): 170.0 (J(P,P)=37), 189.6	[20]
*5	$(C_4H_6)Fe(CO)(C_8H_{12})P(C_6H_5)_3$ (structure)	not isolated	[5]

References on p. 185

Table 5 [continued]

No.	compound	method of preparation (yield in %), properties and remarks	Ref.
*6	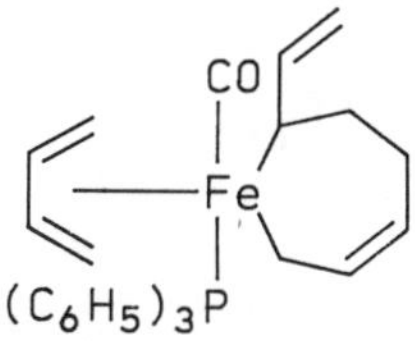	not isolated	[5]
7	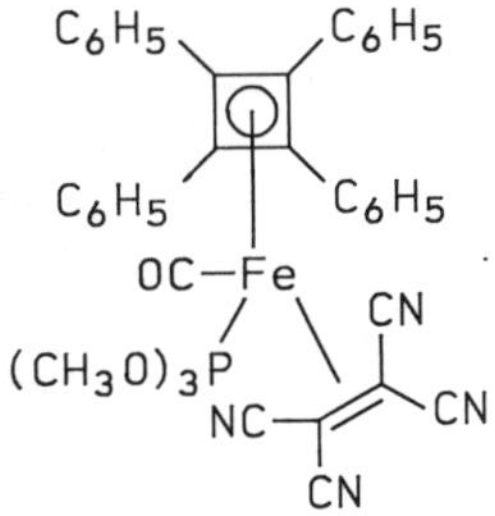 $\cdot$ 1/3 CH_2Cl_2	III d (67) m.p. >120° (dec.), green-black crystals (from CH_2Cl_2/hexane) ^1H NMR ($CDCl_3$): 3.48 (d, CH_3; J(P,H) = 12), 7.28 (m, C_6H_5) IR (CH_2Cl_2): 1981 (CO); 2195, 2215 (CN)	[22]
8	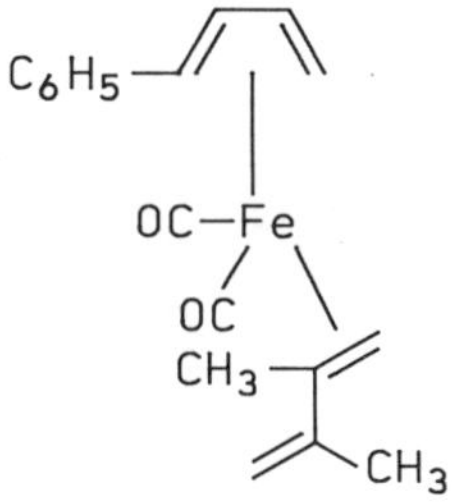	III a isolated	[6]
9	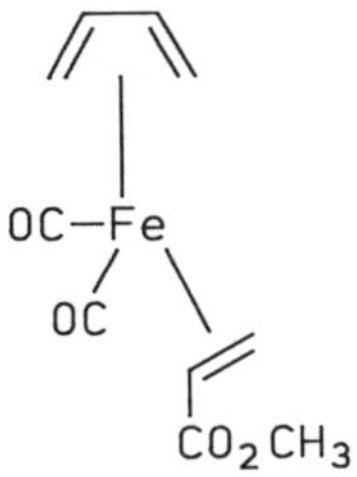	–	[16]
*10	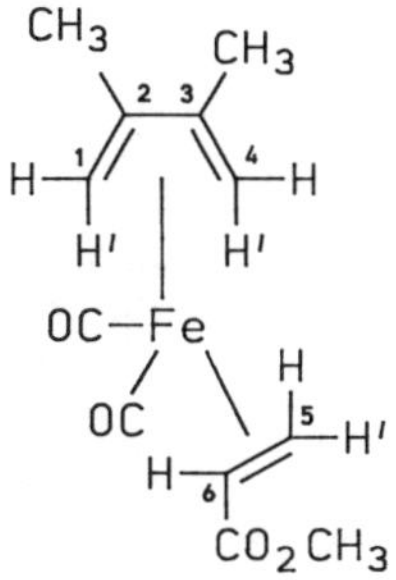	II (27), III a, III e (41) m.p. 24 to 25°, yellow crystals (from hexane at −78°) ^1H NMR (toluene-d_8): −0.40, −0.21 (H-1′,4′); 0.63, 1.09 (H-1,4); 1.83, 1.98 (CH_3-2,3); 2.20 (H-5; J(H-5,5′) ≲ 3), 2.60 (H-6; J(H-5,6) ≈ 7), 3.12 (H-5′; J(H-5′,6) ≈ 11), 3.50 (OCH_3) IR (hexane): 1711, 1724 (C=O); 1968, 1977.5, 1981, 2018, 2023 (CO)	[19, 23, 24]

References on p. 185

Table 5 [continued]

No.	compound	method of preparation (yield in %), properties and remarks	Ref.
11	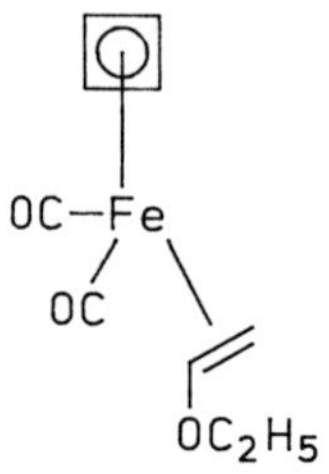 OC—Fe, OC, OC_2H_5	IIIb IR (pentane): 1942, 1995 (CO) mass spectrum: $[M-nCO]^+$ (n = 0 to 2), $[C_4H_4Fe(CO)(CH_2{=}CHOC_2H_5)_2]^+$	[11]
*12	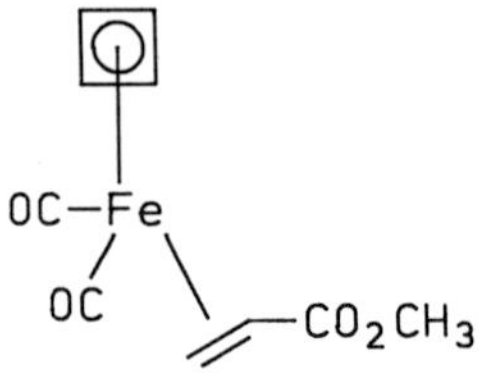 OC—Fe, OC, $—CO_2CH_3$	–	[16]
*13	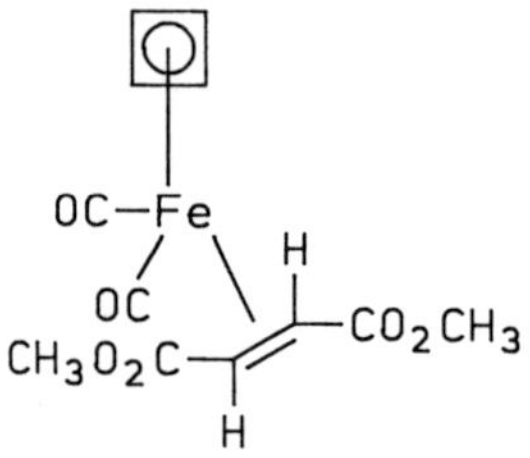 OC—Fe, OC, H, $—CO_2CH_3$, $CH_3O_2C—$, H	IIIb m.p. 26 to 27°, yellow crystals ^{1}H NMR: 3.4 ($2CH_3$), 3.46 (4H), 4.22 (2H) IR: 1990, 2065 (CO)	[1, 4]
*14	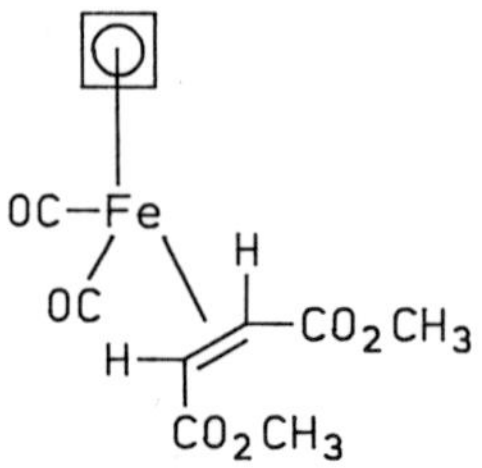 OC—Fe, OC, H, $—CO_2CH_3$, H—, CO_2CH_3	IIIb yellow liquid ^{1}H NMR: 3.35 (s, 4H), 3.37 (s, 2H), 3.47 (s, $2CH_3$) IR: 1980, 2038 (CO)	[1, 3, 12]
*15	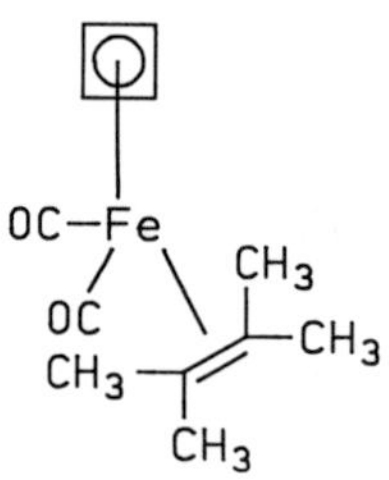 OC—Fe, OC, CH_3, $—CH_3$, $CH_3—$, CH_3	IIIb unstable	[10, 11]

References on p. 185

Table 5 [continued]

No.	compound	method of preparation (yield in %), properties and remarks	Ref.
16	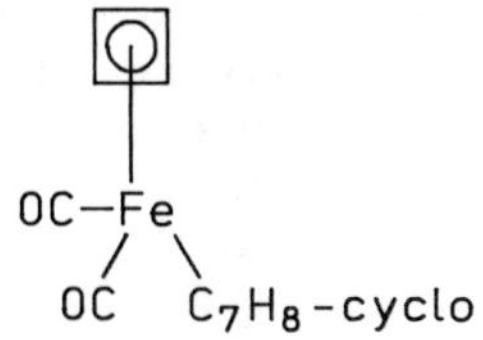 (structure: cyclobutadiene ring, OC–Fe, OC, C_7H_8-cyclo)	not isolated	[8]
*17	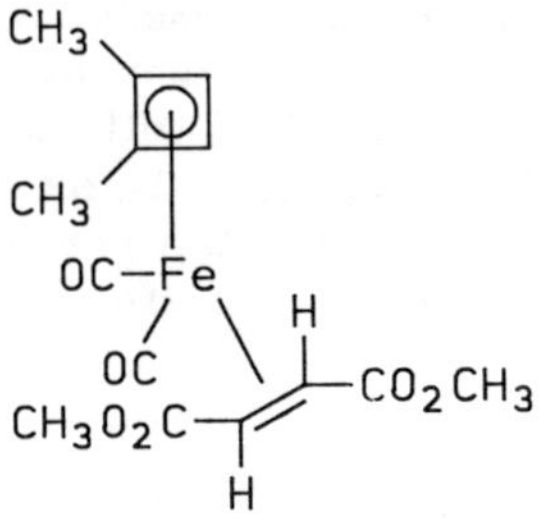 (structure: dimethylcyclobutadiene, OC–Fe, OC, CH_3O_2C–CH=CH–CO_2CH_3)	IIIc (64) b.p. 60°/0.001 Torr, pale yellow oil ^{1}H NMR ($CDCl_3$): 1.72 (s, CH_3), 1.88 (s, CH_3), 3.69 (s, 2 OCH_3), 3.72 (s, 1H), 3.88 (s, 2H), 4.09 (s, 1H) IR (benzene-d_6): 1155; 1710 (C=O); 1968, 2014 (CO)	[14]
*18	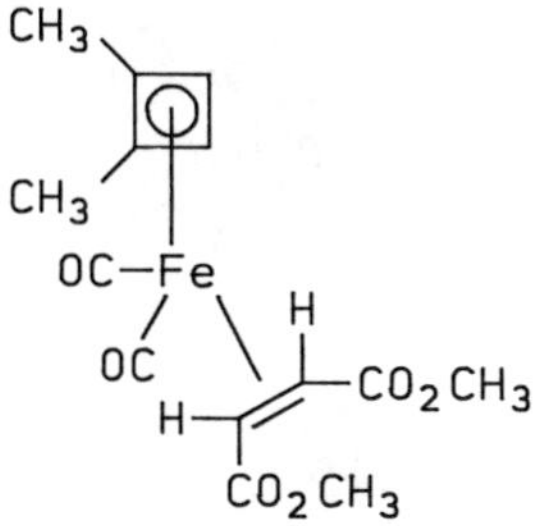 (structure: dimethylcyclobutadiene, OC–Fe, OC, H–C(CO_2CH_3)=CH–CO_2CH_3)	IIIc (31) b.p. 50°/0.001 Torr, pale yellow oil ^{1}H NMR (benzene-d_6): 1.27 (s, 2 CH_3), 3.12 (s, 2H), 3.55 (s, 2H), 3.62 (s, 2 OCH_3) IR (benzene-d_6): 1145; 1710, 1740 (C=O); 1969, 2016 (CO)	[14]
19	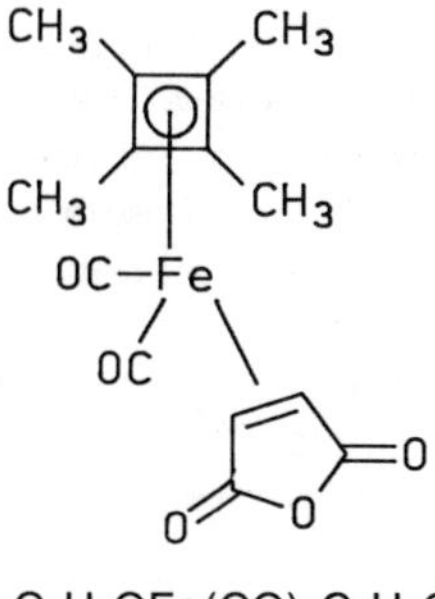 (structure: tetramethylcyclobutadiene, OC–Fe, OC, maleic anhydride)	IIIc stable at room temperature	[2, 4]
20	$C_8H_8OFe(CO)_2C_8H_8O$	IV structure unknown	[7]

* Further information:

$[C_4H_6FeCH_2{=}C{=}O]^+$ (Table **5**, No. **1**) is formulated as one of the products obtained by gasphase reaction of Fe^+ with 3-methylcyclopentanone [21].

$^4LFe(P(OC_3H_7\text{-}i)_3)_2{}^2L$ (Table **5**, Nos. **2** and **4** with $^4L={}^2L$ = 2,3-dimethylbutadiene, or cycloocta-1,3-diene) are not isolated for further characterization, and their ^{1}H NMR spectra

References on p. 185

consist of a singlet at higher temperatures and an AB spin system in the slow exchange limit at −90 °C [20].

$C_4H_6Fe(CO)(P(C_6H_5)_3)C_8H_{12}$ (Table **5**, Nos. **5** and **6**) are suggested as intermediates in the oligomerization of butadiene (= ^4L) catalyzed by $^4L_2FeCO$ in the presence of $P(C_6H_5)_3$ [5].

$(CH_3)_2C_4H_4Fe(CO)_2CH_2{=}CHCO_2CH_3$ (Table **5**, No. **10**). The ^1H NMR resonances are partially split at lower temperatures, and these temperature dependent changes continue below −80 °C. Thus, the compound seems to represent a fluxional system. The IR spectrum shows that the compound exists at least as a mixture of three or four species. There is no change in the intensity of the IR bands in the temperature region of +20 and −50 °C [23]. The electronic spectrum in n-hexane is shown in a figure in [24]. The complex is stable at 20 °C, but very air-sensitive in solution. It reacts in the dark with CO to give a mixture of isomers I, p. 179 [23]. Irradiation in n-hexane in the presence of CO at −45 °C, where the thermal reaction yielding I is almost negligible, gives $^4LFe(CO)_3$ (^4L = 2,3-dimethylbutadiene) [19, 23, 24]. Reaction with $P(OCH_3)_3$ in n-hexane gives IV ($^2D = P(OCH_3)_3$). The analogous product is observed with other ligands like isocyanides, as long as the ligands are not too bulky [23, 24]. However, upon reaction with $P(C_6H_5)_3$, IV ($^2D = P(C_6H_5)_3$) can be isolated at 0 °C, but it is immediately converted to No. 10 in n-hexane at room temperature. The final product is $^4LFe(CO)_2P(C_6H_5)_3$ (^4L = 2,3-dimethylbutadiene) [24].

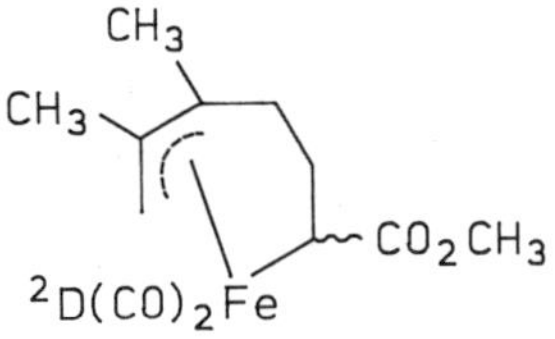

IV 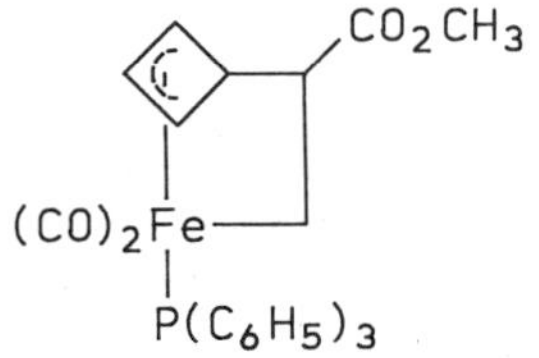 V 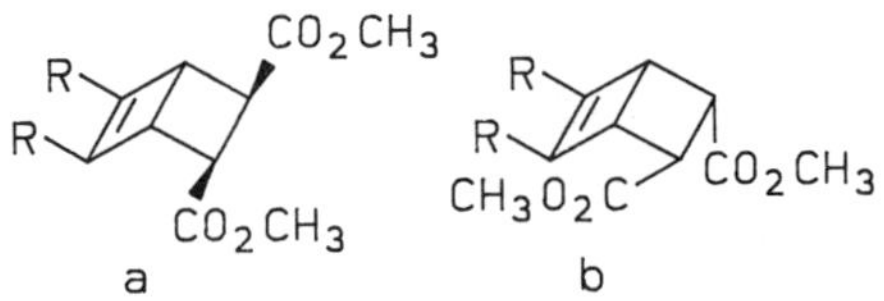 VI (a, b)

$C_4H_4Fe(CO)_2CH_2{=}CHCO_2CH_3$ (Table **5**, No. **12**) gives V with $P(C_6H_5)_3$ [16].

$C_4H_4Fe(CO)_2CH_3O_2CCH{=}CHCO_2CH_3$ (Table **5**, Nos. **13** and **14**). At −78 °C the two olefinic protons of the fumarate ligand in No. 13 appear as separate absorptions indicating restricted rotation about the Fe olefinic bond at these temperatures. Both complexes are stable at 25 °C over extended periods. Examination of the NMR spectra of No. 13, in the presence of a large excess of dimethyl maleate, shows no ligand exchange of the esters, which indicates that there is no tendency to undergo intramolecular Diels-Alder additions. Oxidation of No. 13 with $[NH_4][Ce(NO_3)_6]$ gives bicyclohexene dicarboxylic acid ester VIa (R = H) only. Similar oxidation of No. 14 gives endo-cis-bicyclohexanedicarboxylic acid ester VIb (R = H). Oxidation of No. 13 or No. 14 in the presence of equimolar amounts of either dimethyl fumarate or dimethyl maleate, and $[NH_4]_2[Ce(NO_3)_6]$ gives VIa and b (R = H) in the ratio 25:1 or 50:1, respectively [1].

$C_4H_4Fe(CO)_2(CH_3)_2C{=}C(CH_3)_2$ (Table **5**, No. **15**) gives $(C_4H_4)_2Fe_2(CO)_3$ when treated at low temperature under vacuum [11] or on concentrating its solution [10].

$(CH_3)_2C_4H_2Fe(CO)_2CH_3O_2CH{=}CHCO_2CH_3$ (Table **5**, Nos. **17** and **18**). Oxidative degradation of Nos. 17 and 18 with $[NH_4]_2[Ce(NO_3)_6]$ in acetone at 0 °C gives dimethyl fumarate and VIa (R = CH_3), or dimethyl maleate and VIb (R = CH_3), respectively. Starting the oxidation of No. 18 with addition of a fivefold excess of dimethyl maleate gives an increased yield of VIb. The yield of VIb is independent of the reaction time [14].

References on p. 185

References:

[1] P.C. Reeves, J. Henery, R. Pettit (J. Am. Chem. Soc. **91** [1969] 5888/90). — [2] P.C. Reeves (Diss. Univ. Texas 1969 from [18]). — [3] P.C. Reeves (Diss. Univ. Texas 1969 from [15]). — [4] P.C. Reeves (Diss. Univ. Texas 1969, pp. 1/134; Diss. Abstr. Intern. B **30** [1970] 5436/7). — [5] J.M. Kelly, D.V. Bent, H. Hermann, D. Schulte-Frohlinde, E. Koerner v. Gustorf (J. Organometal. Chem. **69** [1974] 259/69).

[6] R. Pettit (J. Ann. Meeting Chem. Soc. Roy. Inst., Brighton 1971 from [9]). — [7] H. Maltz, G. Deganello (J. Organometal. Chem. **27** [1971] 383/7). — [8] J.S. Ward, R. Pettit (J. Am. Chem. Soc. **93** [1971] 262/4). — [9] J. Buchkremer (Diss. Ruhr-Univ. Bochum 1973). — [10] E.A. Koerner v. Gustorf, I. Fischler, R. Wagner (Proc. 16th Intern. Conf. Coord. Chem., Dublin 1974, Abstr. No. 4.20).

[11] I. Fischler (Diss. Ruhr-Univ. Bochum 1974). — [12] R.E. Davis, P.C. Reeves, R. Pettit (unpublished results from [13]). — [13] R. Pettit (J. Organometal. Chem. **100** [1975] 205/17). — [14] E.K.G. Schmidt (Chem. Ber. **108** [1975] 1609/22). — [15] P.E. Riley, R.E. Davis (Inorg. Chem. **14** [1975] 2507/13).

[16] D. Schultz, F.-W. Grevels, E.A. Koerner v. Gustorf (unpublished results from [17]). — [17] R.C. Kerber, E.A. Koerner v. Gustorf (J. Organometal. Chem. **110** [1976] 345/57). — [18] P.L. Pruitt, E.R. Biehl, P.C. Reeves (J. Organometal. Chem. **134** [1977] 37/45). — [19] T. Akiyama, F.-W. Grevels, I. Salama (9th Intern. Conf. Organometal. Chem., Dijon 1979, Abstr. No. D2). — [20] S.D. Ittel, F.A. Van-Catledge, J.P. Jesson (J. Am. Chem. Soc. **101** [1979] 3874/84).

[21] R.C. Burnier, G.D. Byrd, B.S. Freiser (J. Am. Chem. Soc. **103** [1981] 4360/7). — [22] N.G. Connelly, R.L. Kelly, M.W. Whiteley (J. Chem. Soc. Dalton Trans. **1981** 34/9). — [23] F.-W. Grevels, K. Schneider (Angew. Chem. **93** [1981] 417/8; Angew. Chem. Intern. Ed. Engl. **20** [1981] 410). — [24] T. Akiyama, F.-W. Grevels, J.G.A. Reuvers, P. Ritterskamp (Organometallics **2** [1983] 157/60).

1.4.2.2 Compounds of the Type $^{4}LFe(^{2}L)_2CO$

$C_4H_4Fe[(CH_3)_2C{=}C(CH_3)_2]_2CO$. Irradiation of $C_4H_4Fe(CO)_3$ in $(CH_3)_2C{=}C(CH_3)_2$ at −30 °C gives $C_4H_4Fe[(CH_3)_2C{=}C(CH_3)_2]_2CO$ and $C_4H_4Fe(CO)_2(CH_3)_2C{=}C(CH_3)_2$. The compound decomposes on warming and shows an IR band at 1928 (CO) cm^{-1}.

Reference:

I. Fischler (Diss. Ruhr-Univ. Bochum 1974).

1.4.2.3 Compounds with One ^{4}L Ligand and One ^{3}L Ligand

$RC_4H_5(C_8H_9)Fe(CO)Cl$ (R=H (butadiene), R=1-CH_3 (penta-1,3-diene), R=2-CH_3 (isoprene)) are prepared from $RC_4H_5(C_8H_8)FeCO$ and 0.2 M aqueous HCl in hexane at −78 °C. The main product is $FeCl_2$ and the $^{4}L(^{3}L)Fe(CO)Cl$ compounds formed are impure. The compound with R=H shows in the IR spectrum (KBr) bands at 742, 768, 833, 851, 889, 906, 945, 1010, 1027, 1047, 1068, 1110, 1133, 1168, 1225, 1235, 1375, 1409, 1430, 1460, 1498, and 2002 cm^{-1} and the mass spectrum shows the molecule ion.

Reference:

A. Carbonaro, F. Cambisi (J. Organometal. Chem. **44** [1972] 171/80).

1.4.3 Compounds with Two ^4L Ligands

General References:

I. Omae, Organometallic Intramolecular-Coordination Compounds Containing a Diolefin Donor Ligand, Coord. Chem. Rev. **51** [1983] 1/39.

K. Jonas, Alkali-Metal-Transition Metal π-Complexes, Advan. Organometal. Chem. **19** [1981] 95/122.

P.S. Pregosin, ^{13}C NMR of Group VIII Metal Complexes, Ann. Rept. NMR Spectrosc. A **11** [1981] 227/71.

P.L. Pauson, Nucleophilic Addition to Transition Metal Complexes, J. Organometal. Chem. **200** [1980] 207/21.

M. Herberhold, Komplexchemie mit nackten Metallatomen, Chemie Unserer Zeit **10** [1976] 120/9.

K.J. Klabunde, Organic Chemistry of Metal Vapours, Accounts Chem. Res. **8** [1975] 393/9.

E. Koerner v. Gustorf, O. Jaenicke, O. Wolfbeis, C.R. Eady, The Laser-Evaporation of Metals and Its Application to Organometallic Synthesis, New Syn. Methods **3** [1975] 107/33; Angew. Chem. **87** [1975] 300/9.

R. Ugo, The Contribution of Organometallic Chemistry and Homogeneous Catalysis to the Understanding of Surface Reactions, Catal. Rev. Sci. Eng. **11** [1975] 225/97.

1.4.3.1 Compounds of the Type $^4LFe^4L'$

Some of the compounds described in Table 6 can be prepared by the following methods:

Method I: $FeCl_2$ and a suitable metalorganic derivative of the appropriate ligand react in an inert organic solvent.

a. A sodium naphthalide solution is prepared in tetrahydrofuran, and I (R = R' = H) is added to this solution. The mixture is stirred at room temperature for 4 h. $MgBr_2$ is then added; 2 h later, $FeCl_2$ is added. The reaction medium is stirred again at room temperature for 16 h and then refluxed for 1 h and evaporated. The residue is extracted with C_6H_6, dried, and then reextracted with hexane. At −30 °C the main part of naphthalene crystallizes out. After filtration and evaporation the residual naphthalene is eliminated by sublimation (30 to 40 °C/0.01 Torr). The remaining orange solid is practically pure No. 1. Further purification is achieved by chromatography with hexane/C_6H_6 (4:1). Similar reaction of I (R = R' = CH_3) gives No. 3 [24].

b. 16.7 mmol of I (R = R' = CH_3) is dissolved in tetrahydrofuran (50 mL), Li (353 mg) is added, and the mixture is stirred at 20 °C for 4 h. The reaction mixture is filtered and $AlCl_3$ (5.6 mmol) is added to the resulting equimolar mixture of LiC_6H_5 and lithium 3,4-dimethylphospholyl, cooled at 0 °C. The mixture is stirred for 30 min at 0 °C and then for 30 min at 20 °C. $FeCl_2$ (8.6 mmol) is added to this mixture, and after 12 h at 20 °C, it is hydrolyzed

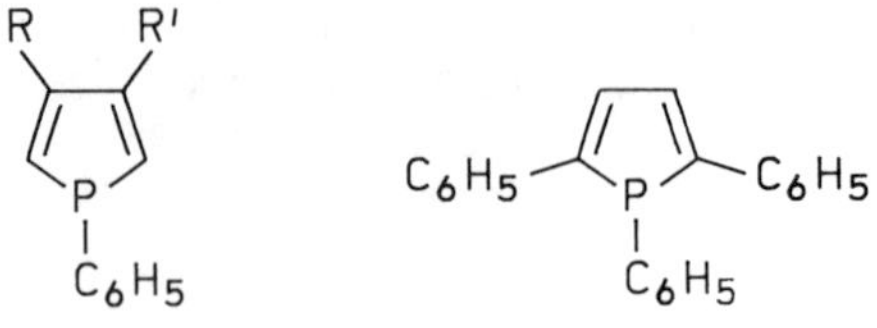

R R'
P
C_6H_5
I

C_6H_5 P C_6H_5
C_6H_5
II

References on pp. 202/3

with 2N HCl and poured into hexane. The organic phase is washed (H_2O), dried (Na_2SO_4), and evaporated. The residue is taken up in hexane, filtered through silica gel, and evaporated to give No 3. Similar reaction of I (R=H, R′=CH_3, 18.4 mmol) with excess Li, $AlCl_3$ (6.1 mmol), and $FeCl_2$ (0.2 mmol) in tetrahydrofuran gives No. 2 [30, 31].

c. I (R=R′=CH_3) reacts with Li in tetrahydrofuran for 4 h at 25 °C. The resulting mixture is treated with t-C_4H_9Cl to destroy the LiC_6H_5. This mixture reacts with 0.5 equivalent of $FeCl_2$. After evaporation the residue is first dissolved in C_6H_6, then in hexane, and chromatographed on silica gel with C_6H_6/hexane (1:4). The crude product contains t-butylphosphole and biphenyl. Treatment with cooled CH_3I precipitates the quaternary salt, and evaporation followed by sublimation of the residue (70 to 80 °C/0.1 Torr) gives crude No. 8. Recrystallization of the sublimate with CH_3OH separates biphenyl. Similar reaction with II gives No. 4 [16].

d. Refluxing III in dimethylether with $FeCl_2$ for 2 h followed by chromatography on silica gel with pentane/C_6H_6 (80:20) gives No. 10 [19]. Refluxing tetramethylthiophene with $FeCl_2$ in the presence of $AlCl_3$ in cyclohexane gives a red oil. Addition of H_2O and $[NH_4]PF_6$ affords No. 12 [4]. Reaction of IV with $FeCl_2$ affords a 33% yield of a 2:1 mixture of No. 11 and 2,2′, 5,5′-tetramethyldistibolyl, separated by fractional sublimation [22].

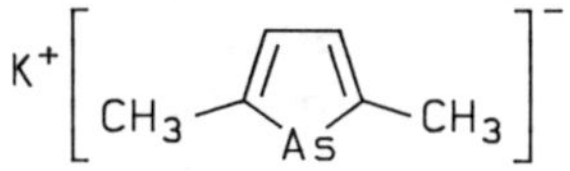

III 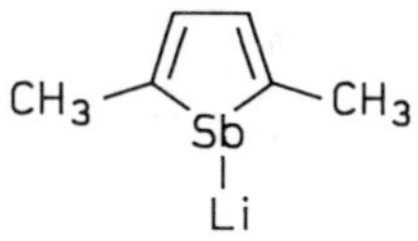 IV

Method II: Fe (vapor) is condensed on a 10% cycloocta-1,5-diene solution in methylcyclohexane at −120 °C [3, 5, 28] or at −190 °C [7]. Washing with cold pentane gives several grams of No. 14 [3, 5, 8, 11, 17, 20, 25]. This compound is not formed below 77 K [14]. Low temperature condensation of Fe and cycloocta-1,3-diene in pentane at $\leqq -120$ °C gives No. 13 [17].

Preliminary electrochemical studies have shown that 1,1′-diphosphaferrocenes are reversibly oxidized to 1,1′-diphosphaferricinium ions less readily than ferrocenes [33]. Attempts to prepare Fe complexes of thiophenes having fewer than four CH_3 substituents did not succeed apparently due to the reduced stability of their respective complex haloaluminate salts to the hydrolysis conditions (cf. Method Id) [4].

References on pp. 202/3

Table 6
Compounds of the Type $^4LFe^4L'$.
Further information on numbers preceded by an asterisk is given at the end of the table, pp. 195/202.
For abbreviations and dimensions see p. VIII.

No.	compound	method of preparation (yield in %), properties and remarks	Ref.
compounds of the type ^4L_2Fe:			
*1	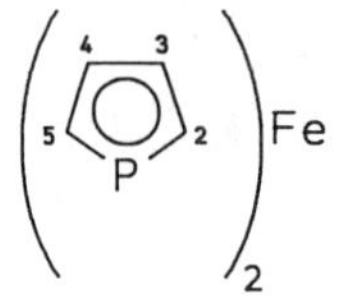 4 3 5 P 2 Fe 2	I a (ca. 30) m.p. ca. 145°, rather stable, orange-red solid ^{1}H NMR ($CDCl_3$): 4.05 (dm, H-2,5; J(H,P) = 36.5), 5.30 (m, H-3,4) ^{13}C NMR ($CDCl_3$): 79.3 (d, C-2,5; J(C,P) = 64.3), 82.0 (C-3,4) ^{31}P NMR ($CDCl_3$): −59 mass spectrum: $[M-nC_4H_4P]^+$ (n = 0, 1)	[24]
2	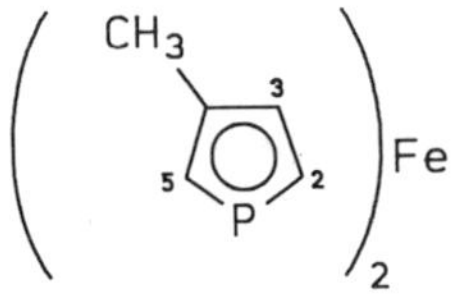 CH_3 3 5 P 2 Fe 2	I b (43) ^{1}H NMR ($CDCl_3$): 2.15 (s, CH_3), 3.68 (d, H-5; ^{2}J(H,P) = 36), 3.70 (dm, H-2; ^{2}J(H,P) = 38), 5.05 (m, H-3) ^{31}P NMR ($CDCl_3$): −61.5 (m) mass spectrum: $[M]^{2+}$, $[M-nCH_3]^+$ (n = 0, 1), $[M-CH_3C_4H_3P]^+$, $[CH_3C_4H_3P]^+$, $[Fe]^+$ and others	[30, 31]
*3	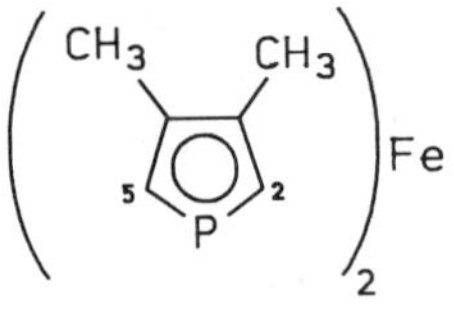 CH_3 CH_3 5 P 2 Fe 2	I a (ca. 60), I b (74), I c (ca. 15), see also Nos. 27, 30, pp. 200 and 201 m.p. 140°, subl. 70 to 80°/0.1 Torr, air-stable, red solid (from CH_3OH) ^{1}H NMR ($CDCl_3$): 2.08 (CH_3), 3.71 (H-2,5; ^{2}J(H,P) = 36.2) ^{13}C NMR ($CDCl_3$): 16.1 (CH_3), 82.1 (C-2,5; ^{1}J(C,P) = 61.6), 97.5 (C-3,4; ^{2}J(C,P) = 7.5) ^{31}P NMR ($CDCl_3$): −72 (^{2}J(P,P) = 10 ± 2) ^{57}Fe-γ (80 K): δ = 0.51(1), Δ = 1.82(1) mass spectrum: $[M]^+$	[1, 16, 24, 29 to 32]
*4	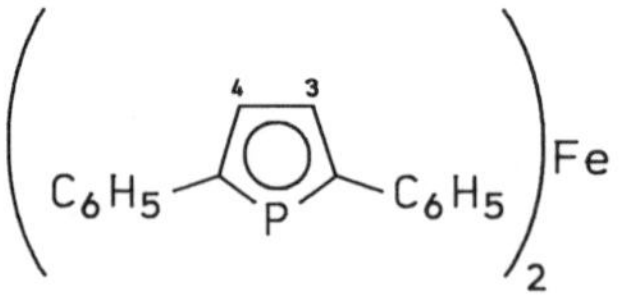 4 3 C_6H_5 P C_6H_5 Fe 2	I c (15 or 53) m.p. 230°, stable red solid (from C_6H_6/CH_3OH) ^{1}H NMR ($CDCl_3$): 5.50 (H-3,4; ^{3}J(P,H) = 4), 6.95 (C_6H_5) ^{31}P NMR (C_6H_6): 62.64 ^{57}Fe-γ (80 K): δ = 0.57(2), Δ = 1.52(2)	[1, 16, 24]

Table 6 [continued]

No.	compound	method of preparation (yield in %), properties and remarks	Ref.
*5	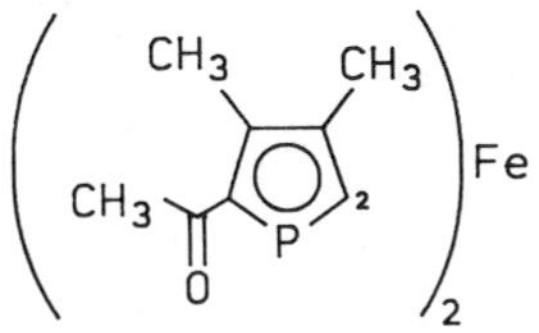 CH_3 CH_3 CH_3 O P 2 Fe 2	see No. 3, p. 197 diastereoisomer a: m.p. 128° (from CS_2) ^1H NMR ($CDCl_3$): 2.0, 2.23 (CH_3); 2.26 ($COCH_3$; J(H, P) = 2); 3.70 (H-2; ^2J(P, H) = 36 ± 1) ^{31}P NMR ($CDCl_3$): −52 diastereoisomer b: m.p. 134° (from CS_2) ^1H NMR ($CDCl_3$): 2.0, 2.20 (CH_3); 2.23 ($COCH_3$; J(H, P) = 2), 4.02 (H-2; ^2J(P, H) = 36 ± 1) ^{31}P NMR ($CDCl_3$): −49 both diastereoisomers: IR (KBr): 1655 (C=O)	[24]
*6	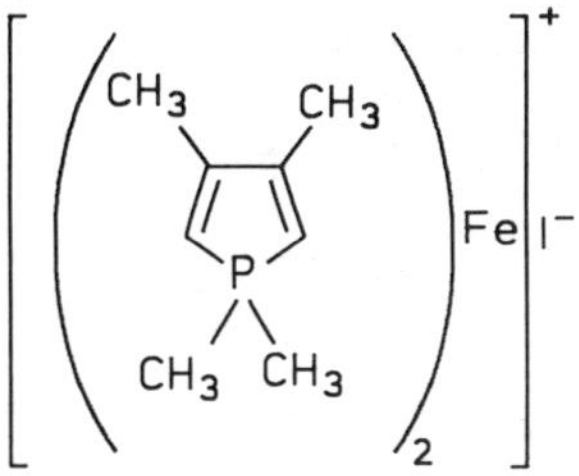 CH_3 CH_3 P CH_3 CH_3 2 Fe I^- +	see No. 3, p. 197 very unstable mass spectrum: shows decomposition	[29]
*7	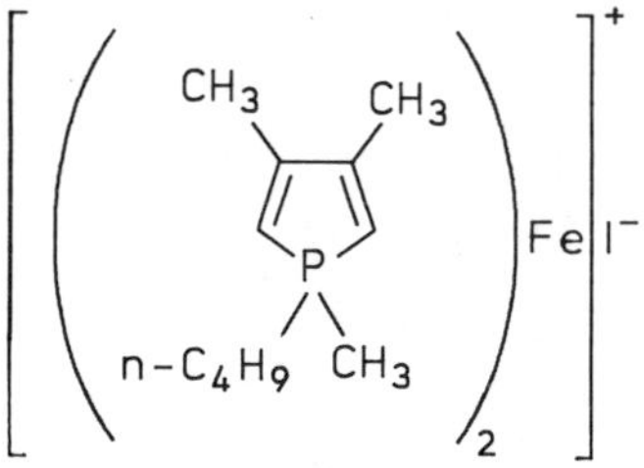 CH_3 CH_3 P n-C_4H_9 CH_3 2 Fe I^- +	see No. 3, p. 197 mass spectrum: $[M-I]^+$, $[M-I-nC_4H_9-nCH_3]^+$ (n = 1, 2)	[29]
*8	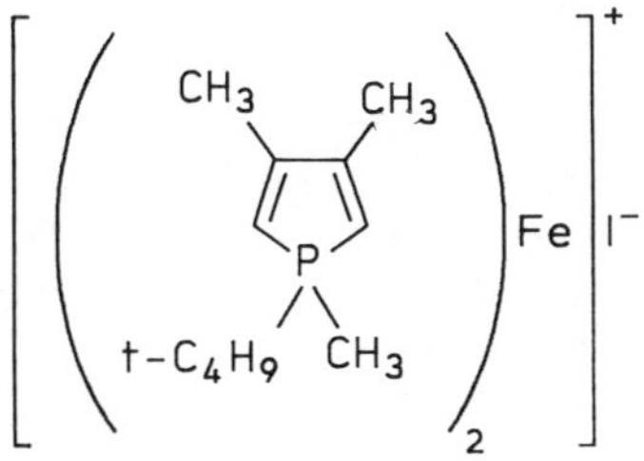 CH_3 CH_3 P t-C_4H_9 CH_3 2 Fe I^- +	see No. 3, p. 197 m.p. 206° (dec.), green powder (from slow evaporation of tetrahydrofuran solutions) mass spectrum: $[M-I]^+$, $[M-I-nC_4H_9-nCH_3]^+$ (n = 1, 2), $[PC_4H_2(CH_3)_2]^+$	[29]

References on pp. 202/3

Table 6 [continued]

No.	compound	method of preparation (yield in %), properties and remarks	Ref.
*9	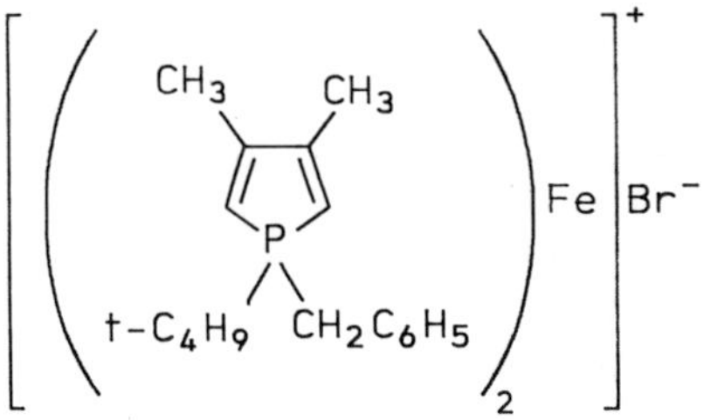 $[(CH_3)_2C_4H_2P(t\text{-}C_4H_9)(CH_2C_6H_5)]_2Fe]^+Br^-$	see No. 3, p. 197 mass spectrum: shows decomposition	[29]
*10	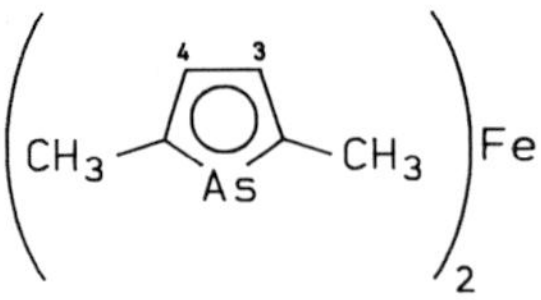 $(2,5\text{-}(CH_3)_2C_4H_2As)_2Fe$	Id (ca. 40) m.p. ca. 34°, red solid ^{1}H NMR ($CDCl_3$): 1.91 (CH_3), 5.09 (H) ^{13}C NMR ($CDCl_3$): 18.90 (CH_3), 86.10 (C-3,4; $^1J(C,H) = 168$), 108.50 (C-1,5) mass spectrum: $[M]^+$	[19, 23]
11	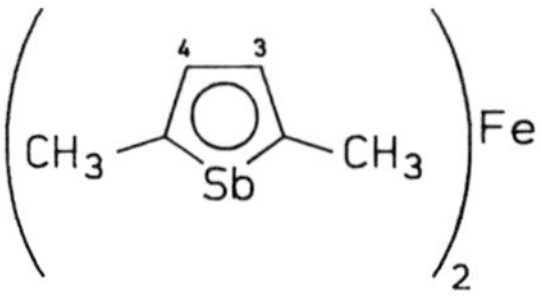 $(2,5\text{-}(CH_3)_2C_4H_2Sb)_2Fe$	Id (22) m.p. 59° (dec.), air-stable, deep red solid ^{1}H NMR (benzene-d_6): 1.8 (CH_3), 5.5 (H) ^{13}C NMR (benzene-d_6): 23.0 (CH_3), 93.5 (C-3,4; $^1J(C,H) = 164$), 113.6 (C-2,5) mass spectrum: $[M]^+$	[22]
*12	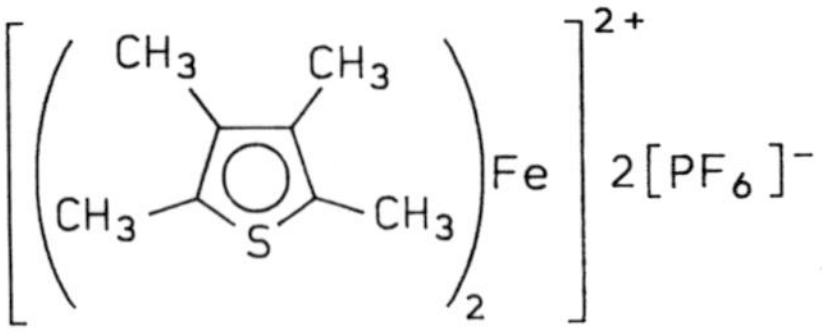 $[((CH_3)_4C_4S)_2Fe]^{2+}\,2[PF_6]^-$	Id (55) dec. 100°, red crystals, diamagnetic ^{1}H NMR (acetone-d_6): 2.50 (s, CH_3), 2.66 (s, CH_3) IR (KBr): 560, 850, 1037, 1092, 1105, 1160, 1250, 1310, 1401, 1457, 1490, 2880, 2958, 3000, 3030	[4]
*13	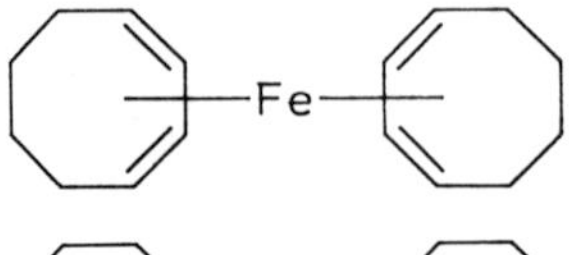 $(C_8H_{12})_2Fe$ (cyclooctadienyl-type ligands, drawn as structure)	II stable at reduced temperature	[17]
*14	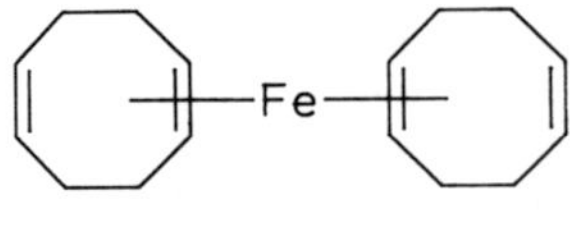 $(C_8H_{12})_2Fe$ (cycloocta-1,5-diene ligands, drawn as structure)	II (30 to 40) dec. −40°, dec. −30°, dec. above −20°, dec. above −25°, brown crystalline solid, pyrophoric brown crystals (from methylcyclohexane)	[3, 5, 7, 8, 11, 14, 15, 17, 20, 25, 28]
*15	$[(C_8H_{12})_2Fe]Li_2$	—	[21, 26]

References on pp. 202/3

Table 6 [continued]

No.	compound	method of preparation (yield in %), properties and remarks	Ref.
compounds of the type $^4LFe^4L'$:			
16	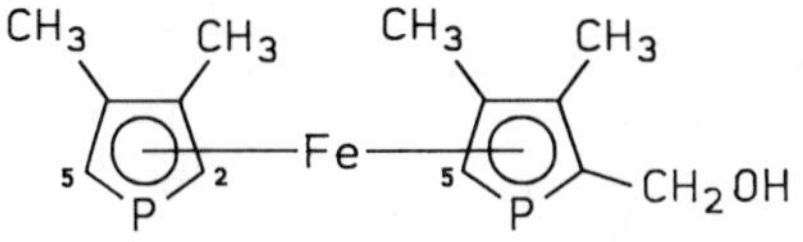 CH_3 CH_3 CH_3 CH_3; 5, 2, P, Fe, 5, P, CH_2OH	see No. 17, p. 200 m.p. 127° ^1H NMR ($CDCl_3$): 1.56 (OH); 2.09, 2.12 (CH_3); 3.58, 3.95 (H-2,5; $^2J(H,P) = 36 \pm 1$); 4.19 (CH_2) ^{31}P NMR ($CDCl_3$): −73.7, −67.0 ($^2J(P,P) = 10 \pm 2$) IR (KBr): 3200 (OH) mass spectrum: $[M - nH_2O]^+$ (n = 0, 1)	[24]
*17	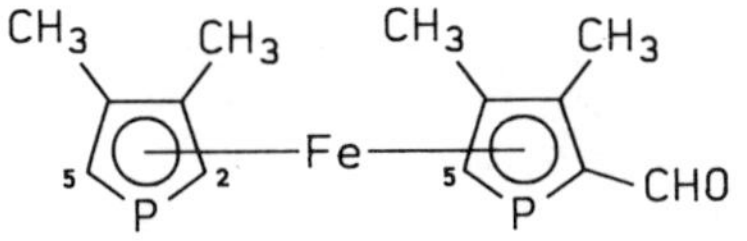 CH_3 CH_3 CH_3 CH_3; 5, 2, P, Fe, 5, P, CHO	see No. 3, p. 197, see also No. 18, p. 200 m.p. 162° ^1H NMR ($CDCl_3$): 2.0, 2.03, 2.10, 2.35 (CH_3); 3.72, 4.10 (H-2,5; $^2J(H,P) = 36 \pm 1$); 9.67 (CHO; J(H,P) = 4.5) ^{31}P NMR ($CDCl_3$): −67, −51 ($^2J(P,P) = 10 \pm 2$) IR (KBr): 1658 (C=O)	
*18	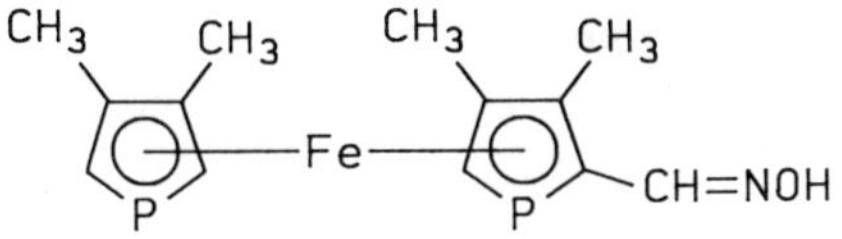 CH_3 CH_3 CH_3 CH_3; P, Fe, P, CH=NOH	see No. 17, p. 200 mixture of two isomers: crystalline solid ^{31}P NMR ($OS(CH_3)_2$): −71.6, −67.3, −66.7, −47 (d's; $^2J(P,P) \approx 7$) mass spectrum: $[M + n]^+$ (n = 0, 1), $[M - H_2O]^+$ and others	[27]
19	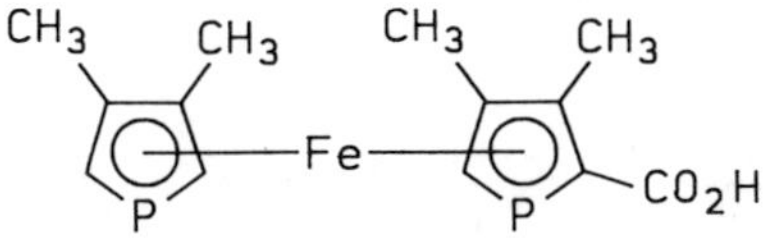 CH_3 CH_3 CH_3 CH_3; P, Fe, P, CO_2H	see No. 20, p. 200 m.p. 190 to 192° (dec.) ^1H NMR ($CDCl_3$): 2.05 (pseudo s, 2CH_3), 2.13 (s, CH_3), 2.37 (s, CH_3), 3.60 (dd, 1H; $^2J(H,P) = 36$), 3.83 (dd, 1H; $^2J(H,P) = 36$), 4.10 (d, 1H; $^2J(H,P) = 36$) ^{31}P NMR ($CDCl_3$): −66.9, −52.5 ($^2J(P,P) \approx 7$) IR (KBr): 1655 (C=O) mass spectrum: $[M + n]^+$ (n = 0, 1), $[M - CO_2 - H_2]^+$, since $[M]^+$ is the base peak even at 190°, it seems that decarboxylation is very difficult	[27]

References on pp. 202/3

Table 6 [continued]

No.	compound	method of preparation (yield in %), properties and remarks	Ref.
*20	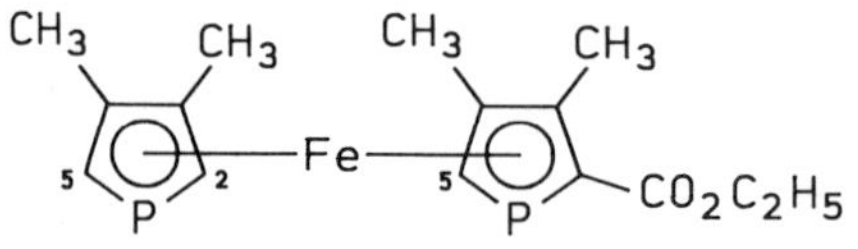 CH_3 CH_3 CH_3 CH_3 — 5 P 2 —Fe— 5 P — $CO_2C_2H_5$	see No. 3, p. 200 m.p. 55° ^{1}H NMR ($CDCl_3$): 1.30 (t, CH_3 in C_2H_5), 2.03 (pseudo s, 2 CH_3), 2.10 (s, CH_3), 2.35 (s, CH_3), 3.57 (dd, 1H; $^2J(H,P)=36$), 3.73 (dd, 1H; $^2J(H,P)=36$), 3.98 (d, 1H; $^2J(H,P)=36$), 4.09 (q, OCH_2) ^{13}C NMR ($CDCl_3$): 13.7 (s, CH_3), 14.5 (s, CH_3+CH_3 in C_2H_5), 16.4 (s, CH_3), 60.0 (s, OCH_2), 82.0 (d, C-2,5; $^1J(C,P)=60.6$), 83.0 (d, C-2,5; $^1J(C,P)=60.7$), 83.3 (d, C-2,5; $^1J(C,P)=58.2$), 84.2 (d, C-2,5; $^1J(C,P)=60.6$), 97.6 (d, C-3,4; $^2J(C,P)\approx 7$), 97.9 (d, C-3,4; $J(C,P)\approx 7$), 98.8 (d, C-3,4; $^2J(C,P)=7.2$), 101.0 (d, C-3,4; $^2J(C,P)=6.8$), 172.2 (d, CO; $^2J(C,P)=18.7$) ^{31}P NMR ($CDCl_3$): −68.1, −55.6 ($^2J(P,P)\approx 7$) IR (KBr): 1688 (C=O) mass spectrum: $[M+n]^+$ (n = 0, 1), $[M-C_2H_4-nCO_2]^+$ (n = 0, 1)	[27]
21	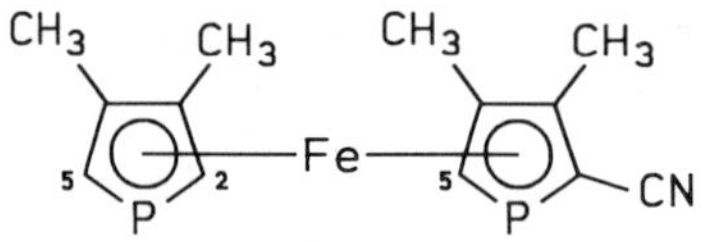 CH_3 CH_3 CH_3 CH_3 — 5 P 2 —Fe— 5 P — CN	see No. 18, p. 200 m.p. 136°, orange solid ^{1}H NMR ($CDCl_3$): 2.05 (pseudo s, 3 CH_3), 2.17 (s, CH_3), 3.63 (d, 1H; $^2J(H,P)=36$), 3.83 (d, 1H; $^2J(H,P)=36$), 3.98 (d, 1H; $^2J(H,P)=37$) ^{13}C NMR ($CDCl_3$): 13.9 (s, CH_3), 14.3 (s, CH_3), 14.8 (s, CH_3), 16.3 (s, CH_3), 83.5 (d, C-2,5; $^1J(C,P)=61.5$), 84.0 (d, C-2,5; $^1J(C,P)=62$), 85.6 (d, C-2,5; $^1J(C,P)=61$) ^{31}P NMR ($CDCl_3$): −64.7, −52.4 ($^2J(P,P)\approx 7$) IR ($CDCl_3$): 2220 (CN) mass spectrum: $[M+n]^+$ (n = 0, 1), $[Fe]^+$ and others	[27]
22	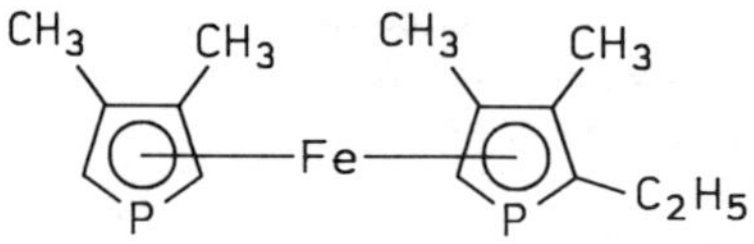 CH_3 CH_3 CH_3 CH_3 — P —Fe— P — C_2H_5	see No. 24, p. 200 orange oil ^{1}H NMR ($CDCl_3$): 1.02 (CH_3 in C_2H_5);	[24]

References on pp. 202/3

Table 6 [continued]

No.	compound	method of preparation (yield in %), properties and remarks	Ref.
		2.02, 2.07 (CH_3); 3.45, 3.57 (CH; $^2J(H,P) = 36 \pm 1$) (the CH_2 protons are masked by the CH_3 peaks)	
23	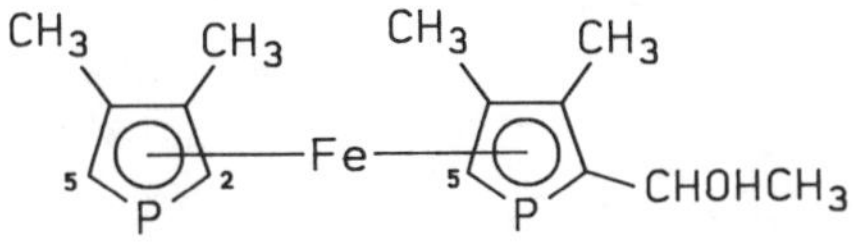 1-(3,4-dimethylphospholyl)-1′-(3,4-dimethyl-2-(CHOHCH₃)phospholyl)iron (structure; ring positions 2, 5 and P marked)	see No. 24, p. 200 isomer a: m.p. 100 to 105° (from CH_3OH) ^1H NMR ($CDCl_3$): 1.29 (CH_3; $^3J(H,H) = 6.1$); 2.01, 2.07, 2.11 (CH_3-3,4); 2.50 (OH); 3.42, 3.53, 4.15 (H-2,5; $^2J(H,P) = 36 \pm 1$), 4.43 (CH) ^{31}P NMR ($CDCl_3$): −79.6 ($^2J(P,P) = 10 \pm 2$) IR (KBr): 3370 (sh), 3485 (OH) isomer b: m.p. 70 to 72° (from CH_3OH) ^1H NMR ($CDCl_3$): 1.26 (OH), 1.48 (CH_3; $^3J(H,H) = 6.3$); 2.05, 2.07, 2.15, 2.24 (CH_3-3,4); 3.53, 3.94 (H-2,5; $^2J(H,P) = 36 \pm 1$), 4.53 (CH) ^{31}P NMR ($CDCl_3$): −72.3, −71.6 ($^2J(P,P) = 10 \pm 2$) IR (KBr): 3250 (OH)	[24]
*24	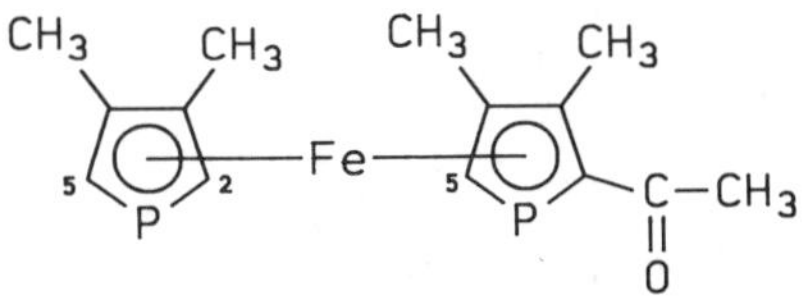 1-(3,4-dimethylphospholyl)-1′-(3,4-dimethyl-2-(C(=O)CH₃)phospholyl)iron (structure; ring positions 2, 5 and P marked)	see No. 3, p. 197 m.p. ca. 71° ^1H NMR ($CDCl_3$): 1.99, 2.06, 2.11 (CH_3); 2.28 ($COCH_3$; $J(H,P) = 2.9$), 2.38 (CH_3); 3.69, 4.02 (H-2,5; $^2J(H,P) = 36 \pm 1$) ^{31}P NMR ($CDCl_3$): −72, −54 ($^2J(P,P) = 10 \pm 2$) IR (KBr): 1650 (C=O)	[24]
25	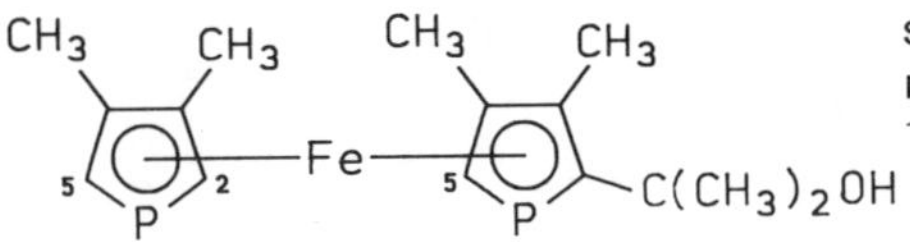 1-(3,4-dimethylphospholyl)-1′-(3,4-dimethyl-2-(C(CH₃)₂OH)phospholyl)iron (structure; ring positions 2, 5 and P marked)	see No. 24, p. 200 m.p. ca. 98° ^1H NMR ($CDCl_3$): 1.41, 1.54 ($C(CH_3)_2$; $J(H,P) = 0$, $J(H,P) = 3.4$); 2.05, 2.07, 2.16, 2.31 (CH_3); 2.43 (OH; $J(H,P) = 7.6$); 3.44, 3.58, 4.29 (H-2,5; $^2J(H,P) = 36 \pm 1$) ^{31}P NMR ($CDCl_3$): −83, −66 ($^2J(P,P) = 10 \pm 2$) mass spectrum: $[M - nH_2O]^+$ (n = 0, 1)	[24]

Table 6 [continued]

No.	compound	method of preparation (yield in %), properties and remarks	Ref.
26	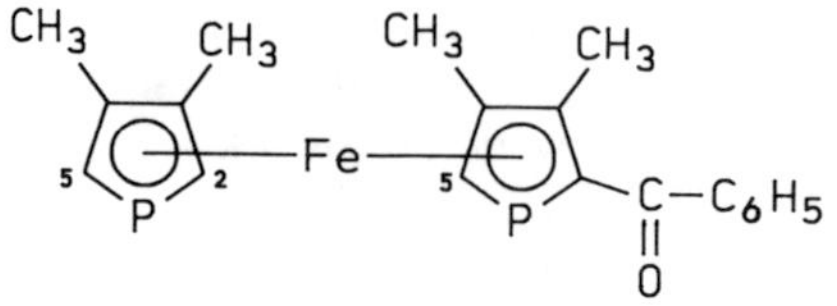 [structure: two 3,4-dimethylphospholyl rings bonded to Fe; one ring with $C(=O)C_6H_5$ at position 2]	see No. 3, p. 197 m.p. ca. 70° ^1H NMR ($CDCl_3$): 2.0, 2.05, 2.13, 2.30 (CH_3); 3.63, 3.93 (H-2,5; $^2J(H,P)=36\pm1$), 7.30 (H-3 to 5 in C_6H_5), 7.63 (H-2,6 in C_6H_5) ^{31}P NMR ($CDCl_3$): −69, −50 ($^2J(P,P)=10\pm2$) IR ($CDCl_3$): 1620 (C=O) mass spectrum: $[M-nCOC_6H_5]^+$ (n=0, 1)	[24]
*27	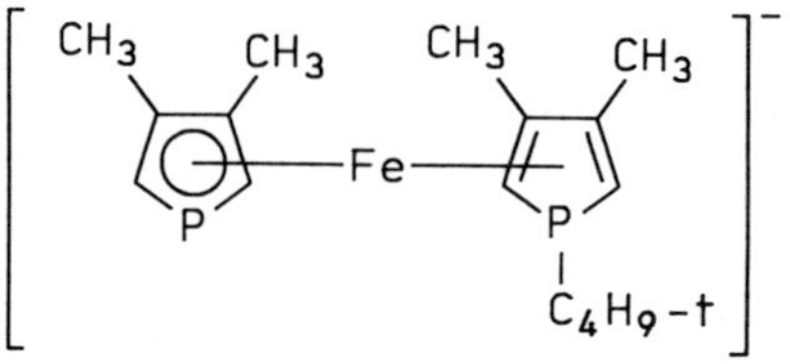 [anion: 3,4-dimethylphospholyl–Fe–1-C_4H_9-t-3,4-dimethylphosphole]$^-$	see No. 3, p. 197 not isolated ^{31}P NMR: −95.4 (P), −21.8 (PC_4H_9-t)	[29, 32]
28	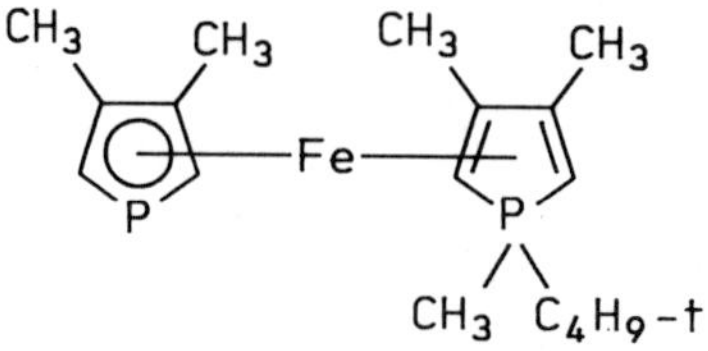 [structure: 3,4-dimethylphospholyl–Fe–phosphole with P(CH_3)(C_4H_9-t)]	see No. 27, p. 200 not isolated	[29]
29	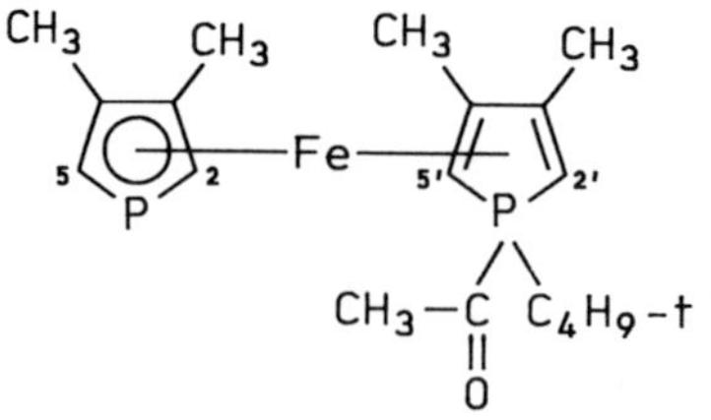 [structure: 3,4-dimethylphospholyl–Fe–phosphole with P($COCH_3$)(C_4H_9-t)]	see No. 3, p. 197 m.p. 80°, red crystals ^1H NMR ($CDCl_3$): 0.70 (d, C_4H_9-t; $^3J(H,P)=13.7$), 1.07 (d, H-2′,5′; $^2J(H,P)=23.9$), 1.77 (s, $2CH_3$), 1.97 (d, $2CH_3$; J=0.9), 2.82 (dd, $COCH_3$; J=1.5, J=4.9), 3.31 (d, H-2,5; $^2J(H,P)=35.9$) ^{31}P NMR: −65.5, −2.4 ($J(P,P)\cong0$) IR (KBr): 1650 (C=O) mass spectrum: $[M-nCOCH_3]^+$ (n=0, 1), $[M-C_4H_9]^+$, $[M-C_4H_9-CH_3]^+$, $[M-C_4H_9-CO]^+$, $[M-COCH_3-C_4H_9]^+$	[32]
*30	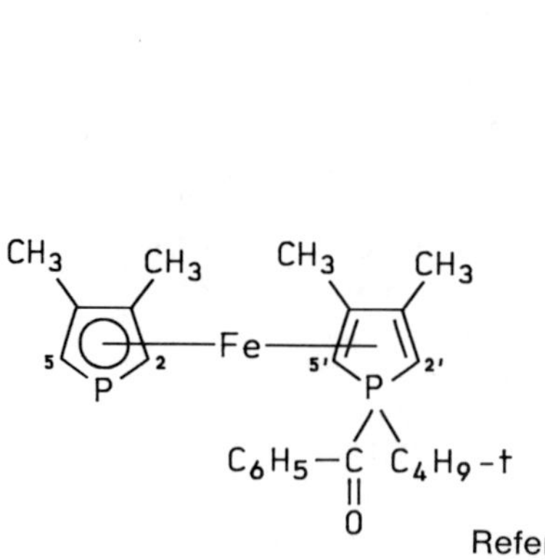 [structure: 3,4-dimethylphospholyl–Fe–phosphole with P(COC_6H_5)(C_4H_9-t)]	see No. 3, p. 197 m.p. 205° (dec.), green crystals (from ether/pentane at low temperature, or from CH_2Cl_2/pentane at −20°) ^1H NMR (CD_2Cl_2): 0.72 (d, C_4H_9-t; $^3J(H,P)=14$), 1.27 (m, H-2,5);	[32]

Table 6 [continued]

No.	compound	method of preparation (yield in %), properties and remarks	Ref.
		1.60, 1.77 (2s, CH_3-3′,4′); 1.97, 2.01 (2s, CH_3-3,4); 2.85 (d, H-5; $^2J(H,P)=36$), 3.01 (d, H-2; $^2J(H,P)=36$), 7.6 (m, H-3 to 5 in C_6H_5), 8.2 (m, H-2,6 in C_6H_5) ^{31}P NMR (benzene-d_6): −63.7, −8.5 (J(P,P)=2.5) IR (KBr): 1592 to 1612 (C=O) UV (ether): $\lambda_{max}=740$ nm mass spectrum: $[M]^+$, $[M-C_4H_9-nCO]^+$ (n=0, 1), $[M-COC_6H_5-C_4H_9]^+$	
*31	CH_3 CH_3 CH_3 CH_3 Fe P P C_6H_5 CH_3 O O $\cdot H_2O$	see No. 3, p. 197 m.p. 188° (dec.), red solid (from $CH_3CO_2C_2H_5$/pentane) ^1H NMR ($CDCl_3$): 1.60 (s, CH_3), 1.68 (s, CH_3), 2.37 (d, CH; J(H,P)=25), 2.85 (d, CH_3P; J(H,P)=14), 3.56 (d, CH; J(H,P)=18), 7.35 (m, C_6H_5) ^{31}P NMR ($CDCl_3$): 1.9, 24.5 IR (KBr): 1208 (P=O)	[29]

*Further information:

$(C_4H_4P)_2Fe$ (Table **6**, No. **1**). The photoelectron spectrum (He I (21.21 eV) and He II (40.41 eV)) gives vertical ionization potentials (in eV): 7.50 ($10a_g$, $6a_g$, $9a_g$), 9.1 ($8b_u$, $8a_g$), 9.5 ($5a_u$, $5b_g$), 10.15 ($7a_g$), 10.35 ($7b_u$), 11.6 to 12.0 ($4a_u$, $6b_u$, $4b_g$), 12.80 to 13.4 ($5b_u$, $3a_u$, $6a_g$, $3b_g$). The intensity of the band at 7.50 eV increases considerably in going from He I to He II excitation. It may thus be unambiguously assigned to the ionization to three orbitals, showing an important participation of the 3d AOs of iron ($10a_g$, $6b_g$, and $9a_g$). On the contrary, the intensity of the fourth band (10.15, 10.35 eV) decreases during the He I/He II transition which results from the ionization of MOs which are highly localized on the phosphorus atoms, or more precisely on their σ lone pair ($7b_u$ and $7a_g$). The two intermediate bands are probably due to the four π MOs localized on the phospholyl rings ($8b_u$, $8a_g$, $5a_u$, and $5b_g$). Thus, there is a general satisfying agreement between the experimental and calculated values, since the theoretical order is $3d_{Fe}$, $4\pi_{ligand}$, $2n_p$. These data are used in conjunction with EHT and MSXα calculations to explain the reactivity of this compound within the frame work of perturbation theory [18].

For a nonparameterized molecular orbital calculation for this compound see [2]. Although the phosphorus lone pairs in the ligands remain localized on this atom, the P atoms are not nucleophilic because the lone pair is not in a frontier molecular orbital. The P atoms are even weakly electrophilic. Nucleophilic substitutions apparently are controlled by orbital and charge effects in concert [2].

References on pp. 202/3

$[(CH_3)_2C_4H_2P]_2Fe$ (Table **6**, No. **3**). The yields of No. 3 in the preparation according to Method Ib, pp. 186/7, vary with the Lewis acid added. Repetition of Method Ib with various Lewis acids gives the following results [30, 31]:

Lewis acid	amount (mmol)	yield of No. 3	Lewis acid	amount (mmol)	yield of No. 3
$AlCl_3$	5.6	74	SbF_3	4.1	51
$AlCl_3$	13.8	59	$SnCl_4$	4.0	48
$ZnCl_2$	8.35	61	$(C_2H_5)_2OBF_3$	5.5	43
$ZnCl_2$	25	62	$TiCl_4$	4.2	32
ZnI_2	25	61			

The photoelectron spectrum (He I (21.21 eV) and He II (40.41 eV)) gives vertical ionization potentials (in eV): 7.05 ($10a_g$, $6a_g$, $9a_g$), 8.35 ($8b_u$), 8.55 ($8a_g$), 8.95 ($5a_u$), 9.2 ($5b_g$), 9.80 ($7a_g$), 9.95 ($7b_u$), 10.90 ($6b_u$), 11.2 to 11.5 ($4a_u$, $4b_g$), 11.9 to 12.2 ($5b_u$, $6a_g$, $3a_u$), 12.8 to 13.1 ($3b_g$, $5a_g$). The intensities of the peaks at 7.05 and 9.2 increase, and those at 9.80 and 9.95 eV decrease when going from He I to the He II spectrum. These data are used in conjunction with EHT (extended Hückel theory) and MSXα (multiple scattering) calculations to explain the reactivity of this compound within the frame work of perturbation theory [18]. Compound No. 3 crystallizes in the monoclinic space group $P2_1/c-C^5_{2h}$ with the unit cell parameters a=22.612(13), b=12.373(12), c=13.725(8) Å, β=99.07(4)°; Z=12 molecules per unit cell. D_{meas}=1.44±0.02 g/cm³, D_{calc}=1.46 g/cm³. The main bonding distances and angles are shown in **Fig. 27**. There are three such molecules in the asymmetric unit. All $(CH_3)_2C_4H_2P$ rings are roughly parallel to the (10$\bar{1}$) plane of the unit cell. The carbon moieties of the two rings are both planar within experimental error, but the P atoms lie out of these carbon mean planes by 0.011(1) to 0.084(1) Å on the opposite side to Fe, with an average value of 0.04(1) Å. The dihedral angles between the mean planes C(1) to C(4) and C(1)-P(1)-C(4) are 0.67°, 0.66°, and 0.67°, and those between mean planes C(7) to C(10) and C(7)-P(2)-C(10) are 3.84°, 3.50°, and 3.48° indicating that for each molecule one phosphoryl ring is more bent than the other.

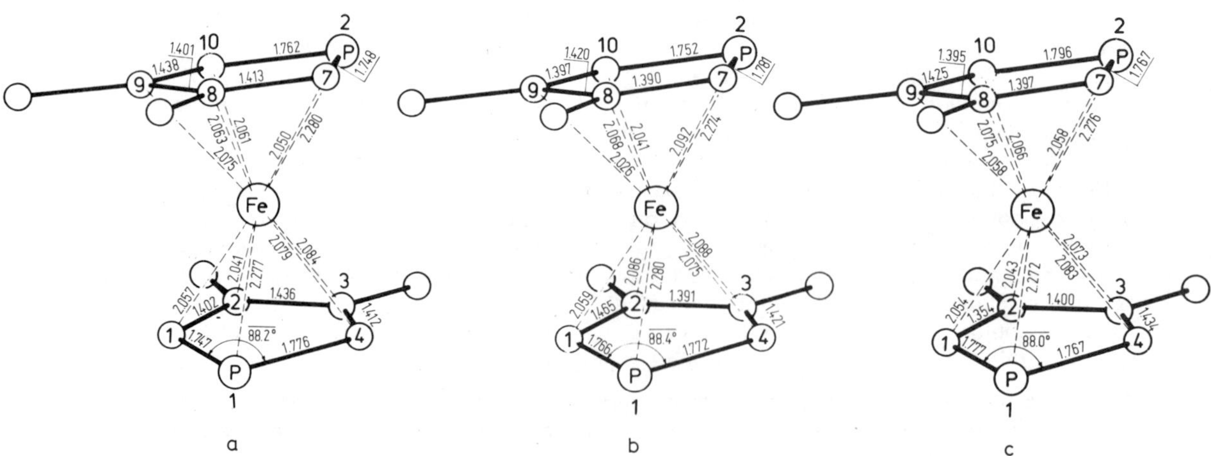

Fig. 27. The molecular structure of the three different molecules of $[(CH_3)_2C_4H_2P]_2Fe$ (No. 3) [24].

References on pp. 202/3

The compound is stable to light [16]. The ^{31}P NMR spectrum of No. 3 in $HOSO_2CF_3$ shows a chemical shift at $\delta = 93.85$ ppm, suggesting that protonation occurs on the C ring (and not on the P atom) which accounts for the very large downfield shift (166 ppm) observed on changing from C_6H_6 to $HOSO_2CF_3$ [1]. Compound No. 3 reacts with $Fe_2(CO)_9$ (ratio 1:1) in tetrahydrofuran in the presence of CO to give V (35% yield). Refluxing with $Fe_2(CO)_9$ (ratio 1:2) in C_6H_6 gives V (74% yield) [27]. Reaction of No. 3 with $OPCl_3/OHCN(CH_3)C_6H_5$ (ratio 1:3:3) in CH_2Cl_2 at room temperature for 1 h and then at 50 °C for 2 h followed by usual workup and chromatography on silica gel (Merck, 70 to 230 mesh) with C_6H_6 gives No. 17 in 58% yield. Compound No. 3 (0.005 mol) reacts with $CH_3COCl/AlCl_3$ (0.0057 mol: 0.0052 mol) in CH_2Cl_2 at 50 °C for 3 h. The mixture is then poured on ice, and the organic layer is washed, dried (Na_2SO_4), and evaporated. Chromatography of the residue as before gives No. 24 in 24% yield. Similar reaction with $C_6H_5COCl/AlCl_3$ at room temperature for 4 h followed by treatment as above gives No. 26 (42%). The acetylation of No. 3 (0.0025 mol) with $CH_3COCl/AlCl_3$ (0.0065 mol: 0.0062 mol) as above, but elution with $C_6H_6/CH_3CO_2C_2H_5$, (95:5) gives a mixture of two diastereoisomers of No. 5 in the ratio of ca. 60:40 (total yield 64%) [24]. Compound No. 3 is dissolved in CS_2 at -5 °C, and $AlCl_3$ is added successively, and then $ClCO_2C_2H_5$ in CS_2 is added within 1 h. At the end of the addition, the mixture is stirred for 40 min and then hydrolyzed with aqueous HCl and ice. The H_2O layer is saturated with NaCl and extracted with $CHCl_3$. The organic layers are washed with a saturated aqueous solution of NaCl, dried (Na_2SO_4), and evaporated. The residue is chromatographed as before but with $C_6H_6/CH_3CO_2C_2H_5$ (90:10). The starting complex No. 3 elutes first and then complex No. 20 (30% yield) is recovered. Heating of No. 3 with $C_6H_5CH_2Br$ at 80 °C for 3 h followed by stirring of the cooled reaction mixture for 1 h gives VII in 72% yield [27]. A stirred solution of No. 3 in tetrahydrofuran is treated at -80 °C with t-C_4H_9Li in pentane. After 15 min CH_3I is added to the solution. Then, after 0.5 h the reaction mixture is hydrolyzed with H_2O at low temperature and extracted with $CH_3CO_2C_2H_5$. The organic layer is evaporated, and the residue is chromatographed as before but with C_2H_5OH as eluent to give No. 8 in 40 to 55% yield. The complexes No. 6, 7, and 9 are prepared in the same way in ca. 10, ca. 20, and ca. 30% yield. Treatment of No. 3 in tetrahydrofuran with C_6H_5Li in ether at -80 °C followed by addition of CH_3I after 15 min and hydrolysis after 0.5 h with H_2O at low temperature gives a mixture which is extracted at room temperature with $CH_3CO_2C_2H_5$. Evaporation and chromatography of the residue as before with C_2H_5OH gives No. 31 in ca. 50% yield [29]. Compound No. 3 in tetrahydrofuran reacts with t-C_4H_9Li in pentane at -80 °C. After 30 min, C_6H_5COCl is added; and after another 1 h the reaction mixture is allowed to come to room temperature. The solvents are evaporated, and the residue is chromatographed with ether. Rechromatography of the green fraction with ether/pentane (10:90) gives No. 30 in 40 to 50% yield. The same procedure with CH_3COCl gives No. 29 (50 to 60%). Compound No. 27 is suggested to be the intermediate in the last two reactions [32].

V VI VII

$[(C_6H_5)_2C_4H_2P]_2Fe$ (Table **6**, No. **4**) shows in the ^{31}P NMR spectrum in $HOSO_2CF_3$ a chemical shift of $\delta = 139.46$ ppm, and the frozen $HOSO_2CF_3$ solid solution gives a ^{57}Fe-γ spectrum of $\delta = 0.55(2)$ and $\Delta = 1.72(2)$ mm/s. Treatment with $DOSO_2CF_3$ gives an exchange in the

References on pp. 202/3

ortho and para positions of the phenyl rings but no exchange of the protons in the five-membered phosphoryl ring. Thus, the protonation clearly occurs in the phenyl rings, and it is most probable that No. 4 on protonation yields a σ complex where the positive charge is stabilized by overlap of e_{2g} with empty p orbitals on the α carbon as shown in VIII [1].

VIII IX X

$(CH_3CO(CH_3)_2C_4HP)_2Fe$ (Table **6**, No. **5**) can be also obtained by reaction of $AlCl_3$ and CH_3COCl with VI at room temperature within 4 h. After hydrolysis, extraction with C_6H_6 and chromatography, No. 5 is isolated in 100% yield [27]. Pure samples of each diastereoisomer are obtained by chromatography on silica gel with $C_6H_6/CH_3CO_2C_2H_5$ (90:10). Diastereoisomer 5a elutes first [24].

$[((CH_3)_2C_4H_2PRCH_3)_2Fe]I$ (Table **6**, Nos. **6** to **8** with R=CH_3, n-C_4H_9, t-C_4H_9) and **$[((CH_3)_2C_4H_2P(C_4H_9\text{-}t)CH_2C_6H_5)_2Fe]Br$** (Table **6**, No. **9**). Compound No. 8 crystallizes in the triclinic space group $P\bar{1}$-C_i^1, S_2^1 with the unit cell parameters a=10.640(1), b=10.865(2), c=24.908(4) Å, α=90.14(1)°, β=95.06(1)°, γ=112.24(1)°; Z=4 molecules per unit cell. D_{meas}= 1.37±0.02 g/cm³, D_{calc}=1.375 g/cm³. There are two different molecules in the asymmetric unit. The main bond distances and angles are shown in **Fig. 28**. The dihedral angles between least-square planes C(1)-C(4) and C(1)-P-C(4) have a mean value of 30.9°. Consequently, the mean value of the Fe-P bond is greater by about 0.4 Å than that observed in No. 3.

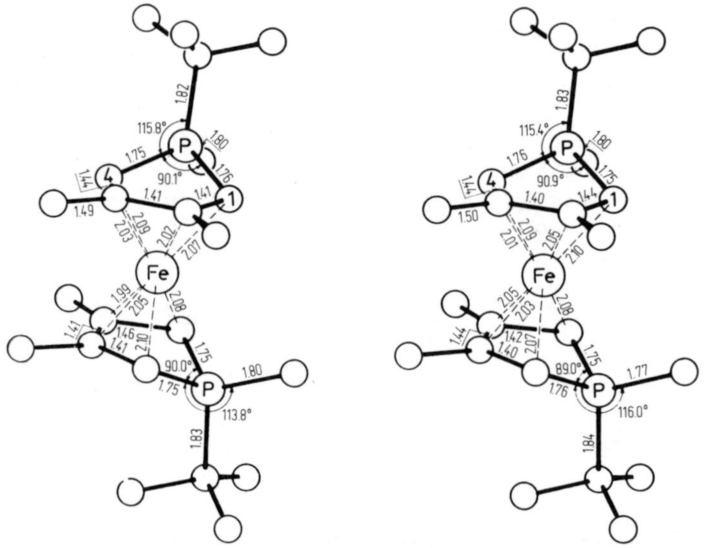

Fig. 28. The molecular structure of the two different molecules of $[((CH_3)_2C_4H_2P(C_4H_9\text{-}t)CH_3)_2Fe]I$ (No. 8) [29].

References on pp. 202/3

Another structural difference of No. 8, by comparison with No. 3, is that the P of one ring superposes with the α-carbon of the other (superposition with the β-carbon in No. 3). In contrast, the Fe-C(ring) mean distances are remarkably similar in Nos. 8 and 3. Both t-C_4H_9 groups in No. 8 are exo. Compound No. 8 is stable, water-soluble, and paramagnetic. The stability decreases in the order No. 8>7>6. Reaction of all four compounds 6 to 9 with HI in H_2O at room temperature gives the appropriate phospholium salts (65% yield in the case of No. 8) [29].

[((CH$_3$)$_2$C$_4$H$_2$As)$_2$]Fe (Table **6**, No. **10**) undergoes deuterium exchange at 25 °C in CF_3COOD, while it can be acetylated with $CH_3COCl/AlCl_3$ in CH_2Cl_2 [22].

[((CH$_3$)$_4$C$_4$S)$_2$Fe][PF$_6$]$_2$ (Table **6**, No. **12**) shows a molar conductance in acetonitrile (5.2×10^{-3} M) of 98.78 $\Omega^{-1} \cdot cm^2 \cdot mol^{-1}$. It decomposes slowly in air or rapidly in solution to give tetramethylthiophene and a purple solid which is not identified. Thermal decomposition gives tetramethylthiophene and iron salts. Cyclic voltammetry in 50% H_2O/acetone (KCl supporting electrolyte) gives $E_{1/2} = -0.27$ V as the first reversible and $E_{1/2} = -1.17$ V as a second irreversible step versus SCE. Possibly $((CH_3)_4C_4S)_2Fe$ may be isolable. Reduction with $NaBH_4$ in tetrahydrofuran gives a brown complex which may be the hydride adduct; it is, however, very unstable and, therefore, not completely characterized [4].

(1,3-C$_8$H$_{12}$)$_2$Fe (Table **6**, No. **13**) gives 1,3-$C_8H_{12}Fe(P(OCH_3)_3)_3$ in the reaction with $P(OCH_3)_3$ in pentane when warmed from −120 °C to 0 °C [17].

(1,5-C$_8$H$_{12}$)$_2$Fe (Table **6**, No. **14**) is presently only obtainable by metal atom techniques (see Method II, p. 187) [9]. Magnetic balance, ^{1}H NMR, and ESR measurements are rendered uncertain by the very facile decomposition of the compound to metallic Fe [3, 4, 28]. The IR spectrum in Nujol at −30 °C resembles that of (1,5-$C_8H_{12})_2Ni$ (see 1.4.2.1.1.2 in "Nickel-Organische Verbindungen" 2, 1974, p. 88) [3]. The UV spectrum of a matrix isolated thin film of No. 14 is given in a figure in [28]. The structure IX is proposed for this compound [17]. A green solution in hexane or pentane quickly deposits an iron mirror above −30 or −20 °C, respectively; but a solution in cycloocta-1,5-diene is slightly more stable (up to 0 °C [5]) [3, 5]. The compound does not react with N_2 but caught fire in air [3]. Reaction with 2,2'-dipyridyl (^{4}D) in ether at −30 °C followed by oxidation by air and addition of $[NH_4]PF_6$ gives two moles cycloocta-1,5-diene and one mole $[Fe(^4D)_3][PF_6]$ [3]. Reaction with PF_3 in n-pentane at −78 °C (or n-hexane at −30 °C [3]) gives 1,5-$C_8H_{12}Fe(PF_3)_3$ [8, 17, 20]. The compound No. 14 reacts in different ways with $P(OR)_3$ and PR_3 nucleophiles. Thus, reaction with excess $P(OR)_3$ (R=CH_3 [7, 8, 13, 17], C_2H_5 [7, 13], or C_3H_7-i [7, 13]) in cold tetrahydrofuran (at −50 °C) [7], or in methylcyclohexane (at −78 °C) [8] gives 1,3-$C_8H_{12}Fe(P(OR)_3)_3$. Similar reaction with $P(OCH_2)_3CR$ (R=C_2H_5, C_3H_7, C_7H_{15}) at −40 °C to room temperature gives $Fe(P(OCH_2)_3CR)_5$ [6]. Reaction of liquid N_2 cooled No. 14 covered with a pentane layer containing $P(OC_6H_5)_3$ followed by warming to room temperature gives $(P(OC_6H_5)_2OC_6H_4)_2Fe(P(OC_6H_5)_3)_2$ (see Formula X) [15, 17]. Reaction with ^{2}D=$P(OCH_3)_2C_6H_5$ or $P(CH_3)_3$ gives 1,3-$C_8H_{12}Fe(^2D)_3$ [13]. Compound No. 14 dissolved in methylcyclohexane yields with ^{4}D=$P(C_6H_5)_2CH_2CH_2P(C_6H_5)_2$ in ether at −20 °C a brown solution (possibly $Fe(^4D)_2$ [13]) decomposing above 0 °C to Fe^0 [8]. Bubbling N_2 through the reaction mixture gives $N_2Fe(^4D)_2$ [8, 11, 13]; with CO, the appropriate $COFe(^4D)_2$ is obtained [11, 13]. Similar reaction with $P(C_6H_5)_3$ results in a very unstable mixture. The brown material obtained gives iron mirrors even at −20 °C. Well resolved ^{1}H NMR spectra are therefore not obtained, but the upfield chemical shift of the second P resonance is consistent with XI [17]. The product of the reaction with CO was first suggested to be $Fe(CO)_5$ and (1,3-$C_8H_{12})_2FeCO$ [7], but on repetition of this reaction in methylcyclohexane at −130 °C there was no evidence for this isomerization, and $Fe(CO)_5$ (traces) and 1,5-$C_8H_{12}Fe(CO)_3$ were the only products [8, 13]. Reaction with CNC_4H_9-t gives 1,5-$C_8H_{12}Fe(CNC_4H_9\text{-t})_3$ [8, 13]. Similar reaction with

References on pp. 202/3

buta-1,3-diene gives $(C_4H_6)_nFe$ (n=2 or 3) [5, 11, 13]. Reaction with buta-1,3-diene in the presence of ^2D=CO, or CNC_4H_9-t gives $(C_4H_6)_2Fe^2D$ [11]. The organic ligand is displaced with C_7H_8 [5]. Reaction with cyclooctatetraene [3, 5, 11, 13, 20] in n-hexane at −30 °C gives $(C_8H_8)_2Fe$ [3, 11, 13, 20] and cycloocta-1,5-diene in quantitative yield [3]. Reaction with C_8H_8 and PF_3 gives $C_8H_8Fe(PF_3)_3$ (trace) [3, 11, 13, 20]. Reaction with bicyclo[6.1.0]nona-2,4,6-triene in n-hexane at −78 °C gives XII [25].

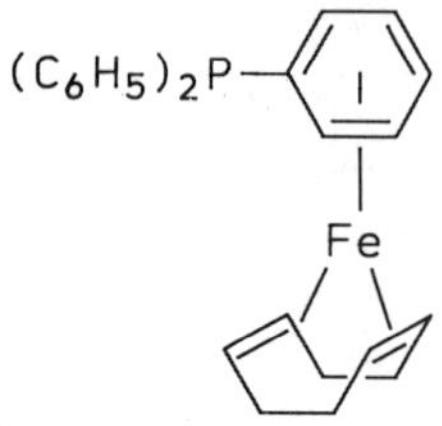

$(C_6H_5)_2P$ Fe

XI

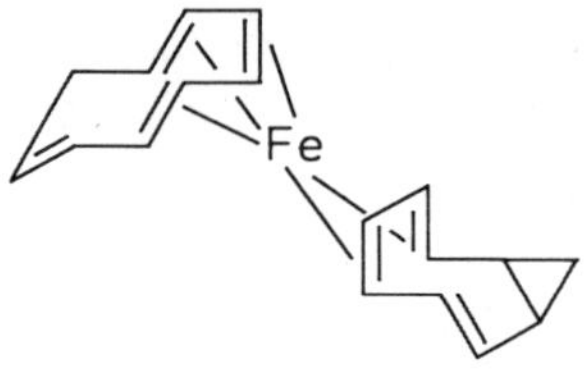

Fe

XII

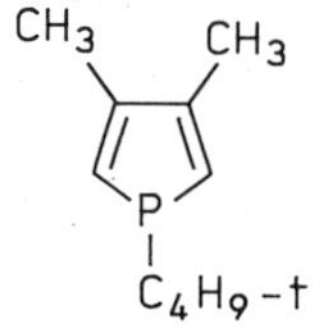

CH_3 CH_3 P C_4H_9-t

XIII

$Li_2[Fe(C_8H_{12})_2]$ (Table **6**, No. **15**) is prepared by the reaction of $Li_2(D)_2Fe(C_2H_4)_4$ with cycloocta-1,5-diene at 40 to 70 °C and isolated as the $Li_2(D)_2Fe(C_8H_{12})_2$ salt (D=dimethoxyethane or N,N,N′,N′-tetramethylethylenediamine [21, 26].

$(CH_3)_2C_4H_2PFePC_4H(CH_3)_2CHO$ (Table **6**, No. **17**) in ether is reduced by dropwise addition of $LiAlH_4$ in tetrahydrofuran. After 1 h at room temperature the reaction mixture is treated with $CH_3CO_2C_2H_5$, and then H_2O is added. The organic phase is washed, dried (Na_2SO_4), and evaporated. Chromatography of the residue with $C_6H_6/CH_3CO_2C_2H_5$ (95:5) gives No. 16 in 53% yield [24]. Refluxing in C_2H_5OH with $[H_3NOH]Cl$ for 1 h followed by evaporation gives a residue which is dissolved in $CHCl_3$, washed with aqueous Na_2CO_3 and H_2O, dried (Na_2SO_4), and evaporated to give No. 18 as a mixture of two isomers best seen on the ^{31}P NMR spectrum (85 to 90% yield) [27].

$(CH_3)_2C_4H_2PFePC_4H(CH_3)_2CH{=}NOH$ (Table **6**, No. **18**) is refluxed in C_6H_6 for 4 h with two equivalents dicyclohexylcarbodiimide. After evaporation, the residue is chromatographed with C_6H_6. Compound No. 21 elutes first (52% yield), and some aldehyde No. 17 is recovered (18%) [27].

$(CH_3)_2C_4H_2PFePC_4H(CH_3)_2CO_2C_2H_5$ (Table **6**, No. **20**) is refluxed for 1 h in 20 cm^3 of 98% formic acid and 5 cm^3 conc. H_2SO_4. The mixture is cooled and extracted with $CHCl_3$. The organic layer is washed with saturated aqueous NaCl, dried (Na_2SO_4), and evaporated. Chromatography of the residue with $C_6H_6/CH_3CO_2C_2H_5$ (90:10) gives No. 19 in 75% yield [27].

$(CH_3)_2C_4H_2PFePC_4H(CH_3)_2COCH_3$ (Table **6**, No. **24**) in CH_3OH is treated at room temperature with small batches of $NaBH_4$. After evaporation the residue is taken up with CH_2Cl_2/H_2O. The organic phase is washed, dried (Na_2SO_4), and evaporated. The crude mixture is chromatographed with $C_6H_6/CH_3CO_2C_2H_5$ (95:5) to give isomer 23a (elutes first) in 52% yield and isomer 23b (29%). Similar reduction with $AlCl_3/LiAlH_4$ in ether at 5 °C for 1 h followed by addition of formic acid and H_2O at 0 °C, work up of the ethereal layer as before, and chromatography with hexane gives No. 22 in 33% yield. Refluxing with CH_3MgI in ether for 1.5 h, followed by hydrolysis and work up of the organic layer as above gives a crude product which is recrystallized from hexane at −20 °C and then chromatographed with C_6H_6/$CH_3CO_2C_2H_5$ (90:10) to give pure No. 25 in 15% yield [24].

$[(CH_3)_2C_4H_2PFe(t\text{-}C_4H_9PC_4H_2(CH_3)_2)]^-$ (Table **6**, No. **27**). Addition of CH_3I in tetrahydrofuran leads only to the isolation of No. 3. A possible explanation lies in the spontaneous decomposition of the expected complex No. 28 [29].

References on pp. 202/3

$(CH_3)_2C_4H_2PFe(t\text{-}C_4H_9(C_6H_5CO)PC_4H_2(CH_3)_2)$ (Table **6**, No. **30**) crystallizes in the monoclinic space group $P2_1/n\text{-}C_{2h}^5$ with the unit cell parameters a=17.564(6), b=12.254(4), c=10.346(4) Å, β=96.39(2)°; Z=4 molecules per unit cell. D_{meas}=1.30±0.02 g/cm³ (by flotation in aqueous KI), D_{calc}=1.32 g/cm³. The main bond distances and angles are shown in **Fig. 29**. The Fe-P(1′) distance is with 2.565(2) Å clearly outside of the normal length. Thus, the structure is certainly partly zwitterionic. Nevertheless, both the folding of the phospholyl ring around the C(2′)-C(5′) axis (20.8°) and the Fe-P(1′) distance are much smaller than those in other known complexes (e.g., Nos. 8 and 31). Thus, it is quite clear that there is a rather strong through-space interaction. A direct interaction between Fe and the carbonyl C is excluded (Fe · · · C=O=3.17 Å). The acyl group possibly increases the positive charge on P, thus creating a stronger Coulombic attraction between Fe and P(1′). Consequently the Fe-P(1) bond (2.289(3) Å) and Fe-ring 1 distance (1.670(1) Å) is weakened. Due to this longer Fe-ring 1 distance, the ring 1 is now strictly planar. Finally, the endo position of the acyl group explains why it migrates easily to Fe in the thermal decomposition of No. 30 (see below).

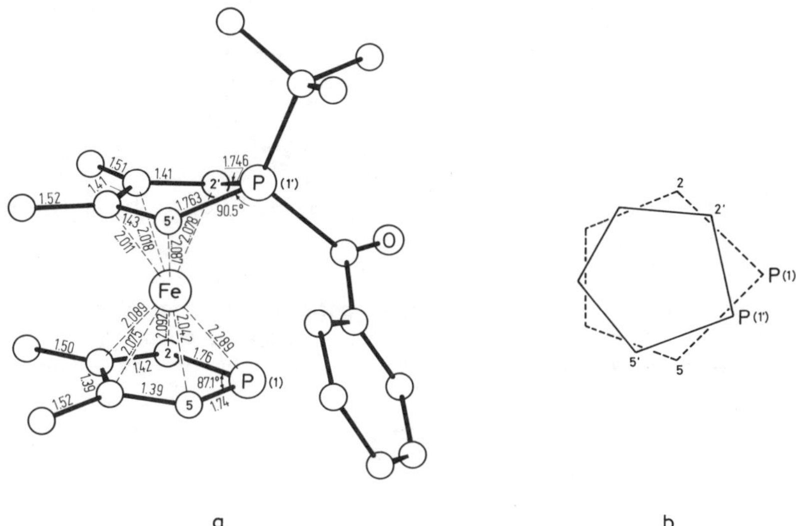

Fig. 29. a) Molecular structure of $(CH_3)_2C_4H_2PFe(t\text{-}C_4H_9(C_6H_5CO)PC_4H_2(CH_3)_2)$ (No. 30). b) Projection of one phosphole ring on the plane of the other in No. 30 [32].

Thermolysis of No. 30 at 170 °C under vacuum gives XIII (36%) which distills slowly. The residue of the thermolysis is recovered in hexane and chromatographed after filtration (eluent: hexane/C_6H_6 (90:10)). The first band is No. 3, and the second red band gives XIV [32].

$[(CH_3)_2C_4H_2P(C_6H_5)CH_3]Fe[O_2PC_4H_2(CH_3)_2] \cdot H_2O$ (Table **6**, No. **31**) crystallizes in the triclinic space group $P\bar{1}\text{-}C_i^1$, S_2^1 with the unit cell parameters a=9.822(1), b=9.845(1), c=10.913(1) Å, α=74.64(1)°, β=67.02(1)°, γ=80.74(1)°; Z=2 molecules per unit cell. D_{meas}=1.49±0.02 g/cm³, D_{calc}=1.493 g/cm³. The main bond distances and angles are shown in **Fig. 30**. An H_2O molecule connects two molecules together, and the oxygen atom in P=O is linked via hydrogen bonds to a third molecule. The phospholium ring is geometrically very similar to the rings of No. 8; the dihedral angle between the mean planes of the two five-membered rings is just slightly larger (34.3° versus 30.9°) and, consequently, the

References on pp. 202/3

Fe–P1 bond length is longer (2.732(1) Å). Quite logically, the C_6H_5 substituent is exo. The bonding of one O atom to Fe causes a drastic decrease of the dihedral angle around the C(2′)–C(5′) axes down to 10.4°. The P atoms P(1) and P(2) lie under one another.

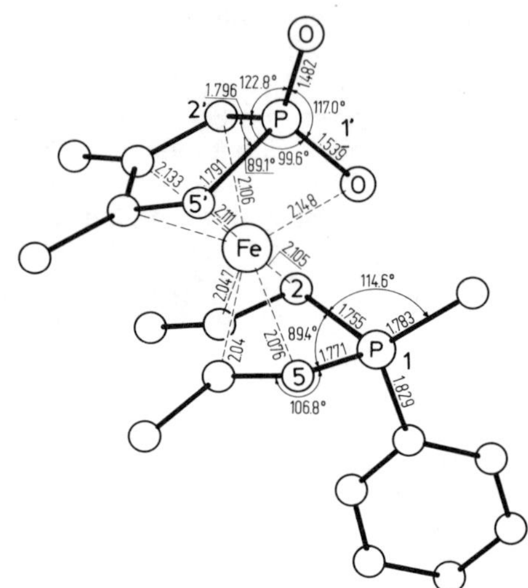

Fig. 30. Molecular structure of $[(CH_3)_2C_4H_2P(C_6H_5)CH_3]Fe[O_2PC_4H_2(CH_3)_2] \cdot H_2O$ (No. 31, H_2O omitted) [29].

An aqueous HI solution is added to a H_2O suspension of compound No. 31 until complete dissolution is achieved. XV is extracted with $CHCl_3$, and the aqueous phase is then saturated with NaCl and reextracted with $CHCl_3$ to give another unstable product, probably XVI [29].

XIV: CH_3, CH_3, CH_3, CH_3, Fe, P, C_6H_5

XV: $[\ldots]^+$ CH_3, CH_3, P, C_6H_5, CH_3, I^-

XVI: CH_3, CH_3, P, O, OH

XIV XV XVI

References:

[1] B. Lukas, R.M.G. Roberts, J. Silver, A.S. Wells (J. Organometal. Chem. **256** [1983] 103/10). — [2] N.M. Kostić, R.F. Fenske (Organometallics **2** [1983] 1008/13). — [3] R. Mackenzie, P.L. Timms (J. Chem. Soc. Chem. Commun. **1974** 650/1). — [4] D.M. Braitsch, R. Kumarappan (J. Organometal. Chem. **84** [1975] C37/C39). — [5] P.L. Timms (Angew. Chem. **87** [1975] 295/9).

[6] J.P. Jesson, C.A. Tolman, E.I. Du Pont de Nemours and Co. (U.S. 3997579 [1976]). — [7] A.D. English, J.P. Jesson, C.A. Tolman (Inorg. Chem. **15** [1976] 1730/2). — [8] R.A. Cable, M. Green, R.E. Mackenzie, P.L. Timms, T.W. Turney (J. Chem. Soc. Chem. Commun. **1976** 270/1). — [9] R.M. Atkins, P.L. Timms, T.W. Turney (unpublished results from [10]). — [10] P.L. Timms, T.W. Turney (J. Chem. Soc. Dalton Trans. **1976** 2021/5).

[11] R.E. McKenzie, P.L. Timms (unpublished results from [12]). — [12] P.L. Timms, T.W. Turney (Advan. Organometal. Chem. **15** [1977] 53/112, 68, 95). — [13] P.L. Timms, T.W. Turney (unpublished results from [12]). — [14] R.M. Atkins (Diss. Univ. Bristol 1977 from [28]). — [15] C.A. Tolman, A.D. English, S.D. Ittel, J.P. Jesson (Inorg. Chem. **17** [1978] 2374/8).

[16] G. De Lauzon, F. Mathey, M. Simalty (J. Organometal. Chem. **156** [1978] C33/C36). — [17] S.D. Ittel, F.A. Van-Catledge, J.P. Jesson (J. Am. Chem. Soc. **101** [1979] 3874/84). — [18] C. Guimon, D. Goubeau, G. Pfister-Guillonzo, G. De Lauzon, F. Mathey (Chem. Phys. Letters **104** [1984] 560/7). — [19] G. Thiollet, F. Mathey, R. Poilblanc (Inorg. Chim. Acta **32** [1979] L67/L68). — [20] T. Saito (Yuki Gosei Kagaku Kyokaishi **37** [1979] 1099/104).

[21] K. Jonas, L. Schieferstein, C. Krüger, Y.-H. Tsay (Angew. Chem. **91** [1979] 590/1). — [22] A.J. Ashe III, T.R. Diephouse (J. Organometal. Chem. **202** [1980] C95/C98). — [23] A.J. Ashe III, T.R. Diephouse (unpublished results from [22]). — [24] G. De Lauzon, B. Deschamps, J. Fischer, F. Mathey, A. Mitschler (J. Am. Chem. Soc. **102** [1980] 994/1000). — [25] E. Meier, R. Roulet, A. Scrivanti (Inorg. Chim. Acta Letters **45** [1980] L267/L268).

[26] K. Jonas, C. Krüger (Angew. Chem. **92** [1980] 513/31). — [27] G. De Lauzon, B. Deschamps, F. Mathey (Nouv. J. Chim. **4** [1980] 683/5). — [28] G.A. Ozin, C.G. Francis, H.X. Huber, M. Andrews, L. Nazar (J. Am. Chem. Soc. **103** [1981] 2453/6). — [29] B. Deschamps, J. Fischer, F. Mathey, A. Mitschler (Inorg. Chem. **20** [1981] 3252/9). — [30] G. De Lauzon, F. Mathey, Soc. Nat. des Poudres et Explosifs (U.S. 4404147 [1981/83]).

[31] G. De Lauzon, F. Mathey, Soc. Nat. des Poudres et Explosifs (Eur. 53987 [1981/82]). — [32] B. Deschamps, J. Fischer, F. Mathey, A. Mitschler, L. Ricard (Organometallics **1** [1982] 312/6). — [33] P. Tordo (unpublished results from [32]).

1.4.3.2 Compounds of the Type $^4L(^4L')Fe^2D$

The compounds listed in Table 7 can be prepared by the following methods:

Method I: Low temperature co-condensation of Fe and the alkadiene followed by addition of ^2D gives the product.

a. Fe is co-condensed with butadiene onto a liquid N_2 cooled surface over 2h. Then a layer of $P(OCH_3)_3$ is added followed by a final layer of ether. The mixture is allowed to warm to room temperature. After 30 min the excess butadiene is evaporated, the suspension is filtered through Celite, and the filtrate is evaporated. The residual oil is taken up in pentane, again filtered through Celite, and chromatographed on alumina to give a mixture of $C_4H_6Fe(P(OCH_3)_3)_3$ and No. 4. Attempted separation of these two complexes by sublimation, crystallization, and conventional chromatography failed [20, 24], cf. [15, 16]. In another experiment the preceding procedure is employed, but the $P(OCH_3)_3$ is placed in the reactor as a pentane solution prior to evaporation of Fe. The C_4H_6-Fe condensate then has to melt and flow down the $P(OCH_3)_3$ before reaction can take place. This gives a mixture of products which is very rich in No. 4. After removal of $P(OCH_3)_3$, No. 4 is crystallized from pentane at −50 °C to give an orange-yellow product. There is a lesser amount of No. 4 left in the pentane filtrate, but it can not be collected without contamination by some of the residual monodiene species [24].

b. Fe is co-condensed with ^4L (^4L = butadiene, penta-1,3-diene, isoprene or hexa-2,3-diene) and ^2D on a liquid N_2 cooled layer of pentane over 3.5 h. A second layer of pentane is condensed, and the mixture is allowed to warm

References on pp. 212/3

to room temperature over 3 h. This procedure gives a mixture of products rich in the $^4LFe(^2D)_3$ species. Work-up involves Celite filtration, removal of the volatiles, and chromatography on alumina. But most of the material coming off the column is a mixture of monodiene species and $^4L_2Fe^2D$. The two species are readily separated by size-exclusion liquid chromatography (on Waters Microstyrogel 100 Å). The elution corresponds to molecular size, e.g., $C_4H_6Fe(CO)_3$ eluting before No. 4 in tetrahydrofuran. Other $^4L_2Fe^2D$ species are prepared and isolated similarly. In the case of nonvolatile ligands ^2D such as $P(OR)_3$ with $R=C_3H_7$-i or C_6H_5 (Nos. 6, 7) it is not practical to simultaneously co-condense the diene and $P(OR)_3$ with the Fe atoms. The $P(OR)_3$ is placed in the reactor as a pentane solution prior to the diene-Fe co-condensation. The mixture is then washed into the $P(OR)_3$ solution with pentane at -120 °C. In some cases, only mixtures are obtained. When the unsymmetrically substituted olefins, e.g. isoprene, are employed, the $^4L_2Fe^2D$ product is a mixture of isomers. The ratio of cis to trans in the isolated products varies with the steric size of ^2D. Thus, for $^2D=P(OC_2H_5)_3$ the ratio of No. 15 to No. 16 is 3:7 and for $^2D=P(OC_6H_5)_3$ the ratio of No. 19 to No. 20 is 2:8. For the ratios of Nos. 13:14 and Nos. 17:18 see "Further information", p. 212 [24].

c. Fe atoms and butadiene are co-condensed at -196 °C. The reaction mixture is warmed to -78 °C, stirred for 1 h, and cooled to -196 °C. PF_3 is added, and the mixture is warmed to room temperature to give No. 3 [1, 2]. Low temperature Fe vapor co-condensation with butadiene in cyclohexane at -90 to -60 °C for 90 to 120 min followed by addition of ^2D gives Nos. 2, 7, 10, 21, and 29 [23].

Method II: An inorganic iron salt is the starting material.

a. $FeCl_2$ in tetrahydrofuran (OC_4H_8) reacts with butadiene and $P(CH_3)_3$ at -196 °C. Warming to 0 °C gives a brown solution. $Mg(C_4H_6)_2 \cdot 2OC_4H_8$ is added dropwise to the ice-cold mixture. After 1 h at 0 °C and then overnight at room temperature the volatiles are evaporated. The residue is dissolved in boiling hexane and filtered through Celite. Concentration of the filtrate gives an oily solid which crystallizes on addition of ether and cooling to -78 °C yields No. 30 [26].

b. The cathodic reduction (Hg cathode, Pt anode) of $Fe(acac)_3$ (acac=acetylacetonate) in the presence of butadiene, $P(C_6H_5)_3$, and $LiCH_3$ in CH_3OH at -15 °C for 10 h gives No. 10 after evaporation and extraction with hexane [4, 5]. Similar reduction in the presence of isoprene gives No. 21 [4].

c. Irradiation of $H_2Fe(P(C_6H_5)_2C_2H_5)_3N_2$ with 2,3-dimethylbutadiene gives traces of No. 25 [9].

Method III: $Fe(CO)_5$, $P(OCH_3)_3$ (ratio 1:1), and excess butadiene (^4L) are irradiated in pentane for 27 h. Evaporation and chromatography of the residue on silica gel with pentane gives $^4L_2FeCO$. Elution with C_6H_6 gives a mixture of $(CO)_3Fe(P(OCH_3)_3)_2$ and No. 4, separated by crystallization from pentane [7].

Method IV: $^4L_2FeCO$ and PR_3 are irradiated in pentane.

a. $^4L_2FeCO$ (^4L=butadiene) and excess PF_3 are irradiated in pentane for 1 h. Evaporation followed by sublimation at 0 °C in vacuum gives a ca. 20:80 mixture of $^4L_2FeCO$ and No. 3. Recrystallization from n-pentane results in orange-red crystals of this mixture (yield of No. 3 ca. 55%). Purification by recrystallization, sublimation, or chromatography on silica gel or alumina failed. Thus, this mixture is again irradiated in the presence of PF_3 as before; but the reaction is stopped, the mixture is cooled to −196 °C, pumped off, and PF_3 is added again. This procedure is repeated three times. Evaporation and chromatography of the residue on silica gel gives pure No. 3 [7]. This reaction proceeds only at rather short wavelengths (λ=254 nm) [7, 13, 14].

b. $^4L_2FeCO$ (^4L=butadiene) and $P(OCH_3)_3$ are irradiated (hν≧254 nm) in pentane for 1 h. Evaporation and chromatography of the residue on silica gel with pentane gives $^4L_2FeCO$. Elution with C_6H_6 gives No. 4 (59.8% yield) [7], cf. [13, 16]. Irradiation (hν≧366 nm) for 14 h followed by similar work-up gives an ca. 80:20 mixture of No. 4 and $^4LFe(P(OCH_3)_3)_2CO$. Further elution with ether gives $^4LFe(P(OCH_3)_3)_2CO$, and elution with CH_3OH gives $^4LFe(P(OCH_3)_3)_3$ [7]. For the effects of conditions on quantum yields see [25], and $(C_4H_6)_2FeCO$ in 1.4.3.3, p. 222.

c. Irradiation of $^4L_2FeCO$ (^4L=butadiene) and $P(C_4H_9\text{-}n)_3$ in pentane for 2.5 h followed by evaporation and chromatography on Florisil (Fluka, 60 to 100 mesh) with pentane (as quickly as possible) gives No. 8. Similar irradiation with $P(C_6H_{11}\text{-cyclo})_3$ in pentane gives No. 9 which decomposes on chromatography. Separation from excess $P(C_6H_{11}\text{-cyclo})_3$ by recrystallization failed [7]. Irradiation of $^4L_2FeCO$ (^4L=cyclohexadiene) and $P(OCH_3)_3$ or ^2D= pyridine at −40 °C gives Nos. 27 or 28 [10].

Method V: $^6LFeC_4H_4(CH_3)_2\text{-}2,3$ (see Formula I) and 2,3-dimethylbutadiene are irradiated [9] for 4 h (λ≧270 nm) [12]. The reaction mixture is evaporated, and the residue is dissolved in pentane to give No. 25 [9], cf. [11].

P(C6H5)(C2H5)
Fe
H3C CH3

I

Some of the compounds listed in Table 7 catalyze the mixed dimerization of butadiene and $CH_2{=}CHCO_2CH_3$. The course of this dimerization and the size of the catalytic activity are influenced by the ligand ^2D [7, 14].

References on pp. 212/3

Table 7
Compounds of the Type $^4L(^4L')Fe^2D$.
Further information on numbers preceded by an asterisk is given at the end of the table, pp. 211/2.
For abbreviations and dimensions see p. VIII.

No.	compound	method of preparation (yield in %), properties and remarks	Ref.
*1	$(C_4H_6)_2Fe(CNC_4H_9\text{-}t)$	–	[21]
2	$(C_4H_6)_2Fe$(2,2'-bipyridine)	Ic (51.4)	[23]
*3	$(C_4H_6)_2FePF_3$ (H-1, H-1', H-4, H-4' labelled)	Ic (3.2 or 17), IVa dec. above 80°, m.p. 185 to 192° (dec.), subl. p. 73°, orange-yellow crystals, yellow-brown needles (from pentane), red product (from pentane at −80°) ^{1}H NMR (benzene-d_6): −0.55 (m, H-1',4'; J(H-1',P) = 18.5), 0.88 (m, H-1,4), 4.37 (H-2,3) ^{57}Fe-γ: δ = 0.284 UV (cyclohexane): $\bar{\nu}_{max}$ (ε) = 25.2 (180), 34 (sh, 1000), 38 (ssh, 23000), 45 (17000) kK mass spectrum: $[M-nPF_3]^+$ (n = 0, 1), $[M-C_4H_6-PF_3]^+$, $[PF_3]^+$, $[Fe]^+$	[1, 2, 7, 13,14]
*4	$(C_4H_6)_2FeP(OCH_3)_3$ (H-1, H-1', H-4, H-4' labelled)	Ia, Ib, III (17), IVb (59.8 or 30) m.p. 78°, dec. above ca. 80°, orange colored crystals (from pentane at −35°) ^{1}H NMR (benzene-d_6): −0.60 (H-1',4'; J = 15.5), 1.08 (H-1,4; J = 9), 3.60 (d, CH_3; J(H,P) = 10.5), 4.50 (H-2,3; J = 2) ^{31}P NMR: 201.7 (s, invariant with temp.) UV (cyclohexane): $\bar{\nu}_{max}$ (ε) = 24.5 (140), 34 (sh, 2000), 43 (ssh, 15000), 47 (28000) kK mass spectrum: $[M-nC_4H_6]^+$ (n = 0 to 2) and other fragments given	[6, 7, 13, 15, 16, 20, 24, 25]
5	$(C_4H_6)_2FeP(OC_2H_5)_3$	Ib ^{31}P NMR: 197.2 (s)	[24]

References on pp. 212/3

Table 7 [continued]

No.	compound	method of preparation (yield in %), properties and remarks	Ref.
6	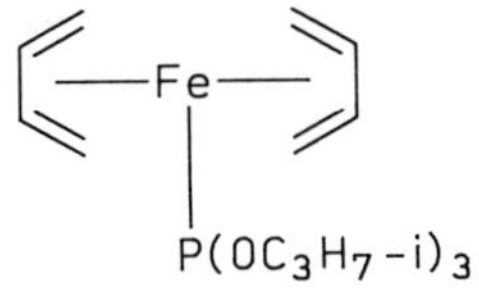 $P(OC_3H_7\text{-}i)_3$	Ib ^{31}P NMR: 188.2 (s)	[24]
*7	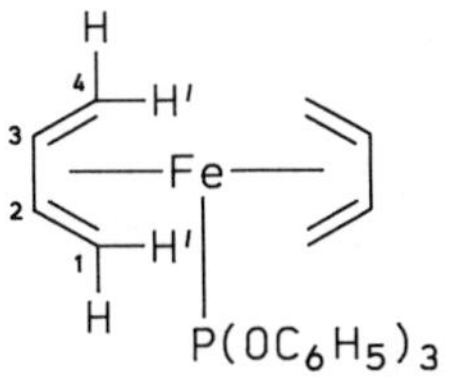 $P(OC_6H_5)_3$	Ib, Ic ^{1}H NMR: −0.43 (dd, H-1′,4′; J(H,H) = 9, J(H,P) = 17), 1.52 (m, H-1,4), 4.36 (t, H-2,3; J(H,H) = 7), 6.86 (t, H-4 in C_6H_5; J(H,H) = 8), 7.02 (t, H-3,5 in C_6H_5; J(H,H) = 8), 7.16 (d, H-2,6 in C_6H_5; J(H,H) = 8) ^{31}P NMR: 181.2 (s)	[23, 24]
*8	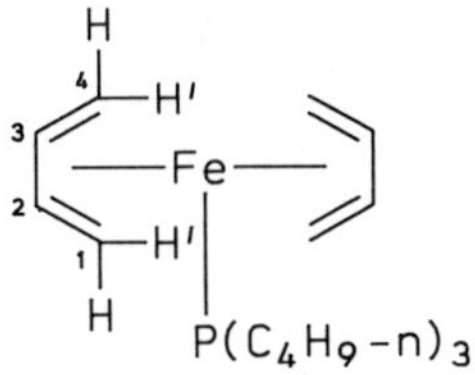 $P(C_4H_9\text{-}n)_3$	IVc (38.5) m.p. 48 to 49°, orange-red needles (from n-pentane at −80°) ^{1}H NMR (benzene-d_6): −1.12 (H-1′,4′; J(H,P) = 13.0), 0.9 to 2.4 (H-1,4, C_4H_9), 4.59 (H-2,3) UV (cyclohexane): $\bar{\nu}_{max}$ (ε) = 28.5 (ssh, 1150), 33 (1850), 46 (21500) kK mass spectrum: $[M]^+$	[7]
9	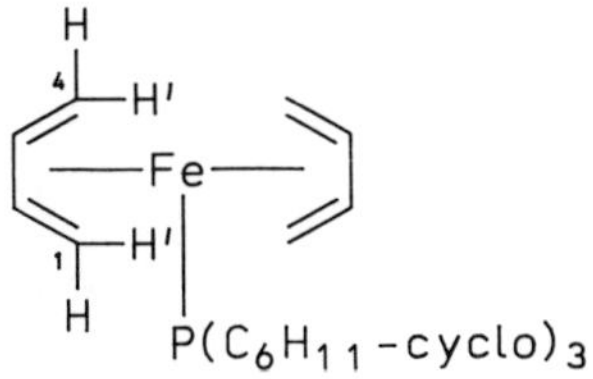 $P(C_6H_{11}\text{-cyclo})_3$	IVc (small amounts) not isolated ^{1}H NMR: −1.1 (H-1′,4′)	[7]
*10	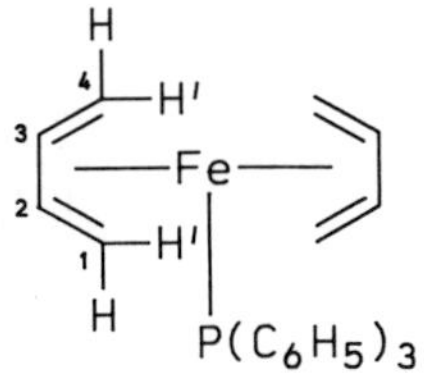 $P(C_6H_5)_3$	Ic (50.2), IIb (30) dec. 105 to 110°, dec. 118 to 125°, very long orange needles (from hexane at −20°) ^{1}H NMR (benzene-d_6): −1.38 (H-1′,4′; J(H-1,1′) = J(H-4,4′) = 1.2, J(H-1′,P) = 13.3), 1.26 (H-1,4; J(H-1,2) = J(H-3,4) = 6.5), 4.57 (H-2,3; J(H-1′,2) = J(H-3,4′) = 8.9) ^{13}C NMR (benzene-d_6): 43.0 (C-1,4; J(P,C) = 10.4), 82.8 (C-2,3), 127.5 (C-3,5 in C_6H_5; J(P,C) = 8.3), 128.7 (C-4 in C_6H_5), 134.6 (C-2,6 in C_6H_5; J(P,C) = 10.4), 140.3 (C-1 in C_6H_5; J(P,C) = 33.3) ^{31}P NMR (benzene-d_6): 88.9 mass spectrum: $[M-nC_4H_6]^+$ (n = 0 to 2), $[M-P(C_6H_5)_3]^+$, $[C_4H_6]^+$, $[P(C_6H_5)_3]^+$	[3, 4, 5, 23]

References on pp. 212/3

Table 7 [continued]

No.	compound	method of preparation (yield in %), properties and remarks	Ref.
11	CH_3 CH_3 Fe $P(OCH_3)_3$	Ib ^{31}P NMR: 192.6	[24]
12	H CH_3 4 H′ 3 Fe 2 1 CH_3 $P(OCH_3)_3$	Ib m.p. 70° ^{1}H NMR: −0.72 (dd, H-4′; J = 9, J(H, P) = 16.5), −0.23 (ddq, H-1′; J = 6, J = 8, J(H, P) = 14), 0.86 (dd, CH_3; J = 6, J(H, P) = 2), 0.92 (m, H-4), 3.61 (d, OCH_3; J(H, P) = 10), 4.22 (m, H-2; J = 5, J = 8, J(H, P) = 1), 4.40 (m, H-3; J = 5, J = 7, J = 9, J(H, P) = 1) ^{31}P NMR: 191.4 (s) mass spectrum: $[M - nC_5H_8]^+$ (n = 0 to 2)	[24]
*13	CH_3 Fe CH_3 $P(OCH_3)_3$	Ib ^{31}P NMR: 205.3	[24]
*14	CH_3 Fe CH_3 $P(OCH_3)_3$	Ib ^{31}P NMR: 206.0	[24]
15	CH_3 Fe CH_3 $P(OC_2H_5)_3$	Ib ^{31}P NMR: 190.4	[24]
16	CH_3 Fe CH_3 $P(OC_2H_5)_3$	Ib ^{31}P NMR: 191.0	[24]

References on pp. 212/3

Table 7 [continued]

No.	compound	method of preparation (yield in %), properties and remarks	Ref.
*17	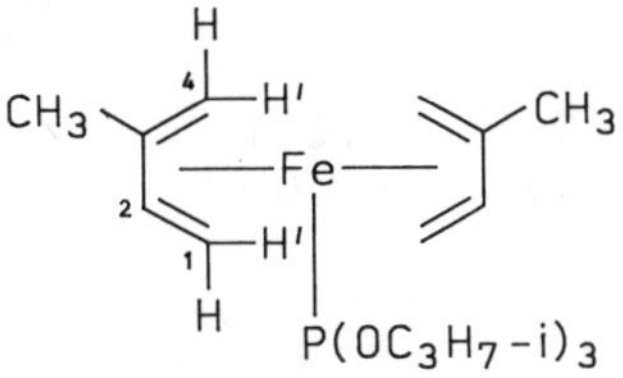 H, H′, CH_3, Fe, CH_3, 4, 2, 1, H′, H, $P(OC_3H_7\text{-}i)_3$	Ib ^{31}P NMR: 183.3 mass spectrum: $[M-nC_5H_8]^+$ (n=0, 1)	[24]
*18	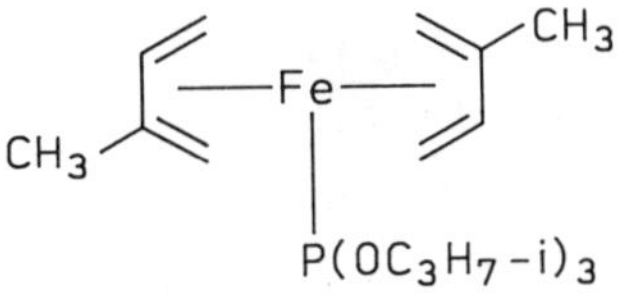 CH_3, Fe, CH_3, $P(OC_3H_7\text{-}i)_3$	Ib ^{31}P NMR: 185.9 mass spectrum: $[M-nC_5H_8]^+$ (n=0, 1)	
19	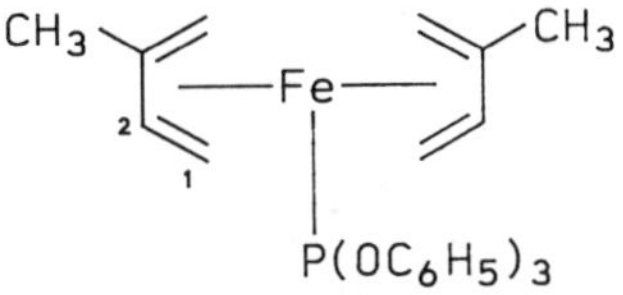 CH_3, Fe, CH_3, 2, 1, $P(OC_6H_5)_3$	Ib ^1H NMR: same as No. 20 except 1.68 (s, CH_3-3), 3.95 (t, H-2; J=9) ^{31}P NMR: 176.9	[24]
20	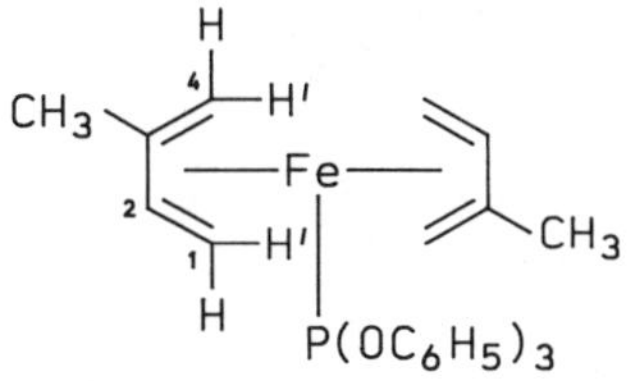 H, H′, CH_3, Fe, 4, 2, 1, H′, H, CH_3, $P(OC_6H_5)_3$	Ib ^1H NMR: −0.58 (dd, H-4′; J(H,P)=18), −0.40 (ddd, H-1′; J=2, J=9, J(H,P)=18), 1.12 (dd, H-1; J=2, J=8), 1.31 (d, H-4), 1.70 (s, CH_3-3), 3.79 (t, H-2; J=9), 6.87 (t, H-4 in C_6H_5; J=8), 7.06 (t, H-3,5 in C_6H_5; J=8), 7.41 (d, H-2,6 in C_6H_5; J=8) ^{31}P NMR: 178.1	[24]
*21	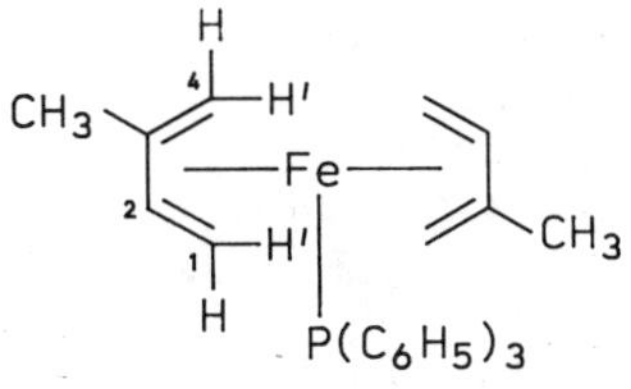 H, H′, CH_3, Fe, 4, 2, 1, H′, H, CH_3, $P(C_6H_5)_3$	Ic (50.5), IIb dec. above 128°, orange-red crystals (from n-hexane at −20°) ^1H NMR (benzene-d_6): −1.88 (H-4′; J(H-4′,P)=16), −1.78 (H-1′; J(H-1,1′)<1, J(H-1′,2)=9, J(H-1′,P)=13), 1.06 (H-4), 1.23 (H-1; J(H-1,2)=7), 1.94 (CH_3-3), 4.07 (H-2) ^{13}C NMR (benzene-d_6): 22.7 (CH_3), 41.2 (C-4; J(C-4,P)=10.0), 42.0 (C-1; J(C-1,P)=12.5), 88.1 (C-2), 94.1 (C-3), 127.4 (C-3,5 in C_6H_5), 128.7 (C-4 in C_6H_5; J(C,P)=9.2), 135.0 (C-2,6 in C_6H_5; J(C,P)=9.6), 140.5 (C-1 in C_6H_5) ^{31}P NMR (benzene-d_6): 90.9 mass spectrum: $[M-nC_5H_8]^+$ (n=0 to 2), $[M-P(C_6H_5)_3]^+$, $[P(C_6H_5)_3]^+$, $[C_5H_8]^+$	[3, 4, 23]

References on pp. 212/3

Table 7 [continued]

No.	compound	method of preparation (yield in %), properties and remarks	Ref.
22	$[(CH_3)_2C_4H_4]_2Fe[P(OCH_3)_3]$	Ib m.p. 75° ^{1}H NMR: 0.74 (d, H-1′,4′; J(H,P) = 20), 0.91 (s, H-1,4), 1.73 (s, CH_3-2,3), 3.43 (d, OCH_3; J(H,P) = 9) ^{31}P NMR: 202.9 mass spectrum: $[M-nC_6H_{10}]^+$ (n = 0, 1), $[M-C_6H_{10}-P(OCH_3)_3]^+$	[24]
23	$[(CH_3)_2C_4H_4]_2Fe[P(OC_2H_5)_3]$	Ib ^{1}H NMR: −0.75 (dd, H-1′,4′; J = 2, J(H,P) = 19.5), 0.95 (d, H-1.4; J = 2), 1.18 (t, CH_3; J = 7), 1.78 (s, CH_3-2,3), 4.03 (quint, CH_2; J = 7, J(H,P) = 7) ^{13}C NMR: 16.4 (d, CH_3; J(C,P) = 6), 19.2 (s, CH_3-2,3), 38.5 (d, CH_2; J(C,P) = 18), 60.2 (d, C-1,4; J(C,P) = 6), 90.0 (d, C-2,3; J(C,P) = 3) ^{31}P NMR: 188.3	[24]
24	$[(CH_3)_2C_4H_4]_2Fe[P(OC_3H_7\text{-}i)_3]$	Ib m.p. 52° ^{1}H NMR: −0.80 (dd, H-1′,4′; J = 2, J(H,P) = 20), 0.95 (d, H-1,4; J = 2), 1.25 (d, $C(CH_3)_2$; J = 6), 1.79 (s, CH_3-2,3), 4.76 (oct, OCH; J = 6, J(H,P) = 6) ^{13}C NMR: 19.3 (s, CH_3-2,3), 24.6 (d, CH_3; J(C,P) = 4), 39.3 (d, OC; J(C,P) = 16), 68.4 (d, C-1,4; J(C,P) = 9), 90.2 (d, C-2,3; J(C,P) = 3) ^{31}P NMR: 187.1	[24]
25	$[(CH_3)_2C_4H_4]_2Fe[P(C_6H_5)_2C_2H_5]$	IIc (traces), V (65) m.p. 134 to 137° (dec.), orange-red crystals (from pentane at −80°), diamagnetic ^{1}H NMR: −1.39 (d, H-1′,4′; J(H-1′,P) = J(H-4′,P) = 18), 0.85 (s, H-1,4; CH_3; J(CH_2, CH_3) = 7.5, J(CH_3, P) = 12), 1.72 (s, CH_3-2,3), 2.56 (CH_2), 7.5 (C_6H_5) mass spectrum: $[M-nC_6H_{10}]^+$ (n = 0 to 2)	[9, 11, 12]
26	$(CH_3C_4H_4CH_3)_2Fe[P(OCH_3)_3]$	Ib m.p. 64° ^{1}H NMR: −0.6 (dqd, H-1,4; J = 6, J = 8, J(H,P) = 15), 0.91 (dd, CH_3-1,4; J = 2, J = 6), 3.75 (d, OCH_3; J(H,P) = 10), 4.42 (dd, H-2,3; J = 2, J = 8)	[24]

References on pp. 212/3

Table 7 [continued]

No.	compound	method of preparation (yield in %), properties and remarks	Ref.
		^{13}C NMR: 18.65 (s, CH_3-1,4), 50.30 (d, OCH_3; J(C,P)=12), 52.48 (d, C-1,4; J(C,P)=7), 79.08 (d, C-2,3; J(C,P)=3) ^{31}P NMR: 187.4 mass spectrum: $[M]^+$	
27	$(C_6H_8)_2Fe$–$P(OCH_3)_3$	IVc	[10]
28	$(C_6H_8)_2Fe$–N (pyridine)	IVc	[10]
29	$(C_6H_8)_2Fe$–N (2,2'-bipyridine)	Ic (49.9)	[23]
supplement:			
*30	$(C_4H_6)_2FeP(CH_3)_3$	IIa (15) dec. p. 126°, subl. p. 33°/vacuum, fine orange needles (from ether at −78°), orange microcrystals (from sublimation) ^1H NMR (benzene-d_6): −1.22 (m, H-1′,4′), 0.92 (m, H-1,4), 1.38 (d, CH_3; J(H,P)=7.6), 4.33 (m, H-2,3) IR (KBr): 402, 465, 540, 648, 668, 710, 753, 838, 845, 900, 925, 942, 1042, 1185, 1208, 1275, 1297, 1418, 1425, 2905, 2970, 3025 mass spectrum: $[M]^+$	[26]

*Further information:

$(C_4H_6)_2FeCNC_4H_9$-t (Table **7**, No. **1**) is prepared by the reaction of ^4L_2Fe (^4L = cycloocta-1,5-diene) with buta-1,3-diene/CNC_4H_9-t [21].

$(C_4H_6)_2FePF_3$ (Table **7**, No. **3**) is monomeric in C_6H_6 (by cryoscopy) [7]. An X-ray structure analysis gives 1.54 Å for the C(2)-C(3) and 1.36 Å for the C(1)-C(2) or C(3)-C(4) distances [8]. The compound does not catalyze the mixed dimerization of butadiene and CH_2=$CHCO_2R$ in the presence of $P(C_6H_5)_3$ [7, 13, 14].

References on pp. 212/3

$(C_4H_6)_2FeP(OCH_3)_3$ (Table **7**, No. **4**) is monomeric in C_6H_6 (by cryoscopy) [7], shows in C_6H_6 a dipole moment of 2.10D [19], and catalyzes at 100 °C [7], in the presence of $P(C_6H_5)_3$ [13, 14], the mixed dimerization of butadiene and CH_2=$CHCO_2CH_3$ to give trans-CH_2=$CH(CH_2)_2$-CH=$CHCO_2CH_3$ [7, 13, 14]. But it does not catalyze the dimerization of butadiene [7].

$(C_4H_6)_2FeP(OC_6H_5)_3$ (Table **7**, No. **7**) does not catalyze the oligomerization of butadiene at 50 °C [23].

$(C_4H_6)_2FeP(C_4H_9$-n$)_3$ (Table **7**, No. **8**) is monomeric in C_6H_6 (by cryoscopy) and catalyzes the mixed dimerization of butadiene and CH_2=$CHCO_2CH_3$ at 100 °C in the presence of $P(C_6H_5)_3$ in a bomb tube. The products are isomeric mixtures of CH_3CH=$CHCH_2CH$=$CHCO_2CH_3$, CH_2=$CH(CH_2)_2CH$=$CHCO_2CH_3$, $CH_3(CH$=$CH)_2CH_2CO_2CH_3$, CH_3O_2CC(=$CH_2)CH_2CH_2CO_2CH_3$, and $CH_3O_2CC_6H_9$ (C_6H_9=cyclohex-3-enyl) [7].

$(C_4H_6)_2FeP(C_6H_5)_3$ (Table **7**, No. **10**) is diamagnetic [3, 4] and shows no temperature dependence in the NMR spectra [3]. Pyrolysis gives mainly butadiene [5]. Reaction with HCl in C_2H_5OH gives buta-1,3-diene and but-1-ene in a ratio of ca. 1:1 [3]. Reaction with CCl_4 gives buta-1,3-diene and cycloocta-1,5-diene (small amounts) [3, 5]. The compound catalyzes the oligomerization of butadiene [23].

$(CH_3C_4H_5)_2FeP(OCH_3)_3$ (Table **7**, Nos. **13** and **14**). The two products are obtained in a ratio of 4:6 (Nos. 13:14) by Method Ib. The product distributions are not at equilibrium since a mixture of crude Nos. 13, 14, and $CH_3C_4H_5Fe(P(OCH_3)_3)_3$ heated at 60 °C for 1 h in the presence of a large excess of $P(OCH_3)_3$ shows a decrease in No. 13 and an increase in No. 14. Neither No. 13 nor 14 reacts with the excess $P(OCH_3)_3$ to give $CH_3C_4H_5Fe(P(OCH_3)_3)_3$ [24].

$(CH_3C_4H_5)_2FeP(OC_3H_7$-i$)_3$ (Table **7**, Nos. **17** and **18**). The two products are obtained in a ratio of 2:8 (Nos. 17:18) by Method Ib. This mixture of products shows in the ^{1}H NMR spectrum chemical shifts at $\delta = -0.9$ (br, H-4′), -0.8 (br, H-1′), 0.8 (br, H-1), 1.0 (br, H-4), 1.25 (br, $C(CH_3)_2$), 1.9 (br, CH_3-3), 3.9 (br, H-2), and 4.80 (br, OCH) ppm [24].

$(CH_3C_4H_5)_2FeP(C_6H_5)_3$ (Table **7**, No. **21**) is diamagnetic and the NMR spectra do not show a temperature dependence. Reaction with HCl in C_2H_5OH gives 3-methylbut-1-ene, isoprene, and 2-methylbut-1-ene [3]. The compound does not catalyze the oligomerization of butadiene [23].

$(C_4H_6)_2FeP(CH_3)_3$ (Table **7**, No. **30**) crystallizes in the monoclinic space group $P2_1/c$-C^5_{2h}. For the ESR spectrum of $(C_4H_6)_2MnP(CH_3)_3$ doped into a single crystal of No. 30 see [26].

References:

[1] D.L. Williams-Smith, L.R. Wolf, P.S. Skell (J. Am. Chem. Soc. **94** [1972] 4042/3). — [2] E. Koerner v. Gustorf, O. Jaenicke, O.E. Polansky (Angew. Chem. **84** [1972] 547/9). — [3] W. Schäfer, A. Zschunke, H.-J. Kerrinnes, U. Langbein (Z. Anorg. Allgem. Chem. **406** [1974] 105/9). — [4] W. Schäfer, H.-J. Kerrinnes, U. Langbein (Z. Anorg. Allgem. Chem. **406** [1974] 101/4). — [5] W. Schäfer, H.-J. Kerrinnes, U. Langbein (Ger. [East] 107862 [1974]).

[6] J. Buchkremer, F.-W. Grevels, O. Jaenicke, P. Kirsch, R. Knoesel, E.A. Koerner v. Gustorf, J. Shields (Intern. Conf. Photochem., Jerusalem 1973, Abstr. No. 7 from [19]). — [7] J. Buchkremer (Diss. Ruhr-Univ. Bochum 1973). — [8] C. Krüger (unpublished results from [7]). — [9] I. Fischler (Diss. Ruhr-Univ. Bochum 1974). — [10] O. Jaenicke, P. Kirsch, R. Rumin, E.A. Koerner v. Gustorf (unpublished results from [9]).

[11] E.A. Koerner v. Gustorf, I. Fischler, R. Wagner (Proc. 16th Intern. Conf. Coord. Chem., Dublin 1974, Abstr. No. 4.20). — [12] I. Fischler, E.A. Koerner v. Gustorf (Z. Naturforsch.

References on pp. 212/3

30b [1975] 291). — [13] J. Buchkremer (Diss. Ruhr-Univ. Bochum 1973 from [17]). — [14] J. Buchkremer (Diss. Ruhr-Univ. Bochum 1973 from [18]). — [15] C.R. Eady, O. Wolfbeis, E.A. Koerner v. Gustorf (unpublished results from [18]).

[16] C.R. Eady, O. Wolfbeis, E.A. Koerner v. Gustorf (unpublished results from [17]). — [17] E.A. Koerner v. Gustorf, O. Jaenicke, O. Wolfbeis, C.R. Eady (New Syn. Methods **3** [1975] 107/33). — [18] E.A. Koerner v. Gustorf, O. Jaenicke, O. Wolfbeis, C.R. Eady (Angew. Chem. **87** [1975] 300/9). — [19] H. Lumbroso, D.M. Bertin (J. Organometal. Chem. **108** [1976] 111/23). — [20] J.R. Blackborow, C.R. Eady, E.A. Koerner v. Gustorf, A. Scrivanti, O. Wolfbeis (J. Organometal. Chem. **111** [1976] C3/C5).

[21] R.E. McKenzie, P.L. Timms (unpublished results from [22]). — [22] P.L. Timms, T.W. Turney (Advan. Organometal. Chem. **15** [1977] 53/112, 95). — [23] V.M. Akhmedov, V.G. Mardanov (U.S.S.R. 695696 [1979]). — [24] S.D. Ittel, F.A. Van-Catledge, J.P. Jesson (J. Am. Chem. Soc. **101** [1979] 3874/84). — [25] O. Jaenicke, R.C. Kerber, P. Kirsch, E.A. Koerner v. Gustorf, R. Rumin (J. Organometal. Chem. **187** [1980] 361/73).

[26] J.M. McCall, J.R. Morton, K.F. Preston (Organometallics **3** [1984] 238/40).

1.4.3.3 Compounds of the Type $^4L(^4L')FeCO$

The compounds treated in this section can be prepared by the following methods:

Method I: Fe atoms and butadiene are co-condensed at −196 °C, and the reaction mixture is warmed to −78 °C. After 1 h at −78 °C the mixture is cooled to −196 °C, and CO is added. Warming to room temperature gives No. 1 [17]. CO addition, before thawing, and chromatography on silica gel with pentane gives No. 1 in better yield [18, 40, 53].

Method II: The Grignard reagent i-C_3H_7MgCl is slowly added at 17 °C to a mixture of $FeCl_3$ and butadiene at −78 °C in ether. After 10 min the reaction mixture is saturated with CO (5 atm) in an autoclave. After 2 h at 0 °C the volatiles are evaporated, the residue is extracted with n-hexane, evaporated again, and the residue is sublimed at 30 °C/0.01 Torr to give No. 1 [1, 14, 20]. Similar reaction of i-C_3H_7MgCl, $FeCl_3$ (molar ratio 3:5), C_8H_8, and the appropriate diolefin for 2 to 8 h at 0 °C (CO pressure 1 to 10 atm) gives the compounds No. 4 to 8. Compound No. 5 can be prepared either from a mixture of (E)- and (Z)-penta-1,3-diene or from pure isomers, but the (Z)-isomer gives the highest yield [20].

Method III: $Fe(CO)_5$ and the appropriate diolefin are irradiated in an inert solvent such as pentane or toluene at low temperatures.

a. $Fe(CO)_5$ and butadiene are irradiated in pentane at −35 °C for 48 h. Inverse filtration at room temperature gives No. 1 (yield 85 to 92%) [10, 12, 33]. Similar irradiation with excess butadiene at 20 °C for 26 h, bubbling Ar through the solution followed by evaporation, and chromatography on silica gel with pentane gives No. 1 in 61% yield [12, 33]. Similar reaction without Ar gives No. 1 in 29% yield [33]. Analogous reaction with cyclohexa-1,3-diene for 22 h at −35 °C gives No. 14 [10, 12]. Photolysis of $Fe(CO)_5$ and isoprene in pentane at −35 °C for 18 h [10, 12] or 48 h [33] followed by evaporation, and chromatography of the residue on Kieselgel with pentane gives No. 7 [10, 12]. The main product is 2-$CH_3C_4H_5Fe(CO)_3$ (2.1% yield) [33]. Similar reaction, but with proportional addition of $Fe(CO)_5$ during the irradiation, gives No. 7 as the main product (51%) and 2-$CH_3C_4H_5Fe(CO)_3$. The latter is separated by distillation, and No. 7 is purified by recrystallization of the residue.

References on pp. 227/9

Similar irradiation with 2,3-dimethylbutadiene followed by cooling gives No. 13 (47% yield); the mother liquor contains 2,3-$(CH_3)_2C_4H_4Fe(CO)_3$ [33]. Analogous irradiation with diethyl muconic acid in pentane at −10 °C followed by filtration gives No. 11. Similarly irradiation with dimethylbutadiene in pentane at −50 °C for 18 h gives No. 13 in 73% yield [33]. Irradiation of $Fe(CO)_5$ with a mixture of (E, E)- and (Z, Z)-hexa-2,4-diene (containing 43.1% of the (E, E)-isomer) in pentane at −50 °C for 24 h, followed by evaporation gives syn,syn-(1,4-$CH_3)_2C_4H_4Fe(CO)_3$. Chromatography of the residue on Kieselgel with pentane gives No. 9 [33].

b. $Fe(CO)_5$, $P(OCH_3)_3$, and butadiene are irradiated in pentane for 27 h. Evaporation and chromatography on silica gel (Serva 50 to 100) with pentane gives No. 1; further elution with C_6H_6 gives $(CO)_3Fe(P(OCH_3)_3)_2$ and $(C_4H_6)_2Fe$-$P(OCH_3)_3$ [33].

c. $Fe(CO)_5$ and $CH_3CH{=}CHCH{=}CHCO_2CH_3$(^{4}L) are irradiated in toluene at −44 °C for 22 h followed by evaporation. The residue is dissolved in ether and chromatographed on Kieselgel [12, 33]. Irradiation for 140 h gives $^4LFe(CO)_3$ as the main product [33].

Method IV: $^4LFe(CO)_3$ compounds (^{4}L = diolefin) and a diolefin are irradiated in pentane or without a solvent at low temperatures.

a. Irradiation of $^4LFe(CO)_3$ with excess butadiene (^{4}L) in pentane at −50 °C for 114 h gives No. 1 [4, 8, 9, 12]. Similar irradiation with ^{4}L = isoprene for 48 h followed by evaporation, and chromatography of the residue on Kieselgel with pentane gives No. 7 [12, 33]. Irradiation with ^{4}L = cyclohexadiene for 95 h followed by filtration and cooling gives No. 14 [12].

b. Irradiation of $^4LFe(CO)_3$ with excess $CH_2{=}C(CH_3)C(CH_3){=}CH_2$ at −35 °C for 18 h followed by filtration and cooling gives No. 13. Similar irradiation with excess $CH_3CH{=}CHCH{=}CHCO_2CH_3$ for 160 h followed by chromatographic work up (Kieselgel, pentane) gives No. 10 [33].

c. Irradiation of $^4LFe(CO)_3$ (^{4}L = butadiene) with cyclohexadiene gives No. 14 [10].

Method V: $[^5L(^4L)FeCO]BF_4$ (^{5}L = cyclohepta-2,4-dienyl, ^{4}L = cyclohexa-1,3-diene) is reduced with $NaBH_4$ (ratio exactly 1:1) at 0 °C in H_2O/ether for 30 min. The organic layer is dried (Na_2SO_4) and evaporated. Recrystallization gives Nos. 17 and 15 in the ratio 9:1. Even at 0 °C small amounts of $^4LFe(CO)_3$ are observed (^{4}L = cyclohexadiene or cycloheptadiene) [36, 49], see also [39]. Similar reduction of $[^5L(^4L)FeCO]BF_4$ (^{5}L = cyclohexa-2,4-dienyl, ^{4}L = cyclohexa-1,3-diene) gives No. 14 [39]. Similar reaction of $[^5L(^4L)FeCO]BF_4$ (^{5}L = cyclohepta-2,4-dienyl, ^{4}L = cyclohexa-1,3-diene) with KCN in H_2O/ether gives No. 18 [39, 49]. Analogous reaction with $LiCH_3$ gives No. 21 and reaction with acetylacetonate gives No. 22 [39]. Analogous reaction of $[C_6H_7(C_8H_8)FeCO]BF_4$ gives No. 20 [39].

Method VI: $(C_8H_8)_2Fe$ and butadiene react in n-hexane at 0 °C overnight. The solution is carbonylated at 0 °C under CO (2 atm) for about 5 h to give No. 4 [1, 13].

Table 8
Compounds of the Type ${}^4L({}^4L')FeCO$.
Further information on numbers preceded by an asterisk is given at the end of the table, pp. 220/7.
For abbreviations and dimensions see p. VIII.

No.	compound	method of preparation (yield in %), properties and remarks	Ref.
*1	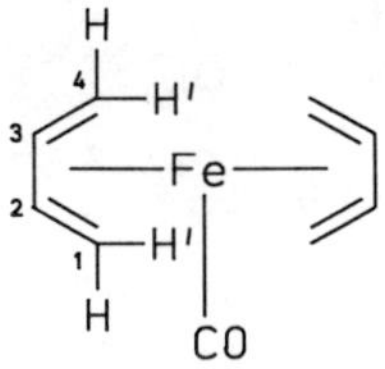	I (ca. 3 to 31), II (30), III a (61 to 92), III b (24.5), IV a (21) m.p. 130° (dec.), m.p. 130 to 135° (dec.), m.p. 130 to 137° (dec.), subl. 30°/0.01 Torr, long orange-red or dark red crystals (from pentane at −78°), shining orange needles, or needle-like tetragonal crystals (from hexane) ^{1}H NMR (benzene-d_6): −0.35 (m, H-1′,4′), 1.04 (m, H-1,4), 4.35 (m, H-2,3) ^{13}C NMR (benzene-d_6): 113, 155 (^{1}J(1,2) = 46.1) ^{57}Fe NMR (C_6H_6, relative to $Fe(CO)_5$): 1411.7 ± 0.5 ^{57}Fe-γ (85.0 ± 1.0 K): $\delta = 0.41 \pm 0.02$, $\Delta = 0.27 \pm 0.05$ (relative to Fe) ^{57}Fe-γ: $\delta = 0.296$, $\Delta = 0.351$ IR (KBr): 415, 492, 537, 545, 584, 660, 777, 795, 898, 922, 950, 1053, 1061, 1192, 1216, 1375, 1434; 1475, 1482 (C_4H_6); 1955 (CO); 2927, 3000, 3049, 3059 IR (n-hexane): 1984.5 (CO) UV (n-hexane): $\bar{\nu}_{max}$ (ε) = 24.7 (180), 32 (sh, 950), 36.5 (sh, 2400), 45 (20000) kK mass spectrum: $[M - n\,CO]^+$ (n = 0, 1), $[M - CO - n\,C_4H_6]^+$ (n = 1, 2)	[1, 2, 4, 8 to 10, 12, 14 to 21, 27, 33, 50, 51, 53, 63, 67 to 69]
*2	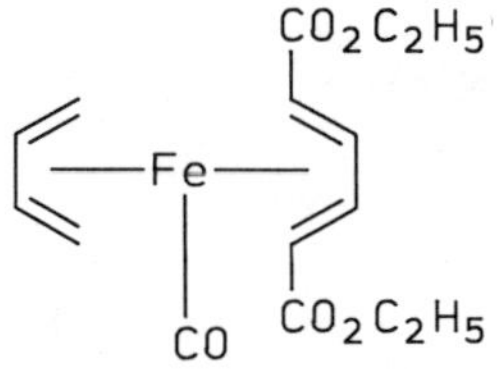	—	[48]
3	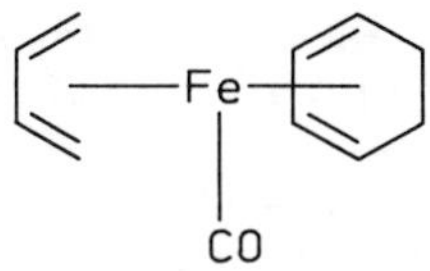	see No. 1, p. 222 IR: 1974 (CO) mass spectrum: $[M]^+$	[33]
*4	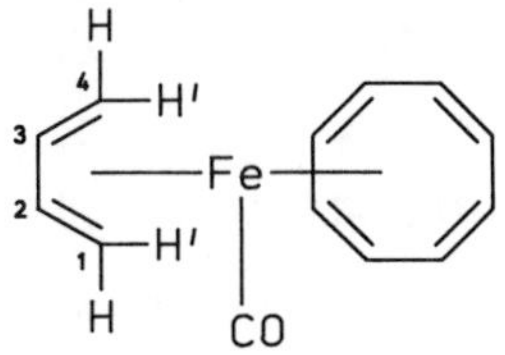	II (ca. 25), VI (ca. 25) m.p. 131°, brown crystals (from hexane at −78°), dark brown needle-like crystals ^{1}H NMR (toluene-d_8): −0.1 (H-1′,4′), 1.7 (H-1,4), 4.3 (H-2,3), 4.5 (s, C_8H_8) ^{57}Fe-γ (85.0 ± 1.0 K): $\delta = 0.42 \pm 0.02$, $\Delta = 0.32 \pm 0.05$	[1, 5, 13, 15, 20]

References on pp. 227/9

Table 8 [continued]

No. compound	method of preparation (yield in %), properties and remarks	Ref.
	IR (KBr): 431, 447, 515, 540, 563, 617, 654, 694, 711, 773, 781, 853, 857, 884, 918, 948, 1051, 1057, 1109, 1189, 1228, 1289, 1302, 1412 to 1426 (complexed C=C), 1474; 1476 (C=C in butadiene), 1485; 1486 (C=C in butadiene), 1571 (noncomplexed C=C), 1922; 1967 (CO), 2982, 3016, 3050, 3069 UV (n-hexane): λ_{max} (log ε) = 205 (4.329), 255 (4.042), 290 (3.974) nm	
*5	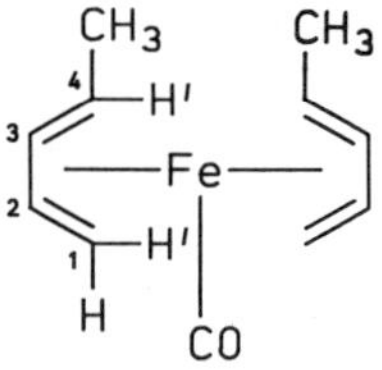 II m.p. ca. 25°, low melting red-orange solid ^{1}H NMR (benzene-d_6): −0.4 (H-1′), 0.20 (H-4′), 0.75 (CH_3), 0.85 (H-1), 4.4 (H-2, 3) IR (KBr): 417, 460, 510, 540, 580, 660, 897 to 922, 948, 990, 1027, 1055, 1193, 1223, 1375, 1446 (C=C); 1482 (C=C); 1960 (CO) UV (n-hexane): λ_{max} (log ε) = 206 (4.857), 232 (4.715) nm mass spectrum: $[M]^+$ and fragments given	[1, 14, 20]
*6	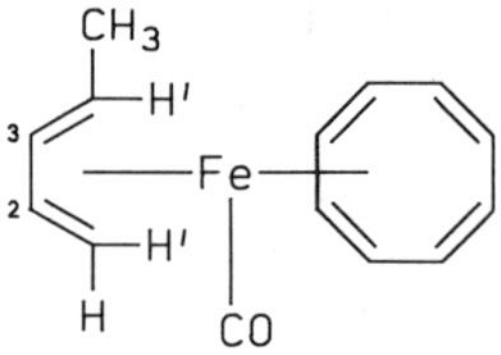 II (20 to 30) m.p. 104°, brown crystals ^{1}H NMR (benzene-d_6): −0.15 (m, H-4′), 0.5 (m, H-1′), 1.1 (m, CH_3), 1.6 (m, H-1), 4.2 (m, H-2, 3), 4.5 (s, C_8H_8) IR (KBr): 653, 693, 709, 775, 860, 882, 898, 957, 1028, 1198, 1235, 1372, 1414, 1431 to 1453, 1475, 1570, 1963 UV (n-hexane): λ_{max} (log ε) = 209 (4.045), 265 (3.767), 292 (3.754) nm mass spectrum: $[M]^+$ and fragments given	[20]
*7	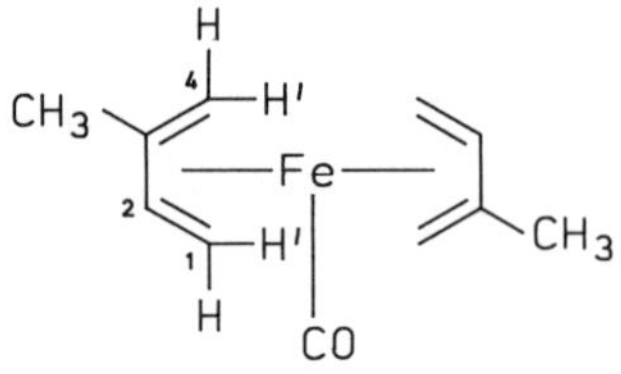 II, IIIa (2.1, 51), IVa (39 to 50) m.p. 99 to 104° (dec.), m.p. 115°, orange needles (from pentane at −78 or −30°) ^{1}H NMR (benzene-d_6): −0.2 (H-1′, 4′), 1.4 (H-1, 4), 1.75 (CH_3), 4.0 (H-2) ^{1}H NMR (benzene-d_6): −0.50 (m, H-4′), −0.38 (m, H-1′; J(H-1′, 2) ≈ 9.5), 0.97 (m, H-4), 1.10 (m, H-1; J(H-1, 1′) ≈ 2), 1.68 (s, CH_3), 3.90 (m, H-2; J(H-1, 2) ≈ 7) ^{57}Fe-γ: δ = 0.348, Δ = 0.371 IR (KBr): 817, 925, 1027, 1198, 1378; 1425, 1470 (C=C); 1965 (CO) IR (hexane): 1980 (CO)	[1, 10, 12, 14, 20, 27, 33, 63, 67]

References on pp. 227/9

Table 8 [continued]

No.	compound	method of preparation (yield in %), properties and remarks	Ref.
		UV (n-hexane): $\tilde{\nu}_{max}$ (ε) = 24.6 (200), 32 (ssh, 1000), 34.5 (ssh, 1500), 44 (sh, 18500), 47.2 (25000) mass spectrum: $[M]^+$ and other fragments given	
*8	CH_3, H, H′, Fe, CO, C_8H_8	II (20 to 30) m.p. 96°, brown needles (from n-hexane at −78°) ^{1}H NMR (benzene-d_6): −0.25 (m, H-1′,4′), 1.5 (m, H-1,4; CH_3), 4.1 (m, H-2), 4.4 (s, 8H) IR (KBr): 693, 709, 777, 818, 862, 1031, 1204, 1380; 1417 to 1424 (C=C in C_8H_8); 1453, 1477 (C=C in isoprene); 1575; 1963 (CO) UV (n-hexane): λ_{max}(log ε) = 207 (4.373), 258 (4.116), 292 (4.078) nm mass spectrum: $[M]^+$ and fragments given	[20]
9	CH_3 CH_3 / Fe / CH_3 CO CH_3	III a (9.5) orange-red crystals (from n-pentane at −78°) IR (n-hexane): 1956 (CO); k_{co} = 15.46 mdyn/Å	[33]
*10	CO_2CH_3 CH_3 / Fe / CH_3 CO CO_2CH_3	III c (23.5 to 24.2), IV b (33.1) m.p. 163.5° (dec.), m.p. 164° (dec.), m.p. 170° (dec.), dark red-brown crystals (from ether at −78°) ^{1}H NMR (benzene-d_6): −0.55 (m, H-4), 0.80 (m, H-1), 1.05 (d, CH_3-1), 3.24 (s, OCH_3), 4.90 (m, H-2), 6.32 (m, H-3) ^{57}Fe-γ: δ = 0.344, Δ = 1.177 IR (n-hexane): 1984 (CO) UV (n-hexane): $\tilde{\nu}_{max}$(ε) = 22 (ssh, 320), 27 (sh, 1700), 38.5 (19500), 44.5 (16500) kK mass spectrum: $[M]^+$	[12, 27, [33]
*11	$C_2H_5O_2C$ $CO_2C_2H_5$ / Fe / $C_2H_5O_2C$ CO $CO_2C_2H_5$	III a (83.5) m.p. 128° (dec.), deep brown-red crystals (from toluene at −30°) ^{1}H NMR (benzene-d_6): 0.24 (m, H-1,4), 1.15 (t, CH_3; J(CH_2, CH_3) = 7.1), 4.10 (q, CH_2), 5.90 (m, H-2,3) ^{57}Fe-γ: δ = 0.397, Δ = 0.961 IR (Nujol): 1690 (C=O), 2032 (CO) IR ($CHCl_3$ at −5°): 2018 (CO)	[27, 33, 55]

References on pp. 227/9

Table 8 [continued]

No.	compound	method of preparation (yield in %), properties and remarks	Ref.
		UV (ether): $\bar{\nu}_{max}(\varepsilon)=21.5$ (ssh, 260), 29 (sh, 2200), 38.4 (33600), 46 (sh, 2400) kK mass spectrum: $[M]^+$	
12	$C_2H_5O_2C$, CH_3, CH_3, Fe, CH_3, $C_2H_5O_2C$, CO, CH_3	see No. 11, p. 225	[46]
*13	H, 4, H′, CH_3, CH_3, Fe, CH_3, CH_3, 1, H′, H, CO	III a (47 to 73), IV b (73) m.p. 136° (dec., from pentane) ^{1}H NMR (benzene-d_6): −0.33 (d, H-1′,4′), 1.14 (d, H-1,4; J(H-1,1′) = J(H-4,4′) ≈ 2), 1.60 (CH_3-2,3) ^{57}Fe-γ: δ = 0.313, Δ = 0.375 IR (n-hexane): 1975 (CO) UV (n-hexane): $\bar{\nu}_{max}(\varepsilon)=24.1$ (320), 32 (850), 36 (ssh, 2500), 42.5 (sh, 15000), 46 (21000) kK mass spectrum: $[M]^+$	[12, 33, 50]
*14	4, 5, 3, Fe, 6, 2, 1, CO	III a (37 to 40), IV a (39.6), IV c, V m.p. 134 to 136° (dec.), orange-red crystals or needles (from pentane at −78°) ^{1}H NMR (benzene-d_6): 1.05 (m), 1.95 (m, H-5,6); 2.54 (m, H-1,4), 4.34 (m, H-2,3) ^{57}Fe-γ: δ = 0.230, Δ = 0.535 IR (n-hexane): 1964.5 (CO) UV (n-hexane): $\nu_{max}(\varepsilon)=25.0$ (130), 32 (1200), 35 to 37 (ssh, 3800), 43.7 (20000), 48 (28500) kK mass spectrum: $[M]^+$	[6, 10, 12, 27, 33, 39, 50]
*15	4, 5, 3, Fe, 6, 2, 1, CO, 4′, 5′, 3′, 6′, 2′, 1′, 7′	V, see No. 17, p. 226 red product ^{1}H NMR (benzene-d_6): 1.10 (m), 1.70 (m), 2.00 (m), 2.40 (m, H-1,4,1′,4′,5,6,5′,6′,7′); 4.15 (m, H-2,3,2′,3′) ^{13}C NMR (n-heptane): 23.8 (t, C-6′; J(C,H) = 130); 24.8 (t), 29.8 (t) (C-5,6,5′,7′; J(C,H) = 115, J(C,H) = 125); 60.1 (d), 61.7 (d, C-1,4,1′,4′; J(C,H) = 140, J(C,H) = 160); 82.4 (d, C-2,3; J(C,H) = 165), 83.7 (d, C-2′,3′; J(C,H) = 165), 230.9 (CO) IR (n-hexane): 1961 (CO) mass spectrum: $[M]^+$	[36, 39, 49]

References on pp. 227/9

Table 8 [continued]

No.	compound	method of preparation (yield in %), properties and remarks	Ref.
16	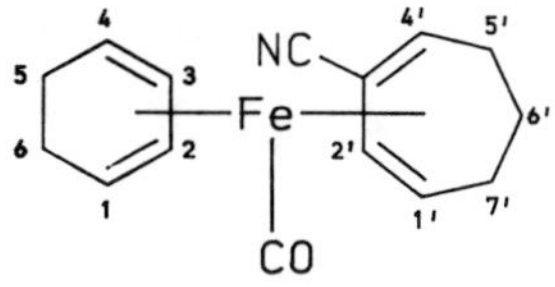 NC, Fe, CO (ring positions 1–6; 1′–7′)	see No. 18, p. 227 red solid (from pentane at −78°) ^{1}H NMR (benzene-d_6): 1.00 (m), 1.70 (m), 2.40 (m, H-1,4,1′,4′,5,6,5′,6′,7′); 4.50 (m, H-2,3,2′) IR (n-hexane): 1977 (CO) mass spectrum: $[M]^+$	[49]
*17	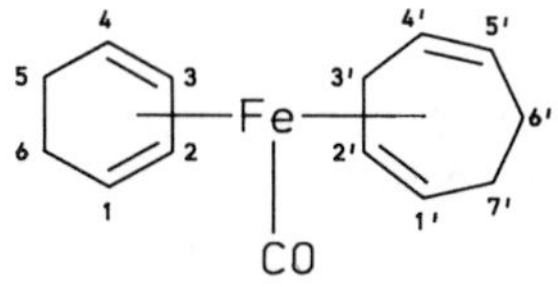 Fe, CO (ring positions 1–6; 1′–7′)	V (59) orange product (from pentane at −78°) ^{1}H NMR (benzene-d_6): 1.15 (m), 1.70 (m), 2.09 (m), 2.29 (m), 2.71 (m, H-1,1′,2′,3′,4,4′,5,5′,6,6′,7′); 5.69 (d of d at 10°, H-2,3) ^{13}C NMR (n-heptane): 22.8 (t, C-3′; J(C,H) = 130); 24.4 (t), 27.6 (t, C-5,6,6′,7′; J(C,H) = 125); 37.1 (d), 52.8 (d, C-1′,2′,4′,5′; J(C,H) = 150, J(C,H) = 165); 71.2 (d, C-1,4; J(C,H) = 180), 89.5 (d, C-2,3; J(C,H) = 170) IR (n-hexane): 1952 (CO) mass spectrum: $[M]^+$	[36, 39, 49]
*18	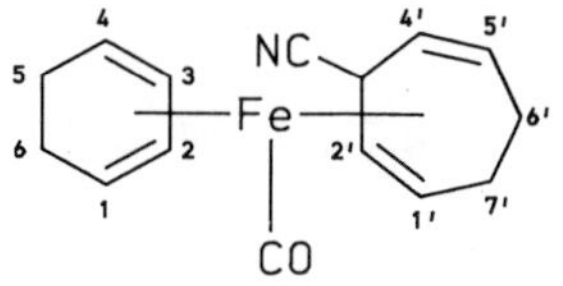 NC, Fe, CO (ring positions 1–6; 1′–7′)	V (30) pink solid ^{1}H NMR (benzene-d_6): 0.93 (m); 1.20 (m), 1.54 (m), 2.25 (m), 2.70 (m, H-1,1′,2′,4,4′,5,5′,6,6′,7′), 3.72 (t, H-3′), 5.37 (d of d at 10°, H-2,3) IR (n-hexane): 1968 (CO) mass spectrum: $[M]^+$	[39, 49]
19	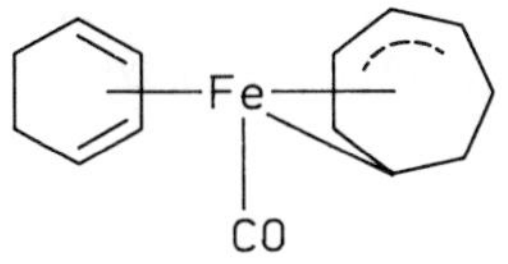 Fe, CO	see No. 15, p. 226, and No. 17, p. 226 not isolated, only in mixtures with No. 15 ^{13}C NMR (n-heptane): 20.3, 29.2, 32.8, 54.7, 73.2, 95.1, 97.9 IR (n-heptane): 1946 (CO)	[36, 49]
*20	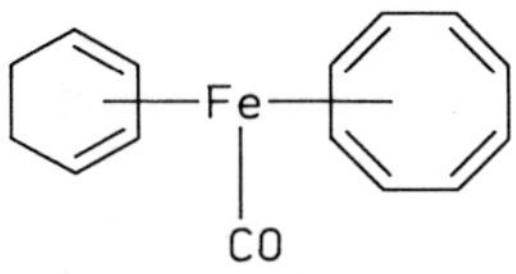 Fe, CO	V	[39]
21	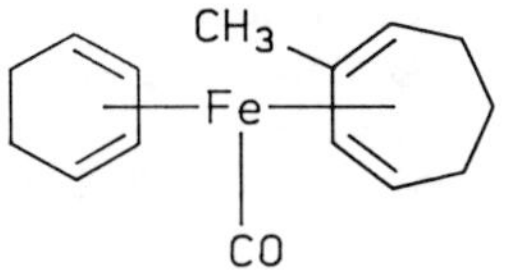 CH_3, Fe, CO	V	[39]

References on pp. 227/9

Table 8 [continued]

No.	compound	method of preparation (yield in %), properties and remarks	Ref.
22	$(C_6H_8)Fe(CO)(C_7H_{10}CH(COCH_3)_2)$ (structure: cyclohexadiene–Fe(CO)–cycloheptadienyl bearing $HC(COCH_3)_2$)	V	[39]

*Further information:

$(C_4H_6)_2FeCO$ (Table **8**, No. **1**) is also formed by a matrix technique. An inert polymer, e.g., polytetrafluoroethylene, is treated with $Fe(CO)_5$ and butadiene and is irradiated. These films can be handled under normal ambient conditions [60, 63, 67]. The compound is also formed by reaction of $(1,5\text{-}C_8H_8)_2Fe$ with C_4H_6/CO [53]. The compound is monomeric in C_6H_6 (by cryoscopy) [8, 33]. The heat of sublimation for crystalline No. 1, measured at 399 to 439 K and corrected to 298 K, is $\Delta H^{298}_{subl} = 18.2$ kcal/mol (76.1 kJ/mol [45, 48, 56]). The enthalpy of formation is $\Delta H^{298}_{f} = -6.2$ kcal/mol (−25.9 kJ/mol) [45, 48, 52, 56], and the enthalpy of disruption is $\Delta H^{298}_{D} < 118.3$ kcal/mol (482 kJ/mol) [48, 56]. The $C_4H_6 \rightarrow Fe$ bond enthalpy contribution is at 298 K <45 kcal/mol (from thermal decomposition), 42.6 kcal/mol (from iodination) [48] or 44.0 kcal/mol (from combustion) [45, 48, 56]. The dipole moment in C_6H_6 at 25 °C is $\mu = 2.15$ D [24, 33, 50]. The IR and Raman spectra of solid No. 1 and of No. 1 in different solvents is shown in the following table [19], cf. [74]:

IR spectrum			Raman spectrum			assignment
solid	CS_2 soln.	C_6D_6 soln.	solid	C_6H_6 soln.	C_6D_6 soln.	
	3066 m		3059 m			CH_2 str. (antisym.)
3052 m	3041 m				3045 w, pol.	CH_2 str. (sym.)
3001 m	3006 m		3001 m		3003 w, pol.	CH str. (sym.) + CH_2 str. (antisym.)
2932 m	2931 s				2934 w	CH str. (antisym.)
					2901 w, ? pol.	
	2658 vw					
2561 vw	2557 vw					
2416 vw	2452 vw					
1951 vs	1976 vs	(1982 vs, C_6H_{12})	1973 w			C≡O str.
	1932 w					
1480 w		1482 w	1483 w		1481 vw	C=C str. (sym.)
1471 m		1474 w				CH_2 scissor (sym.)
1431 m			1435 w			C=C str. (antisym.)
1372 m			1373 w(br)			CH_2 scissor (antisym.)
1216 m	1216 m		1227 w	1231 m, pol.	1225 m	CH bend. or C–C str.
1191	1189 w					CH bend.
1053 m	1053 w		1053 s	1056 m, pol.	1053 m	C–C str. or CH bend.
1021 vw						CH_2 twist

References on pp. 227/9

IR spectrum			Raman spectrum			assignment
solid	CS_2 soln.	C_6D_6 soln.	solid	C_6H_6 soln.	C_6D_6 soln.	
970 vw	969 vw					CH_2 rock
946 w	940 vw					CH_2 twist
916 m	915 w		925 w	925 m, pol.		CH_2 rock+wag
893 m	875 m					CH_2 wag
815 vw						
788 w						
773 m	766 w					CH wag
654 w	651 w					CH wag
580 s	576 s		588 w			δ(FeCO), B_2 or B_1
541 s } 532 s }	532 s		540 w			δ(FeCO), B_1 or B_2
488 m	483 w		478 w		473 w	i.p. skeletal bend
457 w			453 w			i.p. skeletal bend
					?427 vw	o.o.p. skeletal bend (A_2)
402 s		403 w				o.o.p. skeletal bend (B_1)
390 s		394 w				(C_4H_6)-Fe str. (B_2)
			364 vs	361 vs, pol.		Fe-(CO) str. (A_1)
			339 w	335 m, dp.	332 m	(C_4H_6)-Fe tilts
			301 vs	297 vs, pol.		(C_4H_6)-Fe str. (A_1)
			218 w			? torsion

The force constant for the CO group is $k_{co}=15.91$ mdyn/Å [33]. The He(I) photoelectron spectrum shows ionization potentials at 6.95 ($10a_1$), 7.8 ($8b_1$), 8.25 ($6b_2$, $11a_1$), 9.15 ($7b_2$), 9.55 ($5a_2$), 10.63 ($7b_1$), and 11.50 ($9a_1$) eV [66, 71], cf. [38]. The assignment is based on INDO calculations using the ΔSCF procedure and the transition operator model [66]. Compound No. 1 crystallizes in the tetragonal [4, 16, 21] space group $P\bar{4}2_1m$-D^3_{2d}, V^3_d [2 to 4, 7, 16, 21] with the unit cell parameters a=7.748, b=7.221 Å [4], a=b=7.73±0.03, c=7.21±0.02 Å [21], a=7.76, c=7.33 Å [3, 7], or a=7.78(1), c=7.24(1) Å [16]; Z=2 molecules per unit cell [4, 16, 21]. $D_{meas}=1.46$ g/cm³, $D_{calc}=1.48$ g/cm³ [21]. The two types of C-C bonds in the butadiene ligand are equal in length with about 1.45 Å [4]. The main bond distances and angles are shown in **Fig. 31** [16].

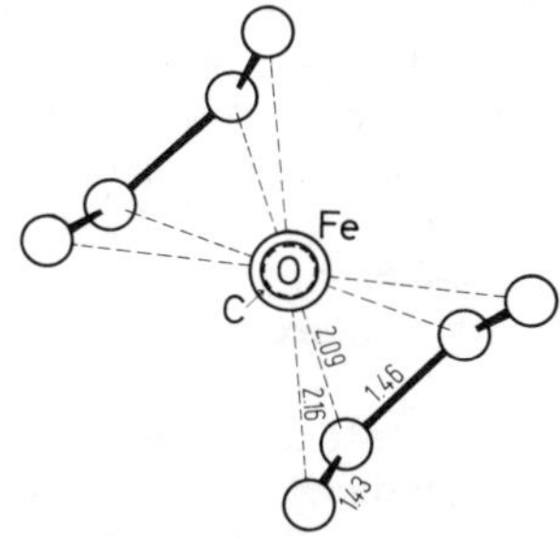

Fig. 31. Molecular structure of $(C_4H_6)_2FeCO$ (No. 1) [16].

References on pp. 227/9

The ground state properties of compound No. 1 are discussed in [66, 71, 72].

Compound No. 1 is a surprisingly fairly [5, 20] thermally [20] stable solid [5, 10, 20]. It is at room temperature fairly stable towards air [10, 16, 20] and H_2O [20], fairly stable in alcoholic solutions [20], and is very soluble in hydrocarbons [5, 20]. It decomposes above the melting point [1, 20, 45, 48], yielding cycloocta-1,5-diene (large amounts), and butadiene (little) [1, 20]. Oxidation with $[NH_4]_2[Ce(NO_3)_6]$ in C_2H_5OH at 0 °C within 2 h gives butadiene (83.6% yield) and 4-vinylcyclohexene (9.7%) [33]. It has been found to efficiently quench the photoreduction of fluorenone by $N(C_2H_5)_3$. The rate constant for the quenching of triplet fluorenone is $k=4.5\times10^{-9}\ M^{-1}\cdot s^{-1}$ [37]. Reaction with aqueous HCl in n-hexane gives mainly $FeCl_2$ (other compounds formed are impure) [20]. Reaction with gaseous HBr in pentane gives $FeBr_2$, 2-bromobutane (6.2%), 1-bromobut-2-ene (2.9%), and an oily residue, possibly containing $(C_4H_7)_2FeBr_2$, $(C_4H_7)_2Fe(CO)Br$, and $C_4H_7Fe(CO)_3Br$ (C_4H_7=butadienyl). Similar reaction in a CO atmosphere gives $FeBr_2$ and $C_4H_7Fe(CO)_3Br$ [33]. The compound No. 1 does not react with 2,2′-dipyridyl at 100 °C [20]. Irradiation with PF_3 in pentane gives $(C_4H_6)_2FePF_3$ [33]. Irradiation in the presence of $P(OCH_3)_3$ in pentane gives different products, dependent of the wavelength [22, 25, 29, 33, 34, 58, 65]. Irradiation at ≧475 nm gives exclusively $C_4H_6Fe(P(OCH_3)_3)_2CO$ [22, 33]. Irradiation at ≧366 nm gives $(C_4H_6)_2FeP(OCH_3)_3$ (30%), $C_4H_6Fe(P(OCH_3)_3)_2CO$ (40%), and $C_4H_6Fe(P(OCH_3)_3)_3$ [25, 29, 33, 34, 58, 65]; but at ≧254 nm, $(C_4H_6)_2FeP(OCH_3)_3$ is obtained in 60% yield [33]. Thus, photolysis results in both carbonyl (Φ_{CO}) and diene (Φ_D) replacement. The quantum yields Φ for several wavelengths are shown in the following table (in cyclohexane):

wavelength (nm)	No. 1 (mmol/L)	$P(OCH_3)_3$ (mol/L)	Φ	Φ_{CO}	Φ_D	Ref.
254	5.7	0.2	—	0.14	0.03	[25, 58]
	5 to 12	0.2	0.15	0.12	0.03	[65]
313	5 to 12	0.2	0.12	0.10	0.02	[65]
366	5 to 12	0.2	0.05	0.04	0.007	[65]
405	5.7	0.2	0.019	0.01	0.01	[25, 58, 65]

The Φ_{CO} values are not directly measurable and are therefore obtained by difference [65]. Irradiation with $P(C_4H_9\text{-}n)_3$ in pentane gives $(C_4H_6)_2FeP(C_4H_9\text{-}n)_3$. Similar photolysis with $P(C_6H_{11}\text{-cyclo})_3$ gives small amounts of $(C_4H_6)_2FeP(C_6H_{11}\text{-cyclo})_3$, not separable from excess $P(C_6H_{11}\text{-cyclo})_3$ [33]. A first report, [20], says that no reaction with $P(C_6H_5)_3$ or $P(C_6H_5)_2CH_2CH_2P(C_6H_5)_2$ occurs at 100 °C [20]. But refluxing compound No. 1 with $P(C_6H_5)_2CH_2CH_2P(C_6H_5)_2$(^{4}D) in C_6H_6 for 19 h gives C_4H_6Fe(^{4}D)CO [47]. Reaction with an excess of CO in toluene at 120 °C gives butadiene and cycloocta-1,5-diene [1]. Reaction with CCl_4 at room temperature gives butadiene (small amounts) and cycloocta-1,5-diene (90 to 95%) [1, 5, 20]. But-2-yne gives red-brown needles containing $(CH_3)_6C_6$ and CO [20]. Reaction with cyclooctatetraene gives brown compounds of variable compositions [20]. Irradiation with cyclohexa-1,3-diene gives a mixture of $(C_6H_8)_2FeCO$ (No. 14) and $C_4H_6(C_6H_8)FeCO$ (No. 3) [10, 33]. Refluxing with cyclohexa-1,3-diene for 7 h gives a similar mixture (not separated) [33]. The catalytic oligomerization of butadiene in solution gives a variety of dimers (e.g., 4-vinylcyclohexene, cycloocta-1,5-diene, 3-methylhepta-1,3,6-triene, octa-1,3,6-triene), trimers (n-dodecatriene), and higher oligomers. The relative amounts of each fraction vary from solvent to solvent [1, 5, 9, 10, 20, 28, 33]. Compound No. 1 induces the polymerization of butadiene in the presence of protic acids such as HCl or CCl_3CO_2H [20]. Octa-1,5-diene is obtained in the presence of $P(C_6H_5)_3$ [9, 10, 33]. For the mixed dimerization of butadiene with $CH_2{=}CHCO_2CH_3$ in the presence of No. 1 and

References on pp. 227/9

$P(C_6H_5)_3$ see [9, 10, 31, 33, 34]. For the oligomerization of cyclohexa-1,3-diene, 5-vinylcyclohexa-1,3-diene, cycloocta-1,3-diene, or cycloocta-1,5-diene in the presence of No. 1 and $P(C_6H_5)_3$ see [33]. For mixed dimerizations of a variety of diolefins with monoolefins see [33].

$C_4H_6(1,4\text{-}(C_2H_5O_2C)_2C_4H_4)FeCO$ (Table **8**, No. **2**). Thermal decomposition at 473 K gives Fe (solid), $(C_4H_6)_2$ (liquid), and syn,syn-$C_2H_5O_2CC_4H_4CO_2C_2H_5Fe(CO)_3$ with ΔH^{298}_{dec} ~3.3 kcal/mol. This enthalpy value is probably too low due to incomplete decomposition at 473 K. Decomposition at 527 to 533 K gives ΔH^{298}_{dec} ~8 kcal/mol. For the idealized decomposition to $C_{10}H_{14}O_4$(gas), butadiene (gas), CO (gas), and Fe (solid), ΔH^{298}_{dec} is estimated to be ca. 44 kcal/mol. Vacuum sublimation at 391 and 403 K gives ΔH^{298}_{subl} ~26.0 ± 1 kcal/mol [48].

$C_4H_6(C_8H_8)FeCO$ (Table **8**, No. **4**). The peak at $\delta = 4.5$ ppm in the ^1H NMR spectrum first broadens and coalesces (from −10 to −20 °C) and then splits (at about −65 °C) into three broad signals centered at about $\delta = 2.35$, 3.47, and 5.84 ppm (ratio 2:2:4). Therefore, the C_8H_8 ligand is fluxional at room temperature [1]. The compound crystallizes in the monoclinic space group $P2_1/m\text{-}C^2_{2h}$ with the unit cell parameters $a = 6.95 \pm 0.02$, $b = 10.82 \pm 0.03$, $c = 7.28 \pm 0.02$ Å, $\beta = 98°38' \pm 20'$; Z = 2 molecules per unit cell. $D_{meas} = 1.43$ g/cm³, $D_{calc} = 1.49$ g/cm³. The main bond distances and angles are shown in **Fig. 32** [21].

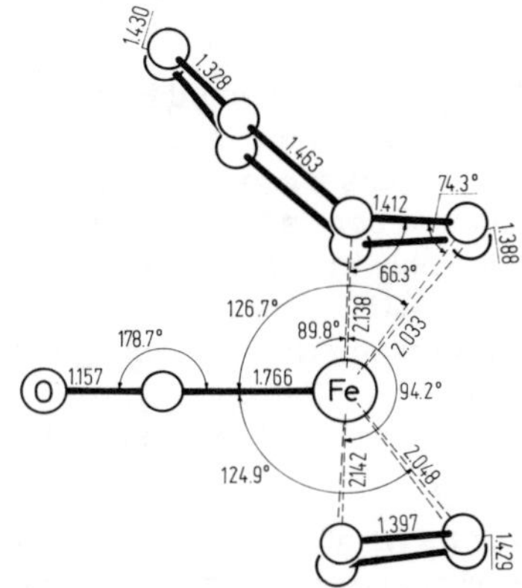

Fig. 32. Molecular structure of $C_4H_6(C_8H_8)FeCO$ (No. 4) [21].

A topological Hückel model is developed which accounts for the bond lengths in this and other η^4-olefin complexes [57].

The compound is fairly stable [1] and decomposes in CCl_4 to give 1:1 adducts of butadiene and C_8H_8 (not characterized) [20]. The compound is an effective catalyst in the oligomerization of butadiene to give 4-vinylcyclohexene and cycloocta-1,5-diene [1, 5, 13], with minor amounts of n-dodecatetraene [20].

$(1\text{-}CH_3C_4H_5)_2FeCO$ (Table **8**, No. **5**) is surprisingly stable, decomposes smoothly above its melting point, but is fairly stable in C_2H_5OH solutions and towards air and H_2O. Reaction with CCl_4/1 wt% n-nonane/1 wt% thiophene gives 3,4-dimethylcycloocta-1,5-diene (50 to 55%) and (E)-penta-1,3-diene (ca. 40%). Reaction in toluene with 0.2 M HCl/H_2O in n-hexane at −78 °C gives mainly $FeCl_2$ and other compounds which are impure and are not isolated [20]. The compound shows a poor catalytic activity towards butadiene. In the presence of protons, butadiene is polymerized [11, 20].

References on pp. 227/9

1-$CH_3C_4H_5(C_8H_8)FeCO$ (Table **8**, No. **6**) is surprisingly stable, decomposes smoothly above its melting point, but is fairly stable in C_2H_5OH solutions and towards air and H_2O. It is very soluble in hydrocarbons. Decomposition in CCl_4 gives 1:1 adducts of C_8H_8 and penta-1,3-diene (not characterized) [20]. Butadiene is oligomerized to vinylcyclohexene and cycloocta-1,5-diene with minor amounts of n-dodecatetraene [11, 20]. But-2-yne and norbornadiene at 70 to 90 °C give $C_6(CH_3)_6$ and polycyclic dimers, respectively. Polymerization of butadiene in the presence of protic acids transforms No. 6 into a stable allylic derivative [20].

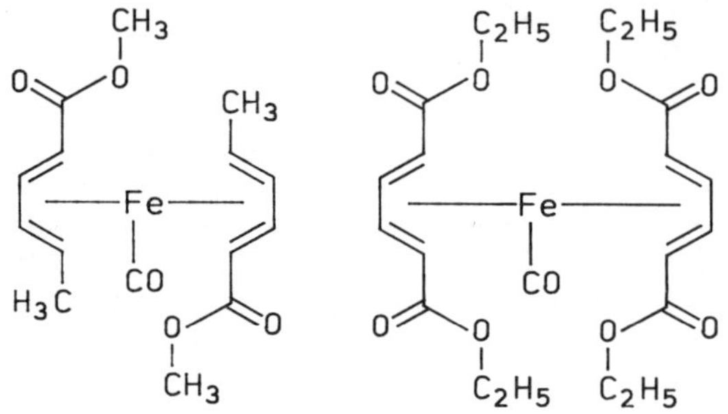

(2-$CH_3C_4H_5)_2FeCO$ (Table **8**, No. **7**) is also formed by a matrix technique. An inert polymer, e.g., polytetrafluoroethylene, is treated with $Fe(CO)_5$ and isoprene and irradiated. These films can be handled under normal ambient conditions [63, 67]. The compound is monomeric in C_6H_6 (by cryoscopy). The CO force constant is 15.84 mdyn/Å [33]. The compound is surprisingly stable, decomposes smoothly above its melting point, is fairly stable in alcoholic solutions and towards air and H_2O, and is very soluble in hydrocarbons. Solutions in chlorinated solvents decompose at room temperature. Decomposition in CCl_4 (containing 1 wt% of n-nonane) at room temperature for 12 h gives 1,5-dimethylcycloocta-1,5-diene in 60 to 90% yield [20]. The compound also readily induces the polymerization of butadiene in the presence of protic acids [11, 20].

2-$CH_3C_4H_5(C_8H_8)FeCO$ (Table **8**, No. **8**) is surprisingly stable, decomposes smoothly above its melting point, is fairly stable in alcoholic solutions and towards air and H_2O, and is very soluble in hydrocarbons. Decomposition in CCl_4 gives 1:1 adducts of C_8H_8 and isoprene (not characterized) [20]. The compound readily induces the polymerization of butadiene in the presence of protic acids [11, 20].

(1-$CH_3(4\text{-}CH_3O_2C)C_4H_4)_2FeCO$ (Table **8**, No. **10**) is monomeric in C_6H_6 (by cryoscopy) [33] and has in C_6H_6 a dipole moment of 1.74 D [24] or 1.78 D. From the possible conformers, I is considered as the predominant one. Thermal decomposition at 473 K gives syn,syn-1-$CH_3(4\text{-}CH_3O_2C)C_4H_4Fe(CO)_3$, and the enthalpy of decomposition is $\Delta H^{298}_{dec} \sim 6.2$ kcal/mol. This value for the decomposition is ca. 9.5 kcal/mol at 513 K. Both values are too low due to incomplete decomposition. The enthalpy for the idealized decomposition according to No. 10 (cryst.) $\rightarrow 2C_7H_{10}O_2$(gas) + CO(gas) + Fe(cryst.) is estimated to $\Delta H^{298}_{dec} \sim 45.5$ kcal/mol [48]. Vacuum sublimation at 414 K gives $\Delta H^{298}_{subl} \sim 28.2$ kcal/mol (118.0 kJ/mol) [48, 56]. The force constant for the CO group is 15.91 mdyn/Å [33]. The trans-configuration shown in Table 8, p. 217, is confirmed by X-ray analysis [23].

(1,4-$(C_2H_5O_2C)C_4H_4)_2FeCO$ (Table **8**, No. **11**) is monomeric in C_6H_6 [33] and shows in C_6H_6 a dipole moment of 2.58 D [24] or 2.65 D [50]. The force constant for the CO group is $k_{CO} = 16.46$ mdyn/Å [33]. A distorted conformation of Type II is roughly in agreement with an X-ray analysis of No. 11 [51]. Reaction with monodentate ligands ^{2}D (^{2}D = pyridine, pyrazine, or quinoline) in toluene for 24 to 25 h gives complexes of the type syn,syn-1,4-$(C_2H_5O_2C)_2C_4H_4Fe(CO)_2{}^2D$. The yields increase up to 75% in a CO atmosphere [70]. Reaction with ^{2}D = PR_3 at 20 °C gives syn,syn-1,4-$(C_2H_5O_2C)_2C_4H_4Fe(PR_3)_2CO$ (R not specified) [46]. Reaction with bidentate ligands ^{4}D (^{4}D = 2,2′-dipyridyl [55, 61], or III (R = H, or OCH_3) [55] gives complexes of the type syn,syn-1,4-$(C_2H_5O_2C)_2C_4H_4Fe(^4D)CO$ [55, 61]. No reaction is observed with 4-$CH_3OC_6H_4N$=CHCH=$NC_6H_4OCH_3$-4 [73]. Reaction with 3,4-dimethylhexa-2,4-diene gives No. 12 [46].

III IV V

(2,3-$(CH_3)_2C_4H_4)_2$FeCO (Table **8**, No. **13**) is monomeric in C_6H_6 (by cryoscopy) [33], and shows in C_6H_6 a dipole moment of 2.96 D [24] or 2.98 D [50]. The force constant for the CO group is k_{CO} = 15.76 mdyn/Å [33]. The He (I) photoelectron spectrum results in the vertical ionization potentials 6.53 (10 a_1), 7.2 (8 b_1), 7.85 (11 a_1, 6 b_2), 8.55 (7 b_2), 8.95 (5 a_1), 9.75 (7 b_1), and 10.45 (9 a_1) eV. These values are compared with those of Nos. 1 and 14. The assignment is based on INDO calculations using the ΔSCF procedure and the "transition operator" model [66].

$(C_6H_8)_2$FeCO (Table **8**, No. **14**). Treatment of $[^5L(^4L)FeCO]BF_4$ (^{5}L = cyclohexadienyl, ^{4}L = cyclohexadiene) with $NaBD_4$ as in Method V gives the monodeuterated compound No. 14 deuterated in exo position at C-5 [39]. The compound No. 14 is monomeric in C_6H_6 (by cryoscopy), and shows in C_6H_6 a dipole moment of 2.04 D [24] or 2.05 D [50]. Thermal decomposition at 417 and 505 K according to No. 14 (cryst.) → Fe (cryst.) + CO (gas) + 2 C_6H_8 (gas) gives ΔH^{298}_{dec} ~ 41.5 kcal/mol. A similar value of ca. 44 kcal/mol is derived from the iodination process at 503 K. Vacuum sublimation at 402 and 414 K (some decomposition at 414 K) gives ΔH^{298}_{subl} = 22.7 ± 0.5 kcal/mol (95.0 kJ/mol) [48, 56]. The force constant for the CO group is k_{CO} = 15.59 mdyn/Å [33]. The He (I) photoelectron spectrum shows vertical ionization potentials at 6.53 (10 a_1), 7.5 (8 b_1), 7.83 (11 a_1, 6 b_2), 8.28 (7 b_2), 8.76 (5 a_2), 9.98 (7 b_1), and 10.90 (9 a_1) eV. These values are compared with those of Nos. 1 and 13. The assignment is based on INDO calculations using the ΔSCF procedure and the "transition operator"

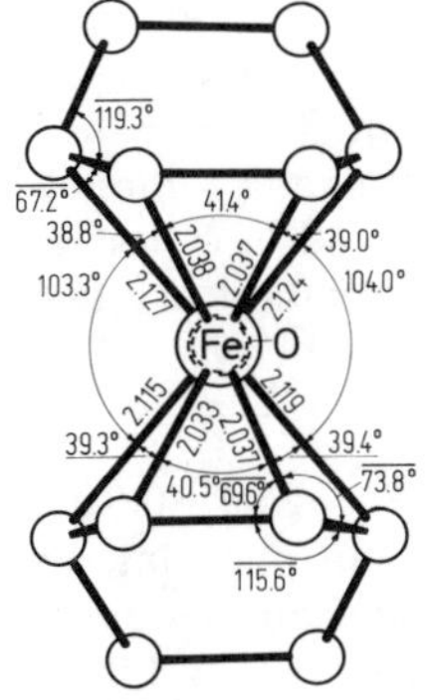

Fig. 33. Molecular structure of $(C_6H_8)_2$FeCO (No. 14) [6, 7].

References on pp. 227/9

model [66]. The compound crystallizes in the orthorhombic space group $P2_12_12_1$-D_2^4, V^4 with the unit cell parameters a = 17.089 ± 0.005, b = 8.495 ± 0.004, c = 7.545 ± 0.004 Å; Z = 4 molecules per unit cell. D_{calc} = 1.48 g/cm³. The main bond distances and angles are shown in **Fig. 33**, p. 225. The rings are arranged around Fe with the saturated part pointing towards the CO group. Although the diene entities are planar, a torsional angle of 3° remains in the strained ring systems. Both rings differ slightly in geometry as well as in their bonding to Fe [6, 7].

Irradiation of No. 14 with N_2 [42, 43] in tetrahydrofuran at −30 °C for 4 h gives a pyrophoric brown precipitate, and 76% of No. 14 is recovered [42]. Similar results are obtained by irradiation with pyridine or CH_3CN/N_2 in toluene overnight or with N_2H_4 [42]. Irradiation with $P(OCH_3)_3$ in cyclohexane results in the formation of $(C_6H_8)_2FeP(OCH_3)_3$ (see 1.4.3.2, Table 7, No. 27, p. 211) and $C_6H_8Fe(P(OCH_3)_3)_2CO$ [64, 65]. Thus, photolysis results in both CO (Φ_{CO}) and diene (Φ_D) replacement. The quantum yields Φ for several wavelengths are shown in the following table (for 5 to 12 mM No. 14 and 0.2 M $P(OCH_3)_3$) [65]:

wavelength (nm)	Φ	Φ_{CO}	Φ_D
254	0.15	0.11	0.04
313	0.10	0.09	0.01
366	0.05	0.05	0.004
405	0.01	0.01	0.003

The Φ_{CO} values are not directly measurable and are therefore obtained by difference [65]. Compound No. 14 oligomerizes butadiene at 100 °C to give 4-vinylcyclohexene and cycloocta-1,5-diene; in the presence of $P(C_6H_5)_3$, cyclododeca-1,5,9-triene and other trimers of butadiene are obtained [10]. No. 14 also dimerizes cyclohexadiene catalytically in the presence of $P(C_6H_5)_3$ at 100 °C in an autoclave [33].

$C_6H_8(C_7H_8O)FeCO$ (Table **8**, No. **15**). The chemical shift at δ = 230.9 (CO) ppm in the ¹³C NMR spectrum is obtained in the presence of 0.04 mol/dm³ $Cr(acac)_3$ (acac = acetylacetonate) as relaxation agent. Heating the compound at 90 °C for several days gives small amounts of No. 19. No reaction occurs with $P(C_6H_5)_3$ in n-heptane at 30 to 50 °C [49]. Reaction with CO (80 atm) at 60 °C gives $C_6H_8Fe(CO)_3$ [36].

$C_6H_8(C_7H_{10})FeCO$ (Table **8**, No. **17**). Reduction of $[^5L(^4L)FeCO]BF_4$ with $NaBD_4$ according to Method V, p. 214, gives No. 17 deuterated in the exo-3′ position [49]. Compound No. 17 isomerizes in the solid state [49] or in solution [36, 49]. Treatment of No. 17-d_1 for 1 h at 60 °C gives No. 15 deuterated at C-2′, and C-3′ by an endo isomerization. Treatment of this deuterated derivative of No. 15 with $[(C_6H_5)_3C]BF_4$ gives undeuterated $[^5L(^4L)FeCO]BF_4$ (^5L = cyclohepta-2,4-dienyl, ^4L = cyclohexa-1,3-diene) [49]. Warming compound No. 17 in n-heptane solutions gives Nos. 15 and 19 in a first-order process [36, 49]. The first-order rate constants for this reaction in n-heptane are shown in the following table (selected) for No. 17 and its monodeuterated derivative No. 17-d_1 (k_{calc} is derived from the activation parameters) [49], see Table 8, p. 227.

The activation parameters are $\Delta H^{\neq}$ = 25.1 ± 0.3 kcal/mol and $\Delta S^{\neq}$ = 3.2 ± 0.9 cal · K^{-1} · mol^{-1} [36, 49]. No. 19 is not isolated but only detected in mixtures with No. 15. The ratio of Nos. 15 to 19 is not markedly temperature dependant being 6 ± 1 : 1 at 50 °C and 9 : 1 at 90 °C. Sublimation of No. 17 at 60 °C within 1 h gives No. 15 (70%). Compound No. 17 is very O_2 sensitive above room temperature [49]. The 1,4-diene ligand is easily replaced

t in °C	No. 17 $k_{obs} \cdot 10^5$ (s^{-1})	$k_{calc} \cdot 10^5$ (s^{-1})	No. 17-d_1 $k_{obs} \cdot 10^5$ (s^{-1})
40.0	9.64	9.90	—
50.0	33.9	35.4	—
50.0	35.2	35.4	—
50.0	39.2	35.4	—
60.0	115	117	112
60.0	117	117	115
60.0	119	117	116
60.0	124	117	112
60.0	127	117	—
70.0	348	362	—
70.0	355	362	—
70.0	372	362	—

by tertiary phosphines or cyclooctatetraene [39]. It reacts in a second-order process with $P(C_6H_5)_3$ in n-heptane. The derived rate constants are shown in the following table (k_1 is obtained from the intercepts of plots of k_{obs} against $[P(C_6H_5)_3]$ and corresponds to the rate of unimolecular isomerization; k_{iso} is the rate constant for the isomerization of No. 17 in the absence of $P(C_6H_5)_3$ calculated from the above given data) [49].

t in °C	$k_2 \cdot 10^3$ ($L \cdot mol^{-1} \cdot s^{-1}$)	$k_1 \cdot 10^4$ (s^{-1})	$k_{iso} \cdot 10^4$ (s^{-1})
30.0	3.02 ± 0.70	0.6 ± 0.4	0.3 ± 0.1
40.0	9.49 ± 1.36	1.8 ± 0.6	1.0 ± 0.1
50.0	26.4 ± 3.6	4.7 ± 0.7	3.5 ± 0.1

The activation parameters for this reaction are $\Delta H^{\neq} = 20.37 \pm 0.68$ kcal/mol and $\Delta S^{\neq} = 3 \pm 2$ cal · K^{-1} · mol^{-1} [49]. Reaction with CO (1 atm) at 20 °C gives $C_6H_8Fe(CO)_3$ [36, 49].

$C_6H_8(NCC_7H_9)FeCO$ (Table **8**, No. **18**). Treatment at 60 °C in C_6H_6 for 1.5 h gives No. 16 (47%), which is consistent with an exo position for the CN-substituent and an endo-metal hydrogen transfer [49]. The complex undergoes a subsequent rearrangement to give the cyclohepta-1,3-diene derivative No. 16. The 1,4-diene ligand is easily replaced by tertiary phosphines or cyclooctatetraene [39].

$C_6H_8(C_8H_8)FeCO$ (Table **8**, No. **20**) shows a temperature dependent ^1H NMR spectrum; C_8H_8 exhibits four resonances at −85 °C. Protonation results in complex IV, and reaction with $[(C_6H_5)_3C]BF_4$ gives V in complete analogy with $C_8H_8Fe(CO)_3$ [39].

References:

[1] A. Carbonaro, A. Greco (J. Organometal. Chem. **25** [1970] 477/81). — [2] R.E. Davis, G.L. Cupper, H.D. Simpson (Am. Cryst. Assoc. Summer Meeting, Ottawa 1970, Abstr. No. N 4 from [10]). — [3] R.E. Davis, G.L. Cupper, H.D. Simpson (Am. Cryst. Assoc. Summer Meeting, Ottawa 1970, Abstr. No. N 4 from [7]). — [4] R.E. Davis, G.L. Cupper, A.D. Simpson (Am. Cryst. Assoc. Summer Meeting, Ottawa 1970, Abstr. No. N 4 from [20]). — [5] A. Carbonaro (Proc. 13th Intern. Conf. Coord. Chem., Cracow 1970, Vol. 2, pp. 347/8).

References on pp. 227/9

[6] C. Krüger, Y.-H. Tsay (J. Organometal. Chem. **33** [1971] 59/67). — [7] C. Krüger, Y.-H. Tsay (Angew. Chem. **83** [1971] 250/1). — [8] E. Koerner v. Gustorf, Z. Pfajfer, F.-W. Grevels (Z. Naturforsch. **26b** [1971] 66/7). — [9] E.A. Koerner v. Gustorf, F.W. Grevels, J. Buchkremer, Z. Pfajfer, M.K. Das (5th Intern. Conf. Organometal. Chem., Moscow 1971, Vol. 2, Abstr. No. 305). — [10] E. Koerner v. Gustorf, J. Buchkremer, Z. Pfajfer, F.-W. Grevels (Angew. Chem. **83** [1971] 249/50).

[11] A. Carbonaro, F. Cambisi (5th Intern. Conf. Organometal. Chem., Moscow 1971, Vol. 2, Abstr. No. 387). — [12] E. Koerner v. Gustorf, J. Buchkremer, Z. Pfajfer, F.W. Grevels, Studienges. Kohle m.b.H. (Ger. 2105627 [1971/72]). — [13] Montedison S.p.A. (Ital. 907325 [1972]). — [14] Montedison S.p.A. (Ital. 907326 [1972]). — [15] P. Mag, L. Korecz, A. Carbonaro, K. Burger (Radiochem. Radioanal. Letters **9** [1972] 137/42).

[16] D.A. Whiting (Cryst. Struct. Commun. **1** [1972] 379/81). — [17] D.L. Williams-Smith, L.R. Wolf, P.S. Skell (J. Am. Chem. Soc. **94** [1972] 4042/3). — [18] E. Koerner v. Gustorf, O. Jaenicke, O.E. Polansky (Angew. Chem. **84** [1972] 547/9). — [19] G. Davidson, D.A. Duce (J. Organometal. Chem. **44** [1972] 365/70). — [20] A. Carbonaro, F. Cambisi (J. Organometal. Chem. **44** [1972] 171/80).

[21] I. Bassi, R. Scordamaglia (J. Organometal. Chem. **37** [1972] 353/9). — [22] E. Koerner v. Gustorf, P. Kirsch (unpublished results from [33]). — [23] C. Krüger (unpublished results from [33]). — [24] H. Lumbroso (unpublished results from [33]). — [25] P. Kirsch, J. Buchkremer, O. Jaenicke, R. Knoesel, R. Rumin, J. Shields, E. Koerner v. Gustorf (Meeting Photochem. Group Ger. Chem. Soc., Göttingen 1973, Abstr. No. 5 from [26]).

[26] C.R. Bock, E.A. Koerner v. Gustorf (Advan. Photochem. **10** [1977] 221/310). — [27] R.H. Herber (unpublished results from [33]). — [28] J. Buchkremer (Diss. Ruhr-Univ. Bochum 1973 from [48]). — [29] J. Buchkremer, F.-W. Grevels, O. Jaenicke, P. Kirsch, R. Knoesel, E.A. Koerner v. Gustorf, J. Shields (Intern. Conf. Photochem., Jerusalem 1973, Abstr. No. 7 from [30]). — [30] R.C. Kerber, E.A. Koerner v. Gustorf (J. Organometal. Chem. **110** [1976] 345/57).

[31] J. Buchkremer (Diss. Ruhr-Univ. Bochum 1973 from [32]). — [32] E. Koerner v. Gustorf, O. Jaenicke, O. Wolfbeis, C.R. Eady (New Syn. Methods **3** [1975] 107/33). — [33] J. Buchkremer (Diss. Ruhr-Univ. Bochum 1973). — [34] J. Buchkremer (Diss. Ruhr-Univ. Bochum 1973 from [35]). — [35] E.A. Koerner v. Gustorf, O. Jaenicke, O. Wolfbeis, C.R. Eady (Angew. Chem. **87** [1975] 300/9).

[36] B.F.G. Johnson, J. Lewis, T.W. Matheson, I.E. Ryder, M.V. Twigg (J. Chem. Soc. Chem. Commun. **1974** 269/70). — [37] A. Gilbert, J.M. Kelly, E. Koerner v. Gustorf (Mol. Photochem. **6** [1974] 225/30). — [38] R. Gleiter, D.O. Cowan (unpublished results from [37]). — [39] J. Ashley-Smith, D.V. Howe, B.F.G. Johnson, J. Lewis, I.E. Ryder (J. Organometal. Chem. **82** [1974] 257/60). — [40] M. Yevitz, P.S. Skell (Intern. Union Cryst. Intercongr. Symp. Intra-Intermol. Forces, University Park, Pa., 1974, Ser. 2, Vol. 2, Abstr. No. 9 from [41]).

[41] K.J. Klabunde, T.O. Murdock (Chem. Technol. [N.Y.] **5** [1975] 624/8). — [42] I. Fischler (Diss. Ruhr-Univ. Bochum 1974). — [43] I. Fischler (Diss. Ruhr-Univ. Bochum 1974 from [44]). — [44] I. Fischler, K. Hildenbrand, E. Koerner v. Gustorf (Angew. Chem. **87** [1975] 35/7). — [45] F.A. Adedeji, D.L.S. Brown, J.A. Connor, M. Leung, M.I. Paz-Andrade, H.A. Skinner (4th Conf. Intern. Thermodyn. Chim. Compt. Rend., Montpellier, France, 1975, Vol. 1, pp. 85/90; C.A. **84** [1976] No. 112518).

[46] F.-W. Grevels, U. Feldhoff, K. Schneider (7th Intern. Conf. Organometal. Chem., Venice 1975, Abstr. p. 192). — [47] C.B. Ungermann, K.G. Caulton (J. Organometal. Chem. **94**

[1975] C9/C13). — [48] D.L.S. Brown, J.A. Connor, M.L. Leung, M.I. Paz-Andrade, H.A. Skinner (J. Organometal. Chem. **110** [1976] 79/89). — [49] B.F.G. Johnson, J. Lewis, I.E. Ryder, M.V. Twigg (J. Chem. Soc. Dalton Trans. **1976** 421/5). — [50] H. Lumbroso, D.M. Bertin (J. Organometal. Chem. **108** [1976] 111/23).

[51] E.A. Koerner v. Gustorf (unpublished results from [50]). — [52] G. Pilcher (unpublished results from [48]). — [53] R.E. McKenzie, P.L. Timms (unpublished results from [54]). — [54] P.L. Timms, T.W. Turney (Advan. Organometal. Chem. **15** [1977] 53/112, 95). — [55] M.-A. De Paoli, H.-W. Frühauf, F.-W. Grevels, E.A. Koerner v. Gustorf, W. Riemer, C. Krüger (J. Organometal. Chem. **136** [1977] 219/33).

[56] J.A. Connor (Top. Current Chem. **71** [1977] 71/110, 100). — [57] D.M.P. Mingos (J. Chem. Soc. Dalton Trans. **1977** 20/5). — [58] P. Kirsch, R.C. Kerber, R. Rumin, E.A. Koerner v. Gustorf (unpublished results from [59]). — [59] C.R. Bock, E.A. Koerner v. Gustorf (Advan. Photochem. **10** [1977] 221/310). — [60] M.-A. De Paoli (9th Intern. Conf. Organometal. Chem., Dijon 1979, Abstr. No. D 6).

[61] H.-W. Frühauf (unpublished results from [62]). — [62] H.-W. Frühauf, F.-W. Grevels, A. Landers (J. Organometal. Chem. **178** [1979] 349/60). — [63] M.-A. De Paoli (Quim. Nova **3** [1980] 112/3). — [64] K. Schneider (unpublished results from [65]). — [65] O. Jaenicke, R.C. Kerber, P. Kirsch, E.A. Koerner v. Gustorf, R. Rumin (J. Organometal. Chem. **187** [1980] 361/73).

[66] M.C. Böhm, R. Gleiter (Chem. Ber. **113** [1980] 3647/55). — [67] M.-A. De Paoli, S.M. De Oliviera, F. Galembeck (J. Organometal. Chem. **193** [1980] 105/10). — [68] T. Jenny, W. von Philipsborn, J. Kronenbitter, A. Schwenk (J. Organometal. Chem. **205** [1981] 211/22). — [69] S. Zobl-Ruh, W. von Philipsborn (Helv. Chim. Acta **64** [1981] 2378/82). — [70] M.-A. De Paoli, M. Makita (J. Organometal. Chem. **216** [1981] 79/85).

[71] M.C. Böhm, R. Gleiter (Theor. Chim. Acta **59** [1981] 153/79). — [72] M.C. Böhm, M. Eckert-Maksić, R.D. Ernst, D.R. Wilson, R. Gleiter (J. Am. Chem. Soc. **104** [1982] 2699/707). — [73] H.-W. Frühauf, G. Wolmershäuser (Chem. Ber. **115** [1982] 1070/82). — [74] I.C. Hisatsune (Spectrochim. Acta A **40** [1984] 391/5).

1.4.3.4 Compounds of the Type $^4L(^4L')Fe^2L$

$(C_4H_6)_2FeC_4H_6$ (C_4H_6 = butadiene) is obtained as a red-brown [10] product by the co-condensation of Fe atoms with butadiene [1 to 3, 9, 11, 12]. It can also be obtained by the reaction of ^4L_2Fe (^4L = cycloocta-1,5-diene) with butadiene [10]. The composition is given as $(C_4H_6)_nFe$ (n = 2 or 3) [10]. It decomposes at +5 °C [9], or −5 °C [10]. The initial spectrum of the co-condensation reaction contains absorption bands of an unstable π-complex with identifiable peak frequencies (cm^{-1}) at 510, 635, 705, 870, 1185, 1480, and 1598. This species, believed to be $(C_4H_6)_2Fe$, decays at about −100 °C into a black polymer formulated as $(FeC_4H_9)_n$ which shows the following bands at −190 °C in the IR spectrum: 310 ν(Fe-C), 355 ν(Fe-C), 424 δ(C=C-C), 445 δ(C=C-C), 492, ca. 525 τ(CH_2), 630, 675 τ(CH_2), 870, 888 ϱ(CH_2), ca. 950 ω(CH), ca. 1040 ω(CH), ca. 1180 ν(C-C), 1220 ϱ(CH), 1335 ϱ(CH), 1373 δ(CH_2), 1430 δ(CH_2), ca. 1465 ν(C=C), 1612 ν(C=C), ca. 2850, ca. 2920 ν(CH), ca. 3000 $\nu_s(CH_2)$,

Fe

I

ca. 3040 ν(CH) cm^{-1}. The polymer is stable to about −15 °C, but it reacts with O_2 at −120 °C and gives butadiene and a brown solid which shows no absorption peaks of organic groups [12]. Structure I is proposed for the product obtained in the matrix [2, 5, 7].

Reaction with PR_3 (R=F [1, 2, 5], OCH_3 [5, 7]) gives $(C_4H_6)_2FePR_3$ [1, 2, 5, 7]. Similar reaction with CO gives $(C_4H_6)_2FeCO$ [1, 3, 5]. Reaction with NCC_4H_9-t gives $(C_4H_6)_2FeNCC_4H_9$-t [10]. Decomposition of the above mentioned $(FeC_4H_9)_n$ in the presence of CO gives a red-brown residue whose IR bands are close to those reported for $(C_4H_6)_2FeCO$ [12].

References:

[1] D.L. Williams-Smith, L.R. Wolf, P.S. Skell (J. Am. Chem. Soc. **94** [1972] 4042/3). — [2] E.A. Koerner v. Gustorf, O. Jaenicke, O.E. Polansky (Angew. Chem. **84** [1972] 547/9). — [3] M. Yevitz, P.S. Skell (Intern. Union Cryst. Intercongr. Symp. Intra-Intermol. Forces, University Park, Pa., 1974, Ser. 2, Vol. 2, Abstr. No. E 9 from [4]). — [4] K.J. Klabunde, T.O. Murdock (Chem. Technol. [N.Y.] **5** [1975] 624/8). — [5] C.R. Eady, O. Wolfbeis, E.A. Koerner v. Gustorf (unpublished results from [6]).

[6] E.A. Koerner v. Gustorf, O. Jaenicke, O. Wolfbeis, C.R. Eady (Angew. Chem. **87** [1975] 300/9). — [7] C.R. Eady, O. Wolfbeis, E.A. Koerner v. Gustorf (unpublished results from [8]). — [8] E.A. Koerner v. Gustorf, O. Jaenicke, O. Wolfbeis, C.R. Eady (New Syn. Methods **3** [1975] 107/33). — [9] R.E. McKenzie, P.L. Timms (unpublished results from [10]). — [10] P.L. Timms, T.W. Turney (Advan. Organometal. Chem. **15** [1977] 53/112, 68, 95).

[11] P.L. Timms, T.W. Turney (unpublished results from [10]). — [12] I.C. Hisatsune (Spectrochim. Acta A **40** [1984] 391/5).

Empirical Formula Index

In the following index the compounds are listed in the order of increasing carbon content. Empirical formulas of ionic compounds are given in brackets; ions as well as components of solvates and adducts are separated by a period.

References to "Organoiron Compounds" B, Mononuclear Compounds 8 to 10 are designated by **B8** to **B10**. Page references are printed in ordinary types, table numbers in boldface, and compound numbers of the tables in italics.

$C_{13}H_{14}FeO_5$	**B10**, 63, **3**, *84*
	B10, 65, **3**, *92*
$C_{13}H_{14}FeO_6$	**B8**, 153, **7**, *9*
	B8, 154, **7**, *16*
	B8, 169
	B8, 213, **9**, *96*
	B8, 237, **10**, *29*
	B8, 242, **11**, *2*
	B8, 243, **11**, *3*
	B9, 55, **2**, *52*
	B10, 48, **3**, *4*
$C_{13}H_{14}FeO_6S$	**B8**, 464, **17**, *18*
$C_{13}H_{14}FeO_7$	**B8**, 362, **13**, *214*
	B10, 10, **1**, *33*
$C_{13}H_{15}ClFeO$	**B10**, 185
$C_{13}H_{15}FeNO_3$	**B8**, 112, **4**, *160*
$C_{13}H_{15}FeNO_3Si$	**B8**, 237, **10**, *26*
$C_{13}H_{15}FeNO_4$	**B8**, 44, **2**, *5*
	B8, 90, **4**, *37*
	B9, 62, **2**, *83*
$C_{13}H_{15}FeNP_2$	**B10**, 192, **6**, *21*
$C_{13}H_{15}FeN_3O_4$	**B8**, 177, **8**, *20*
	B8, 210, **9**, *86*
$C_{13}H_{16}FeGeO_3$	**B9**, 127, **3**, *35*
$C_{13}H_{16}FeO$	**B10**, 218, **8**, *14*
$C_{13}H_{16}FeOP_2$	**B10**, 191, **6**, *17*
$C_{13}H_{16}FeO_2P_2$	**B10**, 191, **6**, *19*
$C_{13}H_{16}FeO_3$	**B8**, 66, **3**, *18*
	B8, 96, **4**, *67*
	B8, 103, **4**, *102*
	B8, 157, **7**, *36*
	B8, 176, **8**, *13*
	B8, 208, **9**, *80*
	B8, 211, **9**, *90*
	B8, 398, **14**, *13*
	B8, 407, **14**, *55*
	B9, 210, **7**, *2*
	B9, 212, **7**, *10*
	B10, 127, **4**, *26*
$C_{13}H_{16}FeO_3Si$	**B9**, 127, **3**, *33*
$C_{13}H_{16}FeO_3Sn$	**B9**, 128, **3**, *38*
$C_{13}H_{16}FeO_4$	**B8**, 64, **3**, *9*
	B8, 85, **4**, *11*
	B8, 94, **4**, *61*
	B8, 174, **8**, *4*
	B8, 202, **9**, *41*
	B8, 206, **9**, *71*
	B9, 46, **2**, *4*
	B9, 70, **2**, *117*
$C_{13}H_{16}FeO_5$	**B8**, 145, **6**, *2*

$C_{17}H_{10}F_4FeO_3$	**B8**, 469, **17**, *41*
	B8, 469, **17**, *42*
$C_{17}H_{10}FeN_2O_9$	**B8**, 464, **17**, *15*
$C_{17}H_{10}FeO_3$	**B8**, 412, **14**, *73*
$C_{17}H_{10}FeO_4$	**B10**, 49, **3**, *10*
$C_{17}H_{11}FeNO_3$	**B10**, 142, **4**, *101*
$C_{17}H_{11}FeNO_4$	**B9**, 10, **1**, *26*
$C_{17}H_{11}FeNO_5$	**B9**, 72, **2**, *126*
	B9, 263, **9**, *32*
$C_{17}H_{12}Br_2FeO_3$	**B10**, 11, **1**, *40*
$C_{17}H_{12}F_6FeO_5$	**B9**, 214, **7**, *19*
$C_{17}H_{12}F_{12}FeO_3$	**B10**, 125, **4**, *22*
$C_{17}H_{12}FeMoO_5$	**B9**, 189, **5**, *32*
$C_{17}H_{12}FeN_2O_3$	**B9**, 258, **9**, *5*
$C_{17}H_{12}FeN_2O_4$	**B8**, 109, **4**, *138*
	B9, 15, **1**, *55*
$C_{17}H_{12}FeN_4O_7$	**B9**, 122, **3**, *12*
$C_{17}H_{12}FeO_3$	**B8**, 268, **12**, *19*
	B9, 158, **4**, *39*
	B9, 160, **4**, *46*
	B9, 263, **9**, *31*
	B9, 269, **9**, *58*
$C_{17}H_{12}FeO_4$	**B9**, 123, **3**, *16*
$C_{17}H_{12}FeO_5$	**B8**, 463, **17**, *14*
$C_{17}H_{13}FeNO_3$	**B8**, 108, **4**, *130*
	B9, 156, **4**, *31*
$C_{17}H_{13}FeNO_4$	**B9**, 60, **2**, *73*
	B9, 156, **4**, *30*
$C_{17}H_{13}FeNO_5$	**B9**, 6, **1**, *6*
	B9, 57, **2**, *64*
$C_{17}H_{14}F_6FeO_3$	**B8**, 470, **17**, *45*
	B8, 470, **17**, *46*
$C_{17}H_{14}FeN_4O_8$	**B8**, 201, **9**, *38*
$C_{17}H_{14}FeO_3$	**B9**, 56, **2**, *58*
	B9, 125, **3**, *26*
	B9, 129, **3**, *47*
	B9, 213, **7**, *18*
$C_{17}H_{14}FeO_4$	**B8**, 368, **13**, *243*
	B9, 67, **2**, *102*
	B9, 122, **3**, *11*
	B9, 203, **6**, *18*
	B9, 203, **6**, *19*
$C_{17}H_{14}FeO_5$	**B8**, 245, **11**, *13*
	B9, 70, **2**, *120*
$C_{17}H_{14}FeO_6$	**B10**, 52, **3**, *27*
	B10, 52, **3**, *28*
$C_{17}H_{14}FeO_6S$	**B8**, 463, **17**, *13*
$C_{17}H_{14}FeO_7$	**B9**, 59, **2**, *69*
$C_{17}H_{16}FeN_2O_4$	**B8**, 353, **13**, *167*
	B8, 415, **14**, *89*
$C_{17}H_{16}FeN_2O_5$	**B8**, 278, **12**, *59*

$C_{21}H_{16}FeO_4$	**B8**, 366, **13**, *231*
$C_{21}H_{17}BFeMoN_6O_5$	**B9**, 188, **5**, *29*
$C_{21}H_{17}FeNO_4$	**B10**, 55, **3**, *47*
	B10, 74, **3**, *129*
	B10, 75, **3**, *138*
$C_{21}H_{17}FeNO_7$	**B9**, 11, **1**, *34*
	B9, 267, **9**, *47*
$C_{21}H_{17}FeO_6P$	**B8**, 91, **4**, *47*
$C_{21}H_{18}F_4FeO_3$	**B8**, 470, **17**, *49*
$C_{21}H_{18}FeO_3$	**B8**, 468, **17**, *37*
$C_{21}H_{18}FeO_{15}$	**B8**, 461, **17**, *1*
$C_{21}H_{19}FeNO_6$	**B8**, 155, **7**, *22*
$C_{21}H_{20}FeN_4O_8$	**B8**, 400, **14**, *23*
$C_{21}H_{20}FeO_4$	**B8**, 103, **4**, *101*
$C_{21}H_{20}FeO_4S$	**B9**, 202, **6**, *13*
$C_{21}H_{20}FeO_7$	**B8**, 313, **13**, *11*
$C_{21}H_{21}FeNO_4$	**B8**, 352, **13**, *164*
$C_{21}H_{22}FeN_4O_8$	**B8**, 156, **7**, *26*
$C_{21}H_{22}FeO_3$	**B10**, 11, **1**, *41*
$C_{21}H_{22}FeO_4S$	**B9**, 200, **6**, *6*
$C_{21}H_{22}FeO_7S$	**B8**, 368, **13**, *245*
$C_{21}H_{23}FeNO_4$	**B8**, 352, **13**, *166*
	B10, 75, **3**, *134*
$C_{21}H_{24}FeO_7S$	**B8**, 155, **7**, *24*
	B8, 343, **13**, *132*
	B8, 361, **13**, *207*
$C_{21}H_{24}FeO_{11}$	**B8**, 347, **13**, *144*
$C_{21}H_{25}FeNO_6$	**B10**, 75, **3**, *137*
$C_{21}H_{26}FeO_7$	**B8**, 325, **13**, *68*
	B8, 357, **13**, *184*
	B8, 413, **14**, *80*
	B8, 416, **14**, *92*
$C_{21}H_{26}FeO_8$	**B8**, 319, **13**, *36*
	B8, 340, **13**, *126*
	B8, 401, **14**, *26*
$C_{21}H_{26}FeO_{10}$	**B8**, 348, **13**, *148*
	B8, 369, **13**, *247*
$C_{21}H_{28}FeO_6S_2$	**B8**, 324, **13**, *64*
	B8, 355, **13**, *180*
$C_{21}H_{28}FeO_8$	**B8**, 339, **13**, *124*
$C_{21}H_{28}FeO_9$	**B10**, 217, **8**, *11*
$C_{21}H_{30}FeO_7Si$	**B8**, 405, **14**, *46*
$C_{21}H_{32}FeO_3$	**B9**, 213, **7**, *16*
$[C_{21}H_{34}FeO_3P]^+ \cdot [BF_4]^-$	**B8**, 134, **5**, *18*
$[C_{21}H_{34}FeO_6P]^+ \cdot [BF_4]^-$	**B8**, 134, **5**, *15*
$C_{21}H_{41}FeO_3P$	**B10**, 210, **7**, *24*
$\mathbf{C}_{22}H_{14}FeN_8O_{11}$	**B9**, 75, **2**, *140*
$C_{22}H_{14}FeO_4$	**B9**, 186, **5**, *20*
$C_{22}H_{17}FeN_3O_7$	**B9**, 63, **2**, *85*
$C_{22}H_{18}FeN_2O_7$	**B8**, 432, **15**, *5*

Ligand Formula Index

Ligands (except CO) are used in the following Ligand Formula Index to locate a compound in the order of increasing carbon content of the respective ligand. The number of identical ligands in a compound and the nature of bonding are not taken into consideration, so that several compounds may be listed at one position, if need be. On the other hand, compounds having two or more different types of carbon-containing ligands occur at two or more positions, respectively. The following examples illustrate the arrangement.

$(CH_3CH{=}CHCH{=}CH_2)_2FeP(OCH_3)_3$:

$C_3H_9O_3P$	C_5H_8	–	–	–	**B10**, 208, **7**, *11*
C_5H_8	$C_3H_9O_3P$	–	–	–	**B10**, 208, **7**, *11*

$(C_6H_5)_4C_4Fe(P(OCH_3)_3)[(NC)_2C{=}C(CN)_2]CO$:

$C_3H_9O_3P$	C_6N_4	$C_{28}H_{20}$	CO	–	**B10**, 181, **5**, *7*
C_6N_4	$C_3H_9O_3P$	$C_{28}H_{20}$	CO	–	**B10**, 181, **5**, *7*
$C_{28}H_{20}$	$C_3H_9O_3P$	C_6N_4	CO	–	**B10**, 181, **5**, *7*

In view of the very large number of carbonyl compounds, CO is not included in the first column; it is given in column 4.

References to "Organoiron Compounds" B, Mononuclear Compounds 8 to 10, are designated by **B8** to **B10**. Page references are printed in ordinary types, table numbers in boldface, and compound numbers of the tables in italics.

C_4H_6					
$C(CH_2)_3$	–	–	CO	–	**B10**, 4, **1**, *1*
$C(CH_2)_3$	C_2H_2O	–	–	–	**B10**, 180, **5**, *1*
$CH_2{=}CHCH{=}CH_2$	–	–	–	–	**B10**, 229
$CH_2{=}CHCH{=}CH_2$	–	–	CO	–	**B10**, 215, **8**, *1*
$CH_2{=}CHCH{=}CH_2$	–	–	–	F_3P	**B10**, 206, **7**, *3*
$CH_2{=}CHCH{=}CH_2$	$C_3H_9O_3P$	–	–	–	**B10**, 206, **7**, *4*
$CH_2{=}CHCH{=}CH_2$	C_3H_9P	–	–	–	**B10**, 211, **7**, *30*
$CH_2{=}CHCH{=}CH_2$	$C_4H_6O_2$	–	CO	–	**B10**, 181, **5**, *9*
$CH_2{=}CHCH{=}CH_2$	C_5H_9N	–	–	–	**B10**, 206, **7**, *1*
$CH_2{=}CHCH{=}CH_2$	C_6H_8	–	CO	–	**B10**, 215, **8**, *3*
$CH_2{=}CHCH{=}CH_2$	$C_6H_{15}O_3P$	–	–	–	**B10**, 206, **7**, *5*
$CH_2{=}CHCH{=}CH_2$	C_8H_8	–	CO	–	**B10**, 215, **8**, *4*
$CH_2{=}CHCH{=}CH_2$	C_8H_9	–	CO	Cl	**B10**, 185
$CH_2{=}CHCH{=}CH_2$	C_8H_{12}	$C_{18}H_{15}$	CO	–	**B10**, 180, **5**, *5* **B10**, 181, **5**, *6*
$CH_2{=}CHCH{=}CH_2$	$C_9H_{21}O_3P$	–	–	–	**B10**, 207, **7**, *6*
$CH_2{=}CHCH{=}CH_2$	$C_{10}H_8N_2$	–	–	–	**B10**, 206, **7**, *2*
$CH_2{=}CHCH{=}CH_2$	$C_{10}H_{14}O_4$	–	CO	–	**B10**, 215, **8**, *2*
$CH_2{=}CHCH{=}CH_2$	$C_{12}H_{27}P$	–	–	–	**B10**, 207, **7**, *8*
$CH_2{=}CHCH{=}CH_2$	$C_{18}H_{15}O_3P$	–	–	–	**B10**, 207, **7**, *7*
$CH_2{=}CHCH{=}CH_2$	$C_{18}H_{15}P$	–	–	–	**B10**, 207, **7**, *10*
$CH_2{=}CHCH{=}CH_2$	$C_{18}H_{33}P$	–	–	–	**B10**, 207, **7**, *9*
2-Butene-1,4-diyl	–	–	CO	H	**B10**, 48, **3**, *1*
$C_4H_6O_2$	C_4H_4	–	CO	–	**B10**, 182, **5**, *12*
$C_4H_6O_2$	C_4H_6	–	CO	–	**B10**, 181, **5**, *9*
$C_4H_6O_2$	C_6H_{10}	–	CO	–	**B10**, 181, **5**, *10*
C_4H_8O	C_4H_4	–	CO	–	**B10**, 182, **5**, *11*
C$_5H_4O_2$	–	–	CO	–	**B8**, 12, **1**, *30*
$C_5H_4O_3$	–	–	CO	–	**B8**, 13, **1**, *31*
C_5H_5N	C_6H_8	–	–	–	**B10**, 211, **7**, *28*
$C_5H_6Cl_2Si$	–	–	CO	–	**B8**, 6, **1**, *6*
$C_5H_6N_2$	–	–	CO	–	**B9**, 12, **1**, *41*
C_5H_6O	–	–	CO	–	**B10**, 5, **1**, *8*
$C_5H_6O_2$					
$C(CH_2)_2CHCO_2CH_3$	–	–	CO	–	**B10**, 7, **1**, *13*
Carbonyloxy[2-butene-1,4-diyl]	–	–	CO	–	**B10**, 61, **3**, *75*
Carbonyloxy[2-methylene-propene-1,3-diyl]	–	–	CO	–	**B10**, 73, **3**, *119*
C_5H_6P	–	–	–	–	**B10**, 188, **6**, *2*
C_5H_7NO	–	–	CO	–	**B10**, 57, **3**, *56*
$C_5H_7N_2$	–	–	CO	–	**B9**, 18, **1**, *70*
$C_5H_7O_2$	–	–	CO	–	**B10**, 63, **3**, *81*
C_5H_8					
$C(CH_2)_2CHCH_3$	–	–	CO	–	**B10**, 4, **1**, *3*
$CH_3CH{=}CHCH{=}CH_2$	–	–	CO	–	**B10**, 216, **8**, *5*
$CH_3CH{=}CHCH{=}CH_2$	$C_3H_9O_3P$	–	–	–	**B10**, 208, **7**, *11* **B10**, 208, **7**, *12*
$CH_3CH{=}CHCH{=}CH_2$	C_8H_8	–	CO	–	**B10**, 216, **8**, *6*
$CH_3CH{=}CHCH{=}CH_2$	C_8H_9	–	CO	Cl	**B10**, 185

$CH_2{=}C(CH_3)CH{=}CH_2$	–	–	CO	–	**B10**, 216, **8**, *7*
$CH_2{=}C(CH_3)CH{=}CH_2$	$C_3H_9O_3P$	–	–	–	**B10**, 208, **7**, *13*
					B10, 208, **7**, *14*
$CH_2{=}C(CH_3)CH{=}CH_2$	$C_6H_{15}O_3P$	–	–	–	**B10**, 208, **7**, *15*
					B10, 208, **7**, *16*
$CH_2{=}C(CH_3)CH{=}CH_2$	C_8H_8	–	CO	–	**B10**, 217, **8**, *8*
$CH_2{=}C(CH_3)CH{=}CH_2$	C_8H_9	–	CO	Cl	**B10**, 185
$CH_2{=}C(CH_3)CH{=}CH_2$	$C_9H_{21}O_3P$	–	–	–	**B10**, 209, **7**, *17*
					B10, 209, **7**, *18*
$CH_2{=}C(CH_3)CH{=}CH_2$	$C_{18}H_{15}O_3P$	–	–	–	**B10**, 209, **7**, *19*
					B10, 209, **7**, *20*
$CH_2{=}C(CH_3)CH{=}CH_2$	$C_{18}H_{15}P$	–	–	–	**B10**, 209, **7**, *21*
2-Pentene-1,5-diyl	–	–	CO	–	**B10**, 56, **3**, *52*
C_5H_8O	–	–	CO	–	**B10**, 5, **1**, *4*
$C_5H_8O_2Si$	–	–	CO	–	**B8**, 6, **1**, *4*
C_5H_8Si	–	–	CO	–	**B8**, 6, **1**, *3*
C_5H_9N	C_4H_6	–	–	–	**B10**, 206, **7**, *1*
$\mathbf{C_6}F_8$	–	–	CO	–	**B8**, 447
$C_6H_4F_4$	–	–	CO	–	**B10**, 120, **4**, *3*
$C_6H_5N_3$	–	–	CO	–	**B9**, 16, **1**, *60*
$C_6H_6Cl_2$					
1,2-Dichloro-1,3-cyclohexadiene	–	–	CO	–	**B8**, 139
2,3-Dichloro-1,3-cyclohexadiene	–	–	CO	–	**B8**, 185
$C_6H_6F_4$					
1,1,2,2-Tetrafluoro-4-methylenepentane-1,5-diyl	–	–	CO	–	**B10**, 73, **3**, *120*
5,5,6,6-Tetrafluoro-2-hexene-1,6-diyl	–	–	CO	–	**B10**, 68, **3**, *107*
C_6H_6O					
2,4-Cyclohexadien-1-one	–	–	CO	–	**B8**, 245, **11**, *15*
Oxepine	–	–	CO	–	**B9**, 20, **1**, *75*
$C_6H_6O_2$	–	–	CO	–	**B10**, 122, **4**, *10*
$C_6H_6O_3$					
3-Methoxy-2*H*-2-pyranone	–	–	CO	–	**B8**, 13, **1**, *32*
6-Methoxy-2*H*-2-pyranone	–	–	CO	–	**B8**, 13, **1**, *37*
C_6H_7Br	–	–	CO	–	**B8**, 84, **4**, *3*
C_6H_7Cl					
1-Chloro-1,3-cyclohexadiene	–	–	CO	–	**B8**, 43, **2**, *1*
2-Chloro-1,3-cyclohexadiene	–	–	CO	–	**B8**, 63, **3**, *1*
5-Chloro-1,3-cyclohexadiene	–	–	CO	–	**B8**, 84, **4**, *2*
$C_6H_7Cl_3Ge$	–	–	CO	–	**B8**, 111, **4**, *150*
$C_6H_7Cl_3Si$	–	–	CO	–	**B8**, 111, **4**, *149*
$C_6H_7Cl_3Sn$	–	–	CO	–	**B8**, 111, **4**, *151*
C_6H_7F					
2-Fluoro-1,3-cyclohexadiene	–	–	CO	–	**B8**, 66, **3**, *16*
5-Fluoro-1,3-cyclohexadiene	–	–	CO	–	**B8**, 83, **4**, *1*
C_6H_7I	–	–	CO	–	**B8**, 84, **4**, *4*
C_6H_7N	–	–	CO	–	**B9**, 5, **1**, *2*

$C_6H_7N_3$	–	–	CO	–	**B8**, 112, **4**, *157*
C_6H_7O	–	–	CO	–	**B10**, 13, **1**, *53*
C_6H_8					
$CH_2{=}CH\text{-}CHC(CH_2)_2$	–	–	CO	–	**B10**, 5, **1**, *6*
1,2-Dimethyl-1,3-cyclo-butadiene	$C_6H_8O_4$	–	CO	–	**B10**, 183, **5**, *17*
					B10, 183, **5**, *18*
2-Methylenecyclopentane-1,3-diyl	–	–	CO	–	**B10**, 11, **1**, *39*
1,3-Cyclohexadiene	–	–	CO	–	**B8**, 21/38
1,3-Cyclohexadiene	–	–	CO	–	**B10**, 218, **8**, *14*
1,3-Cyclohexadiene	–	–	CO	H	**B10**, 122, **4**, *13*
1,3-Cyclohexadiene	$C_3H_9O_3P$	–	–	–	**B10**, 211, **7**, *27*
1,3-Cyclohexadiene	C_4H_6	–	CO	–	**B10**, 215, **8**, *3*
1,3-Cyclohexadiene	C_5H_5N	–	–	–	**B10**, 211, **7**, *28*
1,3-Cyclohexadiene	C_7H_{10}	–	CO	–	**B10**, 218, **8**, *15*
					B10, 219, **8**, *17*
					B10, 219, **8**, *19*
1,3-Cyclohexadiene	C_8H_8	–	CO	–	**B10**, 219, **8**, *20*
1,3-Cyclohexadiene	C_8H_9N	–	CO	–	**B10**, 219, **8**, *16*
					B10, 219, **8**, *18*
1,3-Cyclohexadiene	C_8H_{12}	–	CO	–	**B10**, 219, **8**, *21*
1,3-Cyclohexadiene	$C_{10}H_8N_2$	–	–	–	**B10**, 211, **7**, *29*
1,3-Cyclohexadiene	$C_{12}H_{16}O_2$	–	CO	–	**B10**, 220, **8**, *22*
1,4-Cyclohexadiene	$C_3H_9O_3P$	–	–	–	**B10**, 180, **5**, *3*
C_6H_8As	–	–	–	–	**B10**, 190, **6**, *10*
$C_6H_8N_2$					
1-Methyl-1*H*-1,2-diazepine	–	–	CO	–	**B9**, 14, **1**, *45*
3-Methyl-1*H*-1,2-diazepine	–	–	CO	–	**B9**, 16, **1**, *61*
5-Methyl-1*H*-1,2-diazepine	–	–	CO	–	**B9**, 17, **1**, *62*
$C_6H_8N_2O_2S$	–	–	CO	–	**B9**, 13, **1**, *42*
C_6H_8O					
$CH_3C(O)CHC(CH_2)_2$	–	–	CO	–	**B10**, 6, **1**, *9*
1-Oxo-4-hexene-1,6-diyl	–	–	CO	–	**B10**, 68, **3**, *103*
5-Hydroxy-1,3-cyclo-hexadiene	–	–	CO	–	**B8**, 84, **4**, *5*
$C_6H_8O_2$					
$CH_3O_2C\text{-}CHC(CH_2)_2$	–	–	CO	–	**B10**, 7, **1**, *14*
Carbonyloxy[1-methyl-2-butene-1,4-diyl]	–	–	CO	–	**B10**, 62, **3**, *76*
Carbonyloxy[2-methyl-2-butene-1,4-diyl]	–	–	CO	–	**B10**, 62, **3**, *79*
Carbonyloxy[3-methyl-2-butene-1,4-diyl]	–	–	CO	–	**B10**, 63, **3**, *80*
5,6-Dihydroxy-1,3-cylco-hexadiene	–	–	CO	–	**B8**, 263, **12**, *1*
$C_6H_8O_2P$	$C_{13}H_{16}P$	–	–	–	**B10**, 195, **6**, *31*
$C_6H_8O_3S$	–	–	CO	–	**B8**, 86, **4**, *15*
$C_6H_8O_4$	C_4H_4	–	CO	–	**B10**, 182, **5**, *13*
					B10, 182, **5**, *14*

$C_6H_8O_4$	C_6H_8	–	CO	–	**B10**, 183, **5**, *17*
					B10, 183, **5**, *18*
C_6H_8P	–	–	–	–	**B10**, 188, **6**, *3*
C_6H_8P	C_7H_7NP	–	–	–	**B10**, 192, **6**, *21*
C_6H_8P	C_7H_8OP	–	–	–	**B10**, 191, **6**, *17*
C_6H_8P	$C_7H_8O_2P$	–	–	–	**B10**, 191, **6**, *19*
C_6H_8P	C_7H_9NOP	–	–	–	**B10**, 191, **6**, *18*
C_6H_8P	$C_7H_{10}OP$	–	–	–	**B10**, 191, **6**, *16*
C_6H_8P	$C_8H_{10}OP$	–	–	–	**B10**, 193, **6**, *24*
C_6H_8P	$C_8H_{12}OP$	–	–	–	**B10**, 193, **6**, *23*
C_6H_8P	$C_8H_{12}P$	–	–	–	**B10**, 192, **6**, *22*
C_6H_8P	$C_9H_{12}O_2P$	–	–	–	**B10**, 192, **6**, *20*
C_6H_8P	$C_9H_{14}OP$	–	–	–	**B10**, 193, **6**, *25*
C_6H_8P	$C_{10}H_{17}P$	–	–	–	**B10**, 194, **6**, *27*
C_6H_8P	$C_{11}H_{20}P$	–	–	–	**B10**, 194, **6**, *28*
C_6H_8P	$C_{12}H_{20}OP$	–	–	–	**B10**, 194, **6**, *29*
C_6H_8P	$C_{13}H_{12}OP$	–	–	–	**B10**, 194, **6**, *26*
C_6H_8P	$C_{17}H_{22}OP$	–	–	–	**B10**, 194, **6**, *30*
C_6H_8Sb	–	–	–	–	**B10**, 190, **6**, *11*
C_6H_9BO	–	–	CO	–	**B8**, 5, **1**, *1*
C_6H_9ClSi	–	–	CO	–	**B8**, 7, **1**, *9*
C_6H_9N	–	–	CO	–	**B9**, 7, **1**, *17*
C_6H_9NO	–	–	CO	–	**B10**, 57, **3**, *57*
$C_6H_9N_2$					
2-Methyl-1*H*-1,2-diazepinium	–	–	CO	–	**B9**, 19, **1**, *72*
3-Methyl-1*H*-1,2-diazepinium	–	–	CO	–	**B9**, 19, **1**, *73*
1-Dihydro-3-methyl-1,2-diazepinium	–	–	CO	–	**B9**, 20, **1**, *74*
C_6H_9O	–	–	CO	–	**B10**, 8, **1**, *19*
$C_6H_9O_2P$	–	–	CO	–	**B8**, 91, **4**, *43*
$C_6H_9O_3P$	–	–	CO	–	**B8**, 91, **4**, *44*
C_6H_{10}					
$C_2H_5CHC(CH_2)_2$	–	–	CO	–	**B10**, 5, **1**, *5*
$(CH_3)_2CC(CH_2)_2$	–	–	CO	–	**B10**, 10, **1**, *32*
$CH_2=C(CH_3)C(CH_3)=CH_2$	–	–	CO	–	**B10**, 217, **8**, *9*
$CH_2=C(CH_3)C(CH_3)=CH_2$	–	–	CO	–	**B10**, 218, **8**, *13*
$CH_2=C(CH_3)C(CH_3)=CH_2$	$C_3H_9O_3P$	–	–	–	**B10**, 210, **7**, *22*
$CH_2=C(CH_3)C(CH_3)=CH_2$	$C_4H_6O_2$	–	CO	–	**B10**, 181, **5**, *10*
$CH_2=C(CH_3)C(CH_3)=CH_2$	$C_6H_{15}O_3P$	–	–	–	**B10**, 210, **7**, *23*
$CH_2=C(CH_3)C(CH_3)=CH_2$	$C_9H_{21}O_3P$	–	–	–	**B10**, 180, **5**, *2*
					B10, 210, **7**, *24*
$CH_2=C(CH_3)C(CH_3)=CH_2$	$C_{10}H_{10}$	–	CO	–	**B10**, 181, **5**, *8*
$CH_2=C(CH_3)C(CH_3)=CH_2$	$C_{14}H_{15}P$	–	–	–	**B10**, 210, **7**, *25*
$CH_3CH=CHCH=CHCH_3$	$C_3H_9O_3P$	–	–	–	**B10**, 210, **7**, *26*
2-Hexene-1,6-diyl	–	–	CO	–	**B10**, 67, **3**, *100*
$C_6H_{10}O$					
$HO(CH_3)CHCHC(CH_2)_2$	–	–	CO	–	**B10**, 6, **1**, *10*
$C_2H_5(HO)CC(CH_2)_2$	–	–	CO	–	**B10**, 8, **1**, *20*
$C_6H_{10}OSi$	–	–	CO	–	**B8**, 15, **1**, *46*
C_6H_{12}	C_4H_4	–	CO	–	**B10**, 182, **5**, *15*
$C_6H_{15}O_3P$	C_4H_6	–	–	–	**B10**, 206, **7**, *5*

$C_6H_{15}O_3P$	C_5H_8	–	–	–	**B10**, 208, **7**, *15*
$C_6H_{15}O_3P$	C_5H_8	–	–	–	**B10**, 208, **7**, *16*
$C_6H_{15}O_3P$	C_6H_{10}	–	–	–	**B10**, 210, **7**, *23*
C_6N_4	$C_3H_9O_3P$	$C_{28}H_{20}$	CO	–	**B10**, 181, **5**, *7*
$\mathbf{C_7}F_{10}$	–	–	CO	–	**B9**, 78, **2**, *154*
C_7HF_9	–	–	CO	–	**B9**, 78, **2**, *153*
$C_7H_2F_8$	–	–	CO	–	**B9**, 77, **2**, *152*
$C_7H_4F_6$	–	–	CO	–	**B10**, 120, **4**, *4*
C_7H_5ClO	–	–	CO	–	**B9**, 183, **5**, *4*
$C_7H_5Cl_3N_2O$	–	–	CO	–	**B9**, 15, **1**, *51*
$C_7H_5F_3N_2O$	–	–	CO	–	**B9**, 14, **1**, *50*
$C_7H_6F_6$	–	–	CO	–	**B10**, 69, **3**, *108*
C_7H_6O					
Bicyclo[2.2.1]hepta-2,5-dien-7-one	–	–	CO	–	**B8**, 466, **17**, *28*
2,4,6-Cycloheptatrien-1-one	–	–	CO	–	**B9**, 154, **4**, *26*
Carbonyl[bicyclo[3.1.0]hex-3-ene-2,6-diyl]	–	–	CO	–	**B10**, 122, **4**, *11*
$C_7H_6O_2$	–	–	CO	–	**B9**, 75, **2**, *139*
$C_7H_6O_4$					
3-Acetyloxy-2*H*-pyran-2-one	–	–	CO	–	**B8**, 13, **1**, *33*
5-Methoxycarbonyl-2*H*-pyran-2-one	–	–	CO	–	**B8**, 13, **1**, *36*
$C_7H_7ClN_2O$	–	–	CO	–	**B9**, 15, **1**, *52*
$C_7H_7Cl_3$	–	–	CO	–	**B8**, 100, **4**, *89*
$C_7H_7F_3$	–	–	CO	–	**B10**, 65, **3**, *94*
C_7H_7N					
1,3-Cyclohexadiene-1-carbonitrile	–	–	CO	–	**B8**, 48, **2**, *26*
2,4-Cyclohexadiene-1-carbonitrile	–	–	CO	–	**B8**, 101, **4**, *92*
C_7H_7NO					
4-Hydroxy-2,4-cylohexadiene-1-carbonitrile	–	–	CO	–	**B8**, 196, **9**, *3*
1*H*-Azepine-1-carbaldehyde	–	–	CO	–	**B9**, 21, **1**, *79*
C_7H_7NP	C_6H_8P	–	–	–	**B10**, 192, **6**, *21*
C_7H_7NS					
5-Thiocyanato-1,3-cyclohexadiene	–	–	CO	–	**B8**, 86, **4**, *16*
5-Isothiocyanato-1,3-cyclohexadiene	–	–	CO	–	**B8**, 87, **4**, *21*
C_7H_7NSe					
5-Selenocyanato-1,3-cyclohexadiene	–	–	CO	–	**B8**, 112, **4**, *155*
5-Isoselenocyanato-1,3-cyclohexadiene	–	–	CO	–	**B8**, 112, **4**, *158*
$C_7H_7N_3O$	–	–	CO	–	**B9**, 65, **2**, *95*
C_7H_7O	–	–	CO	–	**B9**, 162, **4**, *52*
C_7H_8					
1,3,5-Cycloheptatriene	–	–	CO	–	**B9**, 102/16

1,3,5-Cycloheptatriene	C_4H_4	–	CO	–	**B10**, 183, **5**, *16*
Bicyclo[4.1.0]hepta-2,4-diene	–	–	CO	–	**B8**, 265, **12**, *10*
Bicyclo[2.2.1]hepta-2,5-diene	–	–	CO	–	**B8**, 461, **17**, *2*
Bicyclo[2.2.1]hepta-2,5-diene	–	–	CO	H	**B8**, 467, **17**, *31*
Bicyclo[2.2.1]hept-5-ene-2,3-diylium	–	–	CO	H	**B10**, 154, **4**, *148*
Bicyclo[3.1.1]hept-3-ene-2,6-diyl	–	–	CO	–	**B10**, 135, **4**, *66*
$C_7H_8Br_2$	–	–	CO	–	**B8**, 96, **4**, *69*
$C_7H_8F_4$					
5,5,6,6-Tetrafluoro-1-methyl-2-hexene-1,6-diyl	–	–	CO	–	**B10**, 70, **3**, *113*
5,5,6,6-Tetrafluoro-2-methyl-2-hexene-1,6-diyl	–	–	CO	–	**B10**, 70, **3**, *111*
5,5,6,6-Tetrafluoro-3-methyl-2-hexene-1,6-diyl	–	–	CO	–	**B10**, 69, **3**, *110*
$C_7H_8N_2O$	–	–	CO	–	**B9**, 15, **1**, *53*
C_7H_8O					
1,3-Cyclohexadiene-1-carbaldehyde	–	–	CO	–	**B8**, 45, **2**, *12*
2,4-Cyclohexadiene-1-carbaldehyde	–	–	CO	–	**B8**, 96, **4**, *70*
3-Methyl-2,4-cyclohexadien-1-one	–	–	CO	–	**B8**, 359, **13**, *197*
4-Methyl-2,4-cyclohexadien-1-one	–	–	CO	–	**B8**, 358, **13**, *190*
5-Methyl-2,4-cyclohexadien-1-one	–	–	CO	–	**B8**, 325, **13**, *73*
2,4-Cycloheptadien-1-one	–	–	CO	–	**B9**, 55, **2**, *55*
3,5-Cycloheptadien-1-one	–	–	CO	–	**B9**, 56, **2**, *60*
7-Hydroxy-1,3,5-cycloheptatriene	–	–	CO	–	**B9**, 126, **3**, *29*
Carbonyl[4-cyclohexene-1,3-diyl]	–	–	CO	–	**B10**, 124, **4**, *17*
2-Oxo-4-cycloheptene-1,3-diyl	–	–	CO	–	**B10**, 129, **4**, *38*
7-Oxo-4-cycloheptene-1,3-diyl	–	–	CO	–	**B10**, 129, **4**, *39*
7-Hydroxybicyclo[2.2.1]hepta-2,5-diene	–	–	CO	–	**B8**, 463, **17**, *10*
C_7H_8OP	C_6H_8P	–	–	–	**B10**, 191, **6**, *17*
$C_7H_8O_2$					
4,6-Dimethyl-2*H*-pyran-2-one	–	–	CO	–	**B8**, 13, **1**, *34*
1,3-Cyclohexadiene-1-carboxylic acid	–	–	CO	–	**B8**, 47, **2**, *22*
1,5-Cyclohexadiene-1-carboxylic acid	–	–	CO	–	**B8**, 65, **3**, *13*
2,4-Cyclohexadiene-1-carboxylic acid	–	–	CO	–	**B8**, 100, **4**, *90*
3-Methoxy-2,4-cyclohexadien-1-one	–	–	CO	–	**B8**, 358, **13**, *193*
4-Methoxy-2,4-cyclohexadien-1-one	–	–	CO	–	**B8**, 353, **13**, *169*

6-Hydroxy-2,4-cyclo-heptadien-1-one	–	–	CO	–	**B9**, 65, **2**, *93*
Carbonyloxy[1-ethenyl-2-butene-1,4-diyl]	–	–	CO	–	**B10**, 62, **3**, *78*
Carbonyloxy[2-cyclohexene-1,4-diyl]	–	–	CO	–	**B10**, 123, **4**, *14*
$C_7H_8O_2P$	C_6H_8P	–	–	–	**B10**, 191, **6**, *19*
$C_7H_8O_4$	–	–	CO	–	**B10**, 48, **3**, *2*
C_7H_9ClO					
1-Chloro-5-methoxy-1,3-cyclohexadiene	–	–	CO	–	**B8**, 167
1-Chloro-6-methoxy-1,3-cyclohexadiene	–	–	CO	–	**B8**, 174, **8**, *1*
2-Chloro-5-methoxy-1,3-cyclohexadiene	–	–	CO	–	**B8**, 196, **9**, *1*
2-Chloro-6-methoxy-1,3-cyclohexadiene	–	–	CO	–	**B8**, 233, **10**, *1*
C_7H_9NO					
1,3-Cyclohexadiene-1-carbaldehyde oxime	–	–	CO	–	**B8**, 46, **2**, *15*
1,3-Cyclohexadiene-1-carboxamide	–	–	CO	–	**B8**, 47, **2**, *25*
2,4-Cyclohexadiene-1-carboxamide	–	–	CO	–	**B8**, 103, **4**, *103*
[2,4-Cyclohexadien-1-yl]-formamide	–	–	CO	–	**B8**, 87, **4**, *23*
C_7H_9NOP	C_6H_8P	–	–	–	**B10**, 191, **6**, *18*
$C_7H_9NO_2$					
1-Methoxycarbonyl-2*H*-pyridine	–	–	CO	–	**B8**, 12, **1**, *26*
5-Nitromethyl-1,3-cyclo-hexadiene	–	–	CO	–	**B8**, 92, **4**, *54*
$C_7H_9NO_2S$	–	–	CO	–	**B9**, 5, **1**, *3*
$C_7H_9NO_3$	–	–	CO	–	**B10**, 61, **3**, *74*
$C_7H_9N_2O$	–	–	CO	–	**B9**, 19, **1**, *71*
$C_7H_9N_3$	–	–	CO	–	**B9**, 47, **2**, *10*
C_7H_9O	–	–	CO	–	**B8**, 467, **17**, *34*
C_7H_{10}					
1-Methyl-1,3-cyclohexadiene	–	–	CO	–	**B8**, 44, **2**, *6*
2-Methyl-1,3-cyclohexadiene	–	–	CO	–	**B8**, 64, **3**, *6*
5-Methyl-1,3-cyclohexadiene	–	–	CO	–	**B8**, 91, **4**, *49*
1,3-Cycloheptadiene	C_6H_8	–	CO	–	**B10**, 218, **8**, *15*
1,4-Cycloheptadiene	C_6H_8	–	CO	–	**B10**, 219, **8**, *17*
2-Cycloheptene-1,6-diyl	–	–	CO	–	**B10**, 128, **4**, *32*
2-Cycloheptene-1,6-diyl	C_6H_8	–	CO	–	**B10**, 219, **8**, *19*
$C_7H_{10}N_2$	–	–	CO	–	**B9**, 14, **1**, *46*
$C_7H_{10}N_2O$	–	–	CO	–	**B9**, 12, **1**, *37*
$C_7H_{10}N_2S$	–	–	CO	–	**B8**, 86, **4**, *17*
$C_7H_{10}O$					
1-Methoxy-1,3-cyclohexadiene	–	–	CO	–	**B8**, 43, **2**, *2*
2-Methoxy-1,3-cyclohexadiene	–	–	CO	–	**B8**, 63, **3**, *3*

5-Methoxy-1,3-cyclohexadiene	–	–	CO	–	**B8**, 84, **4**, *8*
1-(Hydroxymethyl)-1,3-cyclohexadiene	–	–	CO	–	**B8**, 45, **2**, *9*
2-(Hydroxymethyl)-1,3-cyclohexadiene	–	–	CO	–	**B8**, 64, **3**, *8*
5-(Hydroxymethyl)-1,3-cyclohexadiene	–	–	CO	–	**B8**, 92, **4**, *50*
7-Hydroxy-1,3-cycloheptadiene	–	–	CO	–	**B9**, 45, **2**, *1*
6-Hydroxy-1,3-cycloheptadiene	–	–	CO	–	**B9**, 50, **2**, *32*
$C_7H_{10}OP$	C_6H_8P	–	–	–	**B10**, 191, **6**, *16*
$C_7H_{10}O_2$					
1-Methoxycarbonyl-1,3-pentadiene	–	–	CO	–	**B10**, 217, **8**, *10*
5-Hydroxy-2-methoxy-1,3-cyclohexadiene	–	–	CO	–	**B8**, 212, **9**, *93*
Carbonyloxy[1,1-dimethyl-2-butene-1,4-diyl]	–	–	CO	–	**B10**, 63, **3**, *83*
Carbonyloxy[1,4-dimethyl-2-butene-1,4-diyl]	–	–	CO	–	**B10**, 64, **3**, *87*
Carbonyloxy[2,3-dimethyl-2-butene-1,4-diyl]	–	–	CO	–	**B10**, 64, **3**, *89*
$C_7H_{10}S$	–	–	CO	–	**B8**, 86, **4**, *18*
$C_7H_{11}N$	–	–	CO	–	**B8**, 92, **4**, *53*
$C_7H_{11}NO$					
Carbonyl(methylimino)-[2-methyl-2-butene-1,4-diyl]	–	–	CO	–	**B10**, 57, **3**, *58*
Carbonyl(methylimino)-[3-methyl-2-butene-1,4-diyl]	–	–	CO	–	**B10**, 58, **3**, *59*
$C_7H_{11}O$	–	–	CO	–	**B10**, 9, **1**, *22*
$C_7H_{11}O_2$	–	–	CO	–	**B10**, 13, **1**, *48*
$C_7H_{11}O_3P$	–	–	CO	–	**B8**, 91, **4**, *45*
$C_7H_{12}O$	–	–	CO	–	**B10**, 6, **1**, *11*
$C_7H_{12}O$	–	–	CO	–	**B10**, 9, **1**, *23*
$C_7H_{12}O_2$	–	–	CO	–	**B10**, 13, **1**, *49*
$C_7H_{12}Si$	–	–	CO	–	**B8**, 6, **1**, *7*
$\mathbf{C_8}F_8$	–	–	CO	–	**B10**, 134, **4**, *61*
$C_8H_4F_6$					
3-Cyclobutene-1,2-diyl-[2,2-difluoro-1-(tetrafluoroethylidene)-1,2-ethanediyl]	–	–	CO	–	**B10**, 120, **4**, *2*
1,6-Bis(trifluoromethyl)-1,3-cyclohexadiene	–	–	CO	–	**B8**, 180, **8**, *28*
1,2-Bis(trifluoromethyl)-1,4-hexadiene-1,6-diyl	–	–	CO	–	**B10**, 66, **3**, *97*
$C_8H_6O_3$	–	–	CO	–	**B8**, 278, **12**, *61*
C_8H_7Br	–	–	CO	–	**B9**, 258, **9**, *2*
C_8H_7Cl	–	–	CO	–	**B9**, 257, **9**, *1*
$C_8H_7Cl_3N_2O_2$	–	–	CO	–	**B9**, 16, **1**, *57*
$C_8H_7F_3O_2$	–	–	CO	–	**B8**, 84, **4**, *7*

C_8H_7NO					
5-Oxo-2,4-cycloheptadiene-1-carbonitrile	–	–	CO	–	**B9**, 66, **2**, *98*
7-Cyano-2-oxo-5-cycloheptene-1,4-diyl	–	–	CO	–	**B10**, 130, **4**, *40*
$C_8H_7O_2$	–	–	CO	–	**B8**, 467, **17**, *32*
C_8H_8					
7-Methylene-1,3,5-cycloheptatriene	–	–	CO	–	**B9**, 159, **4**, *41*
2-Methylene-4,6-cycloheptadiene-1,3-diyl	–	–	CO	–	**B10**, 12, **1**, *45*
Bicyclo[3.2.1]octa-3,6-diene-2,8-diyl	–	–	CO	–	**B10**, 136, **4**, *69*
Bicyclo[4.2.0]octa-2,4,7-triene	–	–	CO	–	**B8**, 268, **12**, *16*
1,3,5,7-Cyclooctatetraene	–	–	CO	–	**B9**, 241/53
1,3.5,7-Cyclooctatetraene	C_4H_6	–	CO	–	**B10**, 215, **8**, *4*
1,3,5,7-Cyclooctatetraene	C_5H_8	–	CO	–	**B10**, 216, **8**, *6*
					B10, 217, **8**, *8*
1,3,5,7-Cyclooctatetraene	C_6H_8	–	CO	–	**B10**, 219, **8**, *20*
$C_8H_8Br_2$	–	–	CO	–	**B9**, 58, **2**, *66*
$C_8H_8Cl_2$	–	–	CO	–	**B9**, 58, **2**, *65*
$C_8H_8F_4$	–	–	CO	–	**B10**, 124, **4**, *19*
$C_8H_8N_2O$					
1-(Ethenylcarbonyl)-1*H*-1,2-diazepine	–	–	CO	–	**B9**, 15, **1**, *54*
3a,8a-Dihydro-3*H*-cycloheptapyrazol-4-one	–	–	CO	–	**B9**, 72, **2**, *128*
C_8H_8O					
1,3,5-Cycloheptatriene-1-carbaldehyde	–	–	CO	–	**B9**, 123, **3**, *14*
2-Methyl-2,4,6-cycloheptatrien-1-one	–	–	CO	–	**B9**, 183, **5**, *5*
2,4,6-Cyclooctatrien-1-one	–	–	CO	–	**B9**, 223, **8**, *5*
Bicyclo[2.2.1]hepta-2,5-diene-2-carbaldehyde	–	–	CO	–	**B8**, 462, **17**, *5*
Carbonyl[bicyclo[3.1.1]hept-3-ene-6,2-diyl]	–	–	CO	–	**B10**, 135, **4**, *67*
Bicyclo[4.2.0]octa-2,4-dien-7-one	–	–	CO	–	**B8**, 267, **12**, *14*
Bicyclo[5.1.0]octa-3,5-dien-2-one	–	–	CO	–	**B9**, 71, **2**, *123*
7-Oxobicyclo[3.2.1]oct-3-ene-2,6-diyl	–	–	CO	–	**B10**, 138, **4**, *81*
9-Oxabicyclo[4.2.1]nona-2,4,7-triene	–	–	CO	–	**B9**, 21, **1**, *78*
C_8H_8O	C_8H_8O	–	CO	–	**B10**, 183, **5**, *20*
$C_8H_8O_2$	–	–	CO	–	**B8**, 462, **17**, *7*
$C_8H_8O_4$					
1,3-Cyclohexadiene-1,4-dicarboxylic acid	–	–	CO	–	**B8**, 159, **7**, *49*

2,4-Cyclohexadiene-1,2-dicarboxylic acid	–	–	CO	–	**B8**, 181, **8**, *33*
3,5-Cyclohexadiene-1,2-dicarboxylic acid	–	–	CO	–	**B8**, 277, **12**, *58*
C_8H_9	C_4H_6	–	CO	Cl	**B10**, 185
C_8H_9	C_5H_8	–	CO	Cl	**B10**, 185
$C_8H_9Br_2NO_2$	–	–	CO	–	**B8**, 12, **1**, *28*
$C_8H_9ClO_2$					
2-Chloro-6-methoxy-2,4-cyclo-heptadien-1-one	–	–	CO	–	**B9**, 70, **2**, *118*
7-Chloro-6-methoxy-2,4-cyclo-heptadien-1-one	–	–	CO	–	**B9**, 70, **2**, *121*
$C_8H_9F_3$					
4-Methyl-1-(trifluoromethyl)-1,4-hexadiene-1,6-diyl	–	–	CO	–	**B10**, 66, **3**, *95*
5-Methyl-1-(trifluoromethyl)-1,4-hexadiene-1,6-diyl	–	–	CO	–	**B10**, 66, **3**, *96*
2-Cyclohexene-1,4-diyl-4-(1,2,2-trifluoro-1,2-ethanediyl)	–	–	CO	–	**B10**, 124, **4**, *18*
C_8H_9N					
1,6-Cycloheptadiene-1-carbonitrile	C_6H_8	–	CO	–	**B10**, 219, **8**, *16*
2,6-Cycloheptadiene-1-carbonitrile	C_6H_8	–	CO	–	**B10**, 219, **8**, *18*
2-Methyl-2,4-cyclohexadiene-1-carbonitrile	–	–	CO	–	**B8**, 177, **8**, *23*
4-Methyl-2,4-cyclohexadiene-1-carbonitrile	–	–	CO	–	**B8**, 210, **9**, *83*
2,4-Cycloheptadiene-1-carbonitrile	–	–	CO	–	**B9**, 50, **2**, *31*
2-Cyano-4-cycloheptene-1,3-diyl	–	–	CO	–	**B10**, 129, **4**, *34*
C_8H_9NO					
3-Acetyl-1*H*-azepine	–	–	CO	–	**B9**, 6, **1**, *9*
3-Methoxy-2,4-cyclohexa-diene-1-carbonitrile	–	–	CO	–	**B8**, 234, **10**, *6*
4-Methoxy-2,4-cyclohexa-diene-1-carbonitrile	–	–	CO	–	**B8**, 204, **9**, *52*
$C_8H_9NO_2$					
1-Methoxycarbonyl-1*H*-azepine	–	–	CO	–	**B9**, 5, **1**, *4*
1-(2-Nitroethenyl)-1,3-cyclohexadiene	–	–	CO	–	**B8**, 46, **2**, *18*
C_8H_9NS					
5-Thiocyanato-1,3-cycloheptadiene	–	–	CO	–	**B9**, 46, **2**, *6*
5-Isothiocyanato-1,3-cycloheptadiene	–	–	CO	–	**B9**, 47, **2**, *11*
C_8H_9NSe					
5-Selenocyanato-1,3-cycloheptadiene	–	–	CO	–	**B9**, 46, **2**, *9*

4,4-Dimethyl-2,5-cyclohexadien-1-one	–	–	CO	–	**B8**, 453, **16**, *1*
1-(Hydroxymethyl)-1,3,5-cycloheptatriene	–	–	CO	–	**B9**, 122, **3**, *8*
7-Methoxy-1,3,5-cycloheptatriene	–	–	CO	–	**B9**, 126, **3**, *30*
7-(Hydroxymethyl)-1,3,5-cycloheptatriene	–	–	CO	–	**B9**, 128, **3**, *40*
7-Hydroxy-1-methyl-1,3,5-cycloheptatriene	–	–	CO	–	**B9**, 148, **4**, *1*
7-Hydroxy-7-methyl-1,3,5-cycloheptatriene	–	–	CO	–	**B9**, 151, **4**, *13*
7-Hydroxy-5-methylene-1,3-cycloheptadiene	–	–	CO	–	**B9**, 66, **2**, *99*
Carbonyl[4-cycloheptene-1,3-diyl]	–	–	CO	–	**B10**, 128, **4**, *33*
7-Hydroxy-1,3,5-cyclooctatriene	–	–	CO	–	**B9**, 223, **8**, *3*
7-Hydroxy-7-methylbicyclo[2.2.1]hepta-2,5-diene	–	–	CO	–	**B8**, 465, **17**, *22*
5-Hydroxybicyclo[5.1.0]hept-3-ene-2,6-diyl	–	–	CO	–	**B10**, 131, **4**, *44*
6-Hydroxybicyclo[5.1.0]octa-2,4-diene	–	–	CO	–	**B9**, 61, **2**, *77*
7-Hydroxybicyclo[4.2.0]octa-2,4-diene	–	–	CO	–	**B8**, 266, **12**, *12*
$C_8H_{10}OP$	–	–	–	–	**B10**, 189, **6**, *5*
$C_8H_{10}OP$	C_6H_8P	–	–	–	**B10**, 193, **6**, *24*
$C_8H_{10}O_2$					
$CH_3O_2CCH{=}CHCHC(CH_2)_2$	–	–	CO	–	**B10**, 5, **1**, *7*
Carbonyloxy-1-cyclopentyl-2-ylidene-(2-ethanyl-1-ylidene)	–	–	CO	–	**B10**, 64, **3**, *85*
Carbonyloxy(2-ethenylcyclopentane-1,2-diyl)	–	–	CO	–	**B10**, 65, **3**, *91*
1-Methoxycarbonyl-1,3-cyclohexadiene	–	–	CO	–	**B8**, 47, **2**, *23*
2-Methoxycarbonyl-1,3-cyclohexadiene	–	–	CO	–	**B8**, 65, **3**, *14*
5-Methoxycarbonyl-1,3-cyclohexadiene	–	–	CO	–	**B8**, 100, **4**, *91*
4-Methoxy-2,4-cyclohexadiene-1-carbaldehyde	–	–	CO	–	**B8**, 201, **9**, *37*
5-Hydroxycarbonyl-5-methyl-1,3-cyclohexadiene	–	–	CO	–	**B8**, 248, **11**, *25*
6-Methoxy-2,4-cycloheptadien-1-one	–	–	CO	–	**B9**, 65, **2**, *94*
$C_8H_{10}O_3$					
Carbonyl[1-methoxycarbonyl-3-methyl-2-buten-1-ylidene]	–	–	CO	–	**B10**, 50, **3**, *12*

1,3-Dimethyl-1,3-cyclohexadiene	–	–	CO	–	**B 8**, 146, **6**, *11*
1,4-Dimethyl-1,3-cyclohexadiene	–	–	CO	–	**B 8**, 157, **7**, *35*
1,5-Dimethyl-1,3-cyclohexadiene	–	–	CO	–	**B 8**, 168
1,6-Dimethyl-1,3-cyclohexadiene	–	–	CO	–	**B 8**, 176, **8**, *12*
2,5-Dimethyl-1,3-cyclohexadiene	–	–	CO	–	**B 8**, 207, **9**, *73*
2,6-Dimethyl-1,3-cyclohexadiene	–	–	CO	–	**B 8**, 234, **10**, *14*
5,5-Dimethyl-1,3-cyclohexadiene	–	–	CO	–	**B 8**, 243, **11**, *4*
2-Methyl-1,3-cycloheptadiene	C_6H_8	–	CO	–	**B 10**, 219, **8**, *21*
5-Methyl-1,3-cycloheptadiene	–	–	CO	–	**B 9**, 49, **2**, *26*
2-Methyl-4-cycloheptene-1,3-diyl	–	–	CO	–	**B 10**, 129, **4**, *35*
1,3-Cyclooctadiene	–	–	–	–	**B 10**, 190, **6**, *13*
1,3-Cyclooctadiene	–	–	CO	–	**B 9**, 210, **7**, *1*
1,3-Cyclooctadiene	$C_9H_{21}O_3P$	–	–	–	**B 10**, 180, **5**, *4*
1,5-Cyclooctadiene	–	–	–	–	**B 10**, 190, **6**, *14*
1,5-Cyclooctadiene	–	–	CO	–	**B 9**, 211, **7**, *7*
5-Cyclooctene-1,4-diyl	–	–	CO	–	**B 10**, 132, **4**, *50*
$C_8H_{12}B_2F_2$	–	–	CO	–	**B 8**, 14, **1**, *42*
$C_8H_{12}ClNO$	–	–	CO	–	**B 10**, 54, **3**, *42*
$C_8H_{12}N_2O_2$	–	–	CO	–	**B 9**, 12, **1**, *38*
$C_8H_{12}O$					
2,3,4,4-Tetramethyl-1-oxo-2-butene-1,4-diyl	–	–	CO	–	**B 10**, 50, **3**, *13*
1-(1-Hydroxyethyl)-1,3-cyclohexadiene	–	–	CO	–	**B 8**, 45, **2**, *10*
5-(1-Hydroxyethyl)-1,3-cyclohexadiene	–	–	CO	–	**B 8**, 99, **4**, *83*
1-Ethoxy-1,3-cyclohexadiene	–	–	CO	–	**B 8**, 48, **2**, *28*
5-Ethoxy-1,3-cyclohexadiene	–	–	CO	–	**B 8**, 85, **4**, *9*
1-Methoxy-3-methyl-1,3-cyclohexadiene	–	–	CO	–	**B 8**, 145, **6**, *4*
1-Methoxy-4-methyl-1,3-cyclohexadiene	–	–	CO	–	**B 8**, 152, **7**, *4*
1-Methoxy-5-methyl-1,3-cyclohexadiene	–	–	CO	–	**B 8**, 167
1-Methoxy-6-methyl-1,3-cyclohexadiene	–	–	CO	–	**B 8**, 174, **8**, *3*
2-Methoxy-1-methyl-1,3-cyclohexadiene	–	–	CO	–	**B 8**, 139
2-Methoxy-5-methyl-1,3-cyclohexadiene	–	–	CO	–	**B 8**, 199, **9**, *30*
3-Methoxy-1-methyl-1,3-cyclohexadiene	–	–	CO	–	**B 8**, 146, **6**, *8*

3-Methoxy-2-methyl-1,3-cyclohexadiene	–	–	CO	–	**B8**, 186
5-Methoxy-2-methyl-1,3-cyclohexadiene	–	–	CO	–	**B8**, 206, **9**, *70*
6-Methoxy-1-methyl-1,3-cyclohexadiene	–	–	CO	–	**B8**, 176, **8**, *10*
6-Methoxy-2-methyl-1,3-cyclohexadiene	–	–	CO	–	**B8**, 234, **10**, *12*
2-Hydroxy-5,5-dimethyl-1,3-cyclohexadiene	–	–	CO	–	**B8**, 366, **13**, *230*
5-Hydroxy-1,3-dimethyl-1,3-cyclohexadiene	–	–	CO	–	**B8**, 316, **13**, *23*
5-Hydroxy-2,5-dimethyl-1,3-cyclohexadiene	–	–	CO	–	**B8**, 358, **13**, *188*
5-Hydroxy-6,6-dimethyl-1,3-cyclohexadiene	–	–	CO	–	**B8**, 359, **13**, *199*
6-Hydroxy-2,6-dimethyl-1,3-cyclohexadiene	–	–	CO	–	**B8**, 359, **13**, *195*
5-Methoxy-1,3-cycloheptadiene	–	–	CO	–	**B9**, 45, **2**, *2*
?-Hydroxy-1,5-cyclooctadiene	–	–	CO	–	**B9**, 214, **7**, *20*
$C_8H_{12}OP$	C_6H_8P	–	–	–	**B10**, 193, **6**, *23*
$C_8H_{12}O_2$					
2-[2-(Acetyloxy)propylidene]-propane-1,3-diyl	–	–	CO	–	**B10**, 7, **1**, *12*
5-Methoxycarbonyl-2-hexene-1,6-diyl	–	–	CO	–	**B10**, 68, **3**, *102*
6-Methoxycarbonyl-2-hexene-1,6-diyl	–	–	CO	–	**B10**, 67, **3**, *101*
1,3-Dimethoxy-1,3-cyclohexadiene	–	–	CO	–	**B8**, 145, **6**, *1*
1,4-Dimethoxy-1,3-cyclohexadiene	–	–	CO	–	**B8**, 152, **7**, *1*
1,6-Dimethoxy-1,3-cyclohexadiene	–	–	CO	–	**B8**, 174, **8**, *2*
2,5-Dimethoxy-1,3-cyclohexadiene	–	–	CO	–	**B8**, 196, **9**, *4*
2,6-Dimethoxy-1,3-cyclohexadiene	–	–	CO	–	**B8**, 233, **10**, *2*
5-Hydroxy-2-methoxy-5-methyl-1,3-cyclohexadiene	–	–	CO	–	**B8**, 330, **13**, *97*
$C_8H_{12}P$	C_6H_8P	–	–	–	**B10**, 192, **6**, *22*
$C_8H_{12}S$	–	–	–	–	**B10**, 190, **6**, *12*
$C_8H_{13}N$	–	–	CO	–	**B8**, 89, **4**, *36*
$C_8H_{13}NO$					
Carbonyl[(1-methylethyl)-imino][1-methyl-1-propene-1,3-diyl]	–	–	CO	–	**B10**, 54, **3**, *41*
Carbonyl(methylimino)[1,4-dimethyl-2-butene-1,4-diyl]	–	–	CO	–	**B10**, 58, **3**, *60*
$C_8H_{13}O$					
n-C_4H_9(O)CC(CH_2)$_2$	–	–	CO	–	**B10**, 9, **1**, *24*

$C_2H_5(O)CC(CH_2)C(CH_3)_2$	–	–	CO	–	**B10**, 15, **1**, *60*
$C_8H_{13}O_3P$	–	–	CO	–	**B8**, 91, **4**, *46*
C_8H_{14}	$C_{10}H_{14}O_4$	–	CO	–	**B10**, 218, **8**, *12*
$C_8H_{14}O$					
$HO(n\text{-}C_4H_9)CC(CH_2)_2$	–	–	CO	–	**B10**, 9, **1**, *25*
$HO(C_2H_5)CC(CH_2)C(CH_3)_2$	–	–	CO	–	**B10**, 15, **1**, *61*
$C_8H_{14}P$	–	–	–	–	**B10**, 189, **6**, *6*
C_9H_7BrO	–	–	CO	–	**B9**, 266, **9**, *42*
$C_9H_7F_6$	–	–	CO	–	**B8**, 464, **17**, *20*
C_9H_7N	–	–	CO	–	**B9**, 262, **9**, *29*
C_9H_7NO					
5*H*-Pyrrolo[1,2-*a*]azepin-5-one	–	–	CO	–	**B9**, 10, **1**, *28*
9*H*-Pyrrolo[1,2-*a*]azepin-9-one	–	–	CO	–	**B9**, 9, **1**, *25*
$C_9H_7O_2$					
2,5-Cyclopropyl-2,5-cyclohexadiene-1,4-dione	–	–	CO	–	**B8**, 454, **16**, *7*
2,6-Cyclopropyl-2,5-cyclohexadiene-1,4-dione	–	–	CO	–	**B8**, 454, **16**, *6*
C_9H_8	–	–	CO	–	**B8**, 273, **12**, *37*
$C_9H_8AlBr_3O$	–	–	CO	–	**B10**, 141, **4**, *93*
$C_9H_8AlBr_3O$	–	–	CO	$AlBr_3$	**B10**, 141, **4**, *95*
$C_9H_8AlCl_3O$	–	–	CO	–	**B10**, 141, **4**, *92*
$C_9H_8AlCl_3O$	–	–	CO	$AlCl_3$	**B10**, 141, **4**, *94*
$C_9H_8F_6$	–	–	CO	–	**B8**, 104, **4**, *109*
$C_9H_8N_2$	–	–	CO	–	**B8**, 99, **4**, *82*
C_9H_8O					
1,3,5,7-Cyclooctatetraene-1-carbaldehyde	–	–	CO	–	**B9**, 260, **9**, *18*
3a,7a-Dihydro-1*H*-inden-1-one	–	–	CO	–	**B8**, 272, **12**, *36*
Carbonyl[bicyclo[3.2.1]octa-3,6-diene-8,2-diyl]	–	–	CO	–	**B10**, 137, **4**, *73*
Bicyclo[3.2.2]nona-3,6,8-trien-2-one	–	–	CO	–	**B8**, 471, **17**, *54*
Bicyclo[4.2.1]nona-2,4,7-trien-9-one	–	–	CO	–	**B9**, 80, **2**, *166*
7-Oxobicyclo[3.2.2]nona-3,8-diene-2,6-diyl	–	–	CO	–	**B10**, 140, **4**, *91*
$C_9H_8O_2$					
2-Acetyl-2,4,6-cycloheptatrien-1-one	–	–	CO	–	**B9**, 183, **5**, *6*
Carbonyloxy[3,5,7-cyclooctatriene-1,2-diyl]	–	–	CO	–	**B10**, 135, **4**, *64*
C_9H_9N					
3-[1,3-Cyclohexadien-1-yl]-2-propenenitrile	–	–	CO	–	**B8**, 46, **2**, *19*
6-Methylene-2,4-cycloheptadiene-1-carbonitrile	–	–	CO	–	**B9**, 69, **2**, *116*
7-Cyano-2-methylene-5-cycloheptene-1,4-diyl	–	–	CO	–	**B10**, 130, **4**, *41*

7-Cyanobicyclo[3.2.1]-oct-3-ene-1,6-diyl	–	–	CO	–	**B10**, 138, **4**, *79*
C_9H_9NO	–	–	CO	–	**B9**, 224, **8**, *8*
$C_9H_9N_3O_2$	–	–	CO	–	**B9**, 18, **1**, *68*
$C_9H_9O_2$					
2,5-Dipropyl-2,5-cyclo-hexadiene-1,4-dione	–	–	CO	–	**B8**, 454, **16**, *9*
2,6-Dipropyl-2,5-cyclo-hexadiene-1,4-dione	–	–	CO	–	**B8**, 454, **16**, *8*
C_9H_{10}					
1-Ethenyl-1,3,5-cyclo-heptatriene	–	–	CO	–	**B9**, 120, **3**, *3*
7-Ethylidene-1,3,5-cyclo-heptatriene	–	–	CO	–	**B9**, 158, **4**, *37*
1-Methyl-2-methylene-4,6-cycloheptadiene-1,3-diyl	–	–	CO	–	**B10**, 14, **1**, *58*
1-Methyl-1,3,5,7-cyclo-octatetraene	–	–	CO	–	**B9**, 259, **9**, *10*
1,3,5,7-Cyclononatetraene	–	–	CO	–	**B9**, 279/80
3a,7a-Dihydro-1*H*-indene	–	–	CO	–	**B8**, 271, **12**, *31*
Bicyclo[6.1.0]nona-2,4,6-triene	–	–	CO	–	**B9**, 224, **8**, *7*
Bicylo[3.3.1]nona-3,7-diene-2,9-diyl	–	–	CO	–	**B10**, 137, **4**, *75*
Tricyclo[4.3.0.0^{7,9}]nona-2,4-diene	–	–	CO	–	**B8**, 273, **12**, *40*
Tricyclo[3.2.2.0^{2,4}]nona-6,8-diene	–	–	CO	–	**B8**, 468, **17**, *36*
$C_9H_{10}F_6$	–	–	CO	–	**B10**, 72, **3**, *117*
$C_9H_{10}N_2$					
1-[2,4-Cyclohexadien-1-yl]-1*H*-imidazole	–	–	CO	–	**B8**, 112, **4**, *161*
4-[2,4-Cyclohexadien-1-yl]-1*H*-imidazole	–	–	CO	–	**B8**, 108, **4**, *134*
$C_9H_{10}N_2O$	–	–	CO	–	**B9**, 73, **2**, *129*
$C_9H_{10}O$					
1-Methoxycarbonyl-1,3,5-cycloheptatriene	–	–	CO	–	**B9**, 123, **3**, *15*
2-Methoxycarbonyl-1,3,5-cycloheptatriene	–	–	CO	–	**B9**, 124, **3**, *23*
7-Methoxymethylene-1,3,5-cycloheptatriene	–	–	CO	–	**B9**, 157, **4**, *35*
1-Methoxy-1,3,5,7-cyclo-octatetraene	–	–	CO	–	**B9**, 258, **9**, *3*
1-(Hydroxymethyl)-1,3,5,7-cyclooctatetraene	–	–	CO	–	**B9**, 259, **9**, *11*
2,3,3a,7a-Tetrahydro-1*H*-inden-1-one	–	–	CO	–	**B8**, 271, **12**, *28*
3a,7a-Dihydro-1-hydroxy-1*H*-indene	–	–	CO	–	**B8**, 271, **12**, *32*
7-Ethenyl-7-hydroxy-bicyclo[2.2.1]hepta-2,5-diene	–	–	CO	–	**B8**, 465, **17**, *24*

8-Methylbicyclo[5.1.0]octa-3,5-dien-2-one	–	–	CO	–	**B9**, 71, **2**, *124*
Carbonyl[bicyclo[5.1.0]oct-3-ene-6,2-diyl]	–	–	CO	–	**B10**, 130, **4**, *43*
4-Hydroxybicyclo[3.2.2]nona-2,6,8-triene	–	–	CO	–	**B8**, 471, **17**, *51* **B9**, 202, **6**, *14*
9-Hydroxybicyclo[4.2.1]nona-2,4,7-triene	–	–	CO	–	**B9**, 80, **2**, *162* **B9**, 202, **6**, *14*
Bicyclo[3.2.2]nona-6,8-dien-2-one	–	–	CO	–	**B8**, 471, **17**, *50*
Bicyclo[4.2.1]nona-2,4-dien-7-one	–	–	CO	–	**B9**, 60, **2**, *75*
Bicyclo[4.2.1]nona-2,4-dien-9-one	–	–	CO	–	**B9**, 80, **2**, *165*
Bicyclo[4.2.1]nona-3,7-dien-2-one	–	–	CO	–	**B9**, 213, **7**, *17*
7-Oxobicyclo[3.2.2]non-3-ene-2,6-diyl	–	–	CO	–	**B10**, 140, **4**, *89*
$C_9H_{10}O_2$					
5-[(Methoxycarbonyl)methylene]-1,3-cyclohexadiene	–	–	CO	–	**B8**, 246, **11**, *18* **B8**, 247, **11**, *19*
1-Methoxycarbonyl-1,3,5-cycloheptatriene	–	–	CO	–	**B9**, 123, **3**, *17*
7-Methoxycarbonyl-1,3,5-cycloheptatriene	–	–	CO	–	**B9**, 129, **3**, *43*
2-Methoxycarbonylbicyclo-[2.2.1]hepta-2,5-diene	–	–	CO	–	**B8**, 462, **17**, *8*
7-Methoxycarbonylbicyclo-[2.2.1]hepta-2,5-diene	–	–	CO	–	**B8**, 472, **17**, *57*
[5-Hydroxy-6-oxobicyclo-[5.1.0]oct-3-ene-2,6-diyl]	–	–	CO	–	**B10**, 131, **4**, *45*
1,4-Dioxaspiro[4.6]undeca-6,8,10-triene	–	–	CO	–	**B9**, 153, **4**, *19*
$C_9H_{10}O_4$					
1-Methoxycarbonyl-2,4-cyclohexadiene-1-carboxylic acid	–	–	CO	–	**B8**, 244, **11**, *11*
2-Methoxycarbonyl-2,4-cyclohexadiene-1-carboxylic acid	–	–	CO	–	**B8**, 180, **8**, *27*
4-Methoxycarbonyl-1,3-cyclohexadiene-1-carboxylic acid	–	–	CO	–	**B8**, 159, **7**, *50*
$C_9H_{10}O_5$	–	–	CO	–	**B10**, 57, **3**, *55*
$C_9H_{10}O_6$					
5,6-Dimethoxy-3-methoxycarbonyl-2-oxo-2*H*-pyran	–	–	CO	–	**B8**, 14, **1**, *39*
5,6-Dimethoxy-4-methoxycarbonyl-2-oxo-2*H*-pyran	–	–	CO	–	**B8**, 14, **1**, *40*
2-Methoxy-3,4-bis(methoxycarbonyl)-1-oxo-2-butene-1,4-diyl	–	–	CO	–	**B10**, 49, **3**, *5*

3-Methoxy-1,2-bis(methoxy-carbonyl)-1-oxo-2-butene-1,4-diyl	–	–	CO	–	**B10**, 49, **3**, *6*
$C_9H_{10}S$	–	–	CO	–	**B9**, 258, **9**, *4*
C_9H_{11}	–	–	CO	–	**B8**, 467, **17**, *33*
$C_9H_{11}F_3$	–	–	CO	–	**B10**, 67, **3**, *98*
$C_9H_{11}N$					
2,6-Cyclooctadiene-1-carbonitrile	–	–	CO	–	**B9**, 212, **7**, *8*
8-Cyano-5-cyclooctene-1,4-diyl	–	–	CO	–	**B10**, 133, **4**, *55*
$C_9H_{11}NO$					
4-Methoxy-1-methyl-2,4-cyclohexadiene-1-carbonitrile	–	–	CO	–	**B8**, 333, **13**, *108*
3-Methoxycarbonyl-1-methyl-1*H*-azepine	–	–	CO	–	**B9**, 6, **1**, *10*
$C_9H_{11}NO_2$	–	–	CO	–	**B9**, 5, **1**, *5*
$C_9H_{11}N_3O$	–	–	CO	–	**B8**, 462, **17**, *6*
C_9H_{12}					
5-Ethenyl-2-methyl-1,3-cyclohexadiene	–	–	CO	–	**B8**, 208, **9**, *78*
5-(2-Propenyl)-1,3-cyclohexadiene	–	–	CO	–	**B8**, 93, **4**, *57*
5-(1-Propenyl)-1,3-cyclohexadiene	–	–	CO	–	**B8**, 97, **4**, *72*
5-(1-Methylethenyl)-1,3-cyclohexadiene	–	–	CO	–	**B8**, 101, **4**, *95*
6-Ethenyl-1-methyl-1,3-cyclohexadiene	–	–	CO	–	**B8**, 176, **8**, *16*
1,3-Dimethyl-6-methylene-1,3-cyclohexadiene	–	–	CO	–	**B8**, 402, **14**, *31*
2-Ethyl-1,3,5-cycloheptatriene	–	–	CO	–	**B9**, 124, **3**, *22*
1,3,5-Cyclononatriene	–	–	CO	–	**B9**, 279
2,3,4,5-Tetrahydro-1*H*-indene	–	–	CO	–	**B8**, 142
2,3,5,6-Tetrahydro-1*H*-indene	–	–	CO	–	**B8**, 187
2,3,3a,7a-Tetrahydro-1*H*-indene	–	–	CO	–	**B8**, 270, **12**, *27*
Bicyclo[4.2.1]nona-2,4-diene	–	–	CO	–	**B9**, 60, **2**, *74*
$C_9H_{12}N_2$	–	–	CO	–	**B8**, 108, **4**, *135*
$C_9H_{12}N_2O_2$					
1,2-Diacetyl-2,3-dihydro-1*H*-1,2-diazepine	–	–	CO	–	**B9**, 12, **1**, *39*
1-(1-Methylethoxycarbonyl)-1*H*-1,2-diazepine	–	–	CO	–	**B9**, 16, **1**, *59*
$C_9H_{12}O$					
1-(2-Oxopropyl)-1,3-cyclohexadiene	–	–	CO	–	**B8**, 46, **2**, *14*
2-(2-Oxopropyl)-1,3-cyclohexadiene	–	–	CO	–	**B8**, 66, **3**, *17*
5-(2-Oxopropyl)-1,3-cyclohexadiene	–	–	CO	–	**B8**, 93, **4**, *59*

5-(1-Oxopropyl)-1,3-cyclohexadiene	–	–	CO	–	**B8**, 104, **4**, *105*
6-Acetyl-1-methyl-1,3-cyclohexadiene	–	–	CO	–	**B8**, 177, **8**, *19*
5-Acetyl-2-methyl-1,3-cyclohexadiene	–	–	CO	–	**B8**, 210, **9**, *85*
5-Ethylidene-2-methoxy-1,3-cyclohexadiene	–	–	CO	–	**B8**, 354, **13**, *172*
1-(1-Hydroxyethyl)-1,3,5-cycloheptatriene	–	–	CO	–	**B9**, 122, **3**, *9*
5-Ethylidene-7-hydroxy-1,3-cycloheptadiene	–	–	CO	–	**B9**, 66, **2**, *100*
7-Acetyl-5-methylene-1,3-cycloheptadiene	–	–	CO	–	**B9**, 68, **2**, *107*
Carbonyl[5-cyclooctene-1,4-diyl]	–	–	CO	–	**B10**, 135, **4**, *62*
2,3,3a,7a-Tetrahydro-2-methylbenzofuran	–	–	CO	–	**B8**, 269, **12**, *20*
2-(1-Hydroxyethyl)bicyclo-[2.2.1]hepta-2,5-diene	–	–	CO	–	**B8**, 462, **17**, *4*
7-(2-Hydroxyethyl)bicyclo-[2.2.1]hepta-2,5-diene	–	–	CO	–	**B8**, 464, **17**, *17*
7-Ethyl-7-hydroxybicyclo-[2.2.1]hepta-2,5-diene	–	–	CO	–	**B8**, 465, **17**, *23*
4-Methoxybicyclo[3.2.1]-octa-2,6-diene	–	–	CO	–	**B9**, 200, **6**, *2*
7-Hydroxy-7-methylbicyclo[4.2.0]octa-2,4-diene	–	–	CO	–	**B8**, 267, **12**, *15*
9-Hydroxybicyclo[4.2.1]-nona-2,4-diene	–	–	CO	–	**B9**, 61, **2**, *79*
$C_9H_{12}O_2$					
1-(Methoxycarbonylmethyl)-1,3-cyclohexadiene	–	–	CO	–	**B8**, 472, **17**, *56*
5-(Acetyloxymethyl)-1,3-cyclohexadiene	–	–	CO	–	**B8**, 92, **4**, *52*
5-(Methoxycarbonylmethyl)-1,3-cyclohexadiene	–	–	CO	–	**B8**, 94, **4**, *60*
1-Methoxycarbonyl-2-methyl-1,3-cyclohexadiene	–	–	CO	–	**B8**, 140
2-Methoxycarbonyl-1-methyl-1,3-cyclohexadiene	–	–	CO	–	**B8**, 140
3-Methoxycarbonyl-1-methyl-1,3-cyclohexadiene	–	–	CO	–	**B8**, 146, **6**, *12*
1-Methoxycarbonyl-3-methyl-1,3-cyclohexadiene	–	–	CO	–	**B8**, 147, **6**, *14*
6-Methoxycarbonyl-1-methyl-1,3-cyclohexadiene	–	–	CO	–	**B8**, 177, **8**, *22*
1-Methoxycarbonyl-6-methyl-1,3-cyclohexadiene	–	–	CO	–	**B8**, 178, **8**, *25*
3-Methoxycarbonyl-2-methyl-1,3-cyclohexadiene	–	–	CO	–	**B8**, 186

5-Methoxycarbonyl-5-methyl-1,3-cyclohexadiene	–	–	CO	–	**B8**, 244, **11**, *7* **B8**, 248, **11**, *24*
5-Methoxycarbonyl-6-methyl-1,3-cyclohexadiene	–	–	CO	–	**B8**, 264, **12**, *5*
[4-Methoxy-1-methyl-2,4-cyclohexadien-1-yl]carbaldehyde	–	–	CO	–	**B8**, 332, **13**, *103*
4-(1-Methylethoxy)-2,4-cyclohexadien-1-one	–	–	CO	–	**B8**, 357, **13**, *186*
3-(1-Methylethoxy)-2,4-cyclohexadien-1-one	–	–	CO	–	**B8**, 358, **13**, *194*
[2-(Carbonyloxymethyl)-1-cyclohexene-1,2-diyl]-methylene	–	–	CO	–	**B10**, 65, **3**, *90*
[3-Cyclohexene-1,2-diyl]-methoxycarbonylmethyl	–	–	CO	–	**B10**, 123, **4**, *15*
[4-Cyclohexene-1,3-diyl] methoxycarbonylmethyl	–	–	CO	–	**B10**, 123, **4**, *16*
7-Acetyl-5-hydroxy-1,3-cycloheptatriene	–	–	CO	–	**B9**, 59, **2**, *71*
6-Methoxy-2-methyl-2,4-cycloheptadien-1-one	–	–	CO	–	**B9**, 70, **2**, *119*
6-Methoxy-7-methyl-2,4-cycloheptadien-1-one	–	–	CO	–	**B9**, 70, **2**, *122*
2,3,3a,7a-Tetrahydro-3-hydroxymethylbenzofuran	–	–	CO	–	**B8**, 269, **12**, *21*
$C_9H_{12}O_2P$	C_6H_8P	–	–	–	**B10**, 192, **6**, *20*
$C_9H_{12}O_3$					
3-Methoxycarbonyl-2,4,4-trimethyl-1-oxo-2-butene-1,3-diyl	–	–	CO	–	**B10**, 50, **3**, *14*
2-Methoxycarbonyl-3,4,4-trimethyl-1-oxo-2-butene-1,3-diyl	–	–	CO	–	**B10**, 51, **3**, *23*
[5-Ethoxy-3-ethyl-2,3-dihydro-2-oxofuran-3,4-diyl]methylene	–	–	CO	–	**B10**, 15, **1**, *64*
6-(1,1-Dimethylethyloxy)-*2H*-pyran-2-one	–	–	CO	–	**B8**, 14, **1**, *38*
1-Methoxy-2-methoxycarbonyl-1,3-cyclohexadiene	–	–	CO	–	**B8**, 141
1-Methoxy-3-methoxycarbonyl-1,3-cyclohexadiene	–	–	CO	–	**B8**, 145, **6**, *5*
3-Methoxy-1-methoxycarbonyl-1,3-cyclohexadiene	–	–	CO	–	**B8**, 146, **6**, *13*
[4-Methoxy-1,3-cyclohexadien-1-yl]acetic acid	–	–	CO	–	**B8**, 153, **7**, *8*
4-Methoxy-1-methoxycarbonyl-1,3-cyclohexadiene	–	–	CO	–	**B8**, 157, **7**, *33*

5-Methoxy-1-methoxycarbonyl-1,3-cyclohexadiene	–	–	CO	–	**B8**, 168
1-Methoxy-6-methoxycarbonyl-1,3-cyclohexadiene	–	–	CO	–	**B8**, 175, **8**, *8*
1-Acetyloxy-6-methoxy-1,3-cyclohexadiene	–	–	CO	–	**B8**, 181, **8**, *30*
[4-Methoxy-2,4-cyclohexadien-1-yl]acetic acid	–	–	CO	–	**B8**, 213, **9**, *97*
5-Hydroxy-5-(methoxycarbonylmethyl)-1,3-cyclohexadiene	–	–	CO	–	**B8**, 242, **11**, *1*
5-Acetyloxy-6-methoxy-1,3-cyclohexadiene	–	–	CO	–	**B8**, 264, **12**, *4*
C_9H_{13}	–	–	CO	–	**B8**, 48, **2**, *27*
$C_9H_{13}N$	–	–	CO	–	**B9**, 7, **1**, *15*
$C_9H_{13}NO_2$	–	–	CO	–	**B9**, 7, **1**, *18*
$C_9H_{13}N_3O$					
2-[1-[1,3-Cyclohexadien-1-yl]-ethylidene]hydrazine-carboxamide	–	–	CO	–	**B8**, 46, **2**, *17*
2-[1-[1,5-Cyclohexadien-1-yl]-ethylidene]hydrazine-carboxamide	–	–	CO	–	**B8**, 65, **3**, *11*
2-[1-[2,4-Cyclohexadien-1-yl]-ethylidene]hydrazine-carboxamide	–	–	CO	–	**B8**, 101, **4**, *94*
C_9H_{14}					
5-(1-Methylethyl)-1,3-cyclohexadiene	–	–	CO	–	**B8**, 97, **4**, *75*
1,3,5-Trimethyl-1,3-cyclohexadiene	–	–	CO	–	**B8**, 316, **13**, *25*
5,7-Dimethyl-1,3-cycloheptadiene	–	–	CO	–	**B9**, 79, **2**, *159*
2-Ethyl-4-cycloheptene-1,3-diyl	–	–	CO	–	**B10**, 129, **4**, *36*
2,6-Dimethyl-4-cycloheptene-1,3-diyl	–	–	CO	–	**B10**, 129, **4**, *37*
3-Methyl-1,5-cyclooctadiene	–	–	CO	–	**B9**, 212, **7**, *9*
8-Methyl-5-cyclooctene-1,4-diyl	–	–	CO	–	**B10**, 134, **4**, *56*
$C_9H_{14}O$					
1-[(1-Methylethyl)oxy]-1,3-cyclohexadiene	–	–	CO	–	**B8**, 44, **2**, *3*
1-(1-Hydroxy-1-methylethyl)-1,3-cyclohexadiene	–	–	CO	–	**B8**, 45, **2**, *11*
2-[(1-Methylethyl)oxy]-1,3-cyclohexadiene	–	–	CO	–	**B8**, 64, **3**, *4*
5-Propoxy-1,3-cyclohexadiene	–	–	CO	–	**B8**, 85, **4**, *10*
5-(2-Hydroxypropyl)-1,3-cyclohexadiene	–	–	CO	–	**B8**, 93, **4**, *58*

4-Ethyl-1-methoxy-1,3-cyclohexadiene	–	–	CO	–	**B8**, 157, **7**, *39*
6-Ethoxy-2-methyl-1,3-cyclohexadiene	–	–	CO	–	**B8**, 234, **10**, *13*
1-Methoxy-3,5-dimethyl-1,3-cyclohexadiene	–	–	CO	–	**B8**, 315, **13**, *19*
3-Methoxy-1,5-dimethyl-1,3-cyclohexadiene	–	–	CO	–	**B8**, 315, **13**, *20*
1-Methoxy-3,4-dimethyl-1,3-cyclohexadiene	–	–	CO	–	**B8**, 360, **13**, *204*
5-Ethoxy-1,3-cycloheptadiene	–	–	CO	–	**B9**, 45, **2**, *3*
5-Methoxy-1,3-cyclooctadiene	–	–	CO	–	**B9**, 211, **7**, *3*
8-Methoxy-5-cyclooctene-1,4-diyl	–	–	CO	–	**B10**, 133, **4**, *53*
$C_9H_{14}OP$	C_6H_8P	–	–	–	**B10**, 193, **6**, *25*
$C_9H_{14}OS$	–	–	CO	–	**B8**, 237, **10**, *28*
$C_9H_{14}O_2$					
Carbonyloxy[1,1,4-trimethyl-2-pentene-1,4-diyl]	–	–	CO	–	**B10**, 65, **3**, *93*
6-Methoxycarbonyl-3-methyl-2-pentene-1,6-diyl	–	–	CO	–	**B10**, 68, **3**, *104*
6-Methoxycarbonyl-2-methyl-2-pentene-1,6-diyl	–	–	CO	–	**B10**, 68, **3**, *105*
5-[2-Hydroxy-1-(hydroxymethyl)ethyl]-1,3-cyclohexadiene	–	–	CO	–	**B8**, 98, **4**, *78*
4-(2-Hydroxyethyl)-1-methoxy-1,3-cyclohexadiene	–	–	CO	–	**B8**, 153, **7**, *11*
5-Ethoxy-2-methoxy-1,3-cyclohexadiene	–	–	CO	–	**B8**, 196, **9**, *5*
1,3-Dimethoxy-2-methyl-1,3-cyclohexadiene	–	–	CO	–	**B8**, 311, **13**, *1*
5-(Hydroxymethyl)-2-methoxy-5-methyl-1,3-cyclohexadiene	–	–	CO	–	**B8**, 331, **13**, *99*
$C_9H_{15}N$					
2,3-Dihydro-3,5,7-trimethyl-1*H*-azepine	–	–	CO	–	**B9**, 7, **1**, *19*
5-Dimethylamino-1,3-cycloheptadiene	–	–	CO	–	**B9**, 48, **2**, *21*
$C_9H_{15}NO$	–	–	CO	–	**B8**, 197, **9**, *15*
$C_9H_{15}O$	–	–	CO	–	**B10**, 9, **1**, *26*
$C_9H_{16}O$	–	–	CO	–	**B10**, 9, **1**, *27*
$C_9H_{16}OSi$					
1-(Trimethylsilyloxy)-2-methylenecyclopentane-1,3-diyl	–	–	CO	–	**B10**, 13, **1**, *54*
1-(Trimethylsilyloxy)-1,3-cyclohexadiene	–	–	CO	–	**B8**, 44, **2**, *4*
2-(Trimethylsilyloxy)-1,3-cyclohexadiene	–	–	CO	–	**B8**, 63, **3**, *2*

5-(Trimethylsilyloxy)-1,3-cyclohexadiene	–	–	CO	–	**B8**, 84, **4**, *6*
$C_9H_{16}O_2Si$	–	–	CO	–	**B8**, 6, **1**, *5*
$C_9H_{16}O_3P$	–	–	CO	–	**B8**, 133, **5**, *13*
$C_9H_{16}Si$	–	–	CO	–	**B8**, 64, **3**, *5*
$C_9H_{18}OSi$	–	–	CO	–	**B10**, 8, **1**, *21*
$C_9H_{21}O_3P$	C_4H_6	–	–	–	**B10**, 207, **7**, *6*
$C_9H_{21}O_3P$	C_5H_8	–	–	–	**B10**, 209, **7**, *17*
$C_9H_{21}O_3P$	C_5H_8	–	–	–	**B10**, 209, **7**, *18*
$C_9H_{21}O_3P$	C_6H_{10}	–	–	–	**B10**, 180, **5**, *2*
$C_9H_{21}O_3P$	C_6H_{10}	–	–	–	**B10**, 210, **7**, *24*
$C_9H_{21}O_3P$	C_8H_{12}	–	–	–	**B10**, 180, **5**, *4*
C_{10} $H_6F_6O_2$	–	–	CO	–	**B10**, 146, **4**, *116*
$C_{10}H_7NO_2$	–	–	CO	–	**B9**, 10, **1**, *29*
$C_{10}H_8$	–	–	CO	–	**B8**, 412, **14**, *72*
$C_{10}H_8F_6O$	–	–	CO	–	**B10**, 146, **4**, *115*
$C_{10}H_8N_2$	C_4H_6	–	–	–	**B10**, 206, **7**, *2*
$C_{10}H_8N_2$	C_6H_8	–	–	–	**B10**, 211, **7**, *29*
$C_{10}H_8O$	–	–	CO	–	**B9**, 21, **1**, *77*
$C_{10}H_8O_2$	–	–	CO	–	**B9**, 265, **9**, *39*
$C_{10}H_9N$					
1-[(2-Cyano)ethenyl]-1,3,5-cycloheptatriene	–	–	CO	–	**B9**, 121, **3**, *6*
1-(Cyanomethyl)-1,3,5,7-cyclooctatetraene	–	–	CO	–	**B9**, 260, **9**, *17*
7-Cyanobicyclo[3.2.2]nona-3,8-diene-2,6-diyl	–	–	CO	–	**B10**, 140, **4**, *90*
$C_{10}H_9N_3O_3$	–	–	CO	–	**B10**, 140, **4**, *87*
$C_{10}H_{10}$					
$C_6H_5CH{=}CHCH{=}CH_2$	C_6H_{10}	–	CO	–	**B10**, 181, **5**, *8*
$C_6H_5CHC(CH_2)_2$	–	–	CO	–	**B10**, 7, **1**, *17*
1-Ethenyl-1,3,5,7-cyclooctatetraene	–	–	CO	–	**B9**, 261, **9**, *19*
1,3a,4,6a-Tetrahydro-1,4-ethenopentalene	–	–	CO	–	**B9**, 204, **6**, *22*
Bicyclo[3.3.2]deca-3,7,9-triene-2,6-diyl	–	–	CO	–	**B10**, 144, **4**, *109*
$C_{10}H_{10}F_4$	–	–	CO	–	**B10**, 126, **4**, *24*
$C_{10}H_{10}F_6$					
1,2-Bis(trifluoromethyl)-4,5-dimethyl-1,4-hexadiene-1,6-diyl	–	–	CO	–	**B10**, 67, **3**, *99*
1,2-Dimethyl-4,5-bis(trifluoromethyl)-1,3-cyclohexadiene	–	–	CO	–	**B8**, 395, **14**, *1*
$C_{10}H_{10}O$					
5-(2-Furanyl)-1,3-cyclohexadiene	–	–	CO	–	**B8**, 107, **4**, *125*
3-[1-(1-Methyl)ethenyl]-2,4,6-cycloheptatrien-1-one	–	–	CO	–	**B9**, 185, **5**, *16*

1-Acetyl-1,3,5,7-cyclo-octatetraene	–	–	CO	–	**B9**, 262, **9**, *28*
4-Methyl-1,3,5,7-cyclo-octatetraene-1-carbaldehyde	–	–	CO	–	**B9**, 267, **9**, *48*
5-Methyl-1,3,5,7-cyclo-octatetraene-1-carbaldehyde	–	–	CO	–	**B9**, 267, **9**, *50*
9-Carbonyl[bicyclo[3.3.1]-nona-3,6-diene-2,9-diyl]	–	–	CO	–	**B10**, 137, **4**, *76*
8-Methyl-7-oxobicyclo[3.2.2]-nona-3,8-diene-2,6-diyl	–	–	CO	–	**B10**, 142, **4**, *100*
$C_{10}H_{10}O_2$	–	–	CO	–	**B9**, 261, **9**, *21*
$C_{10}H_{10}S$	–	–	CO	–	**B8**, 107, **4**, *127*
$C_{10}H_{11}Cl_2NO_2$	–	–	CO	–	**B9**, 21, **1**, *80*
$C_{10}H_{11}F_3$	–	–	CO	–	**B10**, 126, **4**, *23*
$C_{10}H_{11}F_3O_3$	–	–	CO	–	**B8**, 362, **13**, *213*
$C_{10}H_{11}N$	–	–	CO	–	**B8**, 108, **4**, *129*
$C_{10}H_{11}NO$	–	–	CO	–	**B10**, 144, **4**, *108*
$C_{10}H_{11}NO_3$	–	–	CO	–	**B9**, 6, **1**, *11*
$C_{10}H_{11}N_3O$	–	–	CO	–	**B8**, 91, **4**, *42*
$C_{10}H_{12}$					
7-(2-Propenyl)-1,3,5-cyclo-heptatriene	–	–	CO	–	**B9**, 128, **3**, *41*
7-(1-Methylethylidene)-1,3,5-cycloheptatriene	–	–	CO	–	**B9**, 160, **4**, *44*
1-Ethyl-1,3,5,7-cyclo-octatetraene	–	–	CO	–	**B9**, 260, **9**, *14*
1,3,5,7-Cyclodecatetraene	–	–	CO	–	**B9**, 282
1,4,5,6-Tetrahydronaphthalene	–	–	CO	–	**B8**, 142
2,3-Dihydro-3a,7a-methano-1*H*-indene	–	–	CO	–	**B8**, 407, **14**, *57*
Dispiro[2.0.2.4]deca-7,9-diene	–	–	CO	–	**B8**, 407, **14**, *56*
8-Methyl-tricyclo[4.3.0.0^{7,9}]-nona-2,4-diene	–	–	CO	–	**B8**, 273, **12**, *42*
Tricyclo[4.4.0.0^{2,5}]deca-7,9-diene	–	–	CO	–	**B8**, 274, **12**, *43*
Tricyclo[4.3.1.0^{2,5}]dec-8-ene-7,10-diyl	–	–	CO	–	**B10**, 136, **4**, *70*
$C_{10}H_{12}F_4$					
[1,1,2,2-Tetrafluoroethane-1,2-diyl]-1,2,3,4-tetramethyl-3-cyclobutene-2,1-diyl	–	–	CO	–	**B10**, 121, **4**, *7*
[2-Cyclooctene-1,4-diyl]-1,1,2,2-tetrafluoroethane-2,1-diyl	–	–	CO	–	**B10**, 177
$C_{10}H_{12}N_2O$					
2-Methoxy-5(1-1*H*-pyrazolyl)-1,3-cyclohexadiene	–	–	CO	–	**B8**, 197, **9**, *17*
3a,8a-Dihydro-3,3-di-methyl-3*H*-cycloheptapyrazol-4-one	–	–	CO	–	**B9**, 73, **2**, *130*

$C_{10}H_{12}O$					
1-[3-Oxo-1-butenyl]-1,3-cyclohexadiene	–	–	CO	–	**B8**, 47, **2**, *20*
5-[2-Oxo-3-butenyl]-1,3-cyclohexadiene	–	–	CO	–	**B8**, 95, **4**, *63*
5-[1-(1-Acetyl)ethenyl]-1,3-cyclohexadiene	–	–	CO	–	**B8**, 102, **4**, *96*
7-Cyclopropyl-7-hydroxy-1,3,5-cycloheptatriene	–	–	CO	–	**B9**, 152, **4**, *16*
7-(Ethoxy)methylene-1,3,5-cycloheptatriene	–	–	CO	–	**B9**, 158, **4**, *36*
6-(1-Methylethylidene)-2,4-cyloheptadien-1-one	–	–	CO	–	**B9**, 75, **2**, *141*
1-(Methoxy)methyl-1,3,5,7-cyclooctatetraene	–	–	CO	–	**B9**, 259, **9**, *12*
1-[1-(Hydroxy)ethyl]-1,3,5,7-cyclooctatetraene	–	–	CO	–	**B9**, 261, **9**, *22*
1-[2-(Hydroxy)ethyl]-1,3,5,7-cyclooctatetraene	–	–	CO	–	**B9**, 260, **9**, *16*
5-Methoxy-1,3,5,7-cyclononatetraene	–	–	CO	–	**B9**, 281
9-Methoxy-1,3,5,7-cyclononatetraene	–	–	CO	–	**B9**, 281
3a,7a-Dihydro-1-methoxy-1*H*-indene	–	–	CO	–	**B8**, 272, **12**, *33*
1,3,3a,4,7,7a-Hexahydro-3-methyl-1,4-methanoisobenzofuran-7,8-diyl	–	–	CO	–	**B10**, 139, **4**, *83*
8,8-Dimethylbicyclo-[5.1.0]octa-3,5-dien-2-one	–	–	CO	–	**B9**, 72, **2**, *125*
8-Acetylbicyclo[3.2.1]-oct-3-ene-2,6-diyl	–	–	CO	–	**B10**, 138, **4**, *80*
4-Hydroxy-4-methylbicyclo-[3.2.2]nona-2,6,8-triene	–	–	CO	–	**B8**, 471, **17**, *52*
4-Hydroxy-4-methylbicyclo-[3.2.2]nona-2,6,8-triene	–	–	CO	–	**B9**, 202, **6**, *15*
7-Hydroxybicyclo[3.3.2]-deca-3,9-diene-2,6-diyl	–	–	CO	–	**B10**, 144, **4**, *110*
8-Methoxytricyclo[4.3.0.0^{7,9}]-nona-2,4-diene	– –	–	CO	–	**B8**, 273, **12**, *41*
$C_{10}H_{12}O_2$					
3-[Cyclohexa-2,4-dien-1-yl]-4,5-dihydro-2(3*H*)-furanone	–	–	CO	–	**B8**, 107, **4**, *126*
5-[1-(Methoxycarbonyl)-ethylidene]-1,3-cyclohexadiene	–	–	CO	–	**B8**, 247, **11**, *22* **B8**, 247, **11**, *23*
2,4,5,6-Tetramethyl-2,5-cyclohexadien-1,4-dione	–	–	CO	–	**B8**, 455, **16**, *10*
1-Ethoxycarbonyl-1,3,5-cycloheptatriene	–	–	CO	–	**B9**, 124, **3**, *18*

7-Ethoxycarbonyl-1,3,5-cycloheptatriene	–	–	CO	–	**B9**, 129, **3**, *44*
7-Methoxycarbonyl-7-methyl-1,3,5-cycloheptatriene	–	–	CO	–	**B9**, 151, **4**, *14*
5-(Methoxycarbonyl)methylene-1,3-cycloheptadiene	–	–	CO	–	**B9**, 56, **2**, *57*
1,3,4,5,8,9-Hexahydro-cycloocta[*c*]furan-1-one	–	–	CO	–	**B9**, 213, **7**, *15*
2-Methoxycarbonyl-1,3,5-cyclooctatriene	–	–	CO	–	**B9**, 223, **8**, *2*
2-Ethoxycarbonylbicyclo-[2.2.1]hepta-2,5-diene	–	–	CO	–	**B8**, 463, **17**, *9*
7-(Methoxycarbonyl)methyl-bicyclo[2.2.1]hepta-2,5-diene	–	–	CO	–	**B8**, 464, **17**, *16*
$C_{10}H_{12}O_3$	–	–	CO	–	**B8**, 354, **13**, *173*
$C_{10}H_{12}O_4$					
1,4-Bis(methoxycarbonyl)-1,3-cyclohexadiene	–	–	CO	–	**B8**, 159, **7**, *51*
1,6-Bis(methoxycarbonyl)-1,3-cyclohexadiene	–	–	CO	–	**B8**, 178, **8**, *26*
5,5-Bis(methoxycarbonyl)-1,3-cyclohexadiene	–	–	CO	–	**B8**, 245, **11**, *12*
5,6-Bis(methoxycarbonyl)-1,3-cyclohexadiene	–	–	CO	–	**B8**, 264, **12**, *6*
$C_{10}H_{12}O_5$	–	–	CO	–	**B10**, 52, **3**, *26*
$C_{10}H_{13}F_3$					
[1,2,3,4-Tetramethyl-3-cyclobutene-1,2-diyl]-2,2,2-trifluoroethylidene	–	–	CO	–	**B10**, 120, **4**, *5*
[1,2,3,4-Tetramethyl-3-cyclobutene-1,2-diyl]-1,2,2-trifluoroethane-2,1-diyl	–	–	CO	–	**B10**, 121, **4**, *6*
$C_{10}H_{13}NO$					
1-[(3-Iminio-3-methoxy)-1-propenyl]-1,3-cyclo-hexadiene	–	–	CO	–	**B8**, 47, **2**,*21*
5-[2-Oxo-1*H*-pyrrolidin-1-yl]-1,3-cyclohexadiene	–	–	CO	–	**B8**, 90, **4**, *39*
5-(Cyanomethyl)-2-methoxy-5-methyl-1,3-cyclohexadiene	–	–	CO	–	**B8**, 367, **13**, *234*
$C_{10}H_{13}NO_2$					
2-Ethenyl-1-ethoxycarbonyl-2*H*-pyridine	–	–	CO	–	**B8**, 12, **1**, *29*
1-Ethoxycarbonyl-1,2,5,6-tetrahydro-azocine-2,6-diyl	–	–	CO	–	**B10**, 131, **4**, *46* **B10**, 131, **4**, *47*
1-Ethoxycarbonyl-1,2,7,8-tetrahydro-azocine-2,8-diyl	–	–	CO	–	**B10**, 132, **4**, *48*
1-Ethoxycarbonyl-1,2,3,4-tetrahydro-azocine-2,4-diyl	–	–	CO	–	**B10**, 132, **4**, *49*

2-Ethoxycarbonyl-2-aza-bicyclo[5.1.0]octa-3,5-diene	–	–	CO	–	**B9**, 8, **1**, *21*
$C_{10}H_{13}N_2O$	–	–	CO	–	**B8**, 198, **9**, *21*
$C_{10}H_{14}$					
2-(2-Methyl-1-propenyl)-1,3-cyclohexadiene	–	–	CO	–	**B8**, 65, **3**, *12*
5-(2-Methyl-2-propenyl)-1,3-cyclohexadiene	–	–	CO	–	**B8**, 95, **4**, *62*
5-(1-Methyl-2-propenyl)-1,3-cyclohexadiene	–	–	CO	–	**B8**, 97, **4**, *73*
1-Methyl-4-(2-propenyl)-1,3-cyclohexadiene	–	–	CO	–	**B8**, 157, **7**, *37*
1-Methyl-5-(1-methylethenyl)-1,3-cyclohexadiene	–	–	CO	–	**B8**, 181, **8**, *32*
2-Methyl-5-(1-propenyl)-1,3-cyclohexadiene	–	–	CO	–	**B8**, 208, **9**, *79*
2-Methyl-5-(2-propenyl)-1,3-cyclohexadiene	–	–	CO	–	**B8**, 207, **9**, *76*
2-Methyl-5-(1-methylethenyl)-1,3-cyclohexadiene	–	–	CO	–	**B8**, 214, **9**, *101*
2-Methyl-5-(1-methylethylidene)-1,3-cyclohexadiene	–	–	CO	–	**B8**, 358, **13**, *191*
1-(1-Methylethyl)-1,3,5-cycloheptatriene	–	–	CO	–	**B9**, 120, **3**, *2*
7,7-Dimethyl-1,3,5-cycloheptatriene	–	–	CO	–	**B9**, 185, **5**, *14*
1,2,3,4,5,6-Hexahydronaphtalene	–	–	CO	–	**B8**, 142
1,2,3,4,6,7-Hexahydronaphthalene	–	–	CO	–	**B8**, 187
$C_{10}H_{14}NO$	–	–	CO	–	**B8**, 246, **11**, *17*
$C_{10}H_{14}N_2O_3$	–	–	CO	–	**B9**, 12, **1**, *40*
$C_{10}H_{14}O$					
5-(1-Methyl-2-oxopropyl)-1,3-cyclohexadiene	–	–	CO	–	**B8**, 97, **4**, *76*
5-(1,1-Dimethyl-2-oxoethyl)-1,3-cyclohexadiene	–	–	CO	–	**B8**, 102, **4**, *99*
5-(1-Oxobutyl)-1,3-cyclohexadiene	–	–	CO	–	**B8**, 104, **4**, *106*
2-Methoxy-5-(1-propenyl)-1,3-cyclohexadiene	–	–	CO	–	**B8**, 201, **9**, *39*
2-Methoxy-5-(2-propenyl)-1,3-cyclohexadiene	–	–	CO	–	**B8**, 200, **9**, *33*
2-Methoxy-5-(1-methylethenyl)-1,3-cyclohexadiene	– –	–	CO	–	**B8**, 213, **9**, *98*
2-Methoxy-6-(2-propenyl)-1,3-cyclohexadiene	–	–	CO	–	**B8**, 233, **10**, *3*
2-Methyl-5-(1-methylethyl)-2,4-cyclohexadien-1-one	–	–	CO	–	**B8**, 403, **14**, *36*
3,4,6,6-Tetramethyl-2,4-cyclohexadien-1-one	–	–	CO	–	**B8**, 447

4-Methyl-1-(1-methylethyl)-6-oxo-4-cyclohexene-1,3-diyl	–	–	CO	–	**B10**, 127, **4**, *29*
1-[1-Hydroxy-(1-methylethyl)]-1,3-cycloheptatriene	–	–	CO	–	**B9**, 124, **3**, *19*
7-Hydroxy-7-(1-methylethyl)-1,3,5-cycloheptatriene	–	–	CO	–	**B9**, 152, **4**, *15*
5-Hydroxy-7-(1-methylethylidene)-1,3-cycloheptadiene	–	–	CO	–	**B9**, 67, **2**, *104*
5-Methoxy-7-ethylidene-1,3-cycloheptadiene	–	–	CO	–	**B9**, 68, **2**, *108*
2,6,6-Trimethyl-2,4-cycloheptadien-1-one	–	–	CO	–	**B9**, 75, **2**, *142*
7-(1-Hydroxyethyl)bicyclo-[3.2.1]cycloocta-2,6-diene	–	–	CO	–	**B9**, 200, **6**, *3*
2,3,6,7-Tetrahydro-5-methoxy-1*H*-indene	–	–	CO	–	**B8**, 311, **13**, *2*
2,3,3a,4-Tetrahydro-6-methoxy-1*H*-indene	–	–	CO	–	**B8**, 317, **13**, *30*
9-Methoxy-bicyclo[4.2.1]-nona-2,4-diene	–	–	CO	–	**B9**, 79, **2**, *161*
$C_{10}H_{14}O_2$					
1-Methoxy-4-(2-oxopropyl)-1,3-cyclohexadiene	–	–	CO	–	**B8**, 153, **7**, *7*
2-Methoxy-5-(2-oxopropyl)-1,3-cyclohexadiene	–	–	CO	–	**B8**, 200, **9**, *36*
6-Methoxycarbonyl-1,6-dimethyl-1,3-cyclohexadiene	–	–	CO	–	**B8**, 365, **13**, *227*
6-Methoxycarbonyl-2,6-dimethyl-1,3-cyclohexadiene	–	–	CO	–	**B8**, 370, **13**, *252*
1-[Carbonyloxycyclohexylidene]-1-propene-1,3-diyl	–	–	CO	–	**B10**, 63, **3**, *84*
Carbonyloxy-2-propene-1,3-diylcyclohexylidene	–	–	CO	–	**B10**, 65, **3**, *92*
2,3,3a,7a-Tetrahydro-6-methoxy-3a-methylbenzofuran	–	–	CO	–	**B8**, 404, **14**, *42*
$C_{10}H_{14}O_3$					
4-Methoxycarbonyl-2-(1,1-dimethylethyl)-1-oxo-2-butene-1,4-diyl	–	–	CO	–	**B10**, 48, **3**, *4*
4-Methoxy-1-(methoxycarbonyl]methyl)-1,3-cyclohexadiene	–	–	CO	–	**B8**, 153, **7**, *9*
4-Methoxy-1,3-cyclohexadiene-1-propionic acid	–	–	CO	–	**B8**, 154, **7**, *16*
5-(1-Acetyl-2-oxopropyl)-1-methoxycarbonyl-1,3-cyclohexadiene	–	–	CO	–	**B8**, 169
2-Methoxy-5-(methoxycarbonylmethyl)-1,3-cyclohexadiene	–	–	CO	–	**B8**, 213, **9**, *96*

2-Methoxy-6-(methoxycarbonylmethyl)-1,3-cyclohexadiene	–	–	CO	–	**B8**, 237, **10**, *29*
5-Ethoxycarbonylmethyl-5-hydroxy-1,3-cyclohexadiene	–	–	CO	–	**B8**, 242, **11**, *2*
5-Hydroxy-5-(1-methoxycarbonylethyl)-1,3-cyclohexadiene	–	–	CO	–	**B8**, 243, **11**, *3*
5-Hydroxy-5-(methoxycarbonylmethyl)-1,3-cycloheptadiene	–	–	CO	–	**B9**, 55, **2**, *52*
$C_{10}H_{14}O_3S$	–	–	CO	–	**B8**, 464, **17**, *18*
$C_{10}H_{14}O_4$					
$C_2H_5O_2CCH{=}CHCH{=}CHCO_2C_2H_5$	–	–	CO	–	**B10**, 217, **8**, *11*
$C_2H_5O_2CCH{=}CHCH{=}CHCO_2C_2H_5$	C_4H_6	–	CO	–	**B10**, 215, **8**, *2*
$C_2H_5O_2CCH{=}CHCH{=}CHCO_2C_2H_5$	C_8H_{14}	–	CO	–	**B10**, 218, **8**, *12*
$(C_2H_5O_2C)_2CC(CH_2)_2$	–	–	CO	–	**B10**, 10, **1**, *33*
1,3-Dimethoxy-5-methoxycarbonyl-1,3-cyclohexadiene	–	–	CO	–	**B8**, 362, **13**, *214*
$C_{10}H_{14}S$	–	–	CO	–	**B9**, 126, **3**, *31*
$C_{10}H_{15}N$	–	–	CO	–	**B8**, 112, **4**, *160*
$C_{10}H_{15}NO$					
1-(4-Morpholinyl)-1,3-cyclohexadiene	–	–	CO	–	**B8**, 44, **2**, *5*
5-(4-Morpholinyl)-1,3-cyclohexadiene	–	–	CO	–	**B8**, 90, **4**, *37*
7-[(Dimethylamino)methyl]-2,4-cycloheptadien-1-one	–	–	CO	–	**B9**, 62, **2**, *83*
$C_{10}H_{15}NSi$	–	–	CO	–	**B8**, 237, **10**, *26*
$C_{10}H_{15}N_3O$					
5-[1-[1-(Aminocarbonyl)-hydrazinyl-2-ylidene]ethyl]-2-methyl-1,3-cyclohexadiene	–	–	CO	–	**B8**, 210, **9**, *86*
6-[1-[1-(Aminocarbonyl)-hydrazinyl-2-ylidene]ethyl]-1-methyl-1,3-cyclohexadiene	–	–	CO	–	**B8**, 177, **8**, *20*
$C_{10}H_{16}$					
[4,6,6-Trimethyl-4-cyclohexene-3,1-diyl]methyl	–	–	CO	–	**B10**, 127, **4**, *26*
2-Butyl-1,3-cyclohexadiene	–	–	CO	–	**B8**, 66, **3**, *18*
5-Butyl-1,3-cyclohexadiene	–	–	CO	–	**B8**, 96, **4**, *67*
5-(1,1-Dimethylethyl)-1,3-cyclohexadiene	–	–	CO	–	**B8**, 103, **4**, *102*
1-Methyl-4-(1-methylethyl)-1,3-cyclohexadiene	–	–	CO	–	**B8**, 157, **7**, *36*
1-Methyl-6-(1-methylethyl)-1,3-cyclohexadiene	–	–	CO	–	**B8**, 176, **8**, *13*

2-Methyl-5-(1-methylethyl)-1,3-cyclohexadiene	–	–	CO	–	**B8**, 208, **9**, *80*
5-Methyl-2-(1-methylethyl)-1,3-cyclohexadiene	–	–	CO	–	**B8**, 211, **9**, *90*
1,3,5,5-Tetramethyl-1,3-cyclohexadiene	–	–	CO	–	**B8**, 398, **14**, *13*
5,5,6,6-Tetramethyl-1,3-cyclohexadiene	–	–	CO	–	**B8**, 407, **14**, *55*
2-Ethyl-1,3-cyclooctadiene	–	–	CO	–	**B9**, 210, **7**, *2*
3-Ethyl-1,5-cyclooctadiene	–	–	CO	–	**B9**, 212, **7**, *10*
$C_{10}H_{16}Ge$	–	–	CO	–	**B9**, 127, **3**, *35*
$C_{10}H_{16}O$					
2-(2-Hydroxy-2,2-dimethylethyl)-1,3-cyclohexadiene	–	–	CO	–	**B8**, 64, **3**, *9*
5-(1,1-Dimethylethoxy)-1,3-cyclohexadiene	–	–	CO	–	**B8**, 85, **4**, *11*
5-(2-Hydroxy-2,2-dimethylmethyl)-1,3-cyclohexadiene	–	–	CO	–	**B8**, 94, **4**, *61*
1-Methoxy-6-(1-methylethyl)-1,3-cyclohexadiene	–	–	CO	–	**B8**, 174, **8**, *4*
2-Methoxy-5-(1-methylethyl)-1,3-cyclohexadiene	–	–	CO	–	**B8**, 202, **9**, *41*
2-Methyl-5-(1-methylethoxy)-1,3-cyclohexadiene	–	–	CO	–	**B8**, 206, **9**, *71*
5-(1-Methylethoxy)-1,3-cycloheptadiene	–	–	CO	–	**B9**, 46, **2**, *4*
7-Hydroxy-1,5,5-trimethyl-1,3-cycloheptadiene	–	–	CO	–	**B9**, 70, **2**, *117*
$C_{10}H_{16}O_2$					
Carbonyloxy[1-pentyl-2-butene-1,4-diyl]	–	–	CO	–	**B10**, 62, **3**, *77*
Carbonyloxy[4-pentyl-2-butene-1,4-diyl]	–	–	CO	–	**B10**, 63, **3**, *82*
6-Methoxycarbonyl-2,3-dimethyl-2-hexene-1,6-diyl	–	–	CO	–	**B10**, 68, **3**, *106*
1-Methoxy-3-(1-methylethoxy)-1,3-cyclohexadiene	–	–	CO	–	**B8**, 145, **6**, *2*
3-Methoxy-1-(1-methylethoxy)-1,3-cyclohexadiene	–	–	CO	–	**B8**, 145, **6**, *6*
1-Methoxy-4-(1-methylethoxy)-1,3-cyclohexadiene	–	–	CO	–	**B8**, 152, **7**, *2*
1-Methoxy-4-(2-methoxyethyl)-1,3-cyclohexadiene	–	–	CO	–	**B8**, 153, **7**, *12*
4-(2-Hydroxypropyl)-1-methoxy-1,3-cyclohexadiene	–	–	CO	–	**B8**, 152, **7**, *5*
4-(3-Hydroxypropyl)-1-methoxy-1,3-cyclohexadiene	–	–	CO	–	**B8**, 154, **7**, *19*
5-(2-Hydroxyethyl)-2-methoxy-5-methyl-1,3-cyclohexadiene	–	–	CO	–	**B8**, 331, **13**, *100*
$C_{10}H_{16}O_4$	–	–	CO	–	**B10**, 76, **3**, *140*

$C_{10}H_{16}Si$	–	–	CO	–	**B9**, 127, **3**, *33*
$C_{10}H_{16}Sn$	–	–	CO	–	**B9**, 128, **3**, *38*
$C_{10}H_{17}N$	–	–	CO	–	**B9**, 47, **2**, *13*
$C_{10}H_{17}O$	–	–	CO	–	**B10**, 9, **1**, *28*
$C_{10}H_{17}P$	C_6H_8P	–	–	–	**B10**, 194, **6**, *27*
$C_{10}H_{18}N$	–	–	CO	–	**B9**, 51,2,*35*
$C_{10}H_{18}O$	–	–	CO	–	**B10**, 9, **1**, *29*
$C_{10}H_{18}OSi$					
1-Methyl-3-(trimethylsilyloxy)-1,3-cyclohexadiene	–	–	CO	–	**B8**, 146, **6**, *9*
2-Methyl-3-(trimethylsilyloxy)-1,3-cyclohexadiene	–	–	CO	–	**B8**, 186
$C_{10}H_{18}Si$					
1-Methyl-3-trimethylsilyl-1,3-cyclohexadiene	–	–	CO	–	**B8**, 146, **6**, *10*
5-Methyl-2-trimethylsilyl-1,3-cyclohexadiene	–	–	CO	–	**B8**, 211, **9**, *91*
$C_{11}H_6Cl_4O$	–	–	CO	–	**B9**, 74, **2**, *136*
$C_{11}H_6F_8$	–	–	CO	–	**B10**, 69, **3**, *109*
$C_{11}H_8Cl_4$	–	–	CO	–	**B9**, 58, **2**, *68*
$C_{11}H_8N_2$	–	–	CO	–	**B9**, 11, **1**, *32*
$C_{11}H_8O_3Ru$	–	–	CO	–	**B10**, 154, **4**, *152*
$C_{11}H_9Cl$	–	–	CO	–	**B9**, 264, **9**, *37*
$C_{11}H_9F_3O_2$	–	–	CO	–	**B8**, 272, **12**, *34*
$C_{11}H_9F_6NO_3$	–	–	CO	–	**B10**, 145, **4**, *113*
$C_{11}H_{10}$	–	–	CO	–	**B9**, 74, **2**, *138*
$C_{11}H_{10}F_6$	–	–	CO	–	**B10**, 126, **4**, *25*
$C_{11}H_{10}F_6O$	–	–	CO	–	**B10**, 177
$C_{11}H_{10}N_2O_2S$	–	–	CO	–	**B9**, 13, **1**, *43*
$C_{11}H_{10}O$					
$C_6H_5C(O)CHC(CH_2)_2$	–	–	CO	–	**B10**, 8, **1**, *18*
1,4,4a,5,6,8a-Hexahydro-6-oxo-1,5-methanonaphthalene-4,9-diyl	–	–	CO	–	**B10**, 139, **4**, *86*
Carbonyl[bicyclo[3.3.2]-deca-3,7,9-triene-6,2-diyl	–	–	CO	–	**B10**, 145, **4**, *112*
Carbonyl[bicyclo[4.3.1]-deca-2,4,8-triene-10,7-diyl]	–	–	CO	–	**B10**, 137, **4**, *77*
Carbonyl[tricyclo[4.3.1.0^{2,9}]-diene-10,3-diyl	–	–	CO	–	**B10**, 144, **4**, *107*
Tricyclo[5.4.0.0^{4,9}]undeca-2,5,10-trien-8-one	–	–	CO	–	**B9**, 204, **6**, *23*
8-Oxotricyclo[5.4.0.0^{3,9}]-undeca-5,10-diene-2,4-diyl	–	–	CO	–	**B10**, 142, **4**, *103*
$C_{11}H_{10}O_2$					
Carbonyloxy[4-phenyl-2-butene-1,4-diyl]	–	–	CO	–	**B10**, 76, **3**, *139*
6-Methyl-1,3,5,7-cyclooctatetraene-1,3-dicarbaldehyde	–	–	CO	–	**B9**, 268, **9**, *55*
$C_{11}H_{11}ClN$	–	–	CO	–	**B8**, 133, **5**, *2*

1-[2-(Methoxycarbonyl)-ethenyl]-1,3,5-cycloheptatriene	–	–	CO	–	**B9**, 120, **3**, *4*
[3,5,7-Cyclooctatriene-1,2-diyl](methoxycarbonyl)-methyl	–	–	CO	–	**B10**, 135, **4**, *65*
7-(Ethoxycarbonyl)methylene-bicyclo[2.2.1]hepta-2,5-diene	–	–	CO	–	**B8**, 467, **17**, *30*
$C_{11}H_{12}O_4$	–	–	CO	–	**B9**, 149, **4**, *4*
$C_{11}H_{12}S$					
5-(5-Methylthiophen-2-yl)-1,3-cyclohexadiene	–	–	CO	–	**B8**, 108, **4**, *128*
5,7-Dimethyl-4*H*-cyclo-hepta[*b*]thiophene	–	–	CO	–	**B9**, 185, **5**, *18*
5,7-Dimethyl-4*H*-cyclo-hepta[*c*]thiophene	–	–	CO	–	**B9**, 77, **2**, *149*
$C_{11}H_{12}S_2$	–	–	CO	–	**B9**, 159, **4**, *42*
$C_{11}H_{13}F_3O$	–	–	CO	–	**B10**, 122, **4**, *12*
$C_{11}H_{13}NO_3$	–	–	CO	–	**B9**, 6, **1**, *12*
$C_{11}H_{13}N_3O_2$	–	–	CO	–	**B9**, 211, **7**, *5*
$C_{11}H_{13}O_4$	–	–	CO	–	**B8**, 465, **17**, *21*
$C_{11}H_{14}$					
7-(1-Methylpropylidene)-1,3,5-cycloheptatriene	–	–	CO	–	**B9**, 161, **4**, *49*
1,2,3,3a,4,5,8,8a-Octahydro-4,8-methanoazulene-5,9-diyl	–	–	CO	–	**B10**, 136, **4**, *71*
2,3,3a,3b,7a,7b-Hexahydro-1*H*-cyclopenta[3,4]-cyclobuta[1,2]benzene	–	–	CO	–	**B8**, 274, **12**, *45*
Tricyclo[8.1.0.$0^{3,5}$]undeca-6,8-diene	–	–	CO	–	**B9**, 281
$C_{11}H_{14}O$					
5-(1-Methylene-3-oxobutyl)-1,3-cyclohexadiene	–	–	CO	–	**B8**, 102, **4**, *97*
1-(1-Methoxyethyl)-1,3,5,7-cyclooctatetraene	–	–	CO	–	**B9**, 261, **9**, *23*
1-(1-Hydroxypropyl)-1,3,5,7-cyclooctatetraene	–	–	CO	–	**B9**, 262, **9**, *24*
1,2,5,8-Tetrahydro-3-methoxy-naphthalene	–	–	CO	–	**B8**, 361, **13**, *209*
$C_{11}H_{14}O_2$					
5-(1-Acetyl-2-oxopropyl)-1,3-cyclohexadiene	–	–	CO	–	**B8**, 98, **4**, *79*
Carbonyl[4-methyl-1-(1-methylethyl)-2-oxo-4-cyclohexene-1,3-diyl]	–	–	CO	–	**B10**, 128, **4**, *30*
[3,5-Cyclooctadiene-1,2-diyl]-(methoxycarbonyl)methyl	–	–	CO	–	**B10**, 135, **4**, *63*
3a,3b,7a,7b-Tetrahydro-2,2-dimethylbenzo[3,4]cyclo-buta[1,2-*d*]-1,3-dioxole	–	–	CO	–	**B8**, 274, **12**, *44*

7-[(2-Acetyloxy)ethyl]bicyclo[2.2.1]hepta-2,5-diene	–	–	CO	–	**B8**, 464, **17**, *19*
8-Acetyl-6-methoxybicyclo-[5.1.0]octa-2,4-diene	–	–	CO	–	**B9**, 61, **2**, *78*
8-(Ethoxycarbonyl)bicyclo-[5.1.0]octa-2,4-diene	–	–	CO	–	**B9**, 79, **2**, *156*
8-Acetyl-4-methoxybicyclo-[3.2.1]octa-2,6-diene	–	–	CO	–	**B9**, 201, **6**, *8*
9-(Acetyloxy)bicyclo[4.2.1]-nona-2,4-diene	–	–	CO	–	**B9**, 62, **2**, *80*
$C_{11}H_{14}O_3$					
5-[(1-Acetyl)-2-oxopropyl]-1,3-cyclohexadiene	–	–	CO	–	**B8**, 196, **9**, *2*
5-(2,3,4,5-Tetrahydro-2-oxofuran-3-yl)-2-methoxy-1,3-cyclohexadiene	–	–	CO	–	**B8**, 202, **9**, *45*
3-(Acetyloxymethyl)-2,3,3a,7a-tetrahydrobenzofuran	–	–	CO	–	**B8**, 269, **12**, *22*
$C_{11}H_{14}O_4$					
5-Bis(methoxycarbonyl)methyl-1,3-cyclohexadiene	–	–	CO	–	**B8**, 98, **4**, *80*
5-Methoxycarbonyl-5-(methoxycarbonylmethyl)-1,3-cyclohexadiene	–	–	CO	–	**B8**, 244, **11**, *8* **B8**, 244, **11**, *9*
$C_{11}H_{15}ClO$	–	–	CO	–	**B8**, 102, **4**, *98*
$C_{11}H_{15}NO_2$	–	–	CO	–	**B9**, 6, **1**, *7*
$C_{11}H_{16}$					
3-(2-Propenyl)-1,5-cyclooctadiene	–	–	CO	–	**B9**, 212, **7**, *11*
8-(2-Propenyl)-5-cyclooctene-1,4-diyl	–	–	CO	–	**B10**, 134, **4**, *57*
$C_{11}H_{16}Ge$	–	–	CO	–	**B9**, 259, **9**, *8*
$C_{11}H_{16}NO_2$	–	–	CO	–	**B8**, 353, **13**, *171*
$C_{11}H_{16}N_2O_2$	–	–	CO	–	**B10**, 7, **1**, *16*
$C_{11}H_{16}O$					
5-(1,1-Dimethyl-2-oxopropyl)-1,3-cyclohexadiene	–	–	CO	–	**B8**, 102, **4**, *100*
5-(1-Oxopentyl)-1,3-cyclohexadiene	–	–	CO	–	**B8**, 104, **4**, *107*
2-Methoxy-5-(2-methyl-2-propenyl)-1,3-cyclohexadiene	–	–	CO	–	**B8**, 200, **9**, *34*
2-Methoxy-5-(2-methyl-1-propenyl)-1,3-cyclohexadiene	–	–	CO	–	**B8**, 201, **9**, *40*
2-Methoxy-5-methyl-1-(2-propenyl)-1,3-cyclohexadiene	–	–	CO	–	**B8**, 312, **13**, *6*
2-Methoxy-5-methyl-5-(2-propenyl)-1,3-cyclohexadiene	–	–	CO	–	**B8**, 332, **13**, *102*

2-Methoxy-1-methyl-6-(1-methylethenyl)-1,3-cyclohexadiene	–	–	CO	–	**B8**, 314, **13**, *16*
2-Methoxy-3-methyl-5-(1-methylethenyl)-1,3-cyclohexadiene	–	–	CO	–	**B8**, 326, **13**, *78*
3-Methoxy-2-methyl-5-(1-methylethenyl)-1,3-cyclohexadiene	–	–	CO	–	**B8**, 327, **13**, *79*
5-Ethylidene-2-(1-methylethoxy)-1,3-cyclohexadiene	–	–	CO	–	**B8**, 357, **13**, *187*
Carbonylmethyl[3,5,5-trimethyl-3-cyclohexene-1,2-diyl]	–	–	CO	–	**B10**, 127, **4**, *27*
Carbonyl[4-methyl-1-(1-methylethyl)-cyclohexene-1,3-diyl]	–	–	CO	–	**B10**, 127, **4**, *28*
1-(1-Hydroxy-1-methylpropyl)-1,3,5-cycloheptatriene	–	–	CO	–	**B9**, 132, **3**, *57*
5-Methoxy-7-(1-methylethylidene)-1,3-cycloheptadiene	–	–	CO	–	**B9**, 69, **2**, *112*
6-Acetyl-5-ethyl-1,3-cycloheptadiene	–	–	CO	–	**B9**, 57, **2**, *62*
7-(1,1-Dimethylethoxy)bicyclo-[2.2.1]cyclohepta-2,5-diene	–	–	CO	–	**B8**, 463, **17**, *11*
3,4,5,6,7,8-Hexahydro-2-methoxy-naphthalene	–	–	CO	–	**B8**, 311, **13**, *3*
4,4a,5,6,7,8-Hexahydro-2-hydroxy-4a-methyl-naphthalene	–	–	CO	–	**B8**, 399, **14**, *17*
$C_{11}H_{16}O_2$					
1-[2-Methoxycarbonyl-3,3-dimethylcyclopropyl]-2-methylene-2-propenyl	–	–	CO	–	**B10**, 7, **1**, *15*
3-Acetyl-4,4-dimethyl-2-(1-methylethyl)-1-oxo-2-butene-1,4-diyl	–	–	CO	–	**B10**, 50, **3**, *17*
1-Methoxy-4-(3-oxobutyl)-1,3-cyclohexadiene	–	–	CO	–	**B8**, 154, **7**, *14*
2-Methoxy-5-(1,1-dimethyl-2-oxoethyl)-1,3-cyclohexadiene	–	–	CO	–	**B8**, 204, **9**, *54*
2-Methoxy-6-(1,1-dimethyl-2-oxoethyl)-1,3-cyclohexadiene	–	–	CO	–	**B8**, 234, **10**, *8*
2-Methoxy-5-methyl-5-(2-oxopropyl)-1,3-cyclohexadiene	–	–	CO	–	**B8**, 366, **13**, *232*
6-[(2-Methoxycarbonyl)ethyl]-1,3-cycloheptadiene	–	–	CO	–	**B9**, 50, **2**, *34*
$C_{11}H_{16}O_3$					
3-Methoxycarbonyl-4,4-dimethyl-2-(1-methylethyl)-1-oxo-2-butene-1,4-diyl	–	–	CO	–	**B10**, 51, **3**, *18*

2-Methoxycarbonyl-4,4-dimethyl-3-propyl-1-oxo-2-butene-1,4-diyl	–	–	CO	–	**B10**, 52, **3**, *24*
2-Methoxy-5-(methoxycarbonyl)methyl-5-methyl-1,3-cyclohexadiene	–	–	CO	–	**B8**, 331, **13**, *101*
4-[2-(Acetyloxy)ethyl]-1-methoxy-1,3-cyclohexadiene	–	–	CO	–	**B8**, 154, **7**, *13*
4-[2-(Methoxycarbonyl)ethyl]-1-methoxy-1,3-cyclohexadiene	–	–	CO	–	**B8**, 154, **7**, *17*
[4-Methoxy-1,3-cyclohexadien-1-yl]butyric acid	–	–	CO	–	**B8**, 156, **7**, *28*
$C_{11}H_{16}O_4$	–	–	CO	–	**B8**, 396, **14**, *7*
$C_{11}H_{16}Si$					
1-Trimethylsilyl-1,3,5,7-cyclooctatetraene	–	–	CO	–	**B9**, 258, **9**, *6*
7-Trimethylsilylbicyclo-[4.2.0]octa-2,4,7-triene	–	–	CO	–	**B8**, 268, **12**, *17*
$C_{11}H_{16}Sn$	–	–	CO	–	**B9**, 259, **9**, *9*
$C_{11}H_{17}N$	–	–	CO	–	**B9**, 49, **2**, *23*
$C_{11}H_{17}NO$	–	–	CO	–	**B9**, 65, **2**, *96*
$C_{11}H_{17}NO_2$					
1-Methoxy-3-(4-morpholinyl)-1,3-cyclohexadiene	–	–	CO	–	**B8**, 145, **6**, *3*
1-Methoxy-4-(4-morpholinyl)-1,3-cyclohexadiene	–	–	CO	–	**B8**, 152, **7**, *3*
3-Methoxy-1-(4-morpholinyl)-1,3-cyclohexadiene	–	–	CO	–	**B8**, 146, **6**, *7*
$C_{11}H_{18}$					
1-Methyl-6-(1,1-dimethylethyl)-1,3-cyclohexadiene	–	–	CO	–	**B8**, 176, **8**, *15*
2-Methyl-5-(1,1-dimethylethyl)-1,3-cyclohexadiene	–	–	CO	–	**B8**, 210, **9**, *84*
5-Butyl-2-methyl-1,3-cyclohexadiene	–	–	CO	–	**B8**, 207, **9**, *75*
6-Butyl-1-methyl-1,3-cyclohexadiene	–	–	CO	–	**B8**, 176, **8**, *14*
$C_{11}H_{18}N$	–	–	CO	–	**B9**, 51, **2**, *39*
$C_{11}H_{18}NO_2$					
1-Methoxy-3-(4-morpholinio)-1,3-cyclohexadiene	–	–	CO	–	**B8**, 147
1-Methoxy-4-(4-morpholinio)-1,3-cyclohexadiene	–	–	CO	–	**B8**, 160
3-Methoxy-1-(4-morpholinio)-1,3-cyclohexadiene	–	–	CO	–	**B8**, 147
$C_{11}H_{18}O$					
1-Methoxy-6-(1,1-dimethylethyl)-1,3-cyclohexadiene	–	–	CO	–	**B8**, 175, **8**, *6*
2-Methoxy-5-(1,1-dimethylethyl)-1,3-cyclohexadiene	–	–	CO	–	**B8**, 204, **9**, *53*

Formula / Compound					Reference
2-Methoxy-6-(1,1-dimethylethyl)-1,3-cyclohexadiene	–	–	CO	–	**B8**, 234, **10**, *7*
4-Ethyl-1-(1-methylethoxy)-1,3-cyclohexadiene	–	–	CO	–	**B8**, 159, **7**, *46*
5-(2-Hydroxy-1,2-dimethylpropyl)-1,3-cyclohexadiene	–	–	CO	–	**B8**, 98, **4**, *77*
5-[(2-Methylpentyl)oxy]-1,3-cyclohexadiene	–	–	CO	–	**B8**, 111, **4**, *152*
5-Butyl-2-methoxy-1,3-cyclohexadiene	–	–	CO	–	**B8**, 199, **9**, *32*
6-Butyl-1-methoxy-1,3-cyclohexadiene	–	–	CO	–	**B8**, 175, **8**, *5*
1,2-Dimethyl-4-(1-methylethoxy)-1,3-cyclohexadiene	–	–	CO	–	**B8**, 361, **13**, *210*
$C_{11}H_{18}O_2$					
3-(1-Hydroxyethyl)-4,4-dimethyl-2-(1-methylethyl)-1-oxo-2-butene-1,4-diyl	–	–	CO	–	**B10**, 50, **3**, *16*
1-Methoxy-4-(2-methoxypropyl)-1,3-cyclohexadiene	–	–	CO	–	**B8**, 152, **7**, *6*
1-(2-Hydroxy-2,2-dimethylethyl)-4-methoxy-1,3-cyclohexadiene	–	–	CO	–	**B8**, 153, **7**, *10*
1-(4-Hydroxybutyl)-4-methoxy-1,3-cyclohexadiene	–	–	CO	–	**B8**, 155, **7**, *23*
1-(3-Hydroxypropyl)-4-methoxy-2-methyl-1,3-cyclohexadiene	–	–	CO	–	**B8**, 361, **13**, *206*
5-(2-Hydroxy-2,2-dimethylethyl)-2-methoxy-1,3-cyclohexadiene	–	–	CO	–	**B8**, 200, **9**, *35*
5-Ethyl-5-(2-hydroxyethyl)-2-methoxy-1,3-cyclohexadiene	–	–	CO	–	**B8**, 343, **13**, *131*
$C_{11}H_{18}Si$	–	–	CO	–	**B8**, 267, **12**, *13*
$C_{11}H_{19}N$					
5-Butylamino-1,3-cycloheptadiene	–	–	CO	–	**B9**, 47, **2**, *14*
5-Diethylamino-1,3-cycloheptadiene	–	–	CO	–	**B9**, 49, **2**, *22*
5-(1,1-Dimethylethyl)amino-1,3-cycloheptadiene	–	–	CO	–	**B9**, 47, **2**, *15*
$C_{11}H_{19}O$	–	–	CO	–	**B10**, 10, **1**, *30*
$C_{11}H_{19}O_3P$	–	–	CO	–	**B8**, 325, **13**, *71*
$C_{11}H_{20}N$					
[2,4-Cycloheptadien-1-yl]-butylammonium	–	–	CO	–	**B9**, 51, **2**, *36*
[2,4-Cycloheptadien-1-yl]-diethylammonium	–	–	CO	–	**B9**, 51, **2**, *38*
[2,4-Cycloheptadien-1-yl]-(1,1-dimethylethyl)ammonium	–	–	CO	–	**B9**, 51, **2**, *37*
$C_{11}H_{20}O$	–	–	CO	–	**B10**, 10, **1**, *31*

$C_{11}H_{20}P$	–	–	–	–	**B10**, 189, **6**, *7*
					B10, 189, **6**, *8*
$C_{11}H_{20}P$	C_6H_8P	–	–	–	**B10**, 194, **6**, *28*
$C_{11}H_{20}Si$					
1,1-Dimethyl-4-(1,1-dimethylethyl)-2,4-silacyclohexadiene	–	–	CO	–	**B8**, 7, **1**, *10*
5,5-Dimethyl-2-(trimethylsilyl)-1,3-cyclohexadiene	–	–	CO	–	**B8**, 358, **13**, *192*
$C_{11}H_{21}O_3PSi$	–	–	CO	–	**B8**, 235, **10**, *19*
$C_{11}H_{22}OSi$	–	–	CO	–	**B10**, 15, **1**, *62*
$C_{12}H_6BrF_3$	–	–	CO	–	**B8**, 468, **17**, *39*
$C_{12}H_6F_4$	–	–	CO	–	**B8**, 468, **17**, *38*
$C_{12}H_7CrNO_5$	–	–	CO	–	**B8**, 87, **4**, *22*
$C_{12}H_7F_5$	–	–	CO	–	**B8**, 48, **2**, *29*
$C_{12}H_7F_5S$	–	–	CO	–	**B8**, 86, **4**, *20*
$C_{12}H_8F_8$	–	–	CO	–	**B10**, 71, **3**, *114*
$C_{12}H_8F_{12}$	–	–	CO	–	**B10**, 125, **4**, *20*
$C_{12}H_8N_2$	–	–	CO	–	**B8**, 406, **14**, *52*
$C_{12}H_{10}$					
4a,8b-Dihydro-biphenylene	–	–	CO	–	**B8**, 275, **12**, *47*
6,9-Dihydro-5,9-methano-5*H*-benzocycloheptene-6,10-diyl	–	–	CO	–	**B10**, 136, **4**, *72*
Benzocyclooctene	–	–	CO	–	**B9**, 265, **9**, *40*
$C_{12}H_{10}F_4$	–	–	CO	–	**B10**, 73, **3**, *121*
$C_{12}H_{10}N_2O$	–	–	CO	–	**B9**, 15, **1**, *56*
$C_{12}H_{10}N_4O_4$	–	–	CO	–	**B8**, 246, **11**, *16*
$C_{12}H_{11}Br$	–	–	CO	–	**B8**, 113, **4**, *166*
$C_{12}H_{11}F$					
5-(3-Fluorophenyl)-1,3-cyclohexadiene	–	–	CO	–	**B8**, 113, **4**, *164*
5-(4-Fluorophenyl)-1,3-cyclohexadiene	–	–	CO	–	**B8**, 113, **4**, *165*
$C_{12}H_{11}NO_2S$	–	–	CO	–	**B8**, 112, **4**, *154*
$C_{12}H_{11}N_2$	–	–	CO	–	**B8**, 133, **5**, *6*
$C_{12}H_{12}$					
1-Phenyl-1,3-cyclohexadiene	–	–	CO	–	**B8**, 44, **2**, *8*
2-Phenyl-1,3-cyclohexadiene	–	–	CO	–	**B8**, 65, **3**, *15*
5-Phenyl-1,3-cyclohexadiene	–	–	CO	–	**B8**, 104, **4**, *110*
2-(2,4-Cyclopentadien-1-yl)-1,3,5-cycloheptatriene	–	–	CO	–	**B9**, 124, **3**, *24*
1,4,4a,8a-Tetrahydro-1,4-ethenonaphthalene	–	–	CO	–	**B8**, 275, **12**, *49*
5-Methyl-5*H*-benzocycloheptene	–	–	CO	–	**B9**, 77, **2**, *148*
4a,10a-Dihydrobenzocyclooctene	–	–	CO	–	**B9**, 224, **8**, *6*
9,10-Dimethylbicyclo[6.2.0]-deca-1,3,5,7,9-pentaene	–	–	CO	–	**B9**, 265, **9**, *38*
$C_{12}H_{12}ClN$					
5-(3-Chlorophenyl)amino-1,3-cyclohexadiene	–	–	CO	–	**B8**, 88, **4**, *26*

4-(4-Chlorophenyl)amino-1,3-cyclohexadiene	–	–	CO	–	**B8**, 88, **4**, *27*
$C_{12}H_{12}F_6$					
[1,1-Difluoro-2-(tetrafluoroethylidene)ethane-1,2-diyl]-1,2,3,4-tetramethyl-3-cyclobutene-1,2-diyl	–	–	CO	–	**B10**, 121, **4**, *9*
1,2,3,4-Tetramethyl-5,6-bis(trifluoromethyl)benzene	–	–	CO	–	**B8**, 450
$C_{12}H_{12}N_2$	–	–	CO	–	**B9**, 14, **1**, *48*
$C_{12}H_{12}N_2O_2$	–	–	CO	–	**B8**, 88, **4**, *29*
$C_{12}H_{12}N_2O_2S$					
1-[(4-Methylphenyl)sulfonyl]-1*H*-1,2-diazepine	–	–	CO	–	**B9**, 13, **1**, *44*
3-Methyl-1-(phenylsulfonyl)-1*H*-1,2-diazepine	–	–	CO	–	**B9**, 17, **1**, *63*
$C_{12}H_{12}O$					
5-Phenoxy-1,3-cyclohexadiene	–	–	CO	–	**B8**, 85, **4**, *12*
2,3,4a,8a-Tetrahydronaphto[2,3-*b*]furan	–	–	CO	–	**B8**, 403, **14**, *39*
8,9-Dihydro-1*H*,3*H*-3a,7a-[1',2']-*endo*-cyclobutisobenzofuran	–	–	CO	–	**B8**, 408, **14**, *59*
4a,8a-(Methanoxymethano)-naphthalene	–	–	CO	–	**B8**, 409, **14**, *63*
1,2,3,4-Tetrahydro-2,2-dimethyl-4-oxonaphthalene-1,3-diyl	–	–	CO	–	**B10**, 128, **4**, *31*
1,4,4a,9a-Tetrahydro-1,4-methano-5*H*-benzocyclohepten-5-one	–	–	CO	–	**B9**, 73, **2**, *133*
$C_{12}H_{12}O_2$	–	–	CO	–	**B9**, 268, **9**, *52*
$C_{12}H_{12}O_4$	–	–	CO	–	**B9**, 264, **9**, *36*
$C_{12}H_{12}S$	–	–	CO	–	**B8**, 86, **4**, *19*
$C_{12}H_{12}Se$	–	–	CO	–	**B8**, 112, **4**, *156*
$C_{12}H_{13}B$	–	–	CO	–	**B9**, 4, **1**, *1*
$C_{12}H_{13}N$					
5-(Phenylamino)-1,3-cyclohexadiene	–	–	CO	–	**B8**, 87, **4**, *25*
5-(2-Aminophenyl)-1,3-cyclohexadiene	–	–	CO	–	**B8**, 105, **4**, *112*
5-(4-Aminophenyl)-1,3-cyclohexadiene	–	–	CO	–	**B8**, 105, **4**, *113*
$C_{12}H_{13}NO$					
Carbonyl(methylimino)[1-methyl-3-phenyl-1-propene-1,3-diyl]	–	–	CO	–	**B10**, 54, **3**, *36*
Carbonyl[(phenylmethyl)-imino][2-butene-1,4-diyl]	–	–	CO	–	**B10**, 58, **3**, *61*
Carbonyl(phenylimino)[2-methyl-2-butene-1,4-diyl]	–	–	CO	–	**B10**, 61, **3**, *72*

$C_{12}H_{13}O$	–	–	CO	–	**B10**, 13, **1**, *50*
$C_{12}H_{14}$					
1,2,3,4-Tetrahydro-4a,8a-ethenonaphthalene	–	–	CO	–	**B8**, 408, **14**, *60*
5-(2,4-Cyclopentadien-1-yl)-1,3-cycloheptadiene	–	–	CO	–	**B9**, 50, **2**, *29*
$C_{12}H_{14}N$					
1-[2,4-Cyclohexadien-1-yl]-2-methylpyridinium	–	–	CO	–	**B8**, 133, **5**, *3*
1-[2,4-Cyclohexadien-1-yl]-3-methylpyridinium	–	–	CO	–	**B8**, 133, **5**, *4*
1-[2,4-cyclohexadien-1-yl]-4-methylpyridinium	–	–	CO	–	**B8**, 133, **5**, *5*
1-[2,4-Cycloheptadien-1-yl]-pyridinium	–	–	CO	–	**B9**, 52, **2**, *41*
$C_{12}H_{14}NO$	–	–	CO	–	**B8**, 198, **9**, *22*
$C_{12}H_{14}N_2O$					
5-(Dicyanomethyl)-1-ethyl-4-methoxy-1,3-cyclohexadiene	–	–	CO	–	**B8**, 319, **13**, *40*
5-(Dicyanomethyl)-5-ethyl-2-methoxy-1,3-cyclohexadiene	–	–	CO	–	**B8**, 348, **13**, *149*
5-(Dicyanomethyl)-1,2-dimethyl-4-methoxy-1,3-cyclohexadiene	–	–	CO	–	**B8**, 413, **14**, *77*
6-(Dicyanomethyl)-3-methoxy-1,6-dimethyl-1,3-cyclohexadiene	–	–	CO	–	**B8**, 415, **14**, *88*
$C_{12}H_{14}N_2O_2$	–	–	CO	–	**B9**, 75, **2**, *143*
$C_{12}H_{14}O$					
$HO(C_2H_5)CC(CH_2)CHC_6H_5$	–	–	CO	–	**B10**, 13, **1**, *51*
2-Methyl-5-[5-oxo-1-cyclopentenyl]-1,3-cyclohexadiene	–	–	CO	–	**B8**, 210, **9**, *87*
2-Methyl-6-[5-oxo-1-cyclopentenyl]-1,3-cyclohexadiene	–	–	CO	–	**B8**, 234, **10**, *15*
1,2,3,4,4a,9a-Hexahydro-1,4-methano-5*H*-benzocyclohepten-5-one	–	–	CO	–	**B9**, 74, **2**, *134*
5,6,7-Trimethyl-4*H*-cyclohepta[*b*]furan	–	–	CO	–	**B9**, 186, **5**, *22*
$C_{12}H_{14}O_2$					
2-Methoxy-5-[5-oxo-1-cyclopentenyl]-1,3-cyclohexadiene	–	–	CO	–	**B8**, 204, **9**, *61*
2-Methoxy-5-[5-oxo-1-cyclopentenyl]-1,3-cyclohexadiene	–	–	CO	–	**B8**, 234, **10**, *11*
1-(3-Ethoxy-3-oxo-1-propenyl)-1,3,5-cycloheptatriene	–	–	CO	–	**B9**, 120, **3**, *5*
2-Acetyl-3-(1-methylethyl)-2,4,6-cycloheptatrien-1-one	–	–	CO	–	**B9**, 186, **5**, *19*
7-Acetyl-6-(1-methylethylidene)-2,4-cycloheptadien-1-one	–	–	CO	–	**B9**, 76, **2**, *147*

$C_{12}H_{14}O_3$					
3-Ethoxycarbonyl-3a,7a-dihydro-2-methylbenzofuran	–	–	CO	–	**B8**, 270, **12**, *26*
3-Acetyl-3a,7a-dihydro-6-methoxy-2-methylbenzofuran	–	–	CO	–	**B8**, 329, **13**, *91*
Carbonyl[3a,5,8,8a-tetrahydro-2,2-dimethyl-4,8-methano-4*H*-cyclohepta-1,3-dioxole-9,5-diyl]	–	–	CO	–	**B10**, 137, **4**, *74*
$C_{12}H_{14}S$					
4,5,7-Trimethyl-4*H*-cyclohepta[*b*]thiophene	–	–	CO	–	**B9**, 187, **5**, *24*
5,6,7-Trimethyl-4*H*-cyclohepta[*b*]thiophene	–	–	CO	–	**B9**, 190, **5**, *35*
5,6,7-Trimethyl-4*H*-cyclohepta[*c*]thiophene	–	–	CO	–	**B9**, 82, **2**, *173*
5,7,8-Trimethyl-4*H*-cyclohepta[*b*]thiophene	–	–	CO	–	**B9**, 186, **5**, *23*
$C_{12}H_{15}BSi$	–	–	CO	–	**B8**, 15, **1**, *44*
$C_{12}H_{15}NO$	–	–	CO	–	**B8**, 400, **14**, *24*
$C_{12}H_{15}NO_3$	–	–	CO	–	**B9**, 7, **1**, *13*
$C_{12}H_{16}$					
1,2,3,4,4a,4b,8a,8b-Octahydrobiphenylene	–	–	CO	–	**B8**, 274, **12**, *46*
1,2,7,8-Tetramethylbicyclo-[4.2.0]octa-2,4,7-triene	–	–	CO	–	**B8**, 403, **14**, *38*
1,6,7,8-Tetramethylbicyclo-[4.2.0]octa-2,4,7-triene	–	–	CO	–	**B8**, 407, **14**, *58*
2,4,6,8-Tetramethylbicyclo-[4.2.0]octa-2,4,7-triene	–	–	CO	–	**B8**, 431, **15**, *2*
$C_{12}H_{16}O$					
5-(3-Methyl-2-oxocyclopentyl)-1,3-cyclohexadiene	–	–	CO	–	**B8**, 109, **4**, *140*
5-(2-Oxocyclohexyl)-1,3-cyclohexadiene	–	–	CO	–	**B8**, 109, **4**, *142*
5-(2-Methyl-4-oxo-2-pentenyl)-1,3-cyclohexadiene	–	–	CO	–	**B8**, 113, **4**, *162*
5-(1-Acetyl-2-oxopropyl)-2-methyl-1,3-cyclohexadiene	–	–	CO	–	**B8**, 209, **9**, *81*
1-(1-Methoxypropyl)-1,3,5,7-cyclooctatetraene	–	–	CO	–	**B9**, 262, **9**, *25*
1,2,3,5-Tetrahydro-7-methoxy-1-methyl-naphthalene	–	–	CO	–	**B8**, 400, **14**, *20*
1,2,3,5-Tetrahydro-7-methoxy-4-methyl-naphthalene	–	–	CO	–	**B8**, 399, **14**, *19*
1,2,3,4-Tetrahydro-4a,8a-(methanoxymethano)naphthalene	–	–	CO	–	**B8**, 408, **14**, *62*
[2,3,4,4a,5,6,7,8-Octahydro-7-oxonaphthalene-1,8-diyl]-ethane-1,2-diyl	–	–	CO	–	**B10**, 56, **3**, *54*

Compound					Reference
$C_{12}H_{16}O_2$					
2-Acetyl-5-methyl-3-(1-methylethylidene)-1-oxo-4-hexene-1,2-diyl	–	–	CO	–	**B10**, 51, **3**, *22*
1-Methoxycarbonyl-2-methyl-5-(2-propenyl)-1,3-cyclohexadiene	–	–	CO	–	**B8**, 312, **13**, *8*
2-Methoxy-6-(2-oxocyclopentyl)-1,3-cyclohexadiene	–	–	CO	–	**B8**, 234, **10**, *5*
5-(1-Acetyl-2-oxobutyl)-1,3-cyclohexadiene	–	–	CO	–	**B8**, 99, **4**, *84*
5-(1-Acetyl-2-oxopropyl)-1-methyl-1,3-cyclohexadiene	–	–	CO	–	**B8**, 168
5,6,7,8-Tetrahydro-2-methoxy-4a(*4H*)-naphthalenecarbaldehyde	–	–	CO	–	**B8**, 400, **14**, *22*
5-(1-Acetyl-2-oxopropyl)-1,3-cycloheptadiene	–	–	CO	–	**B9**, 49, **2**, *27*
5-(1-Acetyl-2-oxopropyl)-1,3-cycloheptadiene	C_6H_8	–	CO	–	**B10**, 220, **8**, *22*
$C_{12}H_{16}O_3$					
3-Methoxycarbonyl-4-methyl-2-(2-methyl-1-propenyl)-1-oxo-2-pentene-1,4-diyl	–	–	CO	–	**B10**, 51, **3**, *21*
2-Methoxycarbonyl-5-methyl-3-(1-methylethylidene)-1-oxo-4-hexene-1,2-diyl	–	–	CO	–	**B10**, 52, **3**, *25*
1-Methoxy-6-methoxycarbonyl-4-methyl-1,3-cyclohexadiene	–	–	CO	–	**B8**, 363, **13**, *217*
5-[1-Acetyl-2-ethoxy-2-oxopropyl]-1,3-cyclohexadiene	–	–	CO	–	**B8**, 99, **4**, *85*
5-(1-Acetyl-2-oxopropyl)-2-methoxy-1,3-cyclohexadiene	–	–	CO	–	**B8**, 202, **9**, *42*
6-(1-Acetyl-2-oxopropyl)-1-methoxy-1,3-cyclohexadiene	–	–	CO	–	**B8**, 175, **8**, *7*
6-(1-Acetyl-2-oxopropyl)-2-methoxy-1,3-cyclohexadiene	–	–	CO	–	**B8**, 233, **10**, *4*
$C_{12}H_{16}O_4$					
5-[Bis(methoxycarbonyl)-methyl]-2-methyl-1,3-cyclohexadiene	–	–	CO	–	**B8**, 209, **9**, *82*
6-[Bis(methoxycarbonyl)-methyl]-1-methyl-1,3-cyclohexadiene	–	–	CO	–	**B8**, 177, **8**, *21*
4,5-Bis(methoxycarbonyl)-1,2-dimethyl-1,3-cyclohexadiene	–	–	CO	–	**B8**, 395, **14**, *2*
5-Bis[(methoxycarbonyl)-methyl]-1,3-cycloheptadiene	–	–	CO	–	**B9**, 78, **2**, *155*
$C_{12}H_{16}O_5$	–	–	CO	–	**B8**, 213, **9**, *99*
$C_{12}H_{17}NO$	–	–	CO	–	**B8**, 344, **13**, *133*
$C_{12}H_{17}NO_2$	–	–	CO	–	**B9**, 7, **1**, *16*

$C_{12}H_{18}$					
1,5,9-Cyclodecatriene	–	–	CO	–	**B9**, 284
4a,5,6,7,8,9,10,10a-Octahydrobenzocyclooctene	–	–	CO	–	**B8**, 273, **12**, *38*
"Dihydrobenzocyclooctene"	–	–	CO	–	**B8**, 273, **12**, *39*
$C_{12}H_{18}N$	–	–	CO	–	**B9**, 263, **9**, *33*
$C_{12}H_{18}N_2$	–	–	CO	–	**B9**, 159, **4**, *43*
$C_{12}H_{18}O$					
2,3,4,5,6,6-Hexamethyl-2,4-cyclohexadien-1-one	–	–	CO	–	**B8**, 449
					B8, 450
5-Methoxy-7-(1-methylpropylidene)-1,3-cycloheptadiene	–	–	CO	–	**B9**, 69, **2**, *114*
1,2,3,4,5,6-Hexahydro-7-methoxy-1-methylnaphthalene	–	–	CO	–	**B8**, 312, **13**, *5*
1,2,3,4,7,8-Hexahydro-6-methoxy-1-methylnaphthalene	–	–	CO	–	**B8**, 311, **13**, *4*
1,2,3,4,8,8a-Hexahydro-6-methoxy-1-methylnaphthalene	–	–	CO	–	**B8**, 318, **13**, *32*
$C_{12}H_{18}O_2$					
3-Acetyl-4-methyl-2-(1,1-dimethylethyl)-1-oxo-2-pentene-1,4-diyl	–	–	CO	–	**B10**, 51, **3**, *19*
5-(4-Ethoxy-2-oxobutyl)-1,3-cyclohexadiene	–	–	CO	–	**B8**, 96, **4**, *66*
5-(1-Acetyl-2-ethoxyethyl)-1,3-cyclohexadiene	–	–	CO	–	**B8**, 99, **4**, *86*
1-Methoxy-4-(4-oxopentyl)-1,3-cyclohexadiene	–	–	CO	–	**B8**, 155, **7**, *25*
2-Methoxy-6-(1,1-dimethyl-2-oxopropyl)-1,3-cyclohexadiene	–	–	CO	–	**B8**, 234, **10**, *10*
5-Butoxycarbonyl-5-methyl-1,3-cyclohexadiene	–	–	CO	–	**B8**, 248, **11**, *27*
5-(1-Formylpropyl)-3-methoxy-1-methyl-1,3-cyclohexadiene	–	–	CO	–	**B8**, 316, **13**, *21*
2-[2,4-Cyclohexadien-1-yl]-methoxy-tetrahydro-2*H*-pyran	–	–	CO	–	**B8**, 92, **4**, *51*
1,2,3,4,4a,5-Hexahydro-4a-hydroxymethyl-7-methoxy-naphthalene	–	–	CO	–	**B8**, 400, **14**, *21*
$C_{12}H_{18}O_3$					
4-(3-Acetyloxypropyl)-1-methoxy-1,3-cyclohexadiene	–	–	CO	–	**B8**, 155, **7**, *20*
1-Methoxy-4-(4-methoxy-4-oxobutyl)-1,3-cyclohexadiene	–	–	CO	–	**B8**, 156, **7**, *29*
1-Methoxy-4-[2-methyl-1,4-dioxan-2-yl]-1,3-cyclohexadiene	–	–	CO	–	**B8**, 156, **7**, *32*
5-Ethyl-2-methoxy-5-(methoxycarbonyl)methyl-1,3-cyclohexadiene	–	–	CO	–	**B8**, 343, **13**, *130*

$C_{13}H_9N$	–	–	CO	–	**B8**, 406, **14**, *50*
$C_{13}H_{10}BrN$	–	–	CO	–	**B9**, 156, **4**, *29*
$C_{13}H_{10}ClN$	–	–	CO	–	**B9**, 156, **4**, *28*
$C_{13}H_{10}N_4O_4$					
7-[1-(2,4-Dinitrophenyl)-hydrazinyl-2-ylidene]-1,3,5-cycloheptatriene	–	–	CO	–	**B9**, 157, **4**, *34*
7-[1-(2,4-Dinitrophenyl)-hydrazinyl-2-ylidene]bicyclo-[2,2.1]hepta-2,5-diene	–	–	CO	–	**B8**, 466, **17**, *29*
$C_{13}H_{10}O$					
2-Phenyl-2,4,6-cyclohepta-trien-1-one	–	–	CO	–	**B9**, 183, **5**, *7*
5,9-Dihydro-5,9-etheno-6*H*-benzocyclohepten-6-one	–	–	CO	–	**B9**, 203, **6**, *20*
6,9-Dihydro-11-oxo-5,9-ethano-5*H*-benzocycloheptene-6,10-diyl	–	–	CO	–	**B10**, 142, **4**, *102*
$C_{13}H_{11}Cl_4NO_2$	–	–	CO	–	**B9**, 9, **1**, *24*
$C_{13}H_{11}F_3$	–	–	CO	–	**B8**, 113, **4**, *168*
$C_{13}H_{11}NO$	–	–	CO	–	**B8**, 109, **4**, *137*
$C_{13}H_{11}NO_2$					
1-Phenoxycarbonyl-1*H*-azepine	–	–	CO	–	**B9**, 6, **1**, *8*
3-[3-Oxo-1-butenyl]-5*H*-pyrrolo[1,2-*a*]azepin-5-one	–	–	CO	–	**B9**, 10, **1**, *30*
10-Methyl-9,11-dioxo-4a,8a-(methaniminomethano)-naphthalene	–	–	CO	–	**B8**, 411, **14**, *68*
$C_{13}H_{11}N_3O_2$	–	–	CO	–	**B9**, 131, **3**, *53*
$C_{13}H_{12}$					
5-(Phenylmethylene)-1,3-cyclohexadiene	–	–	CO	–	**B8**, 247, **11**, *20*
1-Phenyl-1,3,5-cyclohepta-triene	–	–	CO	–	**B9**, 124, **3**, *21*
2-Phenyl-1,3,5-cyclohepta-triene	–	–	CO	–	**B9**, 125, **3**, *25*
3-Phenyl-1,3,5-cyclohepta-triene	–	–	CO	–	**B9**, 125, **3**, *28*
7-Phenyl-1,3,5-cyclohepta-triene	–	–	CO	–	**B9**, 129, **3**, *45*
$C_{13}H_{12}N_2$					
5-[1*H*-Benzimidazol-2-yl]-1,3-cyclohexadiene	–	–	CO	–	**B8**, 108, **4**, *136*
7-[1-Phenyl-hydrazinyl-2-ylidene]-1,3,5-cyclo-heptatriene	–	–	CO	–	**B9**, 157, **4**, *33*
$C_{13}H_{12}N_2O_5$	–	–	CO	–	**B9**, 76, **2**, *144*
$C_{13}H_{12}N_4O_4$					
5-[[1-(2,4-Dinitrophenyl)-hydrazinyl-2-ylidene]methyl]-1,3-cyclohexadiene	–	–	CO	–	**B8**, 97, **4**, *74*

2-Methyl-6-[1-(2,4-dinitrophenyl)hydrazinyl-2-ylidene]-1,3-cyclohexadiene	–	–	CO	–	**B8**, 359, **13**, *198*
5-[1-(2,4-Dinitrophenyl)-hydrazinyl-2-ylidene]-1,3-cycloheptadiene	–	–	CO	–	**B9**, 55, **2**, *56*
7-[1-(2,4-Dinitrophenyl)-hydrazinyl-2-ylidene]-1,3-cycloheptadiene	–	–	CO	–	**B9**, 56, **2**, *61*
$C_{13}H_{12}N_4O_5$	–	–	CO	–	**B8**, 353, **13**, *170*
$C_{13}H_{12}O$					
5-Benzoyl-1,3-cyclohexadiene	–	–	CO	–	**B8**, 104, **4**, *108*
7-Hydroxy-7-phenyl-1,3,5-cycloheptatriene	–	–	CO	–	**B9**, 152, **4**, *17*
7-Hydroxy-7-phenylbicyclo-[2.2.1]hepta-2,5-diene	–	–	CO	–	**B8**, 465, **17**, *25*
8,9-Dihydro-8-hydroxy-5,9-etheno-5*H*-benzocycloheptene	–	–	CO	–	**B9**, 203, **6**, *17*
$C_{13}H_{12}OP$	C_6H_8P	–	–	–	**B10**, 194, **6**, *26*
$C_{13}H_{12}O_2$	–	–	CO	–	**B8**, 168
$C_{13}H_{12}O_3$					
5-Methoxy-2-methyl-1,5-dioxo-3-phenyl-2-pentene-1,4-diyl	–	–	CO	–	**B10**, 49, **3**, *7*
5-Methoxy-3-methyl-1,5-dioxo-2-phenyl-2-pentene-1,4-diyl	–	–	CO	–	**B10**, 49, **3**, *8*
$C_{13}H_{12}S$	–	–	CO	–	**B9**, 126, **3**, *32*
$C_{13}H_{13}DN$					
7-Pyridinio-1,3,5-cyclooctatriene-d-8	–	–	CO	–	**B9**, 223, **8**, *4*
6-Pyridiniobicyclo-[5.1.0]octa-2,4-diene-d-8	–	–	CO	–	**B9**, 62, **2**, *82*
$C_{13}H_{13}NO$					
Carbonyl(phenylimino)[2,5-hexadiene-1,4-diyl]	–	–	CO	–	**B10**, 61, **3**, *73*
6-(Phenylamino)-2,4-cyclo-heptadien-1-one	–	–	CO	–	**B9**, 66, **2**, *97*
$C_{13}H_{13}NO_2$					
1,2-Dihydro-10-methyl-9,11-dioxo-4a,8a-(methaniminomethano)naphthalene	–	–	CO	–	**B8**, 410, **14**, *66*
1,4-Dihydro-10-methyl-9,11-dioxo-4a,8a-(methaniminomethano)naphthalene	–	–	CO	–	**B8**, 410, **14**, *67*
$C_{13}H_{14}$					
2-Methyl-5-phenyl-1,3-cyclohexadiene	–	–	CO	–	**B8**, 211, **9**, *88*
5-Methyl-5-phenyl-1,3-cyclohexadiene	–	–	CO	–	**B8**, 243, **11**, *5* **B8**, 243, **11**, *6*

5-(Phenylmethyl)-1,3-cyclohexadiene	–	–	CO	–	**B8**, 96, **4**, *68*
5-(4-Methylphenyl)-1,3-cyclohexadiene	–	–	CO	–	**B8**, 113, **4**, *169*
6,8-Dimethyl-5*H*-benzocycloheptene	–	–	CO	–	**B9**, 77, **2**, *150*
$C_{13}H_{14}ClNO$					
5-[(2-Chlorophenyl)amino]-2-methoxy-1,3-cyclohexadiene	–	–	CO	–	**B8**, 197, **9**, *8*
5-[(3-Chlorophenyl)amino]-2-methoxy-1,3-cyclohexadiene	–	–	CO	–	**B8**, 197, **9**, *9*
5-[(4-Chlorophenyl)amino]-2-methoxy-1,3-cyclohexadiene	–	–	CO	–	**B8**, 197, **9**, *10*
$C_{13}H_{14}NO$	–	–	CO	–	**B9**, 49, **2**, *25*
$C_{13}H_{14}N_2O$	–	–	CO	–	**B8**, 352, **13**, *165*
$C_{13}H_{14}N_2O_2S$	–	–	CO	–	**B9**, 17, **1**, *64*
$C_{13}H_{14}N_2O_3$					
Carbonyl[[(4-nitrophenyl)-methyl]-imino][3-methyl-2-butene-1,4-diyl]	–	–	CO	–	**B10**, 60, **3**, *67*
5-(Dicyanomethyl)-2-methoxy-5-[(methoxycarbonyl)methyl]-1,3-cyclohexadiene	–	–	CO	–	**B8**, 348, **13**, *150*
6-(Dicyanomethyl)-1-methoxy-4-[(methoxycarbonyl)methyl]-1,3-cyclohexadiene	–	–	CO	–	**B8**, 320, **13**, *45*
$C_{13}H_{14}N_2O_4$	–	–	CO	–	**B9**, 63, **2**, *84*
$C_{13}H_{14}O$					
5-(4-Methoxyphenyl)-1,3-cyclohexadiene	–	–	CO	–	**B8**, 104, **4**, *111*
5-Phenoxy-1,3-cycloheptadiene	–	–	CO	–	**B9**, 46, **2**, *5*
5-Hydroxy-5-phenyl-1,3-cycloheptadiene	–	–	CO	–	**B9**, 55, **2**, *53*
$C_{13}H_{14}O_2$	–	–	CO	–	**B8**, 85, **4**, *13*
$C_{13}H_{14}O_3$	–	–	CO	–	**B8**, 330, **13**, *93*
$C_{13}H_{14}O_4$	–	–	CO	–	**B9**, 61, **2**, *76*
$C_{13}H_{14}S$					
1-Methyl-6-phenylthio-1,3-cyclohexadiene	–	–	CO	–	**B8**, 176, **8**, *11*
2-Methyl-5-phenylthio-1,3-cyclohexadiene	–	–	CO	–	**B8**, 207, **9**, *72*
5-[(4-Methylthio)phenyl]-1,3-cyclohexadiene	–	–	CO	–	**B8**, 113, **4**, *167*
5-Phenylthio-1,3-cycloheptadiene	–	–	CO	–	**B9**, 46, **2**, *7*
$C_{13}H_{15}B$	–	–	CO	–	**B8**, 5, **1**, *2*
$C_{13}H_{15}N$					
5-[(2-Methylphenyl)amino]-1,3-cyclohexadiene	–	–	CO	–	**B8**, 88, **4**, *30*
5-[(3-Methylphenyl)amino]-1,3-cyclohexadiene	–	–	CO	–	**B8**, 88, **4**, *31*

5-[(4-Methylphenyl)amino]-1,3-cyclohexadiene	–	–	CO	–	**B8**, 89, **4**, *32*
5-(2-Amino-5-methylphenyl)-1,3-cyclohexadiene	–	–	CO	–	**B8**, 106, **4**, *119*
5-(Phenylamino)-1,3-cycloheptadiene	–	–	CO	–	**B9**, 47, **2**, *16*
7-(Phenylamino)-1,3-cycloheptadiene	–	–	CO	–	**B9**, 50, **2**, *33*
$C_{13}H_{15}NO$					
Carbonyl(methylimino)[1-ethyl-3-phenyl-1-propene-1,3-diyl]	–	–	CO	–	**B10**, 54, **3**, *37*
Carbonyl(ethylimino)[1-methyl-3-phenyl-1-propene-1,3-diyl]	–	–	CO	–	**B10**, 54, **3**, *39*
Carbonyl[(phenylmethyl)imino]-[3-methyl-2-butene-1,4-diyl]	–	–	CO	–	**B10**, 58, **3**, *62*
2-Methoxy-5-(phenylamino)-1,3-cyclohexadiene	–	–	CO	–	**B8**, 197, **9**, *7*
2-Methoxy-5-(2-aminophenyl)-1,3-cyclohexadiene	–	–	CO	–	**B8**, 205, **9**, *66*
2-Methoxy-5-(4-aminophenyl)-1,3-cyclohexadiene	–	–	CO	–	**B8**, 205, **9**, *67*
5-[(4-Methoxyphenyl)amino]-1,3-cyclohexadiene	–	–	CO	–	**B8**, 88, **4**, *28*
5-(2-Amino-5-methoxyphenyl)-1,3-cyclohexadiene	–	–	CO	–	**B8**, 106, **4**, *118*
5-Hydroxy-7-(phenylamino)-1,3-cycloheptadiene	–	–	CO	–	**B9**, 59, **2**, *70*
$C_{13}H_{15}NO_2$	–	–	CO	–	**B8**, 410, **14**, *65*
$C_{13}H_{16}N$					
1-[2,4-Cyclohexadien-1-yl]-2,5-dimethylpyridinium	–	–	CO	–	**B8**, 133, **5**, *8*
1-[2,4-Cyclohexadien-1-yl]-2,6-dimethylpyridinium	–	–	CO	–	**B8**, 133, **5**, *9*
1-[2,4-Cyclohexadien-1-yl]-3,5-dimethylpyridinium	–	–	CO	–	**B8**, 133, **5**, *10*
1-[2,4-Cycloheptadien-1-yl]-2-methylpyridinium	–	–	CO	–	**B9**, 52, **2**, *42*
$C_{13}H_{16}NO$	–	–	CO	–	**B8**, 198, **9**, *23*
$C_{13}H_{16}N_2O$	–	–	CO	–	**B8**, 158, **7**, *41*
$C_{13}H_{16}N_2O_2$					
5-(Dicyanomethyl)-2-methoxy-5-[(2-methoxy)ethyl]-1,3-cyclohexadiene	–	–	CO	–	**B8**, 350, **13**, *158*
5-(Dicyanomethyl)-4-methoxy-1-[(2-methoxy)ethyl]-1,3-cyclohexadiene	–	–	CO	–	**B8**, 321, **13**, *47*
$C_{13}H_{16}O_3$	–	–	CO	–	**B8**, 212, **9**, *94*
$C_{13}H_{16}O_4$					
1-Methoxycarbonyl-2-methyl-5-[2-oxotetrahydrofuran-3-yl]-1,3-cyclohexadiene	–	–	CO	–	**B8**, 313, **13**, *12*

2-Methoxy-5-[2-(1,3-dioxobutyl)oxyethylidene]-1,3-cyclohexadiene	–	–	CO	–	**B8**, 368, **13**, *244*
5-(1-Acetyl-2-oxopropyl)-1-methoxycarbonyl-1,3-cyclohexadiene	–	–	CO	–	**B8**, 169
3-Ethoxycarbonyl-3a,7a-dihydro-6-methoxy-2-methylbenzofuran	–	–	CO	–	**B8**, 329, **13**, *92*
8-Methoxy-2-[(methoxycarbonyl)methylene]-1-oxaspiro-[4.5]deca-7,9-diene	–	–	CO	–	**B8**, 352, **13**, *163*
$C_{13}H_{16}O_5$					
5-[4-Methoxy-2,4-cyclohexadien-1-yl]-2,2-dimethyl-1,3-dioxane-4,6-dione	–	–	CO	–	**B8**, 202, **9**, *46*
3-[[(2-Methoxycarbonyl-1-oxoethyl)oxy]methyl]-2,3,3a,7a-tetrahydrobenzofuran	–	–	CO	–	**B8**, 278, **12**, *60*
$C_{13}H_{16}P$	$C_6H_8O_2P$	–	–	–	**B10**, 195, **6**, *31*
$C_{13}H_{16}Si$	–	–	CO	–	**B8**, 10, **1**, *20*
$C_{13}H_{17}NO$					
1,3,4,5-Tetrahydro-7-methoxy-1-methyl-4a(*2H)*-naphthalenecarbonitrile	–	–	CO	–	**B8**, 401, **14**, *29*
1,3,4,5-Tetrahydro-7-methoxy-4-methyl-4a(*2H)*-napthalenecarbonitrile	–	–	CO	–	**B8**, 401, **14**, *28*
1-[(4-Morpholinyl)methyl]-1,3,5,7-cyclooctatetraene	–	–	CO	–	**B9**, 260, **9**, *13*
$C_{13}H_{17}NO_3$					
5-(1-Cyano-2-methoxy-2-oxoethyl)-1-ethyl-4-methoxy-1,3-cyclohexadiene	–	–	CO	–	**B8**, 319, **13**, *41*
5-(1-Cyano-2-methoxy-2-oxoethyl)-5-ethyl-2-methoxy-1,3-cyclohexadiene	–	–	CO	–	**B8**, 344, **13**, *135*
$C_{13}H_{17}O_4$	–	–	CO	–	**B8**, 354, **13**, *174*
$C_{13}H_{18}$					
1,2,3,4,6,7-Hexahydro-6-(2-propenyl)naphthalene	–	–	CO	–	**B8**, 327, **13**, *81*
1,4,6-Trimethylbicyclo-[3.3.2]deca-3,6,9-triene	–	–	CO	–	**B9**, 203, **6**, *21*
1,4,6-Trimethylbicyclo-[3.3.2]deca-3,6-diene-2,10-diyl	–	–	CO	–	**B10**, 143, **4**, *106*
1,4,7-Trimethylbicyclo-[4.3.1]deca-3,7-diene-2,10-diyl	–	–	CO	–	**B10**, 149, **4**, *129*
$C_{13}H_{18}O$					
1-[(2-Acetyl)ethenyl]-2,6,6-trimethyl-1,3-cyclohexadiene	–	–	CO	–	**B8**, 415, **14**, *86*

5-(3-Methyl-2-oxocyclohexyl)-1,3-cyclohexadiene	-	-	CO	-	**B8**, 110, **4**, *143*
1-[1-Hydroxypentyl]-1,3,5,7-cyclooctatetraene	-	-	CO	-	**B9**, 262, **9**, *26*
$C_{13}H_{18}OS$	-	-	CO	-	**B9**, 201, **6**, *9*
$C_{13}H_{18}O_2$					
2-Methoxy-5-(2-oxocyclohexyl)-1,3-cyclohexadiene	-	-	CO	-	**B8**, 203, **9**, *47*
5-(1-Acetyl-2-oxopropyl)-2,3-dimethyl-1,3-cyclohexadiene	-	-	CO	-	**B8**, 327, **13**, *80*
3-(1-Acetyl-2-oxopropyl)-1,5-cyclooctadiene	-	-	CO	-	**B9**, 213, **7**, *12*
2-(1-Acetyl-2-oxopropyl)-5-cyclooctene-1,4-diyl	-	-	CO	-	**B10**, 134, **4**, *58*
$C_{13}H_{18}O_3$	-	-	CO	-	**B8**, 314, **13**, *14*
$C_{13}H_{18}O_4$					
1-Methoxy-4-[2-(1,3-dioxobutyl)oxyethyl]-1,3-cyclohexadiene	-	-	CO	-	**B8**, 157, **7**, *40*
1-Methoxy-4-(5-methoxy-3,5-dioxopentyl)-1,3-cyclohexadiene	-	-	CO	-	**B8**, 154, **7**, *18*
5-[Bis(ethoxycarbonyl)methyl]-1,3-cyclohexadiene	-	-	CO	-	**B8**, 99, **4**, *81*
5-[Bis(methoxycarbonyl)methyl]-1,4-dimethyl-1,3-cyclohexadiene	-	-	CO	-	**B8**, 365, **13**, *226*
5-(1-Acetyl-2-ethoxy-2-oxoethyl)-2-methoxy-1,3-cyclohexadiene	-	-	CO	-	**B8**, 202, **9**, *43*
5-[1-Ethoxycarbonyl-2-oxopropyl]-2-methoxy-5-methyl-1,3-cyclohexadiene	-	-	CO	-	**B8**, 350, **13**, *156*
6-(1-Acetyl-2-oxopropyl)-1,2-dimethyl-1,3-cyclohexadiene	-	-	CO	-	**B8**, 315, **13**, *17*
$C_{13}H_{18}O_5$					
2-Methoxy-5-[bis(methoxycarbonyl)methyl]-3-methyl-1,3-cyclohexadiene	-	-	CO	-	**B8**, 365, **13**, *228*
3-Methoxy-5-[bis(methoxycarbonyl)methyl]-2-methyl-1,3-cyclohexadiene	-	-	CO	-	**B8**, 365, **13**, *229*
4-Methoxy-5-[bis(methoxycarbonyl)methyl]-1-methyl-1,3-cyclohexadiene	-	-	CO	-	**B8**, 363, **13**, *218*
5-(1-Acetyl-2-oxopropyl)-2-methoxy-5-methyl-1,3-cyclohexadiene	-	-	CO	-	**B8**, 345, **13**, *137*
$C_{13}H_{18}Si$	-	-	CO	-	**B9**, 259, **9**, *7*
$C_{13}H_{20}$					
1,2,3,4,5,5-Hexamethyl-6-methylene-1,3-cyclohexadiene	-	-	CO	-	**B8**, 450

1,2,3,4,4a,5-Hexahydro-3,4,4a-trimethylnaphthalene	–	–	CO	–	**B8**, 326, **13**, *76*
$C_{13}H_{20}OS$	–	–	CO	–	**B9**, 200, **6**, *4*
$C_{13}H_{20}OS_2$	–	–	CO	–	**B10**, 14, **1**, *55*
$C_{13}H_{20}O_3$					
1-Methoxy-4-(5-methoxy-5-oxopentyl)-1,3-cyclohexadiene	–	–	CO	–	**B8**, 156, **7**, *31*
4-[(2-Acetyloxy)ethyl]-1-(1-methylethoxy)-1,3-cyclohexadiene	–	–	CO	–	**B8**, 159, **7**, *48*
5-[(2-Acetyloxy)ethyl]-2-methoxy-5,6-dimethyl-1,3-cyclohexadiene	–	–	CO	–	**B8**, 404, **14**, *40*
$C_{13}H_{21}N$					
1,5,5-Trimethyl-3-(1-pyrrolidinyl)-1,3-cyclohexadiene	–	–	CO	–	**B8**, 397, **14**, *11*
3,5,5-Trimethyl-1-(1-pyrrolidinyl)-1,3-cyclohexadiene	–	–	CO	–	**B8**, 397, **14**, *8*
$C_{13}H_{21}NO$					
1,5,5-Trimethyl-3-(4-morpholinyl)-1,3-cyclohexadiene	–	–	CO	–	**B8**, 398, **14**, *12*
3,5,5-Trimethyl-1-(4-morpholinyl)-1,3-cyclohexadiene	–	–	CO	–	**B8**, 397, **14**, *9*
$C_{13}H_{22}Ge$	–	–	CO	–	**B9**, 127, **3**, *36*
$C_{13}H_{22}OSi$	–	–	CO	–	**B8**, 8, **1**, *12* **B8**, 8, **1**, *13*
$C_{13}H_{22}O_2$	–	–	CO	–	**B8**, 355, **13**, *178*
$C_{13}H_{22}O_3$	–	–	CO	–	**B8**, 234, **10**, *9*
$C_{13}H_{22}Si$					
4-Cyclohexyl-1,1-dimethyl-2,4-silacyclohexadiene	–	–	CO	–	**B8**, 9, **1**, *18*
7-(Triethylsilyl)-1,3,5-cycloheptatriene	–	–	CO	–	**B9**, 127, **3**, *34*
1,2,3,4,4a,5-Hexahydro-7-(trimethylsilyl)naphthalene	–	–	CO	–	**B8**, 318, **13**, *31*
$C_{13}H_{23}NOSi$	–	–	CO	–	**B8**, 235, **10**, *18*
$C_{13}H_{24}P$	–	–	CO	–	**B9**, 52, **2**, *45*
$C_{13}H_{24}Si$	–	–	CO	–	**B8**, 236, **10**, *20*
C_{14} H_7BrN_4	–	–	CO	–	**B10**, 152, **4**, *141*
$C_{14}H_7NO_3$	–	–	CO	–	**B9**, 12, **1**, *36*
$C_{14}H_8F_6N_2$	–	–	CO	–	**B10**, 151, **4**, *135*
$C_{14}H_8F_{12}$	–	–	CO	–	**B10**, 177
$C_{14}H_8N_4$					
1,1,2,2-Tetracyano-1,2,3,8a-tetrahydroazulene	–	–	CO	–	**B9**, 162, **4**, *53*
8,8,9,9-Tetracyanobicyclo-[5.2.1]deca-3,5-diene-2,10-diyl	–	–	CO	–	**B10**, 150, **4**, *134*

$C_{14}H_{12}N_2O$					
5-[5-Phenyl-1,3,4-oxazol-2-yl]-1,3-cyclohexadiene	–	–	CO	–	**B8**, 109, **4**, *138*
1-[(2-Phenylethenyl)carbonyl]-1*H*-diazepine	–	–	CO	–	**B9**, 15, **1**, *55*
$C_{14}H_{12}N_4O_4$	–	–	CO	–	**B9**, 122, **3**, *12*
$C_{14}H_{12}O$	–	–	CO	–	**B9**, 123, **3**, *16*
$C_{14}H_{12}O_2$	–	–	CO	–	**B8**, 463, **17**, *14*
$C_{14}H_{13}N$					
5-(1*H*-Indol-3-yl)-1,3-cyclohexadiene	–	–	CO	–	**B8**, 108, **4**, *130*
7-[(4-Methylphenyl)imino]-1,3,5-cycloheptatriene	–	–	CO	–	**B9**, 156, **4**, *31*
$C_{14}H_{13}NO$					
7-[(4-Methoxyphenyl)imino]-1,3,5-cycloheptatriene	–	–	CO	–	**B9**, 156, **4**, *30*
7-Phenyl-7-azabicyclo[4.2.1]-nona-2,4-dien-8-one	–	–	CO	–	**B9**, 60, **2**, *73*
$C_{14}H_{13}NO_2$					
1-[(Phenylmethoxy)carbonyl]-1*H*-azepine	–	–	CO	–	**B9**, 6, **1**, *6*
8-(4-Nitrophenyl)bicyclo-[5.1.0]octa-2,4-diene	–	–	CO	–	**B9**, 57, **2**, *64*
$C_{14}H_{14}$					
2-(4-Methylphenyl)-1,3,5-cycloheptatriene	–	–	CO	–	**B9**, 125, **3**, *26*
7-(2,4,6-Cycloheptrien-1-yl)-1,3,5-cycloheptatriene	–	–	CO	–	**B9**, 129, **3**, *47*
5-Phenylmethylene-1,3-cycloheptadiene	–	–	CO	–	**B9**, 56, **2**, *58*
5,6,9,10-Tetrahydro-5,10-ethenobenzocyclooctene	–	–	CO	–	**B9**, 213, **7**, *18*
$C_{14}H_{14}F_6$	–	–	CO	–	**B8**, 470, **17**, *45* **B8**, 470, **17**, *46*
$C_{14}H_{14}N_4O_5$	–	–	CO	–	**B8**, 201, **9**, *38*
$C_{14}H_{14}O$					
2-Methoxy-5-(2-phenylethenyl)-1,3-cyclohexadiene	–	–	CO	–	**B8**, 368, **13**, *243*
1-[(Hydroxy)phenylmethyl]-1,3,5-cycloheptatriene	–	–	CO	–	**B9**, 122, **3**, *11*
5-Hydroxy-7-(phenylmethylene)-1,3-cycloheptadiene	–	–	CO	–	**B9**, 67, **2**, *102*
6,9-Dihydro-6-methoxy-5,9-etheno-5*H*-benzocycloheptene	–	–	CO	–	**B9**, 203, **6**, *18*
6,9-Dihydro-6-hydroxy-6-methyl-5,9-etheno-5*H*-benzocycloheptene	–	–	CO	–	**B9**, 203, **6**, *19*
$C_{14}H_{14}O_2$					
5-Methoxycarbonyl-5-phenyl-1,3-cyclohexadiene	–	–	CO	–	**B8**, 245, **11**, *13*

Compound					Reference
6-Methoxy-2-phenyl-2,4-cycloheptadien-1-one	–	–	CO	–	**B9**, 70, **2**, *120*
$C_{14}H_{14}O_3$					
2-Methoxycarbonyl-4-methyl-3-phenyl-1-oxo-2-pentene-1,4-diyl	–	–	CO	–	**B10**, 52, **3**, *27*
3-Methoxycarbonyl-4-methyl-2-phenyl-1-oxo-2-pentene-1,4-diyl	–	–	CO	–	**B10**, 52, **3**, *28*
$C_{14}H_{14}O_3S$	–	–	CO	–	**B8**, 463, **17**, *13*
$C_{14}H_{14}O_4$	–	–	CO	–	**B9**, 59, **2**, *69*
$C_{14}H_{15}P$	C_6H_{10}	–	–	–	**B10**, 210, **7**, *25*
$C_{14}H_{16}$					
1-Methyl-4-(phenylmethyl)-1,3-cyclohexadiene	–	–	CO	–	**B8**, 157, **7**, *38*
2-Methyl-5-(phenylmethyl)-1,3-cyclohexadiene	–	–	CO	–	**B8**, 208, **9**, *77*
1,2,5,8,9,10-Hexahydro-anthracene	–	–	CO	–	**B8**, 143
$C_{14}H_{16}N_2O$					
1,1-Dicyano-8-methoxy-10-methylspiro[4,5]deca-7,9-diene	–	–	CO	–	**B8**, 415, **14**, *89*
7,7-Dicyano-3-methoxy-spiro[5,5]undeca-2,4-diene	–	–	CO	–	**B8**, 353, **13**, *167*
$C_{14}H_{16}N_2O_2$	–	–	CO	–	**B8**, 278, **12**, *59*
$C_{14}H_{16}N_2O_3$					
1-[2-(Acetyloxy)ethyl]-5-(dicyanomethyl)-4-methoxy-1,3-cyclohexadiene	–	–	CO	–	**B8**, 321, **13**, *49*
5-(Dicyanomethyl)-4-methoxy-1-(3-methoxy-3-oxopropyl)-1,3-cyclohexadiene	–	–	CO	–	**B8**, 322, **13**, *51*
5-[2-(Acetyloxy)ethyl]-5-(dicyanomethyl)-2-methoxy-1,3-cyclohexadiene	–	–	CO	–	**B8**, 349, **13**, *152*
5-(Dicyanomethyl)-2-methoxy-5-(3-methoxy-3-oxopropyl)-1,3-cyclohexadiene	–	–	CO	–	**B8**, 351, **13**, *160*
$C_{14}H_{16}O$					
5-[(3,5-Dimethylphenyl)oxy]-1,3-cyclohexadiene	–	–	CO	–	**B8**, 86, **4**, *14*
5-Hydroxy-5-(4-methylphenyl)-1,3-cycloheptadiene	–	–	CO	–	**B9**, 55, **2**, *54*
1,2,5,8,9,10-Hexahydro-2-hydroxyanthracene	–	–	CO	–	**B8**, 362, **13**, *212*
$C_{14}H_{16}O_2$					
5-(2,4-Dimethoxyphenyl)-1,3-cyclohexadiene	–	–	CO	–	**B8**, 106, **4**, *116*
5-(2,5-Dimethoxyphenyl)-1,3-cyclohexadiene	–	–	CO	–	**B8**, 106, **4**, *117*

3,4,5a,9a-Tetrahydro-3,3-dimethyl-1(2*H*)-dibenzofuranone	–	–	CO	–	**B8**, 275, **12**, *48*
$C_{14}H_{16}O_4$	–	–	CO	–	**B9**, 225, **8**, *12*
$C_{14}H_{17}N$					
5-(2-Amino-5-methylphenyl)-2-methyl-1,3-cyclohexadiene	–	–	CO	–	**B8**, 211, **9**, *89*
5-(2-Amino-4,5-dimethylphenyl)-1,3-cyclohexadiene	–	–	CO	–	**B8**, 107, **4**, *123*
5-[4-(Dimethylamino)phenyl]-1,3-cyclohexadiene	–	–	CO	–	**B8**, 105, **4**, *114*
5-(4-Amino-2,5-dimethylphenyl)-1,3-cyclohexadiene	–	–	CO	–	**B8**, 107, **4**, *124*
6-(2-Amino-4-methylphenyl)-1-methyl-1,3-cyclohexadiene	–	–	CO	–	**B8**, 177, **8**, *17*
6-(2-Amino-5-methylphenyl)-2-methyl-1,3-cyclohexadiene	–	–	CO	–	**B8**, 235, **10**, *16*
6,6-Dimethyl-5-(phenylamino)-1,3-cyclohexadiene	–	–	CO	–	**B8**, 359, **13**, *200*
5-[(Methyl)phenylamino]-1,3-cycloheptadiene	–	–	CO	–	**B9**, 49, **2**, *24*
5-[(2-Methylphenyl)amino]-1,3-cycloheptadiene	–	–	CO	–	**B9**, 48, **2**, *19*
5-[(4-Methylphenyl)amino]-1,3-cycloheptadiene	–	–	CO	–	**B9**, 48, **2**, *20*
$C_{14}H_{17}NO$					
Carbonyl(imino)[1-(1,1-dimethylethyl)-3-phenyl-1-propene-1,3-diyl]	–	–	CO	–	**B10**, 73, **3**, *122*
Carbonyl(methylimino)[1-(1-methylethyl)-3-phenyl-1-propene-1,3-diyl]	–	–	CO	–	**B10**, 73, **3**, *124*
Carbonyl(ethylimino)[1-ethyl-3-phenyl-1-propene-1,3-diyl]	–	–	CO	–	**B10**, 54, **3**, *40*
Carbonyl[(1-methylethyl)-imino][1-methyl-3-phenyl-1-propene-1,3-diyl]	–	–	CO	–	**B10**, 55, **3**, *43*
Carbonyl[(1,1-dimethylethyl)-imino][3-phenyl-1-propene-1,3-diyl]	–	–	CO	–	**B10**, 55, **3**, *46*
Carbonyl[(phenylmethyl)-imino][2,3-dimethyl-2-butene-1,4-diyl]	–	–	CO	–	**B10**, 59, **3**, *64*
2-Methoxy-5-[(2-methylphenyl-amino]-1,3-cyclohexadiene	–	–	CO	–	**B8**, 197, **9**, *12*
2-Methoxy-5-[(3-methyl-phenyl)-amino]-1,3-cyclohexadiene	–	–	CO	–	**B8**, 197, **9**, *13*
2-Methoxy-5-[(4-methyl-phenyl)-amino]-1,3-cyclohexadiene	–	–	CO	–	**B8**, 197, **9**, *14*

5-(2-Amino-5-methylphenyl)-2-methoxy-1,3-cyclohexadiene	–	–	CO	–	**B8**, 206, **9**, *69*
5-[(2-Methoxyphenyl)amino]-1,3-cycloheptadiene	–	–	CO	–	**B9**, 48, **2**, *17*
5-[(4-Methoxyphenyl)amino]-1,3-cycloheptadiene	–	–	CO	–	**B9**, 48, **2**, *18*
$C_{14}H_{17}NO_2$	–	–	CO	–	**B8**, 197, **9**, *11*
$C_{14}H_{17}NO_3$	–	–	CO	–	**B8**, 367, **13**, *240*
$C_{14}H_{17}O_2$	–	–	CO	–	**B8**, 110, **4**, *145*
$C_{14}H_{17}O_4$	–	–	CO	–	**B8**, 353, **13**, *168*
$C_{14}H_{18}$	–	–	CO	–	**B8**, 408, **14**, *61*
$C_{14}H_{18}N$					
1-[2,4-Cycloheptadien-1-yl]-2,6-dimethylpyridinium	–	–	CO	–	**B9**, 52, **2**, *43*
1-[2,4-Cyclohexadien-1-yl]-2,4,6-trimethylpyridinium	–	–	CO	–	**B8**, 133, **5**, *11*
$C_{14}H_{18}N_2O$					
1-Ethyl-5-(dicyanomethyl)-4-(1-methylethoxy)-1,3-cyclohexadiene	–	–	CO	–	**B8**, 324, **13**, *65*
1-(4,4-Dicyanobutyl)-4-methoxy-2-methyl-1,3-cyclohexadiene	–	–	CO	–	**B8**, 361, **13**, *208*
1-(5,5-Dicyanopentyl)-4-methoxy-1,3-cyclohexadiene	–	–	CO	–	**B8**, 158, **7**, *43*
5-Ethyl-5-(dicyanomethyl)-2-(1-methylethoxy)-1,3-cyclohexadiene	–	–	CO	–	**B8**, 356, **13**, *181*
5-(Dicyanomethyl)-1,2-dimethyl-4-(1-methylethoxy)-1,3-cyclohexadiene	–	–	CO	–	**B8**, 413, **14**, *79*
6-(Dicyanomethyl)-1,6-dimethyl-3-(1-methylethoxy)-1,3-cyclohexadiene	–	–	CO	–	**B8**, 416, **14**, *91*
$C_{14}H_{18}N_2O_2$					
5-(Dicyanomethyl)-2-methoxy-5-(2-methoxypropyl)-1,3-cyclohexadiene	–	–	CO	–	**B8**, 348, **13**, *151*
5-(Dicyanomethyl)-4-methoxy-1-(2-methoxypropyl)-1,3-cyclohexadiene	–	–	CO	–	**B8**, 323, **13**, *60*
$C_{14}H_{18}O$	–	–	CO	–	**B10**, 150, **4**, *131*
$C_{14}H_{18}O_2$					
5-(2-Hydroxy-4,4-dimethyl-6-oxo-1-cyclohexenyl)-1,3-cyclohexadiene	–	–	CO	–	**B8**, 110, **4**, *144*
Dihydro-3-[2,3,5,6,7,8-hexahydro-2-naphthalenyl]-2(3*H*)-furanone	–	–	CO	–	**B8**, 328, **13**, *84*

Compound					Reference
$C_{14}H_{18}O_5$					
4-Methoxy-1-methyl-5-(2,2-dimethyl-4,6-dioxo-1,3-dioxolan-5-yl)-1,3-cyclohexadiene	–	–	CO	–	**B8**, 318, **13**, *34*
2,3,3a,7a-Tetrahydro-3-[2,2-bis(methoxycarbonyl)-ethyl]benzofuran	–	–	CO	–	**B8**, 270, **12**, *24*
$C_{14}H_{18}O_6$					
5-(1-Acetyl-2-oxopropyl)-1-methoxycarbonyl-2-methyl-1,3-cyclohexadiene	–	–	CO	–	**B8**, 313, **13**, *9*
6-(1-Acetyl-2-oxopropyl)-6-methoxycarbonyl-1-methyl-1,3-cyclohexadiene	–	–	CO	–	**B8**, 326, **13**, *75*
$C_{14}H_{18}O_7$					
6-Bis(methoxycarbonyl)-methyl-1-methoxy-6-methoxy-carbonyl-1,3-cyclohexadiene	–	–	CO	–	**B8**, 326, **13**, *74*
2-Methoxy-1-methoxy-carbonyl-5-bis(methoxy-carbonyl)methyl-1,3-cyclo-hexadiene	–	–	CO	–	**B8**, 312, **13**, *7*
3-Methoxy-1-methoxy-carbonyl-5-bis(methoxy-carbonyl)methyl-1,3-cyclo-hexadiene	–	–	CO	–	**B8**, 363, **13**, *215*
$C_{14}H_{18}P$	–	–	CO	–	**B8**, 135, **5**, *27*
$C_{14}H_{19}NO_3$	–	–	CO	–	**B8**, 158, **7**, *42*
$C_{14}H_{19}O_5$	–	–	CO	–	**B8**, 354, **13**, *175*
$C_{14}H_{20}O$					
1-(1-Methoxypentyl)-1,3,5,7-cyclooctatetraene	–	–	CO	–	**B9**, 262, **9**, *27*
2,3,5,6,7,8-Hexahydro-2-(2-oxo-1,1-dimethylethyl)-naphthalene	–	–	CO	–	**B8**, 329, **13**, *90*
7,7,9,9-Tetramethyl-8-oxobi-cyclo[4.3.1]dec-3-ene-2,10-diyl	–	–	CO	–	**B10**, 149, **4**, *130*
$C_{14}H_{20}OS$	–	–	CO	–	**B9**, 201, **6**, *10*
$C_{14}H_{20}O_2$					
3-Acetyl-2-cyclohexyl-4-methyl-1-oxo-2-pentene-1,4-diyl	–	–	CO	–	**B10**, 51, **3**, *20*
1-Methoxy-4-methyl-6-(2-oxo-cyclohexyl)-1,3-cyclohexadiene	–	–	CO	–	**B8**, 318, **13**, *35*
1-Methoxycarbonyl-2-methyl-5-(1,1-dimethyl-2-propenyl)-1,3-cyclohexadiene	–	–	CO	–	**B8**, 314, **13**, *15*
2-Methoxy-5-(6-methyl-2-oxo-cyclohexyl)-1,3-cyclohexadiene	–	–	CO	–	**B8**, 203, **9**, *48*
2-Methoxy-5-(3-methyl-2-oxo-					

cyclohexyl)-1,3-cyclohexadiene –	–	CO	–	**B8**, 212, **9**, *95*
3-Methoxy-1-methyl-5-(2-oxo-cyclohexyl)-1,3-cyclohexadiene –	–	CO	–	**B8**, 316, **13**, *22*
5-[1-(Hydroxymethyl)-2-cyclo-pentenyl]-2-methoxy-5-methyl-1,3-cyclohexadiene –	–	CO	–	**B8**, 333, **13**, *109*
2,3,5,6-Tetraethyl-2,5-cyclohexadiene-1,4-dione –	–	CO	–	**B8**, 455, **16**, *11*
$C_{14}H_{20}O_3$				
1-Methoxycarbonyl-2-methyl-5-(1-methyl-2-oxobutyl)-1,3-cyclohexadiene –	–	CO	–	**B8**, 313, **13**, *10*
5-[2-(Hydroxymethyl)-6-oxabicyclo[3.1.0]hex-2-yl]-2-methoxy-5-methyl-1,3-cyclohexadiene –	–	CO	–	**B8**, 334, **13**, *111*
$C_{14}H_{20}O_4$				
1-Methoxy-4-[5-methoxy-carbonyl-4-oxopentyl]-1,3-cyclohexadiene –	–	CO	–	**B8**, 156, **7**, *27*
1-Ethyl-4-methoxy-5-[1-meth-oxycarbonyl-2-oxopropyl]-1,3-cyclohexadiene –	–	CO	–	**B8**, 320, **13**, *42*
5-Ethyl-2-methoxy-5-[1-meth-oxycarbonyl-2-oxopropyl]-1,3-cyclohexadiene –	–	CO	–	**B8**, 349, **13**, *154*
4,5-Bis(ethoxycarbonyl)-1,2-dimethyl-1,3-cyclohexadiene –	–	CO	–	**B8**, 395, **14**, *3*
5,6-Bis[2-methoxycarbonyl-ethyl]-1,3-cyclohexadiene –	–	CO	–	**B8**, 265, **12**, *7*
5-Bis(ethoxycarbonyl)methyl-1,3-cycloheptadiene –	–	CO	–	**B9**, 50, **2**, *28*
$C_{14}H_{20}O_5$				
1-Ethyl-4-methoxy-5-[bis(methoxycarbonyl)methyl]-1,3-cyclohexadiene –	–	CO	–	**B8**, 319, **13**, *38*
5-Ethyl-2-methoxy-5-[bis(methoxycarbonyl)methyl]-1,3-cyclohexdiene –	–	CO	–	**B8**, 345, **13**, *138*
3-Methoxy-5-[bis(methoxycar-bonyl)methyl]-1,5-dimethyl-1,3-cyclohexadiene –	–	CO	–	**B8**, 397, **14**, *10*
3-Methoxy-6-[bis(methoxycar-bonyl)methyl]-1,6-dimethyl-1,3-cyclohexadiene –	–	CO	–	**B8**, 415, **14**, *87*
4-Methoxy-5-[bis(methoxycar-bonyl)methyl]-1,2-dimethyl-1,3-cyclohexadiene –	–	CO	–	**B8**, 413, **14**, *76*
6-[Bis(ethoxycarbonyl)methyl]-2-methoxy-1,3-cyclohexadiene –	–	CO	–	**B8**, 238, **10**, *30*
$C_{14}H_{20}O_6$ –	–	CO	–	**B8**, 214, **9**, *102*
$C_{14}H_{22}OS$ –	–	CO	–	**B9**, 201, **6**, *7*

$C_{14}H_{22}O_2Si$	–	–	CO	–	**B8**, 236, **10**, *23*
$C_{14}H_{22}O_3$	–	–	CO	–	**B8**, 362, **13**, *211*
$C_{14}H_{22}O_4Si$	–	–	CO	–	**B8**, 236, **10**, *21*
$C_{14}H_{23}N_2O_3P$	–	–	CO	–	**B8**, 91, **4**, *48*
$C_{14}H_{24}O_3$	–	–	CO	–	**B8**, 369, **13**, *248*
$C_{14}H_{26}Si_2$	–	–	CO	–	**B8**, 330, **13**, *96*
$C_{15}H_8F_{12}$	–	–	CO	–	**B9**, 211, **7**, *6*
$C_{15}H_{10}N_4$					
1,1,2,2-Tetracyano-1,2,3,8a-tetrahydro-3-methylazulene	–	–	CO	–	**B9**, 150, **4**, *8*
8,8,9,9-Tetracyano-2-methylbicyclo[5.2.1]deca-3,5-diene-2,10-diyl	–	–	CO	–	**B10**, 151, **4**, *136*
8,8,9,9-Tetracyano-3-methylbicyclo[5.2.1]deca-3,5-diene-2,10-diyl	–	–	CO	–	**B10**, 152, **4**, *138*
$C_{15}H_{10}N_4O$					
1,2,3,8a-Tetrahydro-1,1,2,2-tetracyano-3-methoxyazulene	–	–	CO	–	**B9**, 149, **4**, *7*
1-Acetyl-7,7,8,8-tetracyanobicyclo[4.2.1]non-4-ene-3,9-diyl	–	–	CO	–	**B10**, 148, **4**, *123*
9-Acetyl-7,7,8,8-tetracyanobicyclo[4.2.1]non-3-ene-2,9-diyl	–	–	CO	–	**B10**, 147, **4**, *121*
7,7,8,8-Tetracyano-9-methoxybicyclo[4.2.2]deca-2,4,9-triene	–	–	CO	–	**B9**, 225, **8**, *11*
8,8,9,9-Tetracyano-4-methoxybicyclo[5.2.1]deca-3,4-diene-2,10-diyl	–	–	CO	–	**B10**, 152, **4**, *140*
$C_{15}H_{11}N_3O_3$	–	–	CO	–	**B10**, 140, **4**, *88*
$C_{15}H_{12}$	–	–	CO	–	**B10**, 53, **3**, *33*
$C_{15}H_{12}F_4$					
5,6,7,8-Tetrafluoro-1,3,9-trimethyl-1,4-ethenonaphthalene	–	–	CO	–	**B8**, 469, **17**, *43*
5,6,7,8-Tetrafluoro-2,3,9-trimethyl-1,4-ethenonaphthalene	–	–	CO	–	**B8**, 469, **17**, *44*
$C_{15}H_{12}O$					
7-(Phenylmethylene)-1,3,5-cycloheptatriene-1-carbaldehyde	–	–	CO	–	**B9**, 184, **5**, *12*
5-Phenyl-1,3,5,7-cyclooctatetraene-1-carbaldehyde	–	–	CO	–	**B9**, 267, **9**, *51*
6-Phenyl-1,3,5,7-cyclooctatetraene-1-carbaldehyde	–	–	CO	–	**B9**, 267, **9**, *49*
7-Phenyl-1,3,5,7-cyclooctatetraene-1-carbaldehyde	–	–	CO	–	**B9**, 267, **9**, *46*
$C_{15}H_{12}O_2$	–	–	CO	–	**B8**, 466, **17**, *27*

$C_{15}H_{13}BrO_3$	–	–	CO	–	**B8**, 113, **4**, *163*
$C_{15}H_{13}ClO$	–	–	CO	–	**B9**, 81, **2**, *170*
$C_{15}H_{13}N$	–	–	CO	–	**B10**, 143, **4**, *104*
$C_{15}H_{13}NO_3$	–	–	CO	–	**B8**, 197, **9**, *16*
$C_{15}H_{13}NO_5$	–	–	CO	–	**B9**, 11, **1**, *31*
$C_{15}H_{14}$					
1-(2-Phenylethenyl)-1,3,5-cycloheptatriene	–	–	CO	–	**B9**, 121, **3**, *7*
7-(2-Phenylethenyl)-1,3,5-cycloheptatriene	–	–	CO	–	**B9**, 129, **3**, *42*
7-(2-Phenylethylidene)-1,3,5-cycloheptatriene	–	–	CO	–	**B9**, 158, **4**, *38*
7-[(4-Methylphenyl)methylene]-1,3,5-cycloheptatriene	–	–	CO	–	**B9**, 159, **4**, *40*
7-(1-Phenylethylidene)-1,3,5-cycloheptatriene	–	–	CO	–	**B9**, 161, **4**, *50*
1-(Phenylmethyl)-1,3,5,7-cyclooctatetraene	–	–	CO	–	**B9**, 260, **9**, *15*
2-[2,4,6-Cycloheptatrien-1-ylidene]bicyclo[5.1.0]-octa-3,5-diene	–	–	CO	–	**B9**, 81, **2**, *167*
$C_{15}H_{14}N_4O_2$	–	–	CO	–	**B8**, 271, **12**, *29*
$C_{15}H_{14}N_4O_4$	–	–	CO	–	**B9**, 122, **3**, *13*
$C_{15}H_{14}O$					
7-[(Methoxy)phenymethylene]-1,3,5-cycloheptatriene	–	–	CO	–	**B9**, 161, **4**, *48*
1,3,3a,4,7,7a-Hexahydro-3-phenyl-1,4-methanoisobenzofuran-7,8-diyl	–	–	CO	–	**B10**, 139, **4**, *85*
5,5a,10a,11-Tetrahydro-6*H*-cyclohepta[*b*]naphthalen-6-one	–	–	CO	–	**B9**, 80, **2**, *164*
3-Hydroxy-3-phenylbicyclo-[3.2.2]nona-2,6,8-triene	–	–	CO	–	**B8**, 471, **17**, *53*
$C_{15}H_{14}O_2$	–	–	CO	–	**B9**, 152, **4**, *18*
$C_{15}H_{15}N$	–	–	CO	–	**B8**, 108, **4**, *131*
$C_{15}H_{15}N$	–	–	CO	–	**B8**, 108, **4**, *132*
$C_{15}H_{15}NO$	–	–	CO	–	**B8**, 204, **9**, *59*
$C_{15}H_{16}$	–	–	CO	–	**B9**, 129, **3**, *46*
$C_{15}H_{16}N_4O_4$	–	–	CO	–	**B8**, 95, **4**, *65*
$C_{15}H_{16}N_4O_5$	–	–	CO	–	**B8**, 332, **13**, *104*
$C_{15}H_{16}O$					
1-(1-Hydroxy-2-phenylethyl)-1,3,5-cycloheptatriene	–	–	CO	–	**B9**, 122, **3**, *10*
1-[(4-Methylphenyl)hydroxymethyl]-1,3,5-cycloheptatriene	–	–	CO	–	**B9**, 132, **3**, *58*
5-Methoxy-7-(phenylmethylene)-1,3-cycloheptadiene	–	–	CO	–	**B9**, 68, **2**, *110*
5-Hydroxy-7-(2-phenylethylidene)-1,3-cycloheptadiene	–	–	CO	–	**B9**, 67, **2**, *101*
5-Hydroxy-7-[(4-methylphenyl)methylene]-1,3-cycloheptadiene	–	–	CO	–	**B9**, 67, **2**, *103*

8-[(Hydroxy)phenylmethyl]-bicyclo[3.2.1]octa-2,6-diene	–	–	CO	–	**B9**, 200, **6**, *5*
$C_{15}H_{16}O_2$					
5-[(Methoxycarbonyl)phenyl-methyl]-1,3-cyclohexadiene	–	–	CO	–	**B8**, 99, **4**, *88*
5-Methoxycarbonyl-5-(phenyl-methyl)-1,3-cyclohexadiene	–	–	CO	–	**B8**, 244, **11**, *10*
5-Benzoyl-7-methoxy-1,3-cyclohexadiene	–	–	CO	–	**B9**, 59, **2**, *72*
$C_{15}H_{16}O_6$	–	–	CO	–	**B9**, 74, **2**, *137*
$C_{15}H_{17}NO$					
Carbonyl[(phenylmethyl)-imino][2-[1-cyclopenten-1-yl]-ethane-1,2-diyl]	–	–	CO	–	**B10**, 59, **3**, *65*
Carbonyl[(phenylmethyl)-imino][2-(ethenyl)cyclo-pentane-1,2-diyl]	–	–	CO	–	**B10**, 59, **3**, *66*
Carbonyl(phenylimino)[5,5-di-methyl-2-cyclohexene-1,4-diyl	–	–	CO	–	**B10**, 125, **4**, *21*
$C_{15}H_{17}N_3O_6$	–	–	CO	–	**B9**, 9, **1**, *23*
$C_{15}H_{18}F_6$	–	–	CO	–	**B8**, 471, **17**, *55*
$C_{15}H_{18}N_2O_3$	–	–	CO	–	**B8**, 323, **13**, *61*
$C_{15}H_{18}N_2O_3$	–	–	CO	–	**B8**, 414, **14**, *82*
$C_{15}H_{18}OS_2$	–	–	CO	–	**B10**, 14, **1**, *56*
$C_{15}H_{18}O_3$					
5-(2,3,4-Trimethoxyphenyl)-1,3-cyclohexadiene	–	–	CO	–	**B8**, 106, **4**, *120*
5-(2,4,5-Trimethoxyphenyl)-1,3-cyclohexadiene	–	–	CO	–	**B8**, 106, **4**, *121*
5-(2,4,6-Trimethoxyphenyl)-1,3-cyclohexadiene	–	–	CO	–	**B8**, 107, **4**, *122*
3,4,5a,9a-Tetrahydro-7-methoxy-3,3-dimethyldibenzo-1(2*H*)-furanone	–	–	CO	–	**B8**, 330, **13**, *94*
3,3a,4,5,5a,9b-Hexahydro-3,5a,9-trimethyl-2*H*,8*H*-naphtho[1,2-*b*]furan-2,8-dione	–	–	CO	–	**B8**, 453, **16**, *2*
$C_{15}H_{18}O_4$	–	–	CO	–	**B8**, 205, **9**, *62*
$C_{15}H_{19}N$	–	–	CO	–	**B8**, 112, **4**, *159*
$C_{15}H_{19}NO$	–	–	CO	–	**B8**, 453, **16**, *2*
Carbonyl(methylimino)[1-butyl-3-phenyl-1-propene-1,3-diyl]	–	–	CO	–	**B10**, 74, **3**, *125*
Carbonyl(methylimino)[1-(1,1-dimethylethyl)-1-propene-1,3-diyl]	–	–	CO	–	**B10**, 74, **3**, *126*
Carbonyl(butylimino)[1-methyl-3-phenyl-1-propene-1,3-diyl]	–	–	CO	–	**B10**, 74, **3**, *131*
Carbonyl[(1-methylethyl)-imino][1-ethyl-3-phenyl-1-propene-1,3-diyl]	–	–	CO	–	**B10**, 74, **3**, *130*

Carbonyl[(2-methylpropyl)-imino][1-methyl-3-phenyl-1-propene-1,3-diyl]	–	–	CO	–	**B10**, 75, **3**, *132*
Carbonyl[(1,1-dimethylethyl)-imino][1-methyl-3-phenyl-1-propene-1,3-diyl]	–	–	CO	–	**B10**, 75, **3**, *133*
2-Methoxy-5-[(1-phenylethyl)-amino]-1,3-cyclohexadiene	–	–	CO	–	**B8**, 197, **9**, *6*
2-Methoxy-5-[4-(dimethyl-amino)phenyl]-1,3-cyclo-hexadiene	–	–	CO	–	**B8**, 206, **9**, *68*
$C_{15}H_{19}NO_3$					
Carbonyl[[(2,4-dimethoxy-phenyl)methyl]imino][3-meth-yl-2-butene-1,4-diyl]	–	–	CO	–	**B10**, 60, **3**, *68*
7-Cyano-3-methoxy-7-methoxycarbonyl-spiro[5.5]undeca-2,4-diene	–	–	CO	–	**B8**, 368, **13**, *241*
$C_{15}H_{19}NO_5$					
5-(1-Cyano-2-methoxy-2-oxo-ethyl)-4-methoxy-1-[2-(meth-oxycarbonyl)ethyl]-1,3-cyclohexadiene	–	–	CO	–	**B8**, 322, **13**, *52*
5-(1-Cyano-2-methoxy-2-oxoethyl)-2-methoxy-5-[2-(methoxycarbonyl)ethyl]-1,3-cyclohexadiene	–	–	CO	–	**B8**, 351, **13**, *161*
$C_{15}H_{19}N_3P$	–	–	CO	–	**B8**, 134, **5**, *17*
$C_{15}H_{19}O$	–	–	CO	–	**B8**, 328, **13**, *85*
$C_{15}H_{20}N$					
[2,4-Cyclohexadien-1-yl]-methyl(1-phenylethyl)am-monium	–	–	CO	–	**B8**, 136, **5**, *31*
1-[2,4-Cycloheptadien-1-yl]-2,4,6-trimethylpyridinium	–	–	CO	–	**B9**, 52, **2**, *44*
$C_{15}H_{20}NO$	–	–	CO	–	**B8**, 197, **9**, *18*
$C_{15}H_{20}NO_3$	–	–	CO	–	**B8**, 454, **16**, *5*
$C_{15}H_{20}O_2$					
2-(1-Acetyl-2-oxopropyl)-2,3,5,6,7,8-hexahydro-naphthalene	–	–	CO	–	**B8**, 327, **13**, *82*
2-[2,4-Cycloheptadien-1-yl]-5,5-dimethylcyclohexane-1,3-dione	–	–	CO	–	**B9**, 50, **2**, *30*
$C_{15}H_{20}O_3$					
2-[4-Methoxy-2,4-cyclohexa-dien-1-yl]-5,5-dimethylcyclo-hexane-1,3-dione	–	–	CO	–	**B8**, 203, **9**, *49*
1-Methoxycarbonyl-3-methyl-5-(2-oxocyclohexyl)-1,3-cyclohexadiene	–	–	CO	–	**B8**, 317, **13**, *27*

2-Methoxy-5-methyl-5-[1-methoxycarbonyl-2-cyclopenten-1-yl]-1,3-cyclohexadiene	–	–	CO	–	**B8**, 336, **13**, *117*
5-(2-Hydroxy-5,5-dimethyl-6-oxo-1-cyclohexen-1-yl)-2-methoxy-1,3-cyclohexadiene	–	–	CO	–	**B8**, 205, **9**, *65*
3a,4,5,5a,6,9b-Hexahydro-8-hydroxy-3,5a,9-trimethylnaphtho[1,2-*b*]-2(3*H*)-furanone	–	–	CO	–	**B8**, 430, **15**, *1*
$C_{15}H_{20}O_4$					
2,3,3a,4a,8a,8b-Hexahydro-6-methoxy-8b-methoxycarbonyl-8a-methyl-1*H*-cyclopenta[*b*]furan	–	–	CO	–	**B8**, 405, **14**, *44*
1-(1-Acetyl-2-oxoethyl)-1,2,5,6,7,8-hexahydronaphthalene	–	–	CO	–	**B8**, 315, **13**, *18*
2-Methoxy-5-[1-methoxycarbonyl-2-oxocyclopentyl]-1,3-cyclohexadiene	–	–	CO	–	**B8**, 337, **13**, *122*
5-[2-[(Formyloxy)methyl]-6-oxabicyclo[3.1.0]hex-2-yl]-methoxy-5-methyl-1,3-cyclohexadiene	–	–	CO	–	**B8**, 335, **13**, *116*
$C_{15}H_{20}O_5$	–	–	CO	–	**B8**, 399, **14**, *18*
$C_{15}H_{20}O_7$					
1-(2-Methoxy-2-oxoethyl)-5-[bis(methoxycarbonyl)methyl]-4-methoxy-1,3-cyclohexadiene	–	–	CO	–	**B8**, 320, **13**, *44*
2-Methoxy-5-[bis(methoxycarbonyl)methyl]-5-(2-methoxy-2-oxoethyl)-1,3-cyclohexadiene	–	–	CO	–	**B8**, 346, **13**, *140*
$C_{15}H_{20}P$	–	–	CO	–	**B9**, 54, **2**, *48*
$C_{15}H_{21}NO_3$					
1-[5-Cyano-5-(methoxycarbonyl)pentyl]-4-methoxy-1,3-cyclohexadiene	–	–	CO	–	**B8**, 158, **7**, *44*
5-(1-Cyano-2-methoxy-2-oxoethyl)-1-ethyl-4-(1-methylethoxy)-1,3-cyclohexadiene	–	–	CO	–	**B8**, 324, **13**, *66*
5-(1-Cyano-2-methoxy-2-oxoethyl)-5-ethyl-2-(1-methylethoxy)-1,3-cyclohexadiene	–	–	CO	–	**B8**, 356, **13**, *183*
$C_{15}H_{22}OSi$	–	–	CO	–	**B10**, 13, **1**, *52*
$C_{15}H_{22}O_3$					
2,3,3a,4a,8a,8b-Hexahydro-6-methoxy-8b-(methoxymethyl)-8a-methyl-1*H*-cyclopenta[*b*]benzofuran	–	–	CO	–	**B8**, 404, **14**, *43*

8,8,9,9-Tetracyano-6-(methoxycarbonyl)bicyclo[5.2.1]-deca-3,5-diene-2,10-diyl	–	–	CO	–	**B10**, 153, **4**, *144*
$C_{16}H_{12}$	–	–	CO	–	**B9**, 266, **9**, *41*
$C_{16}H_{12}CrO_3$	–	–	CO	–	**B8**, 247, **11**, *21*
$C_{16}H_{12}O$					
1-Oxo-2,3-diphenyl-2-butene-1,4-diyl	–	–	CO	–	**B10**, 48, **3**, *3*
9-Acetylanthracene	–	–	CO	–	**B8**, 412, **14**, *74*
$C_{16}H_{12}P$	–	–	–	–	**B10**, 188, **6**, *4*
$C_{16}H_{13}N$					
7-Cyano-8-phenylbicyclo-[3.2.2]nona-3.8-diene-2,6-diyl	–	–	CO	–	**B10**, 142, **4**, *99*
7-Cyano-9-phenylbicyclo-[3.2.2]nona-3,8-diene-2,6-diyl	–	–	CO	–	**B10**, 142, **4**, *98*
$C_{16}H_{13}NO$	–	–	CO	–	**B10**, 73, **3**, *123*
$C_{16}H_{13}N_3O_2$					
5,10-Dihydro-2-phenyl-5,10-etheno-1*H*-[1,2,4]triazolo-[1,2-*a*][1,2]diazocine-1,3(2*H*)-dione	–	–	CO	–	**B9**, 224, **8**, *10*
10,11-Dihydro-2-phenyl-1,3(2*H*)-dioxo-5,11-methano-1*H*,5*H*-[1,2,4]triazolo[1,2-*a*]-[1,2]diazonine-10,13-diyl	–	–	CO	–	**B10**, 150, **4**, *133*
$C_{16}H_{13}N_3O_3$	–	–	CO	–	**B9**, 148, **4**, *3*
$C_{16}H_{14}$					
$(C_6H_5)_2CC(CH_2)_2$	–	–	CO	–	**B10**, 10, **1**, *35*
$C_6H_5CHC(CH_2)CHC_6H_5$	–	–	CO	–	**B10**, 11, **1**, *38*
4a,9,9a,10-Tetrahydro-9,10-ethenoanthracene	–	–	CO	–	**B8**, 276, **12**, *52*
1-(2-Phenylethenyl)-1,3,5,7-cyclooctatetraene	–	–	CO	–	**B9**, 261, **9**, *20*
$C_{16}H_{14}F_4$	–	–	CO	–	**B8**, 470, **17**, *47*
$C_{16}H_{14}N_4$	–	–	CO	–	**B9**, 82, **2**, *172*
$C_{16}H_{14}O$	–	–	CO	–	**B9**, 184, **5**, *13*
$C_{16}H_{14}O_3$	–	–	CO	–	**B10**, 49, **3**, *11*
$C_{16}H_{14}O_4$	–	–	CO	–	**B9**, 268, **9**, *56*
$C_{16}H_{15}N_4$	–	–	CO	–	**B10**, 149, **4**, *128*
$C_{16}H_{16}$					
1-Methyl-7-(2-phenylethenyl)-1,3,5-cycloheptatriene	–	–	CO	–	**B9**, 148, **4**, *2*
7-[1-(4-Methylphenyl)ethylidene]-1,3,5-cycloheptatriene	–	–	CO	–	**B9**, 162, **4**, *51*
4a,4b,5,6,9,10,10a,10b-Octahydro-5,9-ethenobenzo-[3,4]cyclobuta[1,2]cyclooctene-6,10-diyl	–	–	CO	–	**B10**, 145, **4**, *111*
4a,4b,4c,10a,10b,10c-Hexahydro-benzo[3',4']cyclobuta[1',2':3,4]cyclobuta-[1,2]cyclooctene	–	–	CO	–	**B8**, 275, **12**, *50*

4a,4b,5,10,10a,10b-Hexahydro-5,10-ethenobenzo[3,4]cyclobuta[1,2]cyclooctene	–	–	CO	–	**B8**, 275, **12**, *51*
3,3a,3b,3c,7a,7b,8,8a-Octahydro-3,8-ethenobenzo[3,4]-cyclobuta[1,2-*a*]pentalene	–	–	CO	–	**B8**, 276, **12**, *53*
Pentacyclo[9.5.0.0^{2,16}.0^{3,5}.0^{4,10}]hexadeca-6,8,12,14-tetraene	–	–	CO	–	**B9**, 62, **2**, *81*
$C_{16}H_{16}$	–	–	CO	–	**B8**, 277, **12**, *56*
$C_{16}H_{16}Fe$	–	–	CO	–	**B8**, 109, **4**, *141*
$C_{16}H_{16}N_2O_2$	–	–	CO	–	**B9**, 11, **1**, *33*
$C_{16}H_{16}O$					
5-(4-Phenyl-2-oxo-3-butenyl)-1,3-cyclohexadiene	–	–	CO	–	**B8**, 95, **4**, *64*
1-(Hydroxymethyl)-7-[(4-methylphenyl)methylene]-1,3,5-cycloheptatriene	–	–	CO	–	**B9**, 184, **5**, *9*
1-(2-Hydroxy-2-phenylethyl)-methylene-4,6-cycloheptadiene-1,3-diyl	–	–	CO	–	**B10**, 14, **1**, *59*
$C_{16}H_{16}OS$	–	–	CO	–	**B9**, 201, **6**, *11*
$C_{16}H_{16}O_2$	–	–	CO	–	**B8**, 99, **4**, *87*
$C_{16}H_{17}N$	–	–	CO	–	**B8**, 108, **4**, *133*
$C_{16}H_{18}$	–	–	CO	–	**B8**, 276, **12**, *55*
$C_{16}H_{18}O$					
1-[1-Hydroxy-1-(4-methylphenyl)ethyl]-1,3,5-cycloheptatriene	–	–	CO	–	**B9**, 132, **3**, *59*
5-Hydroxy-7-[1-(4-methylphenyl)ethylidene]-1,3-cycloheptadiene	–	–	CO	–	**B9**, 68, **2**, *106*
5-Methoxy-7-(2-phenylethylidene)-1,3-cycloheptadiene	–	–	CO	–	**B9**, 68, **2**, *109*
5-Methoxy-7-[(4-methylphenyl)methylene]-1,3-cycloheptadiene	–	–	CO	–	**B9**, 69, **2**, *111*
$C_{16}H_{18}O_2$	–	–	CO	–	**B8**, 366, **13**, *233*
$C_{16}H_{18}O_3S$	–	–	CO	–	**B9**, 79, **2**, *160*
$C_{16}H_{18}O_4S$	–	–	CO	–	**B8**, 269, **12**, *23*
$C_{16}H_{18}O_5S$					
2-Methoxy-5-[2-methoxy-2-oxo-1-(phenylsulfonyl)ethyl]-1,3-cyclohexadiene	–	–	CO	–	**B8**, 214, **9**, *100*
2-Methoxy-6-[2-methoxy-2-oxo-1-(phenylsulfonyl)ethyl]-1,3-cyclohexadiene	–	–	CO	–	**B8**, 238, **10**, *31*
$C_{16}H_{19}N_5O_4$	–	–	CO	–	**B8**, 89, **4**, *34*
$C_{16}H_{19}N_5O_5$	–	–	CO	–	**B8**, 89, **4**, *35*
$C_{16}H_{19}O_6$	–	–	CO	–	**B8**, 271, **12**, *30*
$C_{16}H_{20}$	–	–	CO	–	**B8**, 276, **12**, *54*

$C_{16}H_{20}F_6$	–	–	CO	–	**B8**, 472, **17**, *56*
$C_{16}H_{20}O_5$	–	–	CO	–	**B8**, 205, **9**, *64*
$C_{16}H_{20}O_8$	–	–	CO	–	**B10**, 175
$C_{16}H_{22}O$					
5-[1,7,7-Trimethyl-2-oxo-bicyclo[2.2.1]hept-3-yl]-1,3-cyclohexadiene	–	–	CO	–	**B8**, 110, **4**, *148*
2,3,5,6,7,8-Hexahydro-2-(2-oxocyclohexyl)naphthalene	–	–	CO	–	**B8**, 328, **13**, *86*
$C_{16}H_{22}O_3$					
1-Methoxycarbonyl-2-methyl-5-(3-methyl-2-oxocyclohexyl)-1,3-cyclohexadiene	–	–	CO	–	**B8**, 313, **13**, *13*
1-Methoxycarbonyl-3-methyl-5-(3-methyl-2-oxocyclohexyl)-1,3-cyclohexadiene	–	–	CO	–	**B8**, 317, **13**, *28*
$C_{16}H_{22}O_4$					
1-Ethyl-4-methoxy-5-[1-methoxycarbonyl-2-oxo-cyclopent-1-yl]-1,3-cyclo-hexadiene	–	–	CO	–	**B8**, 320, **13**, *43*
2-Methoxy-5-[4-methoxy-1-methoxycarbonyl-3-cyclohexen-1-yl]-1,3-cyclo-hexadiene	–	–	CO	–	**B8**, 205, **9**, *63*
2-Methoxy-5-[1-methoxycar-bonyl-2-oxocyclohex-1-yl]-5-methyl-1,3-cyclohexadiene	–	–	CO	–	**B8**, 339, **13**, *125*
5-Ethyl-2-methoxy-5-[1-methoxycarbonyl-2-oxo-cyclopent-1-yl]-1,3-cyclo-hexadiene	–	–	CO	–	**B8**, 344, **13**, *136*
4a,5a,6,7,8,9,9a,9b-Octa-hydro-3-methoxy-9a-methoxycarbonyl-9b-methyldibenzofuran	–	–	CO	–	**B8**, 405, **14**, *45*
$C_{16}H_{22}O_5$					
2-Methoxy-5-[5,5-bis-(methoxycarbonyl)pent-1-ylidene]-1,3-cyclohexadiene	–	–	CO	–	**B8**, 368, **13**, *246*
5-[4-(Ethoxycarbonyl)-4-(methoxycarbonyl)but-1-ylidene]-2-methoxy-1,3-cyclohexadiene	–	–	CO	–	**B8**, 355, **13**, *176*
4,4a,5,6,7,8-Hexahydro-2-methoxy-4a-[bis(methoxy-carbonyl)methyl]naphthalene	–	–	CO	–	**B8**, 401, **14**, *25*
$C_{16}H_{22}O_6$					
2-Methoxy-5-[bis(methoxycar-bonyl)methyl]-5-(3-oxobutyl)-1,3-cyclohexadiene	–	–	CO	–	**B8**, 346, **13**, *141*

2-Methoxy-5-(3-methoxy-3-oxopropyl)-5-[1-methoxycarbonyl-2-oxoethyl]-1,3-cyclohexadiene	–	–	CO	–	**B8**, 351, **13**, *162*
4-Methoxy-1-(3-methoxy-3-oxopropyl)-5-[1-methoxycarbonyl-2-oxopropyl]-1,3-cyclohexadiene	–	–	CO	–	**B8**, 322, **13**, *53*
5-(2-Acetyloxypropyl)-6-[bis(methoxycarbonyl)methyl]-1,3-cyclohexadiene	–	–	CO	–	**B8**, 265, **12**, *9*
$C_{16}H_{22}O_7$					
1-(2-Acetyloxyethyl)-4-methoxy-5-[bis(methoxycarbonyl)-methyl]-1,3-cyclohexadiene	–	–	CO	–	**B8**, 321, **13**, *48*
2-Methoxy-5-(3-methoxy-3-oxopropyl)-5-[bis(methoxycarbonyl)methyl]-1,3-cyclohexadiene	–	–	CO	–	**B8**, 350, **13**, *159*
4-Methoxy-1-(3-methoxy-3-oxopropyl)-5-[bis(methoxycarbonyl)methyl]-1,3-cyclohexadiene	–	–	CO	–	**B8**, 321, **13**, *50*
5-(2-Acetyloxyethyl)-2-methoxy-5-[bis(methoxycarbonyl)methyl]-1,3-cyclohexadiene	–	–	CO	–	**B8**, 345, **13**, *139*
$C_{16}H_{22}P$					
[2,4-Cyclohexadien-1-yl]-diethylphenylphosphonium	–	–	CO	–	**B8**, 136, **5**, *28*
[2,4-Cyclohexadien-1-yl]-methyl(1-methylethyl)-phenylphosphonium	–	–	CO	–	**B8**, 136, **5**, *32*
$C_{16}H_{23}NO_3$					
1-Ethoxycarbonyl-3-(1,1,3-trimethyl-2-oxobutyl)-1*H*-azepine	–	–	CO	–	**B9**, 7, **1**, *14*
8-Ethoxycarbonyl-3,3a,8,8a-tetrahydro-3,3-dimethyl-2-(1-methylethylidene)-2*H*-furo[2,3-*b*]azepine	–	–	CO	–	**B9**, 8, **1**, *22*
$C_{16}H_{24}O_3S_2$					
1-Ethyl-5-[bis[(ethylthio)-carbonyl]methyl]-4-methoxy-1,3-cyclohexadiene	–	–	CO	–	**B8**, 319, **13**, *39*
5-Ethyl-5-[bis[(ethylthio)-carbonyl]methyl]-2-methoxy-1,3-cyclohexadiene	–	–	CO	–	**B8**, 344, **13**, *134*
$C_{16}H_{24}O_4$					
1-Ethyl-4-(1-methylethoxy)-5-[1-methoxycarbonyl-2-oxopropyl]-1,3-cyclohexadiene	–	–	CO	–	**B8**, 324, **13**, *67*

7-[(Chromiumtricarbonyl)-2,4,6-cycloheptatrien-1-yl]-1,3,5-cycloheptatriene	–	–	CO	–	**B9**, 130, **3**, *48*
$C_{17}H_{14}MoO_3$	–	–	CO	–	**B9**, 130, **3**, *49*
$C_{17}H_{14}O_3W$	–	–	CO	–	**B9**, 130, **3**, *50*
$C_{17}H_{15}NO$	–	–	CO	–	**B10**, 54, **3**, *38*
$C_{17}H_{15}N_3O_3$	–	–	CO	–	**B9**, 149, **4**, *6*
$C_{17}H_{16}$	–	–	CO	–	**B10**, 56, **3**, *53*
$C_{17}H_{16}F_4$	–	–	CO	–	**B8**, 470, **17**, *48*
$C_{17}H_{16}N$	–	–	CO	–	**B8**, 133, **5**, *7*
$C_{17}H_{16}O_2$	–	–	CO	–	**B9**, 132, **3**, *56*
$C_{17}H_{16}Si$	–	–	CO	–	**B8**, 7, **1**, *8*
$C_{17}H_{17}NO_2$	–	–	CO	–	**B9**, 72, **2**, *127*
$C_{17}H_{18}O$	–	–	CO	–	**B9**, 76, **2**, *146*
$C_{17}H_{18}O_2$					
3,4-Dihydro-2-[2,4-cyclohexadien-1-yl]-6-methoxy-2*H*-naphthalene-1-one	–	–	CO	–	**B8**, 110, **4**, *146*
7-Methoxycarbonyl-3,4-dimethyl-7-phenyl-1,3,5-cycloheptatriene	–	–	CO	–	**B9**, 186, **5**, *21*
$C_{17}H_{19}NO$	–	–	CO	–	**B8**, 204, **9**, *60*
$C_{17}H_{20}$	–	–	CO	–	**B9**, 131, **3**, *52*
$C_{17}H_{20}N_4O_5$	–	–	CO	–	**B8**, 154, **7**, *15*
$C_{17}H_{20}O$	–	–	CO	–	**B9**, 69, **2**, *115*
$C_{17}H_{20}O_2S$	–	–	CO	–	**B8**, 265, **12**, *8*
$C_{17}H_{20}O_4$	–	–	CO	–	**B8**, 248, **11**, *26*
$C_{17}H_{21}NO$	–	–	CO	–	**B10**, 55, **3**, *49*
$C_{17}H_{21}N_2$	–	–	CO	–	**B8**, 133, **5**, *12*
$C_{17}H_{22}$	–	–	CO	–	**B8**, 328, **13**, *88*
$C_{17}H_{22}N_2O_3$					
5-(Dicyanomethyl)-1-[(2-methoxycarbonyl)ethyl]-2-methyl-4-(1-methylethoxy)-1,3-cyclohexadiene	–	–	CO	–	**B8**, 414, **14**, *84*
6-Dicyanomethyl-6-[(2-methoxycarbonyl)ethyl]-1-methyl-3-(1-methylethoxy)-1,3-cyclohexadiene	–	–	CO	–	**B8**, 416, **14**, *94*
$C_{17}H_{22}OP$	C_6H_8P	–	–	–	**B10**, 194, **6**, *30*
$C_{17}H_{22}O_4S$	–	–	CO	–	**B8**, 155, **7**, *21*
$C_{17}H_{23}NO$	–	–	CO	–	**B10**, 59, **3**, *63*
$C_{17}H_{24}$	–	–	–	–	**B10**, 190, **6**, *9*
$C_{17}H_{24}O$					
3a,7a-Dihydro-2,3,3a,4,5,6,7,7a-octamethyl-1*H*-inden-1-one	–	–	CO	–	**B8**, 448
2,3,5,6,7,8-Hexahydro-2-[3-methyl-2-oxocyclohex-1-yl]-naphthalene	–	–	CO	–	**B8**, 328, **13**, *87*

$C_{17}H_{24}O_2$	–	–	CO	–	**B8**, 203, **9**, *51*
$C_{17}H_{24}O_5$					
5-[4,4-Bis(ethoxycarbonyl)-but-1-ylidene]-2-methoxy-1,3-cyclohexadiene	–	–	CO	–	**B8**, 355, **13**, *177*
1,2,5,6,7,8-Hexahydro-3-methoxy-2-[bis(methoxycarbonyl)-methyl]-8-methylnaphthalene	–	–	CO	–	**B8**, 396, **14**, *4*
1,2,5,6,7,8-Hexahydro-3-methoxy-2-[bis(methoxycarbonyl)-methyl]-5-methylnaphthalene	–	–	CO	–	**B8**, 396, **14**, *5*
$C_{17}H_{24}O_6$	–	–	CO	–	**B8**, 324, **13**, *62*
$C_{17}H_{24}O_7$					
1-[(3-Acetyloxy)propyl]-4-methoxy-5-[bis(methoxycarbonyl)methyl]-1,3-cyclohexadiene	–	–	CO	–	**B8**, 322, **13**, *54*
2-Methoxy-5-[bis(methoxycarbonyl)methyl]-5-[(3-methoxycarbonyl)propyl]-1,3-cyclohexadiene	–	–	CO	–	**B8**, 347, **13**, *146*
4-Methoxy-1-[(2-methoxycarbonyl)ethyl]-5-[bis-(methoxycarbonyl)methyl]-2-methyl-1,3-cyclohexadiene	–	–	CO	–	**B8**, 414, **14**, *81*
4-Methoxy-5-[bis(methoxycarbonyl)methyl]-1-[(3-methoxycarbonyl)propyl]-1,3-cyclohexadiene	–	–	CO	–	**B8**, 322, **13**, *56*
5-[(2-Acetyloxy)ethyl]-2-methoxy-6-[bis(methoxycarbonyl)methyl]-5-methyl-1,3-cyclohexadiene	–	–	CO	–	**B8**, 404, **14**, *41*
5-[(3-Acetyloxy)propyl]-2-methoxy-5-[bis(methoxycarbonyl)methyl]-1,3-cyclohexadiene	–	–	CO	–	**B8**, 346, **13**, *142*
$C_{17}H_{24}P$	–	–	CO	–	**B8**, 136, **5**, *29*
$C_{17}H_{25}NSi$	–	–	CO	–	**B8**, 237, **10**, *27*
$C_{17}H_{26}O_2Si$	–	–	CO	–	**B8**, 237, **10**,
$C_{17}H_{26}O_5$	–	–	CO	–	**B8**, 337, **13**, *121*
$C_{17}H_{26}O_6$					
1-[(2-Methoxy)ethyl]-5-[bis(methoxycarbonyl)methyl]-1-(1-methylethoxy)-1,3-cyclohexadiene	–	–	CO	–	**B8**, 364, **13**, *224*
5-[(2-Methoxy)ethyl]-5-[bis(methoxycarbonyl)methyl]-2-(1-methylethoxy)-1,3-cyclohexadiene	–	–	CO	–	**B8**, 357, **13**, *185*

$C_{17}H_{28}O_4Si$					
5-[Bis(ethoxycarbonyl)methyl]-1-methyl-3-(trimethylsilyl)-1,3-cyclohexadiene	–	–	CO	–	**B8**, 317, **13**, *26*
6-[Bis(ethoxycarbonyl)methyl]-5-methyl-2-(trimethylsilyl)-1,3-cyclohexadiene	–	–	CO	–	**B8**, 330, **13**, *95*
5-[1-Acetyl-3-[[(2-trimethylsilyl)ethyl]oxy]-3-oxopropyl]-2-methoxy-5-methyl-1,3-cyclohexadiene	–	–	CO	–	**B8**, 367, **13**, *239*
$\mathbf{C}_{18}H_6F_8$	–	–	CO	–	**B8**, 469, **17**, *40*
$C_{18}H_{10}N_4$	–	–	CO	–	**B10**, 153, **4**, *146*
$C_{18}H_{12}$	–	–	CO	–	**B8**, 413, **14**, *75*
$C_{18}H_{12}N_2$	–	–	CO	–	**B9**, 269, **9**, *59*
$C_{18}H_{14}N_4O_2$					
5-Cyano-2,3,7,11a-tetrahydro-1,3-dioxo-2-phenyl-1*H*,5*H*-cyclohepta[*c*][1,2,4]triazolo-[1,2-*a*]pyridazine	–	–	CO	–	**B9**, 64, **2**, *87*
1,1,2,2-Tetracyano-3-(ethoxycarbonyl)-2,3,5,9a-tetrahydro-1*H*-benzocycloheptene	–	–	CO	–	**B9**, 64, **2**, *90*
$C_{18}H_{15}NO_2$	–	–	CO	–	**B10**, 53, **3**, *35*
$C_{18}H_{15}O_3P$	C_4H_6	–	–	–	**B10**, 207, **7**, *7*
$C_{18}H_{15}O_3P$	C_5H_8	–	–	–	**B10**, 209, **7**, *19*
					B10 ,209, **7**, *20*
$C_{18}H_{15}P$	C_4H_6	–	–	–	**B10**, 207, **7**, *10*
$C_{18}H_{15}P$	C_4H_6	C_8H_{12}	CO	–	**B10**, 180, **5**, *5*
					B10 ,181, **5**, *6*
$C_{18}H_{15}P$	C_5H_8	–	–	–	**B10**, 209, **7**, *21*
$C_{18}H_{16}$					
5,5-Diphenyl-1,3-cyclohexadiene	–	–	CO	–	**B8**, 245, **11**, *14*
4-[2,4-Cyclohexadien-1-yl]-[1,1'-biphenyl]	–	–	CO	–	**B8**, 105, **4**, *115*
$C_{18}H_{16}O$	–	–	CO	–	**B8**, 366, **13**, *231*
$C_{18}H_{17}BMoN_6O_2$	–	–	CO	–	**B9**, 188, **5**, *29*
$C_{18}H_{17}NO$					
Carbonyl[(phenylmethyl)-imino][1-methyl-3-phenyl-1-propene-1,3-diyl]	–	–	CO	–	**B10**, 55, **3**, *47*
Carbonyl(methylimino)[3-phenyl-1-(4-methylphenyl)-1-propene-1,3-diyl]	–	–	CO	–	**B10**, 74, **3**, *129*
Carbonyl[(phenylmethyl)-imino][1-phenyl-2-butene-1,4-diyl]	–	–	CO	–	**B10**, 75, **3**, *138*

$C_{18}H_{17}NO_4$					
1-Methyl-6-[2-[[(4-nitrophenyl)carbonyl]oxy]ethyl]-1,3,5,7-cyclooctatetraene	–	–	CO	–	**B9**, 267, **9**, *47*
4,5-Bis(ethoxycarbonyl)-azepino[2,1,7-*cd*]indolizine	–	–	CO	–	**B9**, 11, **1**, *34*
$C_{18}H_{17}O_3P$	–	–	CO	–	**B8**, 91, **4**, *47*
$C_{18}H_{18}$	–	–	CO	–	**B8**, 468, **17**, *37*
$C_{18}H_{18}F_4$	–	–	CO	–	**B8**, 470, **17**, *49*
$C_{18}H_{18}O_{12}$	–	–	CO	–	**B8**, 461, **17**, *1*
$C_{18}H_{19}NO_3$	–	–	CO	–	**B8**, 155, **7**, *22*
$C_{18}H_{20}N_4O_5$	–	–	CO	–	**B8**, 400, **14**, *23*
$C_{18}H_{20}O$	–	–	CO	–	**B8**, 103, **4**, *101*
$C_{18}H_{20}OS$	–	–	CO	–	**B9**, 202, **6**, *13*
$C_{18}H_{20}O_4$	–	–	CO	–	**B8**, 313, **13**, *11*
$C_{18}H_{21}NO$	–	–	CO	–	**B8**, 352, **13**, *164*
$C_{18}H_{22}$	–	–	CO	–	**B10**, 11, **1**, *41*
$C_{18}H_{22}N_4O_5$	–	–	CO	–	**B8**, 156, **7**, *26*
$C_{18}H_{22}OS$	–	–	CO	–	**B9**, 200, **6**, *6*
$C_{18}H_{22}O_4S$	–	–	CO	–	**B8**, 368, **13**, *245*
$C_{18}H_{23}NO$					
Carbonyl(cyclohexylimino)-[1-ethyl-3-phenyl-1-propene-1,3-diyl]	–	–	CO	–	**B10**, 75, **3**, *134*
9-Methoxy-1-(phenylmethyl)-1-azaspiro[5.5]undeca-8,10-diene	–	–	CO	–	**B8**, 352, **13**, *166*
$C_{18}H_{24}O_4S$					
4-Methoxy-1-[4-[(4-methylphenylsulfonyl)oxy]butyl]-1,3-cyclohexadiene	–	–	CO	–	**B8**, 155, **7**, *24*
4-Methoxy-2-methyl-1-[3-[(4-methylphenylsulfonyl)oxy]-propyl]-1,3-cyclohexadiene	–	–	CO	–	**B8**, 361, **13**, *207*
5-Ethyl-2-methoxy-5-[2-[(4-methylphenylsulfonyl)oxy]-ethyl]-1,3-cyclohexadiene	–	–	CO	–	**B8**, 343, **13**, *132*
$C_{18}H_{24}O_8$	–	–	CO	–	**B8**, 347, **13**, *144*
$C_{18}H_{25}NO_3$	–	–	CO	–	**B10**, 75, **3**, *137*
$C_{18}H_{26}O_4$					
1-Ethyl-5-[1-methoxycarbonyl-2-oxocyclopent-1-yl]-4-(1-methylethoxy)-1,3-cyclohexadiene	–	–	CO	–	**B8**, 325, **13**, *68*
5-Ethyl-5-[1-methoxycarbonyl-2-oxocyclopent-1-yl]-2-(1-methylethoxy)-1,3-cyclohexadiene	–	–	CO	–	**B8**, 357, **13**, *184*
5-[1-Methoxycarbonyl-2-oxocyclopent-1-yl]-1,2-dimethyl-4-(1-methylethoxy)-1,3-cyclohexadiene	–	–	CO	–	**B8**, 413, **14**, *80*

6-[1-Methoxycarbonyl-2-oxocyclopent-1-yl]-1,6-dimethyl-3-(1-methylethoxy)-1,3-cyclohexadiene	–	–	CO	–	**B8**, 416, **14**, *92*
$C_{18}H_{26}O_5$					
5-[3-Hydroxymethyl-1-methoxycarbonyl-3-methyl-2-oxocyclohex-1-yl]-4-methoxy-1-methyl-1,3-cyclohexadiene	–	–	CO	–	**B8**, 319, **13**, *36*
5-[3-Hydroxymethyl-1-methoxycarbonyl-3-methyl-2-oxocyclohex-1-yl]-2-methoxy-5-methyl-1,3-cyclohexadiene	–	–	CO	–	**B8**, 340, **13**, *126*
4a-[Bis(ethoxycarbonyl)methyl]-4,4a,5,6,7,8-hexahydro-2-methoxynaphthalene	–	–	CO	–	**B8**, 401, **14**, *26*
$C_{18}H_{26}O_7$					
2-Methoxy-5-[4-(methoxycarbonyl)butyl]-5-[bis(methoxycarbonyl)methyl]-1,3-cyclohexadiene	–	–	CO	–	**B8**, 348, **13**, *148*
5-[2-(Methoxycarbonyl)ethyl]-5-[bis(methoxycarbonyl)methyl]-2-(1-methylethoxy)-1,3-cyclohexadiene	–	–	CO	–	**B8**, 369, **13**, *247*
$C_{18}H_{28}O_3S_2$					
1-Ethyl-5-[bis[(ethylthio)carbonyl]methyl]-4-(1-methylethoxy)-1,3-cyclohexadiene	–	–	CO	–	**B8**, 324, **13**, *64*
5-Ethyl-5-[bis[(ethylthio)carbonyl]methyl]-2-(1-methylethoxy)-1,3-cyclohexadiene	–	–	CO	–	**B8**, 355, **13**, *180*
$C_{18}H_{28}O_5$	–	–	CO	–	**B8**, 339, **13**, *124*
$C_{18}H_{30}O_4Si$	–	–	CO	–	**B8**, 405, **14**, *46*
$C_{18}H_{32}$	–	–	CO	–	**B9**, 213, **7**, *16*
$C_{18}H_{33}P$	C_4H_6	–	–	–	**B10**, 207, **7**, *9*
$C_{18}H_{34}O_3P$	–	–	CO	–	**B8**, 134, **5**, *15*
$C_{18}H_{34}P$	–	–	CO	–	**B8**, 134, **5**, *18*
C$_{19}H_{14}N_8O_8$	–	–	CO	–	**B9**, 75, **2**, *140*
$C_{19}H_{14}O$	–	–	CO	–	**B9**, 186, **5**, *20*
$C_{19}H_{17}N_3O_4$	–	–	CO	–	**B9**, 63, **2**, *85*
$C_{19}H_{18}N_2O_4$	–	–	CO	–	**B8**, 432, **15**, *5*
$C_{19}H_{19}NO$	–	–	CO	–	**B10**, 55, **3**, *45*
$C_{19}H_{20}N_2O$	–	–	CO	–	**B10**, 74, **3**, *128*
$C_{19}H_{20}N_2O_3$	–	–	CO	–	**B8**, 403, **14**, *35*

$C_{19}H_{20}N_2O_4$					
4a,8-Dihydro-3,5-dimethoxy-9*H*-8,9c-(iminoethano)-phenanthro[4,5-*bcd*]furan-12-carboxamide	–	–	CO	–	**B8**, 432, **15**, *4*
9,10-Dihydro-4,5-dihydroxy-3,6-methoxy-4*H*-10,4a-(iminoethano)phenanthrene-13-carbonitrile	–	–	CO	–	**B8**, 432, **15**, *6*
$C_{19}H_{21}NO_3$					
3,4-Dihydro-5-hydroxy-6,11-dimethoxy-2-methyl-4a,9a-butadieno-9*H*-indeno[2,1-*c*]-pyridinium	–	–	CO	–	**B8**, 442, **15**, *37*
4a,8-Dihydro-3,5-dimethoxy-12-methyl-9*H*-8,9c-(iminoethano)phenanthro[4,5-*bcd*]furan	–	–	CO	–	**B8**, 431, **15**, *3*
$C_{19}H_{22}O_2$	–	–	CO	–	**B8**, 327, **13**, *83*
$C_{19}H_{22}O_3$	–	–	CO	–	**B8**, 453, **16**, *3*
$C_{19}H_{23}BMoN_4O_2$	–	–	CO	–	**B9**, 189, **5**, *31*
$C_{19}H_{23}NO_3$					
1,2,3,4-Tetrahydro-5-hydroxy-6,11-dimethoxy-2-methyl-4a,9a-butadieno-9*H*-indeno-[2,1-*c*]pyridine	–	–	CO	–	**B8**, 440, **15**, *29*
9,10-Dihydro-5-hydroxy-3,6-dimethoxy-13-methyl-4*H*-10,4a-(iminoethano)phenanthrene	–	–	CO	–	**B8**, 402, **14**, *33*
$C_{19}H_{24}NO_3$	–	–	CO	–	**B8**, 441, **15**, *35*
$C_{19}H_{26}$	–	–	CO	–	**B8**, 329, **13**, *89*
$C_{19}H_{28}O$	–	–	CO	–	**B8**, 363, **13**, *216*
$C_{19}H_{28}O_7$					
1-[2-(Methoxycarbonyl)ethyl]-5-[bis(methoxycarbonyl)methyl]-2-methyl-4-(1-methylethoxy)-1,3-cyclohexadiene	–	–	CO	–	**B8**, 414, **14**, *83*
6-[2-(Methoxycarbonyl)ethyl]-6-[bis(methoxycarbonyl)methyl]-1-methyl-3-(1-methylethoxy)-1,3-cyclohexadiene	–	–	CO	–	**B8**, 416, **14**, *93*
$C_{19}H_{30}O_4S_2$	–	–	CO	–	**B8**, 364, **13**, *225*
$C_{19}H_{32}O$	–	–	CO	–	**B8**, 204, **9**, *55*
$C_{19}H_{32}O_2$	–	–	CO	–	**B8**, 202, **9**, *44*
$C_{19}H_{36}OP$	–	–	CO	–	**B8**, 199, **9**, *25*
$C_{19}H_{36}O_4P$	–	–	CO	–	**B8**, 199, **9**, *24*
$C_{19}H_{36}P$	–	–	CO	–	**B9**, 53, **2**, *47*
$\mathbf{C}_{20}H_{12}N_2$	–	–	CO	–	**B8**, 407, **14**, *54*
$C_{20}H_{12}N_4$					
1,1,2,2-Tetracyano-1,2,3,8a-tetrahydro-3-phenylazulene	–	–	CO	–	**B9**, 150, **4**, *9*

6,6,7,7-Tetracyano-6,7-dihydrobenzo[1,2:3,4]dicycloheptene	–	–	CO	–	**B9**, 151, **4**, *12*
7,7,8,8-Tetracyano-2-(phenylmethylene)bicyclo[4.2.1]non-4-ene-3,9-diyl	–	–	CO	–	**B10**, 148, **4**, *127*
8,8,9,9-Tetracyano-2-phenylbicyclo[5.2.1]deca-3,5-diene-2,10-diyl	–	–	CO	–	**B10**, 151, **4**, *137*
8,8,9,9-Tetracyano-3-phenylbicyclo[5.2.1]deca-3,5-diene-2,10-diyl	–	–	CO	–	**B10**, 152, **4**, *139*
8,8,9,9-Tetracyano-5-phenylbicyclo[5.2.1]deca-3,5-diene-2,10-diyl	–	–	CO	–	**B10**, 153, **4**, *143*
$C_{20}H_{14}$	–	–	CO	–	**B9**, 161, **4**, *47*
$C_{20}H_{14}N_4$	–	–	CO	–	**B9**, 131, **3**, *51*
$C_{20}H_{16}$	–	–	CO	–	**B9**, 160, **4**, *45*
$C_{20}H_{16}N_2O$					
3a,8a-Dihydro-1,3-diphenyl-1*H*-cycloheptapyrazol-4-one	–	–	CO	–	**B9**, 73, **2**, *131*
3a,8a-Dihydro-1,3-diphenyl-1*H*-cycloheptapyrazol-8-one	–	–	CO	–	**B9**, 73, **2**, *132*
$C_{20}H_{17}N$	–	–	CO	–	**B9**, 156, **4**, *32*
$C_{20}H_{17}N_3O_3$	–	–	CO	–	**B8**, 410, **14**, *64*
$C_{20}H_{18}O$					
1-(Hydroxydiphenylmethyl)-1,3,5-cycloheptatriene	–	–	CO	–	**B9**, 124, **3**, *20*
5-Hydroxy-7-[(diphenyl)-methylene]-1,3-cycloheptadiene	–	–	CO	–	**B9**, 67, **2**, *105*
$C_{20}H_{19}NO_2$	–	–	CO	–	**B10**, 55, **3**, *44*
$C_{20}H_{19}N_3O_4$					
5-Ethoxycarbonyl-2,3,7,11a-tetrahydro-1,3-dioxo-2-phenyl-1*H*,5*H*-cyclohepta[*c*]-[1,2,4]triazolo[1,2-*a*]-pyridazine	–	–	CO	–	**B9**, 63, **2**, *86*
4a,9-Dihydro-1,4-bis(methoxycarbonyl)-9-[(4-methylphenyl)imino]-2*H*-cyclohepta-[*d*]pyridazine	–	–	CO	–	**B9**, 76, **2**, *145*
$C_{20}H_{20}$					
5,5-Dimethyl-1,3-diphenyl-1,3-cyclohexadiene	–	–	CO	–	**B8**, 398, **14**, *15*
5,5-Dimethyl-1,4-diphenyl-1,3-cyclohexadiene	–	–	CO	–	**B8**, 403, **14**, *37*
$C_{20}H_{22}ClNO$	–	–	CO	–	**B9**, 153, **4**, *20*
$C_{20}H_{22}N_2O_3$	–	–	CO	–	**B8**, 441, **15**, *32*
$C_{20}H_{22}N_2O_4S$	–	–	CO	–	**B8**, 363, **13**, *219*
$C_{20}H_{22}OP$	–	–	CO	–	**B8**, 215, **9**, *106*

Compound					Reference
$C_{20}H_{22}O_4$	–	–	CO	–	**B8**, 367, **13**, *237*
$C_{20}H_{23}NO_3$	–	–	CO	–	**B8**, 440, **15**, *28*
$C_{20}H_{24}Si_2$	–	–	CO	–	**B8**, 15, **1**, *45*
$C_{20}H_{27}NO$					
Carbonyl(cyclohexylimino)-[1-butyl-3-phenyl-1-propene-1,3-diyl]	–	–	CO	–	**B10**, 75, **3**, *135*
2-Cyano-6,7,8,9,10,11,12,13,14,15,16,17-dodecahydro-3-methoxy-13-methyl-1*H*-cyclopenta[*a*]phenanthrene	–	–	CO	–	**B8**, 414, **14**, *85*
$C_{20}H_{28}O_9$	–	–	CO	–	**B8**, 347, **13**, *147*
$C_{20}H_{32}O_4Si$	–	–	CO	–	**B8**, 398, **14**, *16*
$C_{21}H_{14}N_4$					
1,1,2,2-Tetracyano-2,3,5,9a-tetrahydro-3-phenyl-1*H*-benzocycloheptene	–	–	CO	–	**B9**, 65, **2**, *92*
1,1,2,2-Tetracyano-1,2,3,8a-tetrahydro-3-methyl-3-phenylazulene	–	–	CO	–	**B9**, 150, **4**, *11*
1,1,2,2-Tetracyano-1,2,3,8a-tetrahydro-3-(4-methylphenyl)azulene	–	–	CO	–	**B9**, 162, **4**, *54*
2,2,3,3-Tetracyano-3,3a-dihydrospiro[azulene-1(2*H*)2′-bicyclo[5.1.0]octa-3′,5′-diene	–	–	CO	–	**B9**, 81, **2**, *168*
7,7,8,8-Tetracyano-2-(2-phenylethenyl)bicyclo[4.2.1]-non-4-ene-3,9-diyl	–	–	CO	–	**B10**, 154, **4**, *151*
$C_{21}H_{14}N_4O$	–	–	CO	–	**B9**, 150, **4**, *10*
$C_{21}H_{14}O_2$	–	–	CO	–	**B8**, 14, **1**, *41*
					B8, 412, **14**, *70*
$C_{21}H_{16}N_4O_4$					
15-Methyl-2-phenyl-5*H*,10*H*-5,10-etheno-5a,9a-(methaniminomethano)-1*H*-[1,2,4]triazolo[1,2-*b*]-2*H*-phthalazine-1,3,14,16-tetrone	–	–	CO	–	**B8**, 411, **14**, *69*
1-[1-[(2,4-Dinitrophenyl)-hydrazinyl-2-ylidene]methyl]-7-(phenylmethylene)-1,3,5-cycloheptatriene	–	–	CO	–	**B9**, 184, **5**, *10*
?-[1-[(2,4-Dinitrophenyl)-hydrazinyl-2-ylidene]methyl]-1-phenyl-1,3,5,7-cyclooctatetraene	–	–	CO	–	**B9**, 268, **9**, *53*
$C_{21}H_{16}O$	–	–	CO	–	**B9**, 163, **4**, *55*
$C_{21}H_{16}O_2$	–	–	CO	–	**B8**, 360, **13**, *203*

$C_{21}H_{18}O$					
1-(1-Oxo-2,2-diphenylethyl)-1,3,5-cycloheptatriene	-	-	CO	-	**B9**, 131, **3**, *55*
8-Benzoyl-8-phenylbicyclo-[5.1.0]octa-2,4-diene	-	-	CO	-	**B9**, 79, **2**, *157*
9,9-Diphenylbicyclo[5.2.0]-nona-2,4-dien-8-one	-	-	CO	-	**B9**, 58, **2**, *67*
4,8a-Dihydro-2,3-diphenyl-(3a*H*)-cyclohepta[*b*]furan	-	-	CO	-	**B9**, 79, **2**, *158*
[7-Oxo-8,8-diphenylbicyclo-[4.2.1]non-3-ene-2,9-diyl]	-	-	CO	-	**B10**, 147, **4**, *120*
$C_{21}H_{19}BMoN_8O_2$	-	-	CO	-	**B9**, 188, **5**, *28*
$C_{21}H_{19}N_3O$					
5-(Dicyanomethyl)-1-[3-(1,3-dihydro-1,3-dioxo-2*H*-iso-indol-2-yl)propyl]-4-methoxy-1,3-cyclohexadiene	-	-	CO	-	**B8**, 323, **13**, *58*
5-(Dicyanomethyl)-5-[3-(1,3-dihydro-1,3-dioxo-2*H*-iso-indol-2-yl)propyl]-2-methoxy-1,3-cyclohexadiene	-	-	CO	-	**B8**, 349, **13**, *153*
$C_{21}H_{20}$					
7-[2,4,6-Cycloheptatrien-1-yl]-1-(4-methoxyphenyl)-1,3,5-cyloheptatriene	-	-	CO	-	**B9**, 149, **4**, *5*
7-[2-(4-Methoxyphenyl)-2,4,6-cycloheptatrien-1-yl]-1,3,5-cycloheptatriene	-	-	CO	-	**B9**, 132, **3**, *60*
$C_{21}H_{20}O$	-	-	CO	-	**B9**, 69, **2**, *113*
$C_{21}H_{21}NO_4$	-	-	CO	-	**B10**, 60, **3**, *69*
$C_{21}H_{24}NO_4$	-	-	CO	-	**B8**, 442, **15**, *38*
$C_{21}H_{24}N_2O_4S$	-	-	CO	-	**B8**, 364, **13**, *222*
$C_{21}H_{24}N_4O_4S_2$	-	-	CO	-	**B8**, 466, **17**, *26*
$C_{21}H_{24}O_5$					
5-[1,2,3,4-Tetrahydro-5-methoxy-2-methoxycarbonyl-1-oxonaphthalen-2-yl]-2-methoxy-5-methyl-1,3-cyclo-hexadiene	-	-	CO	-	**B8**, 342, **13**, *128*
5-[1,2,3,4-Tetrahydro-6-methoxy-2-methoxycarbonyl-1-oxonaphthalen-2-yl]-2-methoxy-5-methyl-1,3-cyclo-hexadiene	-	-	CO	-	**B8**, 342, **13**, *129*
$C_{21}H_{24}P$	-	-	CO	-	**B9**, 54, **2**, *49*
$C_{21}H_{24}Si$	-	-	CO	-	**B8**, 7, **1**, *11*
$C_{21}H_{25}NO_4$					
5-(Acetyloxy)-1,2,3,4-tetra-hydro-6,11-dimethoxy-2-methyl-4a,9a-butadieno-9*H*-indeno[2,1-*c*]pyridine	-	-	CO	-	**B8**, 440, **15**, *30*

5-(Acetyloxy)-9,10-dihydro-3,6-dimethoxy-13-methyl-4*H*-10,4a-(iminoethano)phenanthrene	–	–	CO	–	**B8**, 402, **14**, *34*
$C_{21}H_{25}NO_6S$	–	–	CO	–	**B8**, 364, **13**, *220*
$C_{21}H_{26}NO_4$	–	–	CO	–	**B8**, 442, **15**, *36*
$C_{21}H_{26}O_4$					
3-[2-Ethoxy-1-(ethoxycarbonyl)-2-oxo-1-phenylethyl]-1,5-cyclooctadiene	–	–	CO	–	**B9**, 213, **7**, *14*
[2-[2-Ethoxy-1-(ethoxycarbonyl)-2-oxo-1-phenylethyl]-5-cyclooctene-1,4-diyl]	–	–	CO	–	**B10**, 134, **4**, *60*
$C_{21}H_{30}O_4SSi$	–	–	CO	–	**B8**, 405, **14**, *47*
$C_{21}H_{30}O_5S$	–	–	CO	–	**B8**, 369, **13**, *249*
$C_{21}H_{32}O_2$	–	–	CO	–	**B8**, 318, **13**, *33*
$C_{21}H_{34}O$	–	–	CO	–	**B8**, 204, **9**, *57*
$\mathbf{C}_{22}H_{16}Br_2$	–	–	CO	–	**B10**, 12, **1**, *44*
$C_{22}H_{16}N_4$	–	–	CO	–	**B8**, 277, **12**, *57*
$C_{22}H_{16}O$	–	–	CO	–	**B10**, 49, **3**, *9*
$C_{22}H_{18}$					
1,2,3,4-Tetrahydro-2-methylene-4-(1-naphthalenylmethyl)-naphthalene-1,3-diyl	–	–	CO	–	**B10**, 11, **1**, *42*
1,2,3,4-Tetrahydro-2-methylene-4-(2-naphthalenylmethyl)-naphthalene-1,3-diyl	–	–	CO	–	**B10**, 12,, **1**, *43*
$C_{22}H_{18}N_4O_4$	–	–	CO	–	**B9**, 184, **5**, *11*
$C_{22}H_{23}NO$	–	–	CO	–	**B10**, 56, **3**, *50*
$C_{22}H_{25}NO$	–	–	CO	–	**B10**, 56, **3**, *51*
$C_{22}H_{25}NO_4$	–	–	CO	–	**B9**, 11, **1**, *35*
$C_{22}H_{26}NO_4$	–	–	CO	–	**B8**, 442, **15**, *39*
$C_{22}H_{27}NO_4$	–	–	CO	–	**B8**, 441, **15**, *31*
$C_{22}H_{27}NO_5$	–	–	CO	–	**B8**, 441, **15**, *33*
$C_{22}H_{27}NO_6S$	–	–	CO	–	**B8**, 364, **13**, *223*
$C_{22}H_{28}O_6S$	–	–	CO	–	**B8**, 337, **13**, *120*
$C_{22}H_{28}O_8S$					
2-Methoxy-5-[bis(methoxycarbonyl)methyl]-5-[3-[[(4-methylphenyl)sulfonyl]-oxy]propyl]-1,3-cyclohexadiene	–	–	CO	–	**B8**, 346, **13**, *143*
4-Methoxy-5-[bis(methoxycarbonyl)methyl]-1-[3-[[(4-methylphenyl)sulfonyl]-oxy]propyl]-1,3-cyclohexadiene	–	–	CO	–	**B8**, 322, **13**, *55*
$C_{22}H_{30}O_4Si$	–	–	CO	–	**B8**, 367, **13**, *235*

$\mathbf{C}_{23}H_{18}N_4O$					
2,2,3,3-Tetracyano-2,3,4,5-tetrahydro-5-methoxy-1-(4-methylphenyl)-1*H*-benzocycloheptene	–	–	CO	–	**B9**, 182, **5**, *1*
2,2,3,3-Tetracyano-2,3,4,9-tetrahydro-9-methoxy-1-(4-methylphenyl)-1*H*-benzocycloheptene	–	–	CO	–	**B9**, 182, **5**, *2*
$C_{23}H_{18}O$	–	–	CO	–	**B10**, 53, **3**, *31*
$C_{23}H_{19}N_3O_2$	–	–	CO	–	**B9**, 64, **2**, *88*
$C_{23}H_{20}Si$	–	–	CO	–	**B8**, 10, **1**, *21*
$C_{23}H_{22}N_4O_5$	–	–	CO	–	**B8**, 110, **4**, *147*
$C_{23}H_{23}NO_2$	–	–	CO	–	**B10**, 55, **3**, *48*
$C_{23}H_{25}NO_2$	–	–	CO	–	**B10**, 75, **3**, *136*
$C_{23}H_{25}NO_6$					
1-[3-(1,3-Dihydro-1,3-dioxo-2*H*-isoindol-2-yl)propyl]-4-methoxy-5-[1-methoxycarbonyl-2-oxopropyl]-1,3-cyclohexadiene	–	–	CO	–	**B8**, 323, **13**, *59*
5-[3-(1,3-Dihydro-1,3-dioxo-2*H*-isoindol-2-yl)propyl]-2-methoxy-5-[1-methoxycarbonyl-2-oxopropyl]-1,3-cyclohexadiene	–	–	CO	–	**B8**, 349, **13**, *155*
$C_{23}H_{25}NO_7$					
1-[3-(1,3-Dihydro-1,3-dioxo-2*H*-isoindol-2-yl)propyl]-4-methoxy-5-[bis(methoxycarbonyl)methyl]-1,3-cyclohexadiene	–	–	CO	–	**B8**, 323, **13**, *57*
5-[3-(1,3-Dihydro-1,3-dioxo-2*H*-isoindol-2-yl)propyl]-2-methoxy-5-[bis(methoxycarbonyl)methyl]-1,3-cyclohexadiene	–	–	CO	–	**B8**, 347, **13**, *145*
$C_{23}H_{26}Ge$	–	–	CO	–	**B8**, 11, **1**, *25*
$C_{23}H_{26}Si$	–	–	CO	–	**B8**, 10, **1**, *19*
$C_{23}H_{30}NO_4$	–	–	CO	–	**B8**, 441, **15**, *34*
$C_{23}H_{34}O_5$	–	–	CO	–	**B8**, 319, **13**, *37*
$C_{23}H_{34}O_6$	–	–	CO	–	**B8**, 341, **13**, *127*
$\mathbf{C}_{24}H_{18}O_3$	–	–	CO	–	**B10**, 52, **3**, *29*
$C_{24}H_{19}Cl_3P$	–	–	CO	–	**B8**, 135, **5**, *22*
$C_{24}H_{19}F_3P$	–	–	CO	–	**B8**, 135, **5**, *21*
$C_{24}H_{20}N_2$	–	–	CO	–	**B9**, 18, **1**, *69*
$C_{24}H_{22}B_2Fe_2$	–	–	CO	–	**B8**, 15, **1**, *43*
$C_{24}H_{22}O_3P$	–	–	CO	–	**B8**, 134, **5**, *16*
$C_{24}H_{22}P$	–	–	CO	–	**B8**, 134, **5**, *20*

$C_{24}H_{24}O$					
4,7,7-Trimethyl-1-(2,2-diphenyl-1-oxoethyl)-1,3,5-cycloheptatriene	–	–	CO	–	**B9**, 190, **5**, *34*
3,6,6-Trimethyl-9,9-diphenylbicyclo[5.2.0]nona-2,5-dien-8-one	–	–	CO	–	**B9**, 82, **2**, *171*
$C_{24}H_{24}O_4$	–	–	CO	–	**B8**, 396, **14**, *6*
$C_{24}H_{27}N_5O_5$	–	–	CO	–	**B8**, 90, **4**, *38*
$C_{24}H_{29}BMoN_6O_2$	–	–	CO	–	**B9**, 189, **5**, *30*
$C_{24}H_{34}Cl_2Si_2$	–	–	CO	–	**B8**, 11, **1**, *22*
$C_{24}H_{40}P$	–	–	CO	–	**B8**, 134, **5**, *19*
$\mathbf{C}_{25}H_{15}Br_3O$	–	–	CO	–	**B9**, 187, **5**, *26*
$C_{25}H_{18}O$	–	–	CO	–	**B9**, 187, **5**, *25*
$C_{25}H_{20}$					
1,4,7-Triphenyl-1,3,5-cycloheptatriene	–	–	CO	–	**B9**, 183, **5**, *3*
2,5,7-Triphenyl-1,3,5-cycloheptatriene	–	–	CO	–	**B9**, 185, **5**, *15*
$C_{25}H_{20}O$	–	–	CO	–	**B9**, 81, **2**, *169*
$C_{25}H_{20}O_3$	–	–	CO	–	**B10**, 53, **3**, *30*
$C_{25}H_{22}Ge$					
1-(Triphenylgermyl)-1,3,5-cycloheptatriene	–	–	CO	–	**B9**, 119, **3**, *1*
3-(Triphenylgermyl)-1,3,5-cycloheptatriene	–	–	CO	–	**B9**, 125, **3**, *27*
$C_{25}H_{24}OP$	–	–	CO	–	**B8**, 199, **9**, *29*
$C_{25}H_{24}P$	–	–	CO	–	**B9**, 54, **2**, *50*
$C_{25}H_{30}O$	–	–	CO	–	**B9**, 190, **5**, *33*
$C_{25}H_{31}NO_4S$	–	–	CO	–	**B8**, 364, **13**, *221*
$\mathbf{C}_{26}H_{14}N_8$	–	–	CO	–	**B10**, 148, **4**, *124*
$C_{26}H_{20}BrN_3$	–	–	CO	–	**B9**, 153, **4**, *22*
$C_{26}H_{20}ClN_3$	–	–	CO	–	**B9**, 153, **4**, *21*
$C_{26}H_{22}O_2$	–	–	CO	–	**B9**, 74, **2**, *135*
$C_{26}H_{24}P$	–	–	CO	–	**B9**, 80, **2**, *163*
$C_{26}H_{26}P$					
[2,4-Cycloheptadien-1-yl]-triphenylphosphonium	–	–	CO	–	**B9**, 211, **7**, *4*
8-(Triphenylphosphonio)-5-cyclooctene-1,4-diyl	–	–	CO	–	**B10**, 133, **4**, *52*
$C_{26}H_{40}NO$	–	–	CO	–	**B8**, 215, **9**, *104*
$C_{26}H_{40}O$	–	–	CO	–	**B8**, 454, **16**, *4*
$C_{26}H_{41}NO$	–	–	CO	–	**B8**, 215, **9**, *105*
$\mathbf{C}_{27}H_{20}$	–	–	CO	–	**B10**, 53, **3**, *34*
$C_{27}H_{22}$					
1-(Triphenylmethyl)-1,3,5,7-cyclooctatetraene	–	–	CO	–	**B9**, 263, **9**, *30*
7-(Triphenylmethyl)bicyclo-[4.2.0]octa-2,3,7-triene	–	–	CO	–	**B8**, 268, **12**, *18*

$C_{27}H_{23}N_3$	–	–	CO	–	**B9**, 154, **4**, *24*
$C_{27}H_{23}N_3O$	–	–	CO	–	**B9**, 154, **4**, *23*
$C_{27}H_{24}As$	–	–	CO	–	**B9**, 264, **9**, *35*
$C_{27}H_{24}P$	–	–	CO	–	**B9**, 264, **9**, *34*
$C_{27}H_{28}O_3P$					
[2,4-Cyclohexadien-1-yl]-tris(2-methoxyphenyl)phosphonium	–	–	CO	–	**B8**, 135, **5**, *23*
[2,4-Cyclohexadien-1-yl]-tris(4-methoxyphenyl)phosphonium	–	–	CO	–	**B8**, 135, **5**, *24*
$C_{27}H_{28}P$					
[2,4-Cyclohexadien-1-yl]-(2-methyl-2-phenylethyl)-diphenylphosphonium	–	–	CO	–	**B8**, 136, **5**, *33*
[2,4-Cyclohexadien-1-yl]-tris(2-methylphenyl)phosphonium	–	–	CO	–	**B8**, 135, **5**, *25*
[2,4-Cyclohexadien-1-yl]-tris(4-methylphenyl)phosphonium	–	–	CO	–	**B8**, 135, **5**, *26*
$C_{27}H_{29}OP_2$	–	–	CO	–	**B8**, 199, **9**, *26*
$C_{27}H_{44}$					
1,3-Cholestadiene	–	–	CO	–	**B8**, 326, **13**, *77*
2,4-Cholestadiene	–	–	CO	–	**B8**, 360, **13**, *202*
$C_{27}H_{44}O$	–	–	CO	–	**B8**, 432, **15**, *7*
$C_{27}H_{47}O_2$	–	–	CO	–	**B8**, 215, **9**, *103*
C$_{28}H_{20}$	$C_3H_9O_3P$	C_4N_6	CO	–	**B10**, 181, **5**, *7*
$C_{28}H_{20}O$	–	–	CO	–	**B10**, 53, **3**, *32*
$C_{28}H_{22}O$	–	–	CO	–	**B9**, 266, **9**, *45*
$C_{28}H_{24}$					
3a,7a-Dihydro-1-(triphenylmethyl)-1*H*-indene	–	–	CO	–	**B8**, 272, **12**, *35*
7,7a-Dihydro-1-(triphenylmethyl)-1*H*-indene	–	–	CO	–	**B8**, 180, **8**, *29*
$C_{28}H_{27}NO_4$	–	–	CO	–	**B10**, 60, **3**, *70*
$C_{28}H_{36}P$	–	–	CO	–	**B8**, 136, **5**, *34*
$C_{28}H_{42}O$	–	–	CO	–	**B8**, 439, **15**, *26*
$C_{28}H_{43}$	–	–	CO	–	**B8**, 412, **14**, *71*
$C_{28}H_{43}NO$	–	–	CO	–	**B8**, 440, **15**, *27*
$C_{28}H_{44}O$	–	–	CO	–	**B8**, 433, **15**, *8*
$C_{28}H_{46}$	–	–	CO	–	**B8**, 402, **14**, *32*
$C_{28}H_{46}O$	–	–	CO	–	**B8**, 402, **14**, *30*
$C_{28}H_{46}O_2$					
5,7-Ergostadiene-3,22-diol	–	–	CO	–	**B8**, 433, **15**, *9*
5,7-Ergostadiene-3,23-diol	–	–	CO	–	**B8**, 434, **15**, *10*
C$_{29}H_{24}$	–	–	CO	–	**B10**, 153, **4**, *147*
$C_{29}H_{38}OP$	–	–	CO	–	**B8**, 199, **9**, *27*
					B8, 216, **9**, *107*

$C_{29}H_{42}NO_2$	–	–	CO	–	**B8**, 198, **9**, *19*
$C_{29}H_{44}O_4$	–	–	CO	–	**B8**, 367, **13**, *238*
C$_{30}H_{30}Ge$	–	–	CO	–	**B9**, 266, **9**, *44*
$C_{30}H_{30}Si$					
1-(Trimethylsilyl)-3-(triphenylmethyl)-1,3,5,7-cyclooctatetraene	–	–	CO	–	**B9**, 266, **9**, *43*
1-(Trimethylsilyl)-7-(triphenylmethyl)bicyclo[4.2.0]-octa-2,4,7-triene	–	–	CO	–	**B8**, 360, **13**, *201*
$C_{30}H_{42}NO_2$	–	–	CO	–	**B8**, 198, **9**, *20*
$C_{30}H_{44}O_2$	–	–	CO	–	**B8**, 435, **15**, *13*
$C_{30}H_{46}O_2$	–	–	CO	–	**B8**, 434, **15**, *12*
$C_{30}H_{48}O_2$					
2-Carbonyloxy(3-ethenylcholestane-2,3-diyl)	–	–	CO	–	**B10**, 64, **3**, *86*
3-(Acetyloxy)-5,7-ergostadiene	–	–	CO	–	**B8**, 434, **15**, *11*
$C_{30}H_{48}O_3$					
3-(Acetyloxy)-22-hydroxy-5,7-ergostadiene	–	–	CO	–	**B8**, 435, **15**, *14*
3-(Acetyloxy)-23-hydroxy-5,7-ergostadiene	–	–	CO	–	**B8**, 436, **15**, *15*
$C_{30}H_{48}O_4$	–	–	CO	–	**B8**, 436, **15**, *16*
C$_{31}H_{22}O$	–	–	CO	–	**B9**, 188, **5**, *27*
$C_{31}H_{57}BO$	–	–	CO	–	**B8**, 204, **9**, *56*
C$_{32}H_{24}$	–	–	CO	–	**B9**, 269, **9**, *57*
C$_{33}H_{22}F_6N_2$	–	–	CO	–	**B10**, 154, **4**, *150*
$C_{33}H_{22}N_4$	–	–	CO	–	**B10**, 154, **4**, *149*
$C_{33}H_{26}$	–	–	CO	–	**B9**, 268, **9**, *54*
$C_{33}H_{27}N_3$	–	–	CO	–	**B9**, 154, **4**, *25*
$C_{33}H_{55}BO$	–	–	CO	–	**B8**, 204, **9**, *58*
C$_{34}H_{52}O_2$	–	–	CO	–	**B8**, 203, **9**, *50*
C$_{35}H_{37}OP_2$	–	–	CO	–	**B8**, 199, **9**, *28*
$C_{35}H_{48}O_2$	–	–	CO	–	**B8**, 437, **15**, *19*
$C_{35}H_{50}O_2$	–	–	CO	–	**B8**, 437, **15**, *18*
$C_{35}H_{50}O_3$					
3-(Benzoyloxy)-22-hydroxy-5,7-ergostadiene	–	–	CO	–	**B8**, 437, **15**, *20*
3-(Benzoyloxy)-23-hydroxy-5,7-ergostadiene	–	–	CO	–	**B8**, 438, **15**, *21*
$C_{35}H_{56}O_2$	–	–	CO	–	**B8**, 436, **15**, *17*
C$_{36}H_{28}$	–	–	CO	–	**B8**, 279, **12**, *62*

$C_{37}H_{52}O_4$					
22-(Acetyloxy)-3-(benzoyloxy)-5,7-ergostadiene	–	–	CO	–	**B8**, 438, **15**, *22*
23-(Acetyloxy)-3-(benzoyloxy)-5,7-ergostadiene	–	–	CO	–	**B8**, 438, **15**, *23*
$C_{38}H_{58}O_3Si$					
3-(Benzoyloxy)-22-[(trimethylsilyl)oxy]-5,7-ergostadiene	–	–	CO	–	**B8**, 439, **15**, *24*
3-(Benzoyloxy)-23-[(trimethylsilyl)oxy]-5,7-ergostadiene	–	–	CO	–	**B8**, 439, **15**, *25*
$C_{41}H_{25}Br_5O$	–	–	CO	–	**B9**, 284
$C_{41}H_{30}O$	–	–	CO	–	**B9**, 283

Table of Conversion Factors

Following the notation in Landolt-Börnstein [7], values which have been fixed by convention are indicated by a bold-face last digit. The conversion factor between calorie and Joule that is given here is based on the thermochemical calorie, cal_{thch}, and is defined as 4.1840 J/cal. However, for the conversion of the "Internationale Tafelkalorie", cal_{IT}, into Joule, the factor 4.1868 J/cal is to be used [1, p. 147]. For the conversion factor for the British thermal unit, the Steam Table Btu, Btu_{ST}, is used [1, p. 95].

Force	N	dyn	kp
1 N (Newton)	1	10^5	0.1019716
1 dyn	10^{-5}	1	1.019716×10^{-6}
1 kp	9.8066**5**	9.8066**5** $\times 10^5$	1

Pressure	Pa	bar	kp/m²	at	atm	Torr	lb/in²
1 Pa (Pascal) = 1 N/m²	1	10^{-5}	1.019716×10^{-1}	1.019716×10^{-5}	0.986923×10^{-5}	0.750062×10^{-2}	145.0378×10^{-6}
1 bar = 10^6 dyn/cm²	10^5	1	10.19716×10^3	1.019716	0.986923	750.062	14.50378
1 kp/m² = 1 mm H_2O	9.8066**5**	0.98066**5** $\times 10^{-4}$	1	10^{-4}	0.967841×10^{-4}	0.735559×10^{-1}	1.422335×10^{-3}
1 at = 1 kp/cm²	0.98066**5** $\times 10^5$	0.98066**5**	10^4	1	0.967841	735.559	14.22335
1 atm = 76**0** Torr	1.0132**5** $\times 10^5$	1.0132**5**	1.033227×10^4	1.033227	1	760	14.69595
1 Torr = 1 mm Hg	133.3224	1.333224×10^{-3}	13.59510	1.359510×10^{-3}	1.315789×10^{-3}	1	19.33678×10^{-3}
1 lb/in² = 1 psi	6.89476×10^3	68.9476×10^{-3}	703.069	70.3069×10^{-3}	68.0460×10^{-3}	51.7149	1

Work, Energy, Heat	J	kWh	kcal	Btu	MeV
1 J (Joule) = 1 Ws = 1 Nm = 10^7 erg	1	2.778×10^{-7}	2.39006×10^{-4}	9.4781×10^{-4}	6.242×10^{12}
1 kWh	$3.\mathbf{6}\times10^6$	1	860.4	3412.14	2.247×10^{19}
1 kcal	4184.**0**	1.1622×10^{-3}	1	3.96566	2.6117×10^{16}
1 Btu (British thermal unit)	1055.06	2.93071×10^{-4}	0.25164	1	6.5858×10^{15}
1 MeV	1.602×10^{-13}	4.450×10^{-20}	3.8289×10^{-17}	1.51840×10^{-16}	1

1 eV/mol ≙ 23.0578 kcal/mol = 96.473 kJ/mol

Power	kW	PS	kp · m/s	kcal/s
1 kW = 10^{10} erg/s	1	1.35962	101.972	0.239006
1 PS	0.73550	1	**75**	0.17579
1 kp · m/s	9.8066**5** $\times10^{-3}$	0.01333	1	2.34384×10^{-3}
1 kcal/s	4.184**0**	5.6886	426.650	1

References:

[1] A. Sacklowski, Die neuen SI-Einheiten, Goldmann, München 1979. (Conversion tables in an appendix.)
[2] International Union of Pure and Applied Chemistry, Manual of Symbols and Terminology for Physicochemical Quantities and Units, Pergamon, London 1979; Pure Appl. Chem. **51** [1979] 1/41.
[3] The International System of Units (SI), National Bureau of Standards Spec. Publ. No. 330 [1972].
[4] H. Ebert, Physikalisches Taschenbuch, 5th Ed., Vieweg, Wiesbaden 1976.
[5] Kraftwerk Union Information, Technical and Economic Data on Power Engineering, Mülheim/Ruhr 1978.
[6] E. Padelt, H. Laporte, Einheiten und Größenarten der Naturwissenschaften, 3rd Ed., VEB Fachbuchverlag, Leipzig 1976.
[7] Landolt-Börnstein, 6th Ed., Vol. II, Pt. 1, 1971, pp. 1/14.
[8] ISO Standards Handbook 2, Units of Measurement, 2nd Ed., Geneva 1982.

Withdrawn from
University Leicester Library